The Quadratic Formula

The roots of the quadratic equation $ax^2 + bx + c = 0$ are given by $x = -$

Laws of Logarithms (where $\log_b x = y$ means $x = b^y$)

$$\log_b (M \cdot N) = \log_b M + \log_b N, \qquad \log_b \left(\frac{M}{N}\right) = \log_b M - \log_b N$$

$$\log_b (M^x) = x \cdot \log_b M, \qquad \log_b b = 1, \qquad \log_b 1 = 0$$

$$\log_{10} x = \log x, \qquad \log_e x = \ln x, \qquad \ln x = \frac{\log x}{\log e}$$

Determinants

$$\begin{vmatrix} a_1 & b_1 \\ a_2 & b_2 \end{vmatrix} = a_1b_2 - a_2b_1$$

Cramer's rule: The solution to the system $\begin{cases} a_1x + b_1y = c_1 \\ a_2x + b_2y = c_2 \end{cases}$ is

$$x = \frac{\begin{vmatrix} c_1 & b_1 \\ c_2 & b_2 \end{vmatrix}}{\begin{vmatrix} a_1 & b_1 \\ a_2 & b_2 \end{vmatrix}} \qquad y = \frac{\begin{vmatrix} a_2 & c_1 \\ a_2 & c_2 \end{vmatrix}}{\begin{vmatrix} a_1 & b_1 \\ a_2 & b_2 \end{vmatrix}}$$

Business Formulas

Compound Interest Formula: $A_n = P(1 + r)^n$

Present Value Formula: $P = (1 + r)^{-n}A$

Annuity Formula: $V = P\,\dfrac{(1 + R/q)^n - 1}{R/q}$

Mortgage Formula: $M = \dfrac{D(R/q)}{1 - (1 + R/q)^{-n}}$

Polynomials

The Remainder Theorem: If $P(x)$ is divided by the linear factor $x - r$ until a constant remainder R is obtained, then $P(r) = R$.

The Factor Theorem: $P(x)$ has a zero $x = r$ if and only if $x - r$ is a factor of the polynomial.

College Algebra

second edition

Bernard J. Rice
University of Dayton

Jerry D. Strange
University of Dayton

Prindle, Weber & Schmidt
Boston, Massachusetts

Prindle, Weber & Schmidt is a division of Wadsworth, Inc.
Second printing: November 1980

Library of Congress Cataloging in Publication Data

Rice, Bernard J
College algebra.

Includes index.
1. Algebra. I. Strange, Jerry D., joint author.
II. Title.
QA154.2.R53 1980 512.9 79-24838
ISBN 0-87150-288-7

Cover image "Impact" by Mark Rowland. © 1977 by Circle Fine Art Corporation. Used by permission of the publisher. The artist's work is based on uncompromisingly precise mathematical systems, with the compass and protractor as implements.

Cover and text design by Deborah Schneider. Composed in Univers and Times Roman by Doyle Photosetting Ltd. Covers printed by Lehigh Press Lithographers. Text printed and bound by Halliday Lithograph.

Preface

Our major objectives in this edition have changed little since the first edition. While we feel we have made significant improvements, the following overview from the first edition still holds true.

> Besides being a prerequisite subject for the study of calculus and other more advanced mathematics courses, algebra is also a subject that is essential to the understanding of the quantitative nature of our world. While most modern elementary mathematics books at the college level speak to the prerequisite role of algebra, few relate to the student's need to identify mathematical concepts with familiar physical problems. In this book we have attempted to overcome this shortcoming. In writing this book we have followed current pedagogy and made the functional concept the central theme while simultaneously stressing the utilitarian aspects of mathematics by including problems stated in the context of some real world situation.

The material presented in the book presupposes some ability in algebra. However, review topics such as special products, factoring exponents, and radicals are presented early in the text to assist the student in mastering fundamental concepts. In addition, emphasis is placed on graphing throughout. The major changes in the text are:

1) The elementary notion of equation solving has been separated from the more sophisticated concept of a function in order to (a) amplify the importance of the functional concept, and (b) allow equations and inequalities to be treated as review topics if desired. This change has created new chapters (Chapters 1, 2 and 3) and allowed the repositioning of others.

2) The use of examples and exercises as a tool for learning cannot be overstated. The number of examples and exercises has been expanded by more than 20%. There are now 350 examples and over 2800 exercises in *College Algebra.* In many instances these problems and examples are stated in the terminology of an applied field so that the student can see the use of mathematics as a tool. Also, a considerable effort has been made to be certain the exercises are in increasing order of difficulty.

3) New sections have been added that use applications. These include Sections 3.2 and 3.4 on "Applications of Linear Equations" and "Applied Problems that Lead to Quadratic Equations." Strategies for translating verbal problems into mathematical equations are dealt with first. "Interest" and "Annuities and Mortgage Payments" are treated in new Sections 9.5 and 9.6. The instructor or student interested in business applications will also find additional examples in Chapter 7 as well as in other appropriate places in the text.

4) A new Chapter 9 has been added covering probability. The calculator and its uses are given more than just passing attention in this edition. Comments and expository remarks specifically intended to highlight calculator material are included in clearly marked sections of the text  . On the other hand, a calculator is not required for this book. Sections of interpolation are included as are tables of values for logarithms. A complete break with tradition comes slowly.

The following flowchart indicates chapter dependencies as well as those sections (contained in ◺) which are not essential to succeeding material. However, instructors can use their experience and judgment to choose and organize topics to suit their needs.

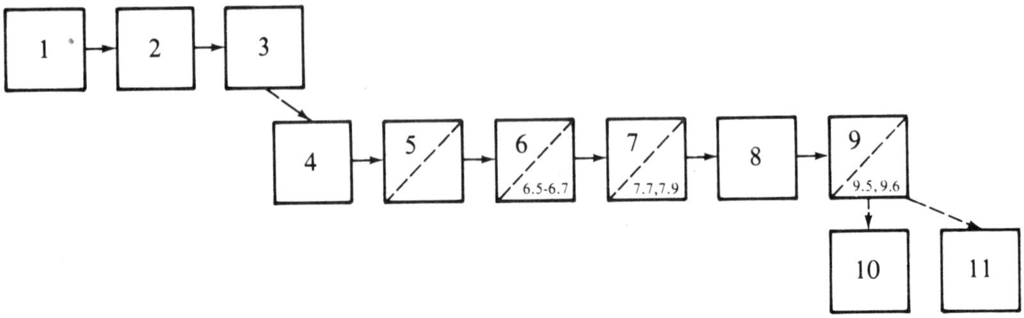

Chapters 6, 7, 8, and 11 would give a sound background for students majoring in sciences. The later sections of Chapter 7, with 7.5 and 7.6 as optional, along with 8, 9, and 10 would provide an equally sound background for business-oriented students.

The answers to the odd-numbered exercises are included in the back of the book and have generally been found with the use of a calculator. However, since decimal approximations to certain irrational numbers such as e and π are often used to varying decimal place accuracy, depending on the method of computation or calculator used, the given answer may differ from your answer. The effect of truncated values on the answers to the exercises can be disconcerting but at this time we know of no easy solution to this accuracy problem.

This edition of *College Algebra* has benefited from critical reviews and comments. Out thanks to the following:

Thomas E. Armstrong (University of Minnesota)
Soo Bong Chae (New College of USF)
James Hall (University of Wisconsin, Madison)
Kendell Hyde (Weber State College)
Ki Woong Kim (Morgan State University)
Helen Kriegsman (Pittsburg State University)
Richard W. Marshall (Eastern Michigan University)
Janet Milles (James Madison University)
Jane Morrison (Thornton Community College)
Karl Rudnick (Texas A & M University)
W. Richard Slinkman (Bemidji State University)
LeRoy V. Stoldt (College of DuPage)
John Vasak (University of Wisconsin, Milwaukee)
Charles Ziegenfus (James Madison University)

A special thanks to Arthur M. Hobbs (Texas A & M University) for his suggestions for Chapter 7.

Finally, it is a pleasure to acknowledge the fine cooperation of the staff of Prindle, Weber & Schmidt, particularly our editor, Theron Shreve, and our production editor, Debbie Schneider. A special thanks goes to Mary LeQuesne for her valuable assistance.

B.J.R., J.D.S.

Contents

5 Elementary Functions, *129*

6 Exponential and Logarithmic Functions, *175*

7 Systems of Equations and Inequalities, *205*

1

Basic Algebra

1.1 Real Numbers

Numbers are the central theme of algebra. Consequently, you should know something about the terminology associated with the system of numbers that we shall use. While there are many different systems of numbers, our current interest is the real number system, which is the system you have worked with in arithmetic. Three important subsets of the real numbers are integer numbers, rational numbers, and irrational numbers.

The most familiar subset of the real numbers is the set of counting numbers 1, 2, 3, 4, 5, ... This set comprises what are also called the **positive integers**, or the **natural numbers**, and is denoted by the letter N. The negative integers -1, -2, -3, -4, -5, . . . together with the positive integers and the number 0 make up the set of **integers** I.

A real number is said to be **rational** if it can be represented as a quotient of the form a/b, where a is any integer and b is any nonzero integer. Numbers such as $-\frac{3}{2}$, $\frac{2}{7}$, 3, and $\frac{17}{11}$ are examples of rational numbers. The set of all rational numbers is denoted by Q. Since each integer n can be written $n/1$, the integers I are a proper subset of the rational numbers Q. This is shown graphically in Figure 1.1.

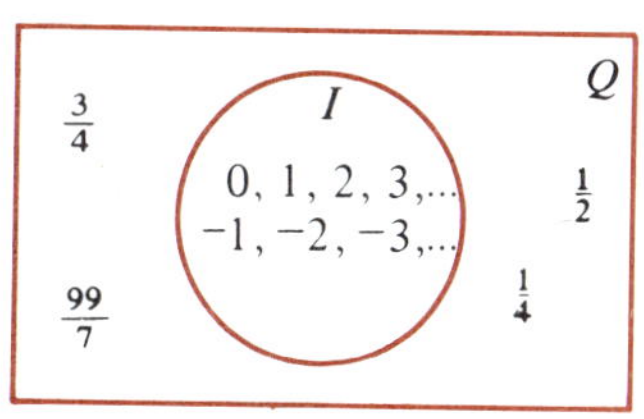

FIGURE 1.1

Rational numbers can be written in *decimal form*. This decimal representation takes one of two forms:

(1) **Terminating decimals,** in which a finite number of places is needed after the decimal point; for example,

$$\frac{1}{2}=0.5, \quad \frac{25}{4}=6.25, \quad \frac{19}{8}=2.375, \text{ or}$$

(2) **Infinitely repeating decimals,** in which infinitely many decimal places are necessary, and a block of digits continually repeats itself; for example,

$$\frac{1}{6}=0.166666\ \ldots, \quad \frac{11}{7}=1.571428571428\ldots$$

Instead of the three dots, a bar, called a vinculum, is often placed over the repeating block to identify it; thus,

$$\frac{1}{6}=0.1\overline{6} \quad \frac{11}{7}=1.\overline{571428}$$

Some real numbers are not rational. There is, for example, no rational number whose square is 2. (The real number whose square is 2 is denoted by $\sqrt{2}$, read "radical two.") Real numbers that cannot be expressed as the ratio of two integers are called **irrational**. Well-known examples of irrational numbers are $\sqrt{2}$ and π.

Irrational numbers have nonterminating, nonrepeating decimal representations. (When we write $\sqrt{2}=1.4142\ldots$ or $\pi=3.14159\ldots$, it is understood that the decimal representation is nonterminating and that no block of digits repeats itself.) Thus, irrational numbers can be thought of as:

(1) Numbers that *cannot* be expressed as the ratio of two integers.

(2) Numbers whose decimal representation is *not* terminating and *not* infinitely repeating; for example, 0.1001000100001 . . . .

The sets of rational and irrational numbers together make up the set of real numbers. The rationals and irrationals are mutually exclusive sets; that is, the set of rationals contains no irrational number, and conversely, the set of irrationals contains no rational number. Figure 1.2 graphically displays the "hierarchy" of real numbers.

You should know the various kinds of real numbers and how they fit in the

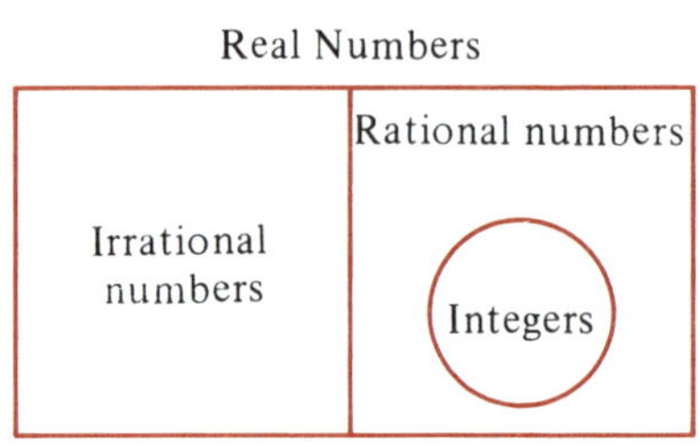

FIGURE 1.2

hierarchy of the real number system. For instance, 3 is at once a natural number, an integer, a rational number, and a real number; $\frac{1}{2}$ is *both* rational and real, but not an integer nor an irrational number; and π is *both* an irrational and a real number, but not an integer nor a rational number.

Computers and calculators use ***truncated*** *numbers, that is numbers rounded off to a certain number of decimal places. The number of digits and the method of representing truncated numbers varies with the particular computer or calculator; some calculators that truncate to seven decimal places will display* $\frac{2}{3}$ *as* 0.6666667, *whereas others will display it as* 0.6666666. *In* 0.6666667, *the number is said to be* ***rounded up***. *Most calculators round up if the truncated digit is* 5 *or greater.*

In the exercises throughout the textbook, you will find problems to be solved with the help of a calculator. A calculator can be used for any arithmetic operation involving real numbers, and we advise you to use one. A calculator does not reason for you; you must do the mathematical reasoning and the precise thinking.

We do not tell you how to use a calculator, except in general terms, but you should gain facility in using your calculator as your mathematical knowledge increases. Since calculators differ in method of operation, consult the operations manual of your calculator for any peculiarities.

1.2 Operations With Numbers

Symbols of Grouping and Order of Operation

Parentheses, brackets, and braces are used to group numbers and to indicate the precise order in which arithmetic operations are to be performed. For instance, $5+(2 \cdot 3)$ indicates that the multiplication $2 \cdot 3$ is performed first and the product is added to 5, giving 11. In contrast, $(5+2) \cdot 3$ means that 5 is added to 2 before multiplying by 3. Thus we have $(5+2) \cdot 3=(7) \cdot 3=21$. Often the multiplication sign is omitted next to a grouping symbol. Thus, the expressions $4 \cdot (3+8)$ and $4(3+8)$ mean the same thing.

Confusion can arise if grouping symbols are omitted or if multiple grouping symbols are used. To avoid confusion, we adopt the following conventions for sequences of arithmetic operations.* These same conventions are used in most computers.

(1) Perform all operations within any grouping symbol before performing other operations. If grouping symbols are contained within one another, begin with the innermost pair.

*Note the distinction between a "law" and a "convention" governing mathematical operations. A law is a direct consequence of the nature of the operation. A convention is merely a convenient, widely used expression of the operation.

Example (a) $(7 \cdot 4)+(8-5)-(16-4)+(42 \div 6)$

$$=28+3-12+7$$
$$=26$$

(b) $3+\{6+[2+(7-2)]\}$

$$=3+\{6+[2+5]\}$$
$$=3+\{6+7\}$$
$$=3+13=16$$

(2) In a sequence of multiplications and divisions, perform the operations in order from left to right.

Example (a) $3 \cdot 18 \div 9=54 \div 9=6$

(b) $24 \div 8 \cdot 5 \div 15=3 \cdot 5 \div 15=15 \div 15=1$

(3) In a sequence of additions, subtractions, multiplications, and divisions, perform the multiplications and divisions first and then the additions and subtractions. Multiplication and division are said to be operations of **higher priority** than addition and subtraction.

Example $5 \cdot 6-3 \cdot 7+24 \div 8=30-21+3=12$

Most scientific calculators can group operations by a [( ] *key and a* [ )] *key. Keystrokes performed after pushing the* [( ] *key and before pushing the* [ )] *key are separated from the sequence of operations outside the grouping symbols. For example,* $3 \cdot (6+4)+7$ *is evaluated by the following sequence:*

3 [×] [( ] 6 [+] 4 [ )] [+] 7 [=] 37

If the grouping keystrokes are not included, the display will show 29, *since the multiplication* $3 \cdot 6$ *will be done first.*

Fundamental Laws

Five fundamental laws govern the arithmetic operations of addition and multiplication. You already use these laws although you may not recognize them as such. We know, for instance, that the sum of two numbers is independent of the order of the numbers. Thus, $2+7=7+2$. This property, which is called the **commutative law for addition**, is valid for all real numbers. Another addition law, called the **associative law for addition**, states that the sum of three or more numbers is the same regardless of how they are grouped for addition; that is, $2+(9+1)=(2+9)+1$.

Using x, y, and z to represent real numbers, we have stated the laws that govern addition and multiplication. You should familiarize yourself with these laws and know how and when to use them.

- **Commutative Law of Addition:** $x+y=y+x$
  For example, $5+2=2+5$.
- **Commutative Law of Multiplication:** $x \cdot y=y \cdot x$
  For example; $2 \cdot 7=7 \cdot 2$.

- **Associative Law of Addition:** $(x+y)+z=x+(y+z)$
  For example, $(7+3)+5=7+(3+5)$.
- **Associative Law of Multiplication:** $x \cdot (y \cdot z)=(x \cdot y) \cdot z$
  For example, $2 \cdot (7 \cdot \pi)=(2 \cdot 7) \cdot \pi$
- **Distributive Law of Multiplication over Addition:** $x \cdot (y+z)=x \cdot y+x \cdot z$
  For example, $2 \cdot (\pi+7)=2 \cdot \pi+2 \cdot 7$.

The distributive law tells us that the grouping symbols can be removed from an expression of the form $x \cdot (y+z)$ simply by multiplying each of the numbers within the parentheses by the coefficient x, giving $x \cdot y + x \cdot z$.

EXAMPLE 1 (a) $3x+2x=(3+2)x=5x$

(b) $\frac{3}{7}+\frac{2}{7}=(3) \cdot \frac{1}{7}+(2) \cdot \frac{1}{7}$

$=(3+2) \cdot \frac{1}{7}=\frac{5}{7}$

(c) $3\sqrt{2}+2\sqrt{2}=(3+2)\sqrt{2}=5\sqrt{2}$

There is a **commutative law** and an **associative law for multiplication** but not for subtraction or division. Since $8 \div 4=2$ and $4 \div 8=\frac{1}{2}$, for example, it is clear that division does not have a commutative law. Likewise, since $8 \div (4 \div 2)=4$ and $(8 \div 4) \div 2=1$, division does not have an associative law.

Signed Numbers

We will review here some of the basic rules for operating with positive and negative numbers because of the confusion arising from the fact that the negative sign designates both the operation of subtraction and the sign of the number.

Grouping Symbols

- If there are grouping symbols around a single number, positive or negative, simply remove the grouping symbols. For instance, $(7)=7$ and $(-7)=-7$.
- If the grouping symbol is preceded by a negative sign remove the grouping symbol by changing the sign of the number. For instance, $-(+2)=-2$ and $-(-2)=2$.

The value of a real number independent of its sign is called the **absolute value** of the number. For instance, the absolute value of -2 is 2; of $-\sqrt{3}$ is $\sqrt{3}$; and of 5 is 5.

Addition

- The sum of two real numbers with *like* signs is the sum of the absolute values of the two numbers preceded by their common sign. For example,

$$(+3)+(+5)=+(3+5)=+8$$
$$(-5)+(-7)=-(5+7)=-12$$

- To add two real numbers with *unlike* signs, subtract the smaller absolute value from the larger. The answer will have the sign of the one with the larger absolute value.
  Thus,

$$(-4)+(+9)=+(9-4)=+5 \quad \text{and} \quad (-8)+(+2)=-(8-2)=-6$$

Subtraction

To subtract two real numbers, change the sign of the one being subtracted and then follow the rules for addition. For example,

$$(+4)-(+9)=(+4)+(-9)=-(9-4)=-5$$
$$(-3)-(-7)=(-3)+(+7)=+(7-3)=+4$$

Multiplication and Division

To multiply (or divide) two real numbers, multiply (or divide) their absolute values. If both numbers have the same sign, the product (or quotient) is positive. If the numbers have unlike signs, the product (or quotient) is negative.

$$(+7)(-6)=-(7\cdot 6)=-42$$
$$(-2)(-3)=+(2\cdot 3)=+6$$
$$\frac{(+5)}{(-9)}=-\left(\frac{5}{9}\right)=-\frac{5}{9}$$

EXAMPLE 2

(a) $(-2)+(-3)=-5$

(b) $3-(-7)=3+7=10$

(c) $9-6=9+(-6)=3$

(d) $(-4)(4-7)=(-4)(-3)=12$

(e) $\dfrac{8-2}{4-(-1)}=\dfrac{8+(-2)}{4+1}=\dfrac{6}{5}$

EXAMPLE 3

(a) $-5(3-\pi)=-15+5\pi$

(b) $-8(\sqrt{5}+\sqrt{2}-4)=-8(\sqrt{5})+(-8)(\sqrt{2})+(-8)(-4)=-8\sqrt{5}-8\sqrt{2}+32$

(c) $-(2\pi-\sqrt{7})=(-1)(2\pi-\sqrt{7})=-2\pi+\sqrt{7}$

In (c), removing the grouping symbol changes the numerical sign of each term within the original grouping. We can generalize this result: **When a grouping is preceded by a negative sign, remove the grouping symbol and change the numerical sign of each term within the grouping.**

1.3 The Real Number Line

The real numbers can be represented geometrically by associating each number with a point on a straight line. This association between real numbers on the one hand and points on the other is critical to the easy description of many algebraic problems. The line is called the **real number line**, or simply the number line. The number line helps in "visualizing" the number system.

Begin by choosing an arbitrary point 0 on the line l, calling it the *origin*. (See Figure 1.3.) With the point 0 we associate the real number 0. Then select any other point to the right of 0 and with it associate the real number 1. The line segment determined by the origin and the point corresponding to 1 is called a *unit distance*.

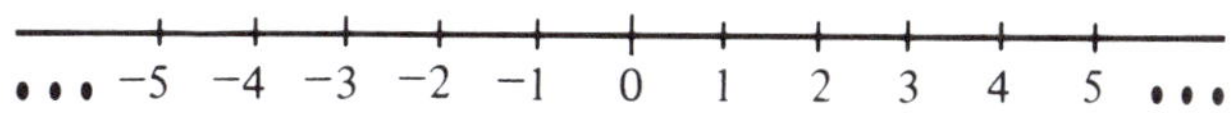

FIGURE 1.3

Proceeding from 0, we lay off 2 units, 3 units, 4 units, and so on. With these points, we associate the numbers 2, 3, 4, ... Thus the entire set of natural numbers is associated with equispaced points on the line. Similarly, unit distances to the left make an association of points with the negative integers.

Figure 1.4 shows another number line with a different unit distance. Although the scale has been expanded, the same numbers are represented.

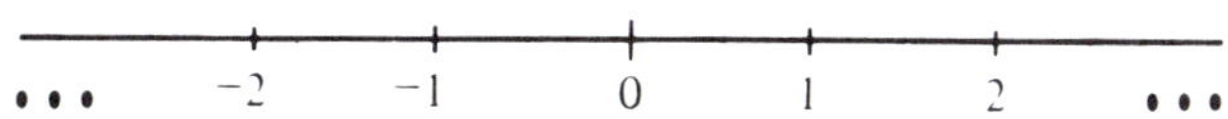

FIGURE 1.4

To locate rational numbers on the number line, divide the unit distance into an appropriate number of equal parts. For example, if the unit distance is divided into thirds, each of the points is labeled as shown in Figure 1.5.

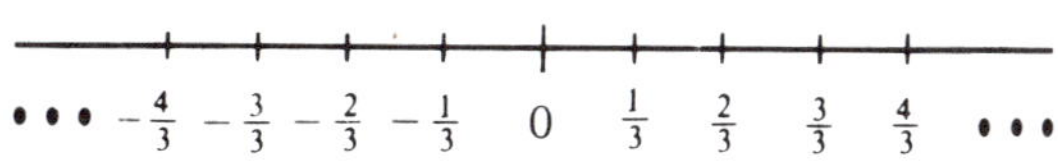

FIGURE 1.5

In locating rational numbers that are represented in decimal form always divide the unit distance by a power of 10. For example, the tenths are divided by making hundredths, and hundredths by making thousandths, as shown in Figure 1.6.

When points on the number line are associated with real numbers, the number is called the **coordinate** of the point and the point is called the **graph** of the number. In precise language, we say the point represents the real number. The two concepts are customarily used interchangeably, however; that is, we say "the point 2" instead of "the point whose coordinate is 2."

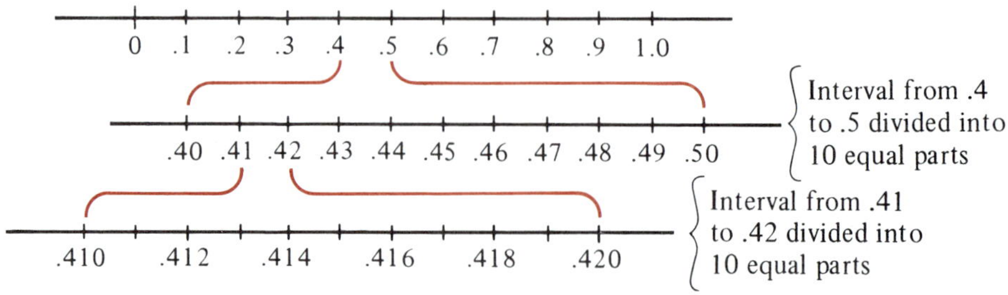

FIGURE 1.6

EXAMPLE 1 Locate the points 1.4 and 1.43 on a number line.

SOLUTION Locate 1.4 by subdividing the portion between 1 and 2 into ten equal parts. Locate 1.43 by subdividing the portion between 1.4 and 1.5 into ten equal parts. See Figure 1.7

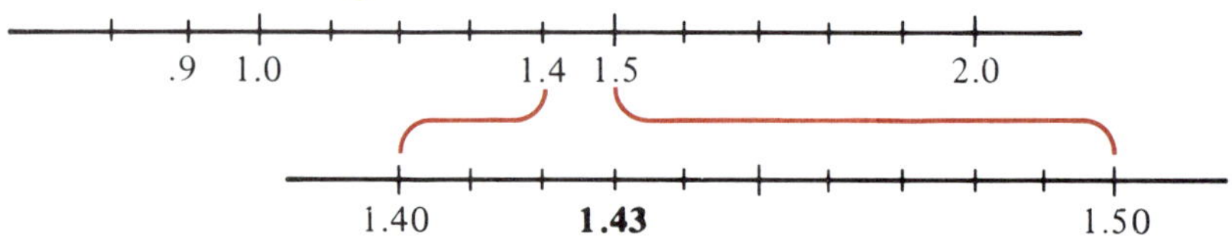

FIGURE 1.7

Real numbers are **ordered** in the sense that for any two distinct real numbers a and b, one is larger than the other. That a is greater than b is denoted by $a>b$ and defined by

$$a>b \quad \text{if } a-b \text{ is positive}$$

where "$a>b$" is read "a is greater than b." Likewise

$$a<b \quad \text{if } b-a \text{ is positive}$$

where "$a<b$" is read "a is less than b."

The symbols $>$ and $<$ are called inequality symbols. The notation $a \geqslant b$ includes the possibility that $a=b$ as well as that $a>b$. A similar statement applies to the notation $a \leqslant b$.

EXAMPLE 2

(a) $9>5$, since $9-5=4$ is positive.
The relation $9>5$ can also be written $5<9$.

(b) $0>-2$, since $0-(-2)=2$ is positive.
The relation $0>-2$ can also be written $-2<0$.

(c) $-1>-8$, since $(-1)-(-8)=(-1)+8=7$ is positive.
The relation $-1>-8$ can also be written $-8<-1$.

From the definition of order, it follows that every positive number is greater than every negative number. Thus, $0.0001>-1000$.

The number line gives an easy and interesting geometric interpretation of ordering. If a and b are real numbers, $a<b$ for any point corresponding to a that is to the left of the point corresponding to b. See Figure 1.8. (a to the left of b means a is less than b.)

FIGURE 1.8

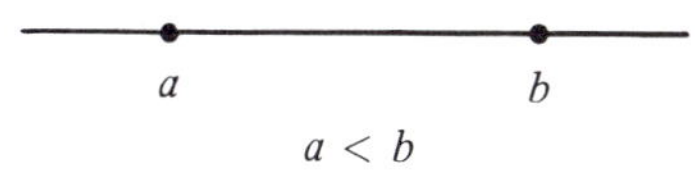

The value of a real number independent of its sign is, we recall, the **absolute value** of the number. The absolute value of a number a is denoted by $|a|$ and defined by

$$|a|=\begin{cases} a & \text{if } a\geqslant 0 \\ -a & \text{if } a<0 \end{cases}$$

From the definition, it follows that the absolute value of every nonzero real number is positive. Thus, if $a>0$, then $|a|=a$, which is positive; and if $a<0$, then $|a|=-a$, which is also positive.

EXAMPLE 3 Evaluate (a) $|19|$; (b) $|-2|$; (c) $|3-\pi|$.

SOLUTION

(a) Since $19>0$, it follows that $|19|=19$.

(b) Since $-2<0$, it follows that $|-2|=-(-2)=2$.

(c) Since $3-\pi<0$, it follows that $|3-\pi|=-(3-\pi)=\pi-3$.

Graphically, the absolute value of a number corresponds to the distance between two points on the coordinate line. Thus, $|a-b|$ represents the number of units between the numbers a and b on the coordinate line. Denoting the distance between a and b by $d(a, b)$, we define it by

$$d(a, b)=|a-b|$$

From the definition of absolute value, it follows that $d(a, b)=d(b, a)$. The distance from the origin of the real line to the point a is given by $d(a, 0)=|a-0|=|a|$, which means that $|a|$ corresponds to the distance from the origin to the point a.

EXAMPLE 4 Find the distance between the indicated points in Figure 1.9.

SOLUTION $d(-3, 2)=|-3-2|=|-5|=5$

FIGURE 1.9

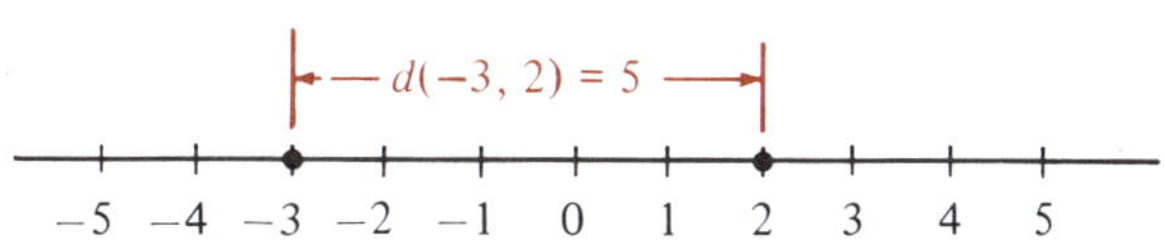

The $\boxed{+/-}$ *button on a calculator can be used to obtain the absolute value of a number. If the displayed number is positive no action is required, but if the displayed number is negative, its absolute value is obtained by pushing the* $\boxed{+/-}$ *button.*

Exercises for Section 1.1 – 1.3

Answer *True* or *False* to Exercises 1–10.

1. An integer is also a rational number.

2. A real number is either positive or negative.

3. An irrational number is the ratio of two rational numbers.

4. If a is to the left of b on the number line, then $a<b$.

5. If a and b are real numbers, then $a+b=b+a$.

6. The real number $-a$ is always negative.

7. A number may be rational and irrational.

8. Every rational number is real.

9. The absolute value of $\sqrt{3}$ is $-\sqrt{3}$.

10. The absolute value of $-a$ is a.

In Exercises 11–20, find the decimal representation of the given rational number and indicate whether it is terminating or nonterminating.

11. 1/7 **12.** 29/3 **13.** 3/8

14. 6/11 **15.** 11/9 **16.** 9/8

17. 11/6 **18.** 12/6 **19.** 6/13

20. 7/13

Write each statement in Exercises 21–44 as a single number. A calculator will be handy for some of them.

21. $(+5)+(-6)$ **22.** $5-(-7)$

23. $-3+7$ **24.** $-15-(+3)$

25. $-9-(-3)$ **26.** $-10-6+20$

27. $(16)(-3)(9)$ **28.** $(-3)(-2)(-6)$

29. $4(5-7)$ **30.** $5+(-4)-(3-4)$

31. $-2-(7-10)(5)$ **32.** $-(-3)(5)$

33. $-[2+3(6-8)][4-(5+2)]$

34. $[3-(6+1)][-2-(-5)]$

35. $[-(-2)-(4-6)][3+(7-3)]$

36. $[3-7(5-6)][-3(-2-5(-4))][-6+(5-7)]$

37. $-2[7.1+(1.2-2.8)-6(2.5-3.2)]$

38. $16.3(3.1-\{6.2-[7.8(4.3+2.1)+8.2]\})$

39. $6[\pi-\frac{2}{3}(6+9)-4.3(\pi-\sqrt{3})+3.1]$

40. $(\pi-\sqrt{2})(\pi+\sqrt{2})-(\sqrt{2}+\pi)(\sqrt{2}+\pi)$

41. $-1395-[748.6-(4096-2716)](10.903)$

42. $[0.395-(1.761-2.059)][5.379+2.13(0.943-1.687)]$

43. $7.5+3.2[1.5-6.6(8.1-4.5)]$

44. $4.07-\{5.87-[0.95+2.53(1.65-3.02)]\}$

Locate on the number line the indicated numbers in Exercises 45–54.

45. 2.3 **46.** -3.5 **47.** 1.67

48. 4.16 **49.** 3.14 **50.** $\frac{3}{8}$

51. $\frac{11}{8}$ **52.** $-\frac{9}{4}$ **53.** -0.017

54. 1.414

Indicate the correct inequality symbol for each set of numbers in Exercises 55–64.

55. 25 17 **56.** 0.01 0.009 **57.** $-\frac{7}{6}$ $\frac{9}{7}$

58. -10 -2 **59.** -0.1 0.002 **60.** 100 -200

61. $\frac{99}{12}$ $-\frac{99}{7}$ **62.** $-\frac{1}{9}$ $-\frac{1}{7}$ **63.** $\sqrt{2}$ π

64. $-\sqrt{5}$ $-\sqrt{7}$

In Exercises 65–74, evaluate the absolute value.

65. $|7|$ **66.** $|-3|$ **67.** $|6-9|$

68. $|3-1|$ **69.** $5+|-9|$ **70.** $3+|-2|$

71. $3+|-3|$ **72.** $6-|9|$ **73.** $1-|3|$

74. $|5+(-9)|$

Find the distance between the indicated points on the number line in Exercises 75–84.

75. $-3, 1$ **76.** $5, 9$ **77.** $-5, -1$

78. $-2, 7$ **79.** $-0.1, 7$ **80.** $-0.5, -0.3$

81. $-6, 0$ **82.** $-8, -1$

83. $-0.125, 0.176$ **84.** $-0.385, 4.671$

85. Show by example that $|a+b| \neq |a|+|b|$. When does $|a+b|=|a|+|b|$?

1.4 Exponential Notation; Algebraic Expressions

A special notation, called **exponential notation**, is used for repeated multiplications of the *same* number. Ordinarily, the product $x \cdot x$ is written x^2 (read "x squared" or "x to the second power"). The product $x \cdot x \cdot x$ is written x^3 (read "x cubed" or "x to the third power").* In general, if n is a positive integer, x^n (read "x to the nth power") is defined by

$$x^n = \overbrace{x \cdot x \cdot x \cdot \cdots \cdot x}^{n \text{ times}}$$

The letter x is called the **base** and n the **exponent** of x. The quantity x^n is called the ***n*th power** of x.

EXAMPLE 1 (a) $2 \cdot 2 \cdot 2 \cdot 2 \cdot 2 \cdot 2 = 2^6 = 64$ (2 is the base and 6 is the exponent)

(b) $(\frac{1}{3})(\frac{1}{3})(\frac{1}{3})(\frac{1}{3}) = (\frac{1}{3})^4$

(c) $(-\pi)(-\pi)(-\pi) = (-\pi)^3$

(d) $(5x+4)(5x+4)(5x+4) = (5x+4)^3$

Squares of numbers are found with a calculator by using the [x^2] *key. Higher powers are found by using the* [y^x] *key. Thus,* $(3.1)^9$ *is found by using the following keystrokes.*

3.1 [y^x] 9 [=] 26439.6

Any grouping of numbers and symbols generated by the elementary arithmetic operations, or by taking powers or extracting roots, is called an **algebraic expression**. Symbols that represent arbitrary elements of a set are usually called **variables**; and unless it has been stated otherwise, the variables are real numbers. The set of all values of a variable for which the expression itself is a real number is called the **domain** of the expression. The domain of the expression $\sqrt{x-1}$, for instance, is the set of all x such that $x \geq 1$ since when $x < 1$, $\sqrt{x-1}$ is not real. The domain of an expression is also referred to as the set of **permissible values**.

*The terms *squared* and *cubed* for x^2 and x^3 come from the early historic perspective of the unknown quantity representing a length; a length times itself is a **square** (geometrically), and a length times a square represents a **cube**.

Terms are groupings of symbols within an algebraic expression that are separated from other groupings by a plus or a minus sign. Thus, x^3, $5x^2y^2$, $2(x-3y)$, and 8 are terms of the expression $x^3+5x^2y^2-2(x-3y)+8$. Likewise, x and $3y$ are terms of the expression $x-3y$.

When a term is composed of a product of numbers and symbols, each number and symbol is called a **factor** of the term. For example, 5, x^2, and y^2 are factors of the term $5x^2y^2$. Similarly, 2 and $(x-3y)$ are factors of the term $2(x-3y)$.

Algebraic expressions are frequently classified according to the number of terms in the expression. An expression with one term is called a **monomial**; with two terms, a **binomial**; and with three terms, a **trinomial**. An expression of more than three terms is called a **multinomial**. A **polynomial** is any expression involving only constant terms or terms in which the variables have positive integer exponents. Each expression in the next example is a polynomial.

EXAMPLE 2

(a) Monomials: $2x$, $5x^2y^2$, $3(x+y)(x-y)$

(b) Binomials: $3x+2y$, $9(a+b)+4(a+b+c)$

(c) Trinomials: $5+x-w$, $r+2s-3(r+s)$

In a term expressed as the product of two or more factors, each factor is called a **coefficient** of the other factors. For example, in the term $5xy$, we have 5 as the coefficient of xy, and x as the coefficient of $5y$, and y as the coefficient of $5x$. In practice, "the coefficient" usually means the numerical factor of a term. Thus, 5 is the coefficient of xy in the term $5xy$. If no numerical coefficient has been expressed, the number 1 is understood. In the expression $2x^2+y-6y^2$, we have 2 as the coefficient of x^2, and 1 the coefficient of y, and -6 the coefficient of y^2.

An important distinction exists between a term of the form ax^n and the term $(ax)^n$. In ax^n, the number x is raised to the nth power, and the result multiplied by the coefficient a. In the second, the product of a and x is raised to the nth power.

EXAMPLE 3

(a) $3 \cdot 2^4 = 3 \cdot (16) = 48$

(b) $(3 \cdot 2)^4 = 6^4 = 1296$

In general, an exponent applies only to the factor to which it is affixed. When it is to apply to more than one factor, as shown in Example 3 (b), grouping symbols must be used.

The distinction between ax^2 and $(ax)^2$ must be observed in working with a calculator. To find $3 \cdot 2^2$, one must use the following keystrokes.

3 [×] 2 [x^2] [=] 12

In computing $(3 \cdot 2)^2$, the multiplication is done before the squaring. Procedurally, this involves interchanging the last two keystrokes in the previous sequence. Try it on your own calculator. The display should read 36.

Terms that differ only in their numerical coefficients are said to be **similar**. For instance, $5xy^2$, $7xy^2$, and $\frac{1}{8}xy^2$ are similar terms, since they have the same literal parts xy^2. Recognizing similar terms is an important part of algebraic simplification. We add, or "combine," two or more similar terms by adding their coefficients. Example 4 shows that combining terms is simply an application of the distributive law.

EXAMPLE 4

(a) $8x^3+3x^3-2x^3=(8+3-2)x^3=9x^3$

(b) $3x^2y^2-(-7x^2y^2)=[3-(-7)]x^2y^2=10x^2y^2$

(c) $5(8x+3)^7-3(8x+3)^7=(5-3)(8x+3)^7=2(8x+3)^7$

It is possible to combine only similar terms by adding coefficients. Thus, the sum of $2x^4$ and $5y^4$, dissimilar terms, can only be indicated, and we write $2x^4+5y^4$; the sum cannot be written any more simply. The sum of two or more algebraic expressions is **simplified** if similar terms have been combined.

EXAMPLE 5 Simplify $(3y^2-5xy^3+7)+(2x^2+3xy^3-7x+2)$.

SOLUTION Since $-5xy^3$ and $3xy^3$ are similar terms, and so are the constant terms 7 and 2, the simplified form is

$$3y^2+2x^2+(-5+3)xy^3-7x+(7+2)=3y^2+2x^2-2xy^3-7x+9$$

EXAMPLE 6 Simplify $3x^2-xy+5x^2-2xy^2+9-4x^2+3xy^2-17-xy^2$.

SOLUTION Combining similar terms, we have

$$(3+5-4)x^2+(-2+3-1)xy^2-xy+(9-17)$$

Since the coefficient of xy^2 reduces to 0, we get

$$4x^2-xy-8$$

EXAMPLE 7 Simplify $(3a^2+ab-9)-(2a^2-5ab-2)-(5ab+6)$.

SOLUTION First, we remove the parentheses from each grouping. The sign of each term in the second and third groupings changes when the parentheses are removed, since each grouping has a negative sign fixed. Hence,

$$\begin{aligned}(3a^2+ab-9)-(2a^2-5ab-2)-(5ab+6) &= 3a^2+ab-9-2a^2+5ab+2-5ab-6\\ &=(3-2)a^2+(1+5-5)ab+(-9+2-6)\\ &=a^2+ab-13\end{aligned}$$

In applications, we sometimes combine expressions of terms that are themselves combinations of terms.

EXAMPLE 8 Let $A=(2x+y-7)$ and $B=(x-3y)$. Find and simplify the expression $3A-2B-(B+2A-1)$.

SOLUTION To remove the parentheses around $B+2A-1$, we must change the sign of each term. Thus,

$$\begin{aligned}3A-2B-(B+2A-1)&=3A-2B-B-2A+1\\&=A-3B+1\end{aligned}$$

Now we let A and B be the given values:

$$\begin{aligned}A-3B+1&=(2x+y-7)-3(x-3y)+1\\&=2x+y-7-3x+9y+1\\&=-x+10y-6\end{aligned}$$

When expressions are grouped within other expressions, we remove the innermost pair of grouping symbols first and then combine like terms. The next example illustrates the technique.

EXAMPLE 9 Simplify $x^2-\{y-x^2+2[x-x^2+5(y-x^2)+3x]-3x\}-3x$.

SOLUTION Remove the innermost parentheses first; the expression is

$$=x^2-\{y-x^2+2[x-x^2+5y-5x^2+3x]-3x\}-3x$$

Collecting like terms within the bracket gives

$$=x^2-\{y-x^2+2[4x-6x^2+5y]-3x\}-3x$$

and removing the brackets

$$=x^2-\{y-x^2+8x-12x^2+10y-3x\}-3x$$

We collect like terms in the inner group of terms,

$$=x^2-\{11y-13x^2+5x\}-3x$$

and then remove the last set of parentheses:

$$=x^2-11y+13x^2-5x-3x$$

Collecting like terms gives

$$=14x^2-11y-8x$$

Exercises for Section 1.4

In 1–16, perform the indicated operation.

1. $(x-3y+7)+(2x+y-6)$

2. $(2p-5q-8)+(p+q)$

3. $(x^2+y^2+x-y)-(y+x-3y^2)$

4. $(xy^2+x^2+7x+2)-(1-y-x-y^2-x^2-2xy^2)$

5. $(a^2-y^2)-(a^2+y^2)$

6. $(z+2w-7)-(2z-w-7)$

7. $(x-y-x^2)+(3x^2-x+y)$

8. $(2y^2+y^4-y^6)+(1-3y^2-2y^4-y^6)$

9. $(x^2y^2+2xy+5)-(x^2+y^2-7-xy)$

10. $(b+b^2+b^3)+(-b+3b^2+b^3)+(b-b^3)$

11. $(3x+2y^2-4x^3-8)+(xy^3-xy^2+x)$
$-(y^2-2x+8)$

12. $(x^4-y^4)+(x^4+x^3+x^2-y-y^3-y^4)$
$-(2x^4+2y^4-x^3-x^2+y+y^3)$

13. $5(x^2+y^2)-2(7x^2+2-y^2)+(9y^2-x^2)$

14. $(c-d)+5(c-d)-(c-d)+18c-18d$

15. $3(m+2n)+2(m-n)-5(2m-3n)$

16. $3(x^2-y^2)+2(x-y)-(8+2x^2+3y^2+2x-3y)$

In 17–24, let $A=x^2-y^2$ and $B=x^2-y^2+2$. Find and simplify the following.

17. $A-B$

18. $2A+3B$

19. $B-A-2$

20. $6A+7B+3$

21. $2A-B+7(A-B)$

22. $A+(A-B)+A-(2A+B)$

23. $B+A-4A-2$

24. $A+8-B$

In 25–30, remove the symbols of grouping and simplify.

25. $\{2x-[x+3(x-7)+4(x-2)+2]-2x\}+3$

26. $2x+y-[x-y-2(x-y)-(1-x)+2]-2y$

27. $4+y+\{x-[1-y+2(x-y)+2x]-3x\}+2x$

28. $x+\{1-[1-(1-x)]\}$

29. $x-\{2x-3y+3[3x-8y-(2x-y)+2(x-3y)]\}$

30. $4a-(3b-1)-2\{a-[a-2(b-a)]\}$

31. In finding the current through an electric current circuit, an electrical engineering student is confronted with the expression

$$(i_1-i_2)-(3i_1-2i_2)+(i_2-i_1)$$

How could the student simplify this expression?

32. In computing the slope of a roadway, an engineer gets the following expression:

$$(m-3)+2m-(3m+6)-(2-5m)$$

What is its simplest expression?

33. Simplify this expression from a chemistry problem:

$$T(3T+2)-[(4-T)T-(T^2-3)]$$

1.5 Multiplication of Algebraic Expressions

An algebraic expression written as a product of other algebraic expressions is said to be in a **factored form**. The individual expressions are called the **factors**. An algebraic expression written without factors is said to be in **expanded form**.

EXAMPLE 1 (a) The algebraic expression $x^3-yx^2-xy^2+y^3$ is in expanded form. One of its factored forms is $(x-y)(x^2-y^2)$. Another of its factored forms is $(x-y)^2(x+y)$.

(b) The expression $3x(x+2y-4z)$ is in a factored form, whereas the expanded form of the same expression is $3x^2+6xy-12xz$.

To expand an expression in factored form, we use the distributive law. For example,

$$x^2(3y+x^5)=x^2(3y)+x^2(x^5)$$

The second term in the expanded form of the expression can be simplified further because

$$x^2 \cdot x^5 = (x \cdot x)(x \cdot x \cdot x \cdot x \cdot x) = x^7$$

Thus, the expanded form of the given expression is $3x^2y + x^7$. In the second term, the exponent equals the sum of the two given exponents; that is, $7 = 5 + 2$. The rule for multiplying exponential factors having the same base is generalized for positive integer exponents as follows:

$$x^m \cdot x^n = \overbrace{(x \cdot x \cdot x \cdot \cdots \cdot x)}^{m \text{ factors}}\overbrace{(x \cdot x \cdot \cdots \cdot x)}^{n \text{ factors}}$$

Since there is a total of $m+n$ factors on the right-hand side, we conclude that

(1.1) $\quad x^m \cdot x^n = x^{m+n}$

EXAMPLE 2

(a) $x^4 \cdot x = x^{4+1} = x^5$

(b) $y^3y^5 = y^8$

(c) $a^4(a^2 - 3b^3) = a^4(a^2) - a^4(3b^3) = a^6 - 3a^4b^3$

Another rule for positive integer exponents is

(1.2) $\quad (x \cdot y)^n = x^ny^n$

The validity of this rule is established by observing that

$$(x \cdot y)^n = \overbrace{(x \cdot y)(x \cdot y)(x \cdot y) \cdots (x \cdot y)}^{n \text{ factors}}$$

Grouping all the x factors together and also all the y factors, we can write

$$(x \cdot y)^n = \overbrace{(x \cdot x \cdot x \cdot \cdots \cdot x)}^{n \text{ factors}} \cdot \overbrace{(y \cdot y \cdot y \cdot \cdots \cdot y)}^{n \text{ factors}}$$

from which we conclude that

$$(x \cdot y)^n = x^ny^n$$

EXAMPLE 3

(a) $(2x)^3 = 2^3x^3 = 8x^3$

(b) $(5xy)^2 = 5^2x^2y^2 = 25x^2y^2$

(c) $64a^3 = 4^3a^3 = (4a)^3$

To read mathematics, you must know how to express algebraic expressions in either their factored form or their equivalent expanded form. The next three examples show how to expand a product using the distributive law.

EXAMPLE 4 Expand $3x(x+2x^2-4y)$.

SOLUTION Using the distributive law, we have

$$\begin{aligned} 3x(x+2x^2-4y) &= 3x(x)+3x(2x^2)-3x(4y) \\ &= 3x^2+6x^3-12xy \end{aligned}$$

EXAMPLE 5 Expand the product $(x-y)(x^3-y^3)$.

SOLUTION Applying the distributive law, we have

$$\begin{aligned} (x-y)(x^3-y^3) &= x(x^3-y^3)-y(x^3-y^3) \\ &= x^4-xy^3-yx^3+y^4 \end{aligned}$$

EXAMPLE 6 Expand $(x^2+5x-3)(x^3-x^2-7)$.

SOLUTION Applying the distributive law, we have

$$\begin{aligned} &(x^2+5x-3)(x^3-x^2-7) \\ &\quad =(x^2+5x-3)x^3+(x^2+5x-3)(-x^2)+(x^2+5x-3)(-7) \end{aligned}$$

Applying the distributive law again, we get

$$=x^5+5x^4-3x^3-x^4-5x^3+3x^2-7x^2-35x+21$$

Finally, collecting like terms, we get

$$=x^5+4x^4-8x^3-4x^2-35x+21$$

Algebraic expressions are sometimes expanded by using a format like that used in arithmetic to multiply numbers. As the next example shows, however, this scheme is just another way of using the distributive law.

EXAMPLE 7 The product in the previous example can also be expanded by the following scheme. Write one expression under the other and then multiply each term in the bottom row by every term in the top row. The resulting products are aligned so that like terms appear in the same column. Finally, add like terms.

$$\begin{array}{llllll}
x^2 & +\ 5x & -3 & & & \\
x^3 & -\ x^2 & -7 & & & \\
\hline
x^5+5x^4-3x^3 & & & & & \leftarrow x^3(x^2+5x-7) \\
\quad -\ x^4-5x^3+3x^2 & & & & & \leftarrow -x^2(x^2+5x-7) \\
\qquad\qquad -7x^2-35x+21 & & & & & \leftarrow -7(x^2+5x-7) \\
\hline
x^5+4x^4-8x^3-4x^2-35x+21 & & & & &
\end{array}$$

EXAMPLE 8 Multiply ax^2+bx+c by $dx+e$, where a, b, c, d, and e are constants.

SOLUTION Using the same format as in Example 7, we have

$$\begin{array}{lllll} ax^2 & +bx & +c & & \\ & dx & +e & & \\ \hline adx^3+bdx^2 & & +dcx & & \leftarrow dx(ax^2+bx+c) \\ & aex^2 & +bex & +ce & \leftarrow e(ax^2+bx+c) \\ \hline \multicolumn{4}{l}{adx^3+(bd+ae)x^2+(dc+be)x+ce} & \end{array}$$

Special Products

An algebraic expression involving the sum or difference of two terms is called a *binomial*. The product of two binomials follows a pattern that can be identified by expanding the general product $(ax+by)(cx+dy)$. Thus,

$$\begin{aligned}(ax+by)(cx+dy) &= ax(cx+dy)+by(cx+dy)\\ &= acx^2+adxy+bcxy+bdy^2\end{aligned}$$

$$\mathbf{(ax+by)(cx+dy)=acx^2+(ad+bc)xy+bdy^2}$$

This result can be applied to the product of any two binomials. The final form of the expansion has the following structure:

- The first term is the product of the two first terms of the given binomials.
- The middle term is the sum of the products obtained by multiplying the first term in each binomial by the second term in the other.
- The third term is the product of the two second terms of the given binomials.

EXAMPLE 9 Expand $(2x+3)(5x-4)$.

SOLUTION

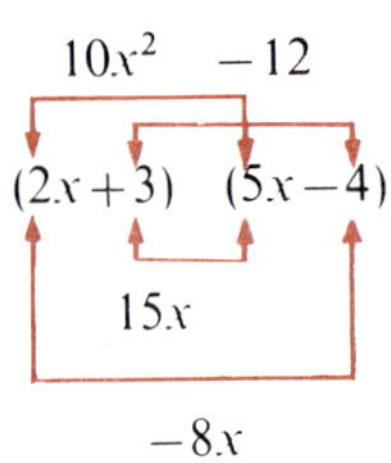

Referring to the diagram shows that

$$\begin{aligned}(2x+3)(5x-4) &= 10x^2+15x-8x-12\\ &= 10x^2+7x-12\end{aligned}$$

EXAMPLE 10 Expand $(a-4b)(8a-3b)$.

SOLUTION

$$\begin{aligned}(a-4b)(8a-3b) &= (1)(8)a^2+[(1)(-3)+(-4)(8)]ab+(-3)(-4)b^2\\ &= 8a^2-35ab+12b^2\end{aligned}$$

Certain products are so frequent in algebraic expressions that learning the equivalent expanded form is advantageous. The following list contains some of these *special products* in factored and expanded form. The validity of each of these expansions is easily checked by the fundamental rules.

(1) $(x+y)^2=x^2+2xy+y^2$
(2) $(x-y)^2=x^2-2xy+y^2$
(3) $(x+y)(x-y)=x^2-y^2$
(4) $(x+y)^3=x^3+3x^2y+3xy^2+y^3$
(5) $(x-y)^3=x^3-3x^2y+3xy^2-y^3$
(6) $(x+y)(x^2-xy+y^2)=x^3+y^3$
(7) $(x-y)(x^2+xy+y^2)=x^3-y^3$

EXAMPLE 11 Expand $(x-7)(x+7)$.

SOLUTION Using the above Expression (3),

$$(x-7)(x+7)=x^2-7^2=x^2-49$$

EXAMPLE 12 Find the square of $(3z+5)$.

SOLUTION Let $x=3z$ and $y=5$ in Expression (1).

$$\begin{aligned}(3z+5)^2 &= (3z)^2+2(3z)(5)+(5)^2\\ &= 9z^2+30z+25\end{aligned}$$

EXAMPLE 13 Expand $(2x-3y)^3$.

SOLUTION According to Expression (5), but replacing x by $2x$ and y by $3y$, we get

$$\begin{aligned}(2x-3y)^3 &= (2x)^3-3(2x)^2(3y)+3(2x)(3y)^2-(3y)^3\\ &= 8x^3-36x^2y+54xy^2-27y^3\end{aligned}$$

EXAMPLE 14 Expand $(a+5b)(a^2-5ab+25b^2)$.

SOLUTION Using Expression (6) with $x=a$ and $y=5b$, we have

$$(a+5b)(a^2-5ab+25b^2)=a^3+(5b)^3=a^3+125b^3$$

Exercises for Section 1.5

Use the distributive law to expand the products in Exercises 1–20.

1. $x^2(x^3+2)$
2. $(x^2+3x-5)x^3$
3. $x^3(x^2-5x)$
4. $x^9(x^3+x^5)$
5. $(x+2y)(x-2y)$
6. $(x-1)(x^2-1)$
7. $(x^2-1)(x^2+1)$
8. $(x+3)(x^2-2x+5)$
9. $(x^3-x^2+x-1)(x-1)$
10. $(x^2-y)^2$
11. $(xy+1)(xy-1)$
12. $(x^2+1)(x^2-2)$
13. $(x+y^2)^3$
14. $(2x+4)^2(3x-2)$
15. $(x+y)(x-y)(x^2-y)$
16. $(x^2-y)(x-y^2)$
17. $(x+y+z)^2$
18. $(x+y)^2(x^2-y^2)$
19. $(3x+4y-z)(2x-3y)$
20. $(x^2+2x-3)^2(x+3)$

Use the table of special products for assistance in expanding the following products.

21. $(x+4)^2$
22. $(x-2)^2$
23. $(x+15)^2$
24. $(a-9)^2$
25. $(b+12)(b-12)$
26. $(y+7)(y-7)$
27. $(2x+5)^2$
28. $(3a+2)^2$
29. $(9y-2)^2$
30. $(2x+3)(2x-3)$
31. $(4x+5)(4x-5)$
32. $(7x+9)(7x-9)$
33. $(x+2)^3$
34. $(x-4)^3$
35. $(5a+b)^3$
36. $(4x+3)^3$
37. $(2x-5)^3$
38. $(a+3b)^3$
39. $(x+2)(x^2-2x+4)$
40. $(x^2-5x+25)(x+5)$
41. $(x-3)(x^2+3x+9)$
42. $(x+1)(x^2-x+1)$
43. $(2x+1)(4x^2-2x+1)$
44. $(3x-2)(9x^2+6x+4)$

Use the general rule for expanding the product of two binomials in the following exercises.

45. $(x+5)(x+9)$
46. $(y+3)(y+7)$
47. $(2x-8)(3x+5)$
48. $(x-4)(3x+9)$
49. $(2y+5)(y+2)$
50. $(8x-3)(3x-2)$
51. $(2x-3y)(x+4y)$
52. $(6m+1)(7m-3)$
53. $(12x-5)(3x-8)$
54. $(x+2y)(5x-4y)$

55. Simplify the expression $\dfrac{(2+h)^3-8}{h}$.

56. Simplify the expression $\dfrac{(h-1)^2-1}{h}$.

57. In calculating the temperature variation of a certain oven, the expression $(3T^3+4)(T^2-2T+5)$ occurs. Write this expression in expanded form.

58. A student encounters the indicated expression while solving a statistics problem: $(2x+4)(x^2-2x+5)$. What is its expanded form?

1.6 Division of Algebraic Expressions

The division of two numbers x and y is written x/y; this is called the quotient of the two numbers. Of particular interest at this time are quotients of the form x^m/x^n, where m and n are positive integers and $m>n$. Consider, for example, the division of x^5 by x^2. For the quotient, we have

$$\frac{x^5}{x^2}=\frac{x\cdot x\cdot x\cdot x\cdot x}{x\cdot x}$$

Two of the x's cancel from numerator and denominator, and hence

$$\frac{x^5}{x^2}=x\cdot x\cdot x=x^3$$

More generally, the quotient x^m/x^n can be determined by first noting that $m=m-n+n$ and that therefore, by (1.1), $x^m=x^{m-n+n}=x^{m-n}\cdot x^n$. Hence,

$$\frac{x^m}{x^n}=\frac{x^{m-n}\cdot x^n}{x^n}=x^{m-n}\left(\frac{x^n}{x^n}\right)=x^{m-n}$$

We have demonstrated the validity of the following rule:

(1.3) $$\frac{x^m}{x^n}=x^{m-n}$$

EXAMPLE 1 (a) $\dfrac{r^4}{r^2}=r^{4-2}=r^2$

(b) $\dfrac{(ab)^7}{(ab)^4}=(ab)^{7-4}=(ab)^3=a^3b^3$

(c) $\dfrac{(x^2y)^{10}}{(x^2y)^7}=(x^2y)^{10-7}=(x^2y)^3=(x^2)^3y^3=x^6y^3$

Division of a multinomial by a monomial is accomplished by dividing the monomial into each term of the multinomial.

EXAMPLE 2 $$\frac{5a^3b^5-7a^4}{20a^2b^2}=\frac{5a^3b^5}{20a^2b^2}-\frac{7a^4}{20a^2b^2}=\frac{ab^3}{4}-\frac{7a^2}{20b^2}$$

EXAMPLE 3 $$\frac{4x^2y^3-3x^3y^4+x^4}{2xy^2}=\frac{4x^2y^3}{2xy^2}-\frac{3x^3y^4}{2xy^2}+\frac{x^4}{2xy^2}=2xy-\frac{3}{2}x^2y^2+\frac{x^3}{2y^2}$$

Suppose that we want to divide one polynomial by another, for example,

$$\frac{2x^3+3x^2-x+5}{x^2-6}$$

To find quotients of this type, we proceed as follows:

(1) Arrange both the dividend and the divisor in descending powers of the variable. Leave space between given terms for the missing powers of x.

(2) Divide the first term of the dividend by the first term of the divisor. The result is the first term of the quotient.

(3) Multiply the divisor by the first term of the quotient and subtract the product from the dividend.

(4) Divide the first term of this difference by the first term of the divisor. This result is the second term of the quotient; it is multiplied by the divisor and subtracted from the first difference.

(5) Repeat these steps until the remainder is either zero or of lower degree than the divisor.*

EXAMPLE 4 Divide $4x^3+5x-9$ by $2x-3$.

SOLUTION

(1) Since both dividend and divisor are already in descending powers of x, there is nothing to do in this step.

(2) $4x^3$ is the first term of the dividend and $2x$ is the first term of the divisor. Since $4x^3 \div 2x = 2x^2$, the term $2x^2$ is the first term of the quotient.

(3) Each term of the divisor is multiplied by $2x^2$:

$$2x^2(2x-3)=2x^2(2x)+2x^2(-3)=4x^3-6x^2$$

This last expression is subtracted from the dividend:

$$\begin{aligned}(4x^3+5x-9)-(4x^3-6x^2)&=4x^3+5x-9-4x^3+6x^2\\&=6x^2+5x-9\end{aligned}$$

(4) The first term of this expression, $6x^2$, is divided by $2x$, to get $3x$, the second term of the quotient. The divisor is multiplied by the second term of the quotient, to get $6x^2-9x$, and this is subtracted from the $6x^2+5x-9$, to get $14x-9$.

(5) Finally, $14x$ divided by $2x$ is 7, which gives the third term of the quotient. Multiplying 7 by the divisor and subtracting from $14x-9$, we get $+12$. Since this is of lower degree than the divisor, $+12$ is the remainder.

We arrange the above work in the following compact format:

$$\begin{array}{lrl l}
 & & 2x^2+3x+7 & \text{Quotient}\\
\text{Divisor} & 2x-3\,) & 4x^3 \qquad\quad +5x-9 & \text{Dividend}\\
 & & 4x^3-6x^2 & \\
 & & \qquad 6x^2+5x-\ 9 & \\
 & & \qquad 6x^2-9x & \\
 & & \qquad\qquad 14x-\ 9 & \\
 & & \qquad\qquad 14x-21 & \\
 & & \qquad\qquad\quad +12 & \text{Remainder}
\end{array}$$

EXAMPLE 5 Divide $x^3+3x^4-2+x^2+x^5$ by $3-x+x^2$.

SOLUTION We first arrange the dividend and divisor in descending powers of x. The problem restated is then: Divide $x^5+3x^4+x^3+x^2-2$ by x^2-x+3.

*The degree of a polynomial in *one* variable is the value of the largest power of the variable in the expression. Thus, the degree of $2x^3+3x^2-x+5$ is 3; of x^2-6 is 2; of $2x-3$ is 1; and of -30 is 0.

The division is carried out as shown:

$$
\begin{array}{lrl}
 & x^3+4x^2+2x-9 & \text{Quotient} \\
\text{Divisor} \quad x^2-x+3\,\big) & x^5+3x^4+x^3+x^2\qquad-2 & \text{Dividend} \\
 & x^5-x^4+3x^3 & \\
 & 4x^4-2x^3+x^2 & \\
 & 4x^4-4x^3+12x^2 & \\
 & 2x^3-11x^2 & \\
 & 2x^3-2x^2+6x & \\
 & -9x^2-6x-2 & \\
 & -9x^2+9x-27 & \\
 & -15x+25 & \text{Remainder}
\end{array}
$$

Synthetic Division

Much of the work in dividing one polynomial by another lies in writing the variable over and over. If you study Examples 4 and 5 just completed, you will see that much of what we did involved repeating the variable. The division can be shortened considerably by eliminating redundant notation. A particularly useful method when the divisor is a linear expression of the form $x-r$ is called **synthetic division**. The idea is to eliminate all the unnecessary parts of the division process. The method of synthetic division is developed in the context of the next example.

EXAMPLE 6 (a) Divide $3x^3-7x^2-x-2$ by $x-3$.

(b) Use this problem to develop synthetic division.

SOLUTION (a) Using long division, we have

$$
\begin{array}{rr}
 & 3x^2+2x+5 \\
x-3\,\big) & 3x^3-7x^2-x-2 \\
 & 3x^3-9x^2 \\
 & 2x^2-x \\
 & 2x^2-6x \\
 & 5x-2 \\
 & 5x-15 \\
 & 13
\end{array}
$$

(b) Synthetic division is long division with all redundancy removed, that is, all unnecessary operations stripped away. In long division, the coefficients are the only important quantities. Since the powers of x are not needed, we omit them and write the indicated division as

$$
\begin{array}{rr}
 & 3\quad 2\quad 5 \\
-3\,\big) & \textcircled{3}\;-7\;-1\;-2 \\
 & 3\;-9 \\
 & \textcircled{2}\;-1 \\
 & 2\;-6 \\
 & \textcircled{5}\;-2 \\
 & 5\;-15 \\
 & 13
\end{array}
$$

You will notice that the three circled coefficients are identical to the coefficients of the quotient, so we eliminate the quotient and other identical terms.

$$
\begin{array}{rrrrr}
-3\,) & \textcircled{3} - 7 & -1 & -2 & \\
 & -9 & & & \\
\hline
 & \textcircled{2} & & & \\
 & & -6 & & \\
\cline{3-3}
 & & \textcircled{5} & & \\
 & & & -15 & \\
\cline{4-4}
 & & & 13 &
\end{array}
$$

Now the numbers below the dividend can be written on two lines,

$$
\begin{array}{rrrrr}
-3\,) & \textcircled{3} & -7 & -1 & -2 \\
 & & -9 & -6 & -15 \\
\hline
 & & \textcircled{2} & \textcircled{5} & 13
\end{array}
$$

All coefficients of the quotient appear below the line except the first. We therefore write this one below the line. It is always the same as the first coefficient of the dividend.

$$
\begin{array}{rrrrr}
-3\,) & 3 & -7 & -1 & -2 \\
 & & -9 & -6 & -15 \\
\hline
 & \nearrow & \nearrow & \nearrow & \\
 & 3 & 2 & 5 & 13
\end{array}
$$

Notice that after bringing down the first coefficient, $3(-3)=-9$; subtracting -9 from -7 yields 2. Next, $2(-3)=-6$; subtracting -6 from -1 yields 5. Finally, $5(-3)=-15$; subtracting -15 from -2 gives 13 as the remainder. In each of these operations, we subtracted the second line from the first. By changing the sign of the divisor, we can add the terms instead of subtracting them; hence, the entire outline takes on a false, or *synthetic*, appearance. The original division problem can now be solved and written as

$$
\begin{array}{rrrrr}
3\,) & 3 & -7 & -1 & -2 \\
 & & 9 & 6 & 15 \\
\hline
 & \nearrow & \nearrow & \nearrow & \\
 & 3 & 2 & 5 & 13
\end{array}
$$

The meaning of the various coefficients relative to the original problem is indicated next.

$$
\begin{array}{rrrrrl}
x-3 & 3x^3 & -7x^2 & -x & -2 & \\
3\,) & 3 & -7 & -1 & -2 & \\
 & & 9 & 6 & 15 & \\
\hline
 & 3 & 2 & 5 & 13 & = \text{Remainder}
\end{array}
$$

$$\text{Quotient} = 3x^2 + 2x + 5$$

To use synthetic division for dividing a polynomial by $x-r$, proceed as follows:

(1) Arrange the coefficients of the polynomial in order of descending powers of x, remembering that missing terms have the coefficient zero.

(2) Replace $x-r$ by r.

(3) Bring down the coefficient of the highest power of x, multiply it by r, and add the result to the coefficient of the next highest power of x. Multiply this sum by r and add to the next coefficient. Continue until a product is added to the constant term.

(4) The last number in the bottom row is the remainder. The numbers to the left of the remainder are the respective coefficients of the quotient. Notice that the quotient is lower by one degree than the dividend.

EXAMPLE 7 Divide $5x^4-18x^2-6x+3$ by $x+2$, using synthetic division.

SOLUTION The divisor is $x+2$, so $r=-2$. Thus,

$$\begin{array}{r|rrrrr} -2 & 5 & 0 & -18 & -6 & 3 \\ & & -10 & 20 & -4 & 20 \\ \hline & 5 & -10 & 2 & -10 & 23 \end{array}$$

The quotient is $5x^3-10x^2+2x-10$, with a remainder of 23.

The limitations on the use of synthetic division must be clearly understood. The dividend must be a polynomial (that is, the terms must include only nonnegative powers of x) and the divisor must be of the form $x-r$. Thus, a divisor such as $2x-3$ must initially be written $2(x-\frac{3}{2})$ before proceeding.

Exercises for Section 1.6

Perform the indicated divisions.

1. $\dfrac{5a^2b^3c^4}{10ab^2c^2}$

2. $\dfrac{3x^5y^3z^2}{7x^3y^2z}$

3. $\dfrac{16rs^2t^3}{4rst^2}$

4. $\dfrac{125p^5q^7}{25p^4q^3v^2}$

5. $\dfrac{4x^3y^2-3}{12x^2y}$

6. $\dfrac{3x^3y^5+15y^3}{6x^2y^2}$

7. $\dfrac{a^2b^5-a^3b^4+a}{ab^3}$

8. $\dfrac{2fL^5-mfL^6}{3fL^5}$

9. $\dfrac{x^3y^2+4-2x^2y^3}{4x^2y}$

10. $\dfrac{u^3v+w+u^4v^3w}{u^3w}$

11. $\dfrac{2x^3-9x^2+11x-6}{x-3}$

12. $\dfrac{y^2-5y+3}{y+4}$

13. $\dfrac{3y^2-y^3+2y-7}{y+1}$

14. $\dfrac{x^3+3x-4}{x+3}$

15. $\dfrac{2x^3-7x^2+9x-3}{x^2-3x+3}$

16. $\dfrac{8x+2x^3-3x^2-2}{x^2-x+2}$

17. $\dfrac{3x^3+19x^2+16x-20}{3x-2}$

18. $\dfrac{3x^3-5x^2-3x-1}{x^2-x-1}$

Perform the divisions in Exercises 19–30, using synthetic division.

19. $(x^2-5x+4)\div(x-4)$

20. $(x^2+7x+12)\div(x-3)$

21. $(x^3-2x^2+3x-1)\div(x-2)$

22. $(x^3+2x^2-3x+4)\div(x+1)$

23. $(x^3-3x+9)\div(x+\frac{1}{2})$

24. $(x^3-1)\div(x+1)$

25. $(x^4-x^3+2x^2-72)\div(x-3)$

26. $(2x^3-4x^2+x-1)\div(x+2)$

27. $(8+7x-3x^5)\div(x-2)$

28. $(4x^3+x^2-16x-4)\div(x+\frac{1}{2})$

29. $(3x^3-20x^2+23x+10)\div(x+\frac{1}{3})$

30. $(3x^3+2x^2-4x+1)\div(x-\frac{1}{3})$

31. The indicated expression occurs when solving for the equivalent resistance of three resistors connected in parallel:

$$\frac{R_1R_2+R_1R_3+R_2R_3}{R_1R_2R_3}$$

Perform the indicated division.

1.7 Factoring

Expanding a product is usually easy since it is a straightforward application of the distributive property or of some well-known formula. Factoring, on the other hand, requires much more ingenuity. We can often use "factoring formulas," the reverse of the product formulas of Section 1.5.

(1) $x^2+2xy+y^2=(x+y)^2$

(2) $x^2-2xy+y^2=(x-y)^2$

(3) $x^2-y^2=(x+y)(x-y)$

(4) $x^3+3x^2y+3xy^2+y^3=(x+y)^3$

(5) $x^3-3x^2y+3xy^2-y^3=(x-y)^3$

(6) $x^3+y^3=(x+y)(x^2-xy+y^2)$

(7) $x^3-y^3=(x-y)(x^2+xy+y^2)$

Sometimes factoring involves simply recognizing common factors, as in the following example.

EXAMPLE 1 Factor $6x^5+12x^2-2x$.

SOLUTION Each term contains $2x$ as a factor; therefore, by the distributive law, we get

$$\begin{aligned}6x^5+12x^2-2x&=2x\cdot 3x^4+2x\cdot 6x-2x\cdot 1\\&=2x(3x^4+6x-1)\end{aligned}$$

EXAMPLE 2 Factor (a) $9x^2-25y^2$; (b) w^2-5.

SOLUTION Since each expression is the difference of two squares, we can use Formula (3):

(a) $9x^2-25y^2=(3x)^2-(5y)^2=(3x-5y)(3x+5y)$

(b) $$\begin{aligned} w^2-5&=w^2-(\sqrt{5})^2\\ &=(w-\sqrt{5})(w+\sqrt{5}) \end{aligned}$$

EXAMPLE 3 Factor $9x^2+24x+16$.

SOLUTION Observe that $9x^2=(3x)^2$ and $16=4^2$. What can we say about the x term? Since $24x=2(3x)(4)$, we can use Formula (1), replacing x by $3x$ and y by 4, to get

$$9x^2+24x+16=(3x+4)^2$$

EXAMPLE 4 Factor $2x^2-x-1$.

SOLUTION If this expression can be factored, it will be of the form $(ax+b)(cx+d)$. Expanding this general form gives

$$(ax+b)(cx+d)=acx^2+(ad+bc)xy+bdy^2$$

This means that $ac=2$ and $bd=-1$. Since the signs of b and d must be opposite for their product to be negative, the only possible factors are

$$(2x-1)(x+1) \quad \text{and} \quad (2x+1)(x-1)$$

Which one is correct? The clue is the coefficient of the x term. Which of these possibilities gives the middle term $-x$ when expanded? The second one is correct, since

$$2x(-1)+(x)(1)=-2x+x=-x$$

The correct factorization is therefore

$$2x^2-x-1=(2x+1)(x-1)$$

EXAMPLE 5 Factor $5x^2+13x-6$.

SOLUTION As in the previous example, if the expression can be factored, it will be of the form $(ax+b)(cx+d)$. Here, $ac=5$ and $bd=-6$. The factors of 5 are $+5$, -5, $+1$, -1. The factors of 6 are $+6$, -6, $+1$, -1, $+2$, -2, $+3$, -3. We know that since the sign of bd is negative, the factors will be of the form

$$(ax+b)(cx-d) \quad \text{or} \quad (ax-b)(cx+d)$$

We can also investigate the x term. In the general form, the coefficient is given by $ad+bc$. Here the x term coefficient is $+13$. This means that any possible factorizations must yield a middle term $+13x$ when they are expanded.

The following are possible factors and their expanded forms. We can ignore factors having $-5x$ and $-x$ as the first terms in the two factors. (Why can we do this?)

$$\begin{aligned} &(5x+6)(x-1)\rightarrow 5x^2+x-6\\ &(5x-6)(x+1)\rightarrow 5x^2-x-6\\ &(5x+1)(x-6)\rightarrow 5x^2-29x-6 \end{aligned}$$

$$(5x-1)(x+6) \rightarrow 5x^2+29x-6$$
$$(5x+2)(x-3) \rightarrow 5x^2-13x-6$$
correct $(5x-2)(x+3) \rightarrow 5x^2+13x-6$
$$(5x+3)(x-2) \rightarrow 5x^2-7x-6$$
$$(5x-3)(x+2) \rightarrow 5x^2+7x-6$$

Thus the correct factorization is $(5x-2)(x+3)$.

EXAMPLE 6 Factor (a) $15a^2-8ab-12b^2$; (b) $6x^2-29xy+20y^2$; (c) $(m+n)^2-2(m+n)-8$.

SOLUTION (a) $15a^2-8ab-12b^2=(3a+2b)(5a-6b)$

(b) $6x^2-29xy+20y^2=(6x-5y)(x-4y)$

(c) $(m+n)^2-2(m+n)-8=[(m+n)-4][(m+n)+2]=[m+n-4][m+n+2]$

EXAMPLE 7 Factor x^3+27.

SOLUTION From Formula (6), we find that $(x+y)(x^2-xy+y^2)=x^3+y^3$; and since $x^3+27=x^3+3^3$, we can write

$$x^3+27=(x+3)(x^2-3x+9)$$

EXAMPLE 8 Factor $27r^3-54r^2s+36rs^2-8s^3$.

SOLUTION $27r^3=(3r)^3$ and $8s^3=(2s)^3$, the expression can be written

$$(3r)^3-3(3r)^2(2s)+3(3r)(2s)^2-(2s)^3$$

This looks like Formula (5) if we let $x=3r$ and $y=2s$. Therefore

$$27r^3-54r^2s+36rs^2-8s^3=(3r-2s)^3$$

Factoring by Grouping

Sometimes algebraic expressions can be factored by grouping terms appropriately. Example 9 illustrates the technique.

EXAMPLE 9 Factor $a+2ab-3c-6cb$.

SOLUTION Inspection shows that a is a common factor in the first two terms and that $-3c$ is common to the last two. Therefore, we write

$$a+2ab-3c-6cb=a(1+2b)-3c(1+2b)$$

We can now see, however, that $(1+2b)$ is a common factor, and we have completely factored the expression. Thus

$$a+2ab-3c-6cb=(1+2b)(a-3c)$$

Completing the Square

The expression x^2+6x+9 is a **perfect square** since it factors into $(x+3)^2$. Likewise, $x^2-3x+\frac{9}{4}$ is a perfect square since it equals $(x-\frac{3}{2})^2$. From the preceding examples, can you guess what value of c will make x^2+8x+c a perfect square? How about $3x^2-7x+c$? For expressions that do not factor easily, we use the method called **completing the square**. The process is based on the rule*

$$(x+k)^2=x^2+(2k)x+(k)^2$$

The constant term k^2 on the right side of the equation is exactly the square of one-half the coefficient of x, where we assume that the coefficient of x^2 is 1. Thus, the expression x^2+8x+c is a perfect square if $c=[\frac{1}{2}(8)]^2=16$. Completing the square is explained in the next two examples.

EXAMPLE 10 Factor x^2+6x+4 by completing the square.

SOLUTION We can make x^2+6x a perfect square by adding $[\frac{1}{2}(6)]^2=9$. Therefore, we add and subtract 9 in the given expression to get

$$\begin{aligned} x^2+6x+4 &= x^2+6x+9-9+4 \\ &= \underbrace{x^2+6x+9}-5 \\ &= (x+3)^2-(\sqrt{5})^2 \leftarrow \text{difference of two squares} \\ &= (x+3-\sqrt{5})(x+3+\sqrt{5}) \end{aligned}$$

EXAMPLE 11 Factor $4x^2-12x-1$ by completing the square.

SOLUTION The "trick" in this problem is to factor 4 from the terms $4x^2-12x$. Thus,

$$4x^2-12x-1=4(x^2-3x+\quad)-1$$

Now complete the square inside the parentheses by adding $(\frac{3}{2})^2=\frac{9}{4}$ to x^2-3x. Since adding $\frac{9}{4}$ inside the parentheses is equivalent to adding 9 to the given expression, we

*We can represent the process geometrically by drawing a square of $x+k$ units on a side, as shown in the accompanying figure. Then x^2+2kx is represented by the shaded area. To complete the area of the square we need to add the area k^2 to the area x^2+2kx.

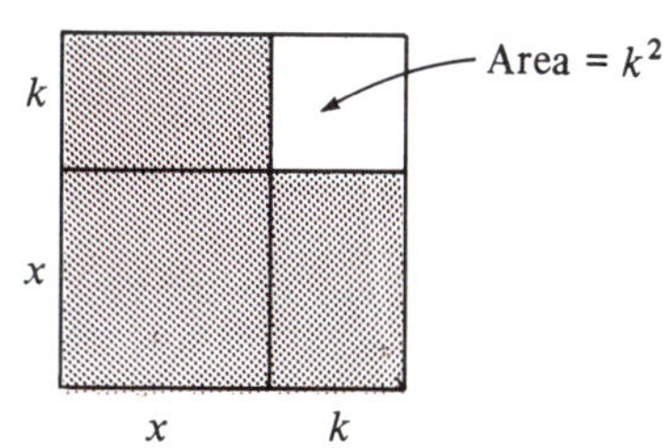

must also subtract 9. The factorization is written

$$\begin{aligned}4x^2-12x-1&=4(x^2-3x+\tfrac{9}{4})-1-9\\&=4(x-\tfrac{3}{2})^2-10\\&=[2(x-\tfrac{3}{2})]^2-(\sqrt{10})^2\\&=[2(x-\tfrac{3}{2})-\sqrt{10}][2(x-\tfrac{3}{2})+\sqrt{10}]\\&=[2x-3-\sqrt{10}][2x-3+\sqrt{10}]\end{aligned}$$

In summary, possible factoring approaches are:

(1) Look for common factors.

(2) Look for special factors (such as the difference of two squares).

(3) Determine whether you can factor by grouping.

(4) Factor by completing the square.

(5) If all these fail, ask a friend for help.

Exercises for Section 1.7

Factor the expressions in Exercises 1–56.

1. $3x^3-5x^2+15x$ **2.** $x^{10}+16x^5$

3. $13y^7-27y^3$ **4.** $y^2+14y+49$

5. $z^2-24z+144$ **6.** $x^2+50x+625$

7. $9x^2+6x+1$ **8.** $25x^2-10x+1$

9. $\frac{1}{4}x^2+x+1$ **10.** x^2-25

11. x^2-169 **12.** x^2-400

13. $4m^2-64$ **14.** $9y^2-25$

15. x^2-7 **16.** $16x^2-13$

17. $\frac{1}{4}x^2-\frac{1}{9}$ **18.** $\frac{4}{9}x^2-\frac{1}{25}$

19. $9x^2+12x+4$ **20.** $25x^2-30x+9$

21. $\frac{1}{9}x^2+\frac{2}{9}x+\frac{1}{9}$ **22.** x^2-x-6

23. x^2-x-12 **24.** a^2-a-2

25. $y^2+13y+42$ **26.** m^2+m-42

27. $x^2-3x-54$ **28.** $b^2-4b-21$

29. $2x^2+7x+3$ **30.** $2x^2+5x+3$

31. $5x^2-28x+15$ **32.** $5y^2-3y-2$

33. $6(a+b)^2-11(a+b)+5$

34. $6(h+k)^2+29(h+k)-5$

35. $3x^2+4xy-15y^2$ **36.** $x^2-xy-12y^2$

37. $3x^2-4xy-4y^2$ **38.** $12a^2+13ab-35b^2$

39. $8x^2-20xy+8y^2$ **40.** x^3-27

41. x^3+64 **42.** y^3-125

43. $8x^3+1$ **44.** $27y^3-8$

45. $a^3+\frac{1}{8}$ **46.** $x^3+6x^2+12x+8$

47. $y^3-9y^2+27y-27$

48. $a^3-6a^2b+12ab^2-8b^3$

49. $8x^3+12x^2+6x+1$

50. $8x^3-24x^2+24x-8$

51. $a^2+a^3b+ab^3+b^2$

52. x^2+y-y^2-x **53.** $z^2-w^2-3z+3w$

54. $r^5+4r^3+3r^2+12$

55. $a^2-c^2+2cd-d^2$

56. $9-x^2-4xy-4y^2$

In Exercises 57–66, factor the expressions by completing the square.

57. x^2+4x+1

58. y^2-6y-2

59. a^2-8a-5

60. $x^2+12x+21$

61. $x^2+3x+\frac{3}{4}$

62. $w^2-2w+\frac{1}{2}$

63. $4y^2-8y-8$

64. $3x^2+12x-7$

65. $2b^2+6b-3$

66. $4z^2-4z-1$

1.8 Fractional Expressions

If both the numerator and the denominator of a fraction are multiplied or divided by the same nonzero quantity, the value of the fraction is unchanged. Thus,

$$\frac{-a}{b}=\frac{a}{-b}=-\frac{a}{b}$$

An algebraic expression is in simplest form when the numerator and the denominator have no common factors. To simplify a fraction, write the numerator and the denominator in factored form and then divide each by every factor common to both. Thus,

$$\frac{ak}{bk}=\frac{a}{b} \quad \text{when } k\neq 0$$

EXAMPLE 1 Reduce to its simplest form

$$\frac{x^2+3x+2}{x^2-1}$$

SOLUTION Factoring, we have

$$\frac{x^2+3x+2}{x^2-1}=\frac{(x+2)(x+1)}{(x-1)(x+1)}$$

Dividing the numerator and the denominator by $x+1$ gives

$$\frac{x+2}{x-1}$$

as the simplest form of the given fraction. We note that this form is valid only if $x\neq -1$. Because the factor $x+1$ effectively has been cancelled, we call this procedure *cancellation* of terms.

Multiplication and Division of Algebraic Fractions

The product of two fractions is the product of the two numerators divided by the product of the two denominators. If a/b and c/d are the given fractions,

$$\frac{a}{b}\cdot\frac{c}{d}=\frac{a\cdot c}{b\cdot d}=\frac{ac}{bd}$$

Any common factors in the numerator and the denominator of the product should be cancelled.

EXAMPLE 2 Simplify $\frac{3x^2y^3}{5a^4} \cdot \frac{15a^3}{16x^5y^5} \cdot \frac{xy}{a^2}$.

SOLUTION

$$\frac{3x^2y^3}{5a^4} \cdot \frac{15a^3}{16x^5y^5} \cdot \frac{xy}{a^2} = \frac{45a^3x^3y^4}{80a^6x^5y^5} = \frac{9}{16a^3x^2y}$$

EXAMPLE 3 Simplify $\frac{3(x^2-y^2)}{x^2+x-6} \cdot \frac{x^2+3x}{x+y}$.

SOLUTION

$$\frac{3(x^2-y^2)}{x^2+x-6} \cdot \frac{x^2+3x}{x+y} = \frac{3(\cancel{x+y})(x-y)}{(x-2)(\cancel{x+3})} \cdot \frac{x(\cancel{x+3})}{\cancel{x+y}} = \frac{3x(x-y)}{x-2}$$

The method learned in arithmetic for simplifying the quotient of two fractions is to invert the denominator fraction and then follow the rules of multiplication. This same rule is followed in dividing two algebraic fractions. If a/b and c/d are given fractions,

$$\frac{a}{b} \div \frac{c}{d} = \frac{a}{b} \cdot \frac{d}{c} = \frac{ad}{bc} \left(\frac{a}{b} \div \frac{c}{d} \text{ is also written } \frac{a/b}{c/d} \right)$$

EXAMPLE 4 Simplify: $\frac{x^2-3x-4}{x^2-1} \div \frac{x^2-16}{x-1}$.

SOLUTION

$$\frac{x^2-3x-4}{x^2-1} \div \frac{x^2-16}{x-1} = \frac{x^2-3x-4}{x^2-1} \cdot \frac{x-1}{x^2-16}$$

$$= \frac{(\cancel{x-4})(\cancel{x+1})}{(\cancel{x+1})(\cancel{x-1})} \cdot \frac{\cancel{x-1}}{(x+4)(\cancel{x-4})}$$

$$= \frac{1}{x+4}$$

Addition and Subtraction of Fractions

Summing algebraic fractions is like summing arithmetic fractions. The general procedure for adding fractions can be outlined as follows:

(1) Factor the numerator and the denominator of each fraction.

(2) Reduce the fractions to lowest terms.

(3) Find the lowest common denominator (LCD).*

(4) Multiply the numerator and the denominator of each fraction by the quotient obtained by dividing the specific denominator into the LCD.

(5) Add the numerators of the fractions and place over the LCD.

(6) Reduce the resulting fraction to lowest terms.

$$\frac{a}{b}+\frac{c}{d}=\frac{ad}{bd}+\frac{bc}{bd}=\frac{ad+bc}{bd}$$

EXAMPLE 5 Simplify $\dfrac{x}{x+1}-\dfrac{x}{x-2}$.

SOLUTION The LCD is $(x+1)(x-2)$. Thus,

$$\begin{aligned}\frac{x}{x+1}-\frac{x}{x-2}&=\frac{x(x-2)}{(x+1)(x-2)}-\frac{x(x+1)}{(x-2)(x+1)}\\&=\frac{x(x-2)-x(x+1)}{(x+1)(x-2)}\\&=\frac{x^2-2x-x^2-x}{(x+1)(x-2)}\\&=\frac{-3x}{(x+1)(x-2)}\end{aligned}$$

EXAMPLE 6 Simplify $\dfrac{2a}{3b^2c^2}-\dfrac{3b}{5a^2c^5}-\dfrac{c}{2ab^3}$.

SOLUTION The LCD is $30a^2b^3c^5$. Therefore, for the first fraction, multiply the numerator and the denominator by $10a^2bc^3$; for the second, by $6b^3$; and for the third, by $15ac^5$.

$$\begin{aligned}\frac{2a}{3b^2c^2}-\frac{3b}{5a^2c^5}-\frac{c}{2ab^3}&=\frac{2a}{3b^2c^2}\cdot\frac{10a^2bc^3}{10a^2bc^3}-\frac{3b}{5a^2c^5}\cdot\frac{6b^3}{6b^3}-\frac{c}{2ab^3}\cdot\frac{15ac^5}{15ac^5}\\&=\frac{20a^3bc^3}{30a^2b^3c^5}-\frac{18b^4}{30a^2b^3c^5}-\frac{15ac^6}{30a^2b^3c^5}\\&=\frac{20a^3bc^3-18b^4-15ac^6}{30a^2b^3c^5}\end{aligned}$$

EXAMPLE 7 Simplify $\dfrac{a-4}{a^2-7a+12}+\dfrac{5}{a^2-9}-\dfrac{3a-6}{a^2+a-6}$.

*The LCD of a fraction is the simplest expression that is exactly divisible by each of the given denominators.

SOLUTION Factor the denominators before finding the LCD. The given fractions can then be rewritten:

$$\frac{\cancel{a-4}}{(a-3)(\cancel{a-4})}+\frac{5}{(a+3)(a-3)}-\frac{3(\cancel{a-2})}{(\cancel{a-2})(a+3)}=\frac{1}{a-3}+\frac{5}{(a+3)(a-3)}-\frac{3}{a+3}$$

From this, we see that a^2-9 is the LCD. Therefore, the original expression equals

$$\frac{1}{a-3}\cdot\frac{a+3}{a+3}+\frac{5}{(a+3)(a-3)}-\frac{3}{a+3}\cdot\frac{a-3}{a-3}=\frac{a+3}{a^2-9}+\frac{5}{a^2-9}-\frac{3a-9}{a^2-9}$$

$$=\frac{(a+3)+5-(3a-9)}{a^2-9}$$

$$=\frac{a+3+5-3a+9}{a^2-9}=\frac{-2a+17}{a^2-9}$$

EXAMPLE 8 Simplify the fraction $\dfrac{3-\dfrac{x+3y}{x+y}}{1+\dfrac{y}{x-y}}$

SOLUTION Before dividing, combine the fractions in the numerator and the denominator. The numerator becomes

$$\frac{3(x+y)-(x+3y)}{x+y}=\frac{2x}{x+y}$$

The denominator becomes

$$\frac{(x-y)+y}{x-y}=\frac{x}{x-y}$$

Thus, the original fraction can be written

$$\frac{3-\dfrac{x+3y}{x+y}}{1+\dfrac{y}{x-y}}=\frac{\dfrac{2x}{x+y}}{\dfrac{x}{x-y}}=\frac{2x}{x+y}\cdot\frac{x-y}{x}=\frac{2(x-y)}{x+y}$$

Exercises for Section 1.8

Perform the indicated multiplications and divisions in Exercises 1–24.

1. $\dfrac{b}{3a}\cdot\dfrac{5c}{4ab^3}$

2. $\dfrac{x^2y}{5xy^2}\cdot\dfrac{3xy^2}{x^3y^5}$

3. $\dfrac{3x}{4y^3}\div\dfrac{9x^4}{2y}$

4. $\dfrac{xyz}{5x^2y^3z}\div\dfrac{4}{25x^3y^2z^2}$

5. $\dfrac{9b^2}{4c^3}\cdot\dfrac{24ac^2}{18b^5}\div\dfrac{c}{2a^3b}$

6. $\dfrac{3x+6}{x+3}\cdot\dfrac{x^2+3x}{5x+10}$

7. $\dfrac{x^2+x-6}{x^2-9}\cdot\dfrac{x^2-6x+9}{2x-4}$

8. $\dfrac{2x^2+11x+15}{x^2+2x-3}\cdot\dfrac{x^2-x}{4x^2+20x+25}$

9. $\dfrac{4xy^2}{4x^2-9} \div \dfrac{10x^2y}{6x+9}$

10. $\dfrac{15ab^2}{a^2+a-12} \div \dfrac{25a^2b}{a-3}$

11. $\dfrac{x^2-xy}{y(x+y)} \cdot \dfrac{x^2+xy}{y(x-y)} \div \dfrac{x^2-x}{xy-2y}$

12. $\dfrac{xy^2+xy}{x^2-4} \cdot \dfrac{x^2-2x}{y^2+y} \div \dfrac{x}{y(x+2)}$

13. $\dfrac{2y^2+11y+5}{2y^2+7y+6} \div \dfrac{y^2+4y-5}{2y^2+y-3}$

14. $\dfrac{6m^2+m-1}{2m^2+5m+2} \div \dfrac{3m^2-7m+2}{m^2-m-6}$

15. $\dfrac{2x^2+x-3}{3x^2+10x-8} \cdot \dfrac{3x^2-2x}{x^2-1} \div \dfrac{x+6}{x^2+5x+4}$

16. $\dfrac{2x^2+3x-9}{6x^2-7x+2} \cdot \dfrac{3x-2}{4x^2-8x+3} \div \dfrac{x^2+3x}{2x^2+13x-7}$

17. $\dfrac{(s-2)s-3}{s-2} \div \dfrac{s^2-9}{(s+1)(s-2)}$

18. $\dfrac{x(x+3)+2}{x+5} \cdot \dfrac{x(x+3)+5(x+3)}{x^2+5x+6}$

19. $\dfrac{ax+ay+bx+by}{ax-ay+bx-by} \div \dfrac{x+y}{x-y}$

20. $\dfrac{x(x-2)-3(x-2)}{x^2-4} \cdot \dfrac{x+2}{x(x-3)+(x-3)}$

21. $\dfrac{2y^2-5y-3}{y-4} \div \left[\dfrac{(y-3)^2}{y^2-16} \cdot \dfrac{1}{3-y}\right]$

22. $\dfrac{x^5+x^4-x-1}{x-1} \div \dfrac{x+1}{2x}$

23. $\dfrac{a^2+300a+20{,}000}{1600-a^2} \cdot \dfrac{a^2+360a-16{,}000}{2a+400}$

24. $\dfrac{(2x+3)x+1}{(2x+3)(x-1)} \cdot \dfrac{(x+1)x-2}{x(x+1)+2(x+1)}$

Perform the indicated additions and subtractions in Exercises 25–54.

25. $\dfrac{2}{b}+\dfrac{1}{c}-\dfrac{3}{a}$

26. $\dfrac{a}{bc}-\dfrac{b}{ac}$

27. $\dfrac{2}{3s}-\dfrac{1}{t}+\dfrac{4}{12st}$

28. $\dfrac{1}{x^2}-\dfrac{1}{y^2}$

29. $\dfrac{3}{xy^2}+\dfrac{2}{x}-\dfrac{3}{y}$

30. $\dfrac{s+1}{s}+\dfrac{s}{2}-\dfrac{1}{s}$

31. $\dfrac{3}{x+1}+\dfrac{2}{x-3}$

32. $\dfrac{5}{x+3}-\dfrac{3}{x-2}$

33. $\dfrac{x}{x-5}+\dfrac{2}{x+3}$

34. $\dfrac{3x}{2x+3}+\dfrac{x}{x-2}$

35. $\dfrac{3}{x}-\dfrac{2}{x+2}-\dfrac{1}{x(x+2)}$

36. $\dfrac{x-y}{x}+\dfrac{y}{x+y}+\dfrac{y^2}{x^2+xy}$

37. $\dfrac{2}{a}+\dfrac{3}{2a+3}-\dfrac{2}{a(2a+3)}$

38. $\dfrac{5}{2x-1}+\dfrac{x}{4x^2-1}-\dfrac{7}{2x+1}$

39. $4-\dfrac{9x-16}{3x+4}$

40. $x+\dfrac{4-x^2}{x-3}$

41. $\dfrac{2}{p}+\dfrac{2}{p^2+2p+1}-\dfrac{1}{p+1}$

42. $\dfrac{xy}{x-y}+\dfrac{x^2}{y-x}+x$

43. $\dfrac{1}{(x+3)(x-2)}+\dfrac{1}{(x+1)(x+3)}-\dfrac{1}{(x+1)(x-2)}$

44. $\dfrac{3}{x(x+2)}-\dfrac{1}{(x+2)(x-1)}+\dfrac{4}{x(x-1)}$

45. $\dfrac{5x+6}{x^2+x-6}+\dfrac{x+1}{x^2-4x+4}-\dfrac{x^2}{x^3-2x^2}$

46. $\dfrac{2-k}{k^2+k-6}+\dfrac{5}{k^2-9}-\dfrac{4-k}{k^2-5k+6}$

47. $\dfrac{x+2}{(x-5)x+6}+\dfrac{x+2}{(x-7)x+12}-\dfrac{1-x}{(6-x)x-8}$

48. $\dfrac{x}{(x-3)(x+2)}-\dfrac{2x}{x^2+6x+9}+\dfrac{x}{x^2-9}$

49. $\dfrac{x-\dfrac{1}{x}}{1-\dfrac{1}{x}}$

50. $\dfrac{1-\dfrac{3}{x}+\dfrac{2}{x^2}}{\dfrac{1}{x}-\dfrac{2}{x^2}}$

51. $\dfrac{\dfrac{2}{x-1}-x-2}{1-x+\dfrac{2}{x+2}}$

52. $\dfrac{\dfrac{1}{x}+\dfrac{2}{x+1}-x}{1-\dfrac{1}{x^2+2x+1}}$

53. $\dfrac{\dfrac{x}{x-4}+\dfrac{1}{x-1}}{\dfrac{x}{x-1}+\dfrac{2}{x-3}}$

54. $\dfrac{x^2-4y^2}{1-\dfrac{2x+y}{x-y}}$

55. An exothermal process is one that releases heat. To calculate how much heat is released by a new process, a chemist uses the expression

$$\frac{1}{3}+\frac{2}{4t}-\frac{5}{2t^3}$$

Combine the terms and simplify.

56. In calculating the center of gravity of a certain concrete beam, the difference

$$\frac{1}{x}-\frac{1}{x(2x+1)}$$

is encountered. Simplify this expression.

57. A marketing student uses the expression

$$\frac{3}{s^2+s}+\frac{2}{s^2+4s+3}$$

while analyzing past trends in the sales of a certain product. Simplify this expression.

1.9 Permissible Values; Complex Numbers

We have defined the domain, or set of permissible values of a variable, as the set of real numbers for which the expression itself is a real number (Section 1.4). Although this definition suffices for most of the problems we treat, it is sometimes necessary to limit the domain of a variable to the rational numbers, the integer numbers, or some other subset of the real numbers. A variable that represents the number of people at a ball game, for instance, is restricted to the positive integers, since there cannot be 3254.72 people, or -215 people, at a ball game.

Sometimes, restrictions on the domain of a variable affect how we simplify an expression. For example, x^2-2 factors as $(x-\sqrt{2})(x+\sqrt{2})$ if the domain of x is the set of real numbers. If we are limited to the set of rational numbers, however, $\sqrt{2}$ is disallowed; and we should conclude that x^2-2 cannot be factored.* Similarly, the expression $x^2-\frac{9}{4}$, which factors as $(x-\frac{3}{2})(x+\frac{3}{2})$ in the setting of the real numbers or the rational numbers, is not factorable if the domain is limited to the set of integers. Limitations on the domain of a variable to some subset of the real numbers is not the only possibility; in some instances, it is convenient to extend the domain to include numbers outside the real number system.

Complex Numbers

Problems arise in mathematics that cannot be solved if the domain is restricted to the set of real numbers. For instance, $\sqrt{-4}$ is not a real number, since there is no real number whose square is -4. To allow for considering and describing nonreal numbers we introduce the number i with the property

$$i^2=-1$$

or

$$i=\sqrt{-1}$$

Numbers of the form bi, where b is a real number, are called **imaginary numbers**.

*This is sometimes described by saying that x^2-2 is not factorable over the rational numbers.

Note that $(2i)^2 = 4i^2 = -4$ and hence

$$\sqrt{-4} = 2i$$

This development is extended one step further to define a new system of numbers. By adding a real number and an imaginary number, we get a number called a **complex number**. Thus, complex numbers are numbers of the form $a+bi$, where a and b are real. The real number a is called the **real part** of the complex number, and the real number b is called the **imaginary part** of the complex number.

EXAMPLE 1 (a) Imaginary numbers: $3i$, $5i$, $2i$, $-7i$, i

(b) Complex numbers: $2+5i$, $3-i$, $-4-3i$, $-4+3i$

Notice that a complex number $a+bi$ yields a real number if $b=0$ and an imaginary number if $a=0$. Thus, the sets of real numbers and of imaginary numbers are subsets of the set of complex numbers.

EXAMPLE 2 Some important sets of numbers to which mathematical discussions are sometimes limited are given here.

Positive integers: 1, 2, 3, ...

Integers: All the previously listed numbers, and also numbers like 0, -1, -2, -3, ...

Rational numbers: All the previously listed numbers, and also numbers like $\frac{2}{3}$, $-\frac{1}{2}$, $0.\overline{178}$.

Real numbers: All the previously listed numbers, and also numbers like π, $\sqrt{2}$, $3+\pi$, $\sqrt{2}-\pi$.

Complex numbers: All the previously listed ones, and also numbers like $2+i$, $\frac{1}{2}-\pi i$, $\sqrt{2}+\frac{3}{7}i$

Complex numbers that differ only in the sign of their imaginary part are said to be **conjugate pairs**. The general form of a conjugate pair is $a \pm bi$. Thus, $3-2i$ and $3+2i$ are conjugate numbers.

Two complex numbers are equal if and only if their real parts are equal and their imaginary parts are equal. Thus, $a+bi$ and $c+di$ are equal if and only if $a=c$ and $b=d$.

Combinations of complex numbers obey the ordinary algebraic rules for real numbers. Thus, the sum, the difference, the product, or the quotient of two complex numbers is found in the same way as the sum, the difference, the product, or the quotient of two real binomials—with the restriction that $i^2 = -1$.

EXAMPLE 3 Find the sum and difference of $3+5i$ and $-9+2i$.

SOLUTION (a) $(3+5i)+(-9+2i)=(3-9)+(5+2)i=-6+7i$

(b) $(3+5i)-(-9+2i)=(3+9)+(5-2)i=12+3i$

EXAMPLE 4 Find the product of $(3-2i)(4+i)$.

SOLUTION

$$\begin{aligned}(3-2i)(4+i)&=12+3i-8i-2i^2\\&=12-5i+2\\&=14-5i\end{aligned}$$

To find the quotient of two complex numbers, we use the following scheme: *Multiply the numerator and the denominator of the given quotient by the conjugate of the denominator*. Example 5 shows how.

EXAMPLE 5 Find the quotient $\dfrac{2+3i}{4-5i}$.

SOLUTION

$$\begin{aligned}\frac{2+3i}{4-5i}&=\frac{(2+3i)(4+5i)}{(4-5i)(4+5i)}\\&=\frac{8+(12+10)i+15i^2}{16-25i^2}\\&=\frac{-7+22i}{16+25}\\&=\frac{-7+22i}{41}\end{aligned}$$

Or more commonly, this is written

$$-\frac{7}{41}+\frac{22}{41}i$$

EXAMPLE 6 Factor each of the following expressions and indicate the domain for which the factorization is valid: (a) x^2-3x-4; (b) x^2-2x-1; (c) x^2+4.

SOLUTION (a) $x^2-3x-4=(x-4)(x+1)$. This factorization is valid for any set of numbers for which the integers are a subset.

(b) The expression x^2-2x-1 is factorable by completing the square; that is,

$$\begin{aligned}x^2-2x-1&=x^2-2x+1-2\\&=(x-1)^2-(\sqrt{2})^2\\&=(x-1-\sqrt{2})(x-1+\sqrt{2})\end{aligned}$$

Consequently, we observe, x^2-2x-1 is factorable over the irrational numbers but not over the rational numbers.

(c) The expression x^2+4 is not factorable if the domain is limited to the real numbers. If the domain is the set of complex numbers, however, then

$$x^2+4=(x-2i)(x+2i)$$

Throughout the remainder of the book you should be alert to any restrictions on the permissible values of a variable. Unless it is specified otherwise, you should assume that the domain of the variable is the set of real numbers. Thus, the domain of $\sqrt{x-1}$ is understood to be $x \geqslant 1$, since $\sqrt{x-1}$ is not a real number for $x<1$. Similarly, the expression x^2+1 is not factorable (unless the domain is specified as the set of complex numbers).

Exercises for Section 1.9

In Exercises 1–10, state the domain of numbers for which the given expression is factorable and factor it.

1. x^2-3 **2.** x^2-8

3. x^2+4x+3 **4.** x^2-6x-7

5. x^2-3x-5 **6.** x^2-x-3

7. $x^2-4x-10$ **8.** x^2+6

9. x^2+3 **10.** $2x^2-x-4$

Perform the indicated operations in Exercises 11–34, expressing all answers in the form $a+bi$.

11. $(3+2i)+(4+3i)$

12. $(6+3i)+(5-i)$

13. $(5-2i)+(-7+5i)$

14. $(-1+i)+(2-i)$

15. $(1+i)+(3-i)$ **16.** $7-(5+3i)$

17. $(3+5i)-4i$ **18.** $(3+2i)+(3-2i)$

19. $(2+3i)(4+5i)$ **20.** $(7+2i)(-1-i)$

21. $(5-i)(5+i)$ **22.** $(6-3i)(6+3i)$

23. $(4+\sqrt{3}\,i)^2$ **24.** $(5-2i)^2$

25. $6i(4-3i)$ **26.** $3i(-2-i)$

27. $\dfrac{3+2i}{1+i}$ **28.** $\dfrac{4i}{2+i}$

29. $\dfrac{3}{2-3i}$ **30.** $\dfrac{7-2i}{6-5i}$

31. $\dfrac{1}{5i}$ **32.** $\dfrac{-3+i}{-2-i}$

33. $\dfrac{-1-3i}{4-\sqrt{2}\,i}$ **34.** $\dfrac{i}{2+\sqrt{5}\,i}$

35. Show that the sum of a complex number and its conjugate is a real number.

36. Show that the product of a complex number and its conjugate is a real number.

Test 1 for Chapter 1

Answer *True* or *False* to Exercises 1–10.

1. $x^2 \cdot x^3=x^6$ **2.** $\dfrac{x+y}{x}=1+\dfrac{y}{x}$

3. $(a+b)^3=a^3+b^3$ **4.** $-4y^2=16y^2$

5. $\dfrac{a}{b+c}=\dfrac{a}{b}+\dfrac{a}{c}$

6. $-3-2i$ and $-3+2i$ are conjugate pairs of complex numbers.

7. $(a-b)(a+b)=a^2-b^2$

8. The domain of $\sqrt{2+x}$ is the set of all x such that $x \geqslant 2$.

9. $\dfrac{-x}{x-y}=\dfrac{x}{y-x}$

10. The expression $3(x+y)-2xyz+3x(y-2z)$ is a trinomial.

11. Simplify $2\{-3x+2[5x-7(2x+5)]\}$.

12. Multiply m^3+2m^2-5m by m^2-3m-3.

13. Factor (a) $r^2-9r+14$; (b) $8a^3-27c^3$; (c) $3x^2-16xy-12y^2$.

14. Divide x^3+3x^2-x+7 by $x+2$, using synthetic division.

15. Simplify $\dfrac{1-\dfrac{6}{x}+\dfrac{5}{x^2}}{\dfrac{1}{x}-\dfrac{5}{x^2}}$.

16. Simplify $\dfrac{2}{x^2-2x-3}-\dfrac{5}{x^2+3x+2}+\dfrac{4}{x^2-x-6}$.

17. Evaluate each of the following: (a) $3i-(2+i)$; (b) $(-2+i)(3-4i)$.

Test 2 for Chapter 1

1. Simplify $3xy+2\{x+y-3[x-y(2+x)]\}$.

2. Write $(x+2y)(x^3-5y^2+2)$ in expanded form.

3. Write x^2-6x+9 in factored form.

4. Write $6x^2+7x-3$ in factored form.

5. Divide $2r^3-5r^2+9$ by $r-3$, using synthetic division.

6. Write $3x-6y+x^2-2xy$ in factored form.

Perform the indicated operations in Exercises 7–10, and write in simplest form.

7. $\dfrac{x^2-3x+2}{3x-x^2} \div \dfrac{x^2+2x-8}{x-3}$

8. $\dfrac{2}{ab}+\dfrac{3}{ac^2}-\dfrac{5}{b^3c}$

9. $3+\dfrac{5}{x+1}-\dfrac{x}{x^2-2x-3}$

10. $\dfrac{\dfrac{1}{x^2y^2}}{\dfrac{1}{x^2}-\dfrac{1}{y^2}}+\dfrac{1}{x^2}$

11. Divide $4+3i$ by $-1-2i$.

12. (a) Write $x-3(x+2)$ in expanded form.
(b) Write $(x-3)(x+2)$ in expanded form.

2

Exponents and Radicals

2.1 Integral Exponents

In Section 1.4 we introduced the concept of an exponent, and used the following basic rules.

RULE 1 $$a^m \cdot a^n = a^{m+n}$$

RULE 2 $$(a^m)^n = a^{mn}$$

RULE 3 $$(ab)^n = a^n b^n$$

$$\left(\frac{a}{b}\right)^n = \frac{a^n}{b^n}$$

RULE 4 $$\frac{a^m}{a^n} = a^{m-n}, \quad m > n$$

$$= \frac{1}{a^{n-m}}, \quad m < n$$

Another rule, easily derivable from Rules 2 and 3 is

RULE 5 $$(a^n b^m)^p = a^{np} b^{mp}$$

Examples 1–4 show how to use these rules for positive integer exponents.

EXAMPLE 1 (a) $x^3 x^4 = x^{3+4} = x^7$ (b) $y^2 y = y^{2+1} = y^3$

EXAMPLE 2 (a) $(2^3)^2 = 2^6 = 64$ (b) $(a^5)^4 = a^{20}$

EXAMPLE 3 (a) $(3xy)^2 = 3^2x^2y^2 = 9x^2y^2$ (b) $\left(\frac{2}{3}\right)^4 = \frac{2^4}{3^4} = \frac{16}{81}$

EXAMPLE 4 (a) $\frac{a^5}{a^3} = a^{5-3} = a^2$ (b) $\frac{r^7}{r^{11}} = \frac{1}{r^{11-7}} = \frac{1}{r^4}$

In passing, we note the difference between $(a^m)^n$ and $(a)^{m^n}$. The first of these equals a^{mn} (by Rule 2); the value of the second quantity, however, is obtained by first taking the power of the exponent and then using this as the power to which a must be taken. Thus

$$(2^3)^2 = 2^6 = 64 \qquad \text{but} \qquad (2)^{3^2} = 2^9 = 512$$

We sometimes say that exponentiation is *not* an associative process.

The concept of a positive integer exponent is used to introduce exponential notation, since it is easy to interpret. However, exponents are not limited to positive integers only. Our present purpose is to extend exponential notation to include 0 and negative integers.

To give meaning to 0 as an exponent, observe that $a^2 = a^{2+0}$. If Rule 1 is to continue to hold, $a^{2+0} = a^2 \cdot a^0$. Thus, for a^2 to equal $a^2 \cdot a^0$, we make the following definition:

RULE 6 $$a^0 = 1, \qquad a \neq 0$$

To interpret a negative integer exponent, note that

$$1 = a^0 = a^{n-n}$$

Under the assumption that Rule 1 can be extended to negative integers, we have

$$a^{n-n} = a^n \cdot a^{-n} = 1$$

A natural consequence of this result is Rule 7.

RULE 7 $$a^{-n} = \frac{1}{a^n}, \qquad a \neq 0$$

The last two rules follow from the others.

RULE 8 $$a^n = \frac{1}{a^{-n}}$$

RULE 9 $$\left(\frac{a}{b}\right)^{-n} = \left(\frac{b}{a}\right)^n$$

EXAMPLE 5 Simplify $\dfrac{xy^{-2}}{3xb^{-3}}$.

SOLUTION Eliminating the negative exponents first gives

$$\frac{xy^{-2}}{3xb^{-3}}=\frac{x\dfrac{1}{y^2}}{3x\dfrac{1}{b^3}}=\frac{\dfrac{1}{y^2}}{\dfrac{3}{b^3}}=\frac{1}{y^2}\cdot\frac{b^3}{3}=\frac{b^3}{3y^2}$$

EXAMPLE 6 Simplify $\dfrac{x^0+y^0}{(x+y)^0}$.

SOLUTION $\dfrac{x^0+y^0}{(x+y)^0}=\dfrac{1+1}{1}=\dfrac{2}{1}=2$

EXAMPLE 7 Simplify $x^{-2}-y^{-3}$.

SOLUTION $x^{-2}-y^{-3}=\dfrac{1}{x^2}-\dfrac{1}{y^3}=\dfrac{y^3}{x^2y^3}-\dfrac{x^2}{x^2y^3}=\dfrac{y^3-x^2}{x^2y^3}$

EXAMPLE 8 Simplify $\dfrac{x^{-1}+y^{-1}}{(x+y)^{-1}}$.

SOLUTION

$$\frac{x^{-1}+y^{-1}}{(x+y)^{-1}}=\frac{\dfrac{1}{x}+\dfrac{1}{y}}{\dfrac{1}{x+y}}=\frac{\dfrac{y+x}{xy}}{\dfrac{1}{x+y}}=\frac{x+y}{xy}\cdot\frac{x+y}{1}=\frac{(x+y)^2}{xy}$$

Scientists must often express positive real numbers in an abbreviated manner, called *scientific notation*. The real number is expressed as the product of a number between 1 and 10 and a power of 10. That is, if x is a positive real number, then in scientific notation,

$$x=m\cdot 10^c$$

where m is a number between 1 and 10 and c is an integer.

Scientific notation is especially valuable for very large and very small numbers. The number

$$0.000000000000000000000053$$

for example, represents the weight of an oxygen molecule. In scientific notation this value equals 5.3×10^{-23}, since the decimal point is moved to the right by 23 places to get the number 5.3, which is between 1 and 10.

EXAMPLE 9 Write in scientific notation the numbers 5278; 37,910,000; 0.000172; and 0.24.

SOLUTION

$$5278 = 5.278 \times 10^3$$
$$37{,}910{,}000 = 3.791 \times 10^7$$
$$0.000172 = 1.72 \times 10^{-4}$$
$$0.24 = 2.4 \times 10^{-1}$$

Most scientific calculators have a button that you can push to display the numbers in scientific notation. The number between 1 and 10 appears with the usual 8-to-10-digit accuracy. To the right appears a two-digit number for the exponent. Such a calculator can handle any number between 10^{-99} *and* 10^{99}.

Your calculator displays the number 3.6513×10^5 *as* 3.6513 05. *Similarly, the number* 2.0715×10^{-4} *is displayed as* 2.0715 -04.

Exercises for Section 2.1

In Exercises 1–54, simplify the given expression.

1. 3^5 **2.** 4^3 **3.** $(-2)^4$

4. $(-10)^3$ **5.** x^6x^8 **6.** b^3b^{10}

7. $(2x^3)(3x^5)$ **8.** $(\frac{1}{2}a^8)(6a^2)$ **9.** $(c^{12})^4$

10. $(y^4)^7$ **11.** $(3^2)^3$ **12.** $(2^4)^2$

13. $(xy)^5$ **14.** $(3z)^4$ **15.** $(2x^3y^7)^3$

16. $(-4a^2b^5)^3$ **17.** $5(99^0)$ **18.** $3(25x^3y)^0$

19. $(\frac{2}{7})^2$ **20.** $(-\frac{3}{4})^3$ **21.** $\dfrac{7x^2y^5}{27xy}$

22. $\dfrac{a^5b^8}{a^0b^{11}}$ **23.** $\left(\dfrac{x^2}{y^5}\right)^6$ **24.** $\left(\dfrac{2c^3}{d^6}\right)^3$

25. $(8)^{-1}$ **26.** $(3)^{-4}$ **27.** $5x^{-2}$

28. $3z^{-3}$ **29.** $(x^{-2}y^5)^4$ **30.** $(r^5s^{-1})^6$

31. $\left(\dfrac{2x}{y}\right)^{-3}$ **32.** $\left(\dfrac{6}{5}\right)^{-1}$ **33.** $\left(\dfrac{a^{-3}b^7}{a^2b^4}\right)^{-2}$

34. $\left(\dfrac{xy^9}{x^{-4}y^{-1}}\right)^{-1}$ **35.** $\left(\dfrac{x^3y^{-6}}{x^7y^{-4}}\right)^{-2}$

36. $\left(\dfrac{m^{-2}n^{-4}}{m^{-6}n^{-9}}\right)^0$ **37.** $(x+y)^{-1}$ **38.** $(x+y)^0$

39. $x^0+(2y)^0$ **40.** $[(2xyz)^7]^0$ **41.** $2^{-1}-3^{-1}$

42. $5^{-1}+2^{-2}$ **43.** $(x+y^{-1})^{-1}$ **44.** $(a^{-1}-b^{-1})^{-1}$

45. $[(3^{-1})^{-2}]^{-1}$ **46.** $[(2^{-2})^{-2}]^{-1}$

47. $ab^{-1}+a^{-1}b$ **48.** $x^2y^{-2}-x^{-1}y^3$

49. $(x-y)^{-1}+(x+y)^{-1}$ **50.** $5(x-4)^{-1}-3(x+2)^{-1}$

51. $\dfrac{(xy)^{-1}}{x^{-1}-y^{-1}}$ **52.** $\dfrac{2^{-1}+3^{-1}}{2^{-2}-3^{-2}}$

53. $\dfrac{(x+1)^{-2}}{(x+1)^{-1}+(x-3)^{-1}}$ **54.** $\dfrac{(a+3)(a-5)^{-1}}{(a-5)^{-1}+(a+3)^{-1}}$

Write the entries of Exercises 55–60 in scientific notation.

55. 8,234,400,000 **56.** 93,002

57. 0.000000000052

58. 0.00786

59. 46 followed by 38 zeros

60. 53 followed by 13 zeros

Write the entries of Exercises 61–66 without scientific notation.

61. 3.4852×10^9 **62.** 1.423×10^{19}

63. 9.385×10^{-7} **64.** 8.99×10^{-11}

65. 2.222×10^{-2} **66.** 7.003×10^{20}

67. The speed of light is approximately 30 billion cm/sec. Express this number in scientific notation.

68. A television rating service estimates that 28.5 million people watch a particular program. Express this number in scientific notation.

69. An economist estimates that 1.5×10^6 people will be unemployed next year. Write this number without scientific notation.

2.2 Radicals

The statement $5^2 = 25$ means that 25 is the square of 5, or equivalently, 5 is the square root of 25. Similarly, in the expression $(-2)^3 = -8$, the term -2 is the cube root of -8; in the expression $3^4 = 81$, the term 3 is the fourth root of 81; and in the expression $4^5 = 1024$, the term 4 is the fifth root of 1024.

DEFINITION

If r is a real number and n is a positive integer such that $r^n = a$, then r is an ***n*th root of *a*.**

The nth root of a number is not unique. Both 5 and -5, for example, are square roots of 25. To avoid confusion, we use one root, the **principal root**. The principal nth root of a number a is denoted by

$$\sqrt[n]{a}$$

where $\sqrt{\ }$ is called a *radical sign*, n the *index* of the radical, and a, the radicand. The meaning attached to $\sqrt[n]{a}$ depends on the values of a and n and is governed by the definition of the principal nth root.

DEFINITION

- If n is even and $a > 0$, then $\sqrt[n]{a}$ is the positive nth root.
- If n is odd and $a < 0$, then $\sqrt[n]{a}$ is the negative nth root.
- If n is odd and $a > 0$, then $\sqrt[n]{a}$ is the positive nth root.
- If n is even and $a < 0$, the principal nth root is not a real number.

The radical sign without an index is understood to be the principal square root. This operation is included on most calculators; simply enter the number and then push the [$\sqrt{\ }$] *key.*

EXAMPLE 1 Find the principal root of (a) $\sqrt[4]{81}$; (b) $\sqrt[5]{-32}$; (c) $\sqrt{25}$.

SOLUTION

(a) $\sqrt[4]{81} = 3$ (*Note*: The answer is not ± 3.)

(b) $\sqrt[5]{-32} = -2$

(c) $\sqrt{25} = 5$

Since $\sqrt[n]{a} = r$ means $r^n = a$, it follows that

$$(\sqrt[n]{a})^n = a$$

For example, $(\sqrt[3]{5.14})^3 = 5.14$. Also, since a principal square root is a unique number, we can write $\sqrt{a^2} = |a|$. Thus, $\sqrt{(-2)^2} = 2$, not -2.

The following rules are useful in working with radical expressions.

RULE 1 $\sqrt[n]{a}\sqrt[n]{b}=\sqrt[n]{ab}$

RULE 2 $\dfrac{\sqrt[n]{a}}{\sqrt[n]{b}}=\sqrt[n]{\dfrac{a}{b}}$

These two rules tell how to multiply or divide numbers taken to the same root. *No corresponding rule exists for combining radicals by addition.*

EXAMPLE 2 Simplify $\sqrt{16x^3y^6}$; that is, remove any perfect square roots.

SOLUTION $\sqrt{16x^3y^6}=\sqrt{16x^2y^6x}=\sqrt{16x^2y^6}\sqrt{x}=4xy^3\sqrt{x}$

EXAMPLE 3 Simplify $\sqrt[3]{-16x^5}$.

SOLUTION $\sqrt[3]{-16x^5}=\sqrt[3]{(-8x^3)(2x^2)}=\sqrt[3]{-8x^3}\cdot\sqrt[3]{2x^2}=-2x\sqrt[3]{2x^2}$

EXAMPLE 4 Simplify $\sqrt{2}\sqrt{3}$ and $\sqrt{\frac{1}{9}}$.

SOLUTION $\sqrt{2}\sqrt{3}=\sqrt{2\cdot 3}=\sqrt{6}$ and $\sqrt{\dfrac{1}{9}}=\dfrac{\sqrt{1}}{\sqrt{9}}=\dfrac{1}{3}$

Writing a fraction without using radicals in the denominator is called **rationalization**. Thus, the rationalized form of $\sqrt{\dfrac{7}{9}}$ is $\dfrac{\sqrt{7}}{3}$. When a perfect root does not appear in the denominator, we multiply both denominator and numerator by the root. To rationalize $\sqrt{\dfrac{3}{5}}$, for example, we write

$$\sqrt{\frac{3}{5}}=\frac{\sqrt{3}}{\sqrt{5}}=\frac{\sqrt{3}}{\sqrt{5}}\cdot\frac{\sqrt{5}}{\sqrt{5}}=\sqrt{\frac{3\cdot 5}{5\cdot 5}}=\frac{\sqrt{15}}{5}$$

EXAMPLE 5 (a) $\dfrac{2}{\sqrt{3}}=\dfrac{2}{\sqrt{3}}\cdot\dfrac{\sqrt{3}}{\sqrt{3}}=\dfrac{2\sqrt{3}}{3}$

(b) $\dfrac{x}{\sqrt[4]{x+y}}=\dfrac{x}{\sqrt[4]{x+y}}\cdot\dfrac{\sqrt[4]{(x+y)^3}}{\sqrt[4]{(x+y)^3}}=\dfrac{x\sqrt[4]{(x+y)^3}}{\sqrt[4]{(x+y)^4}}=\dfrac{x\sqrt[4]{(x+y)^3}}{x+y}$

(c) $\dfrac{1}{\sqrt[3]{2}}=\dfrac{1}{\sqrt[3]{2}}\cdot\dfrac{\sqrt[3]{4}}{\sqrt[3]{4}}=\dfrac{\sqrt[3]{4}}{\sqrt[3]{8}}=\dfrac{\sqrt[3]{4}}{2}$

EXAMPLE 6 Simplify $\sqrt{xy^{-1}+x^{-1}y+2}$.

SOLUTION

$$\sqrt{xy^{-1}+x^{-1}y+2}=\sqrt{\frac{x}{y}+\frac{y}{x}+2}=\sqrt{\frac{x^2+y^2+2xy}{xy}}$$

$$=\sqrt{\frac{(x+y)^2}{xy}}$$

$$=\frac{(x+y)}{\sqrt{xy}}\cdot\frac{\sqrt{xy}}{\sqrt{xy}}$$

$$=\frac{(x+y)\sqrt{xy}}{xy}$$

Sometimes you can extract the "exact" square root by first expressing the radicand as a perfect square.

EXAMPLE 7 Write $\sqrt{4+2\sqrt{3}}$ as an expression with a single radical.

SOLUTION

$$\sqrt{4+2\sqrt{3}}=\sqrt{3+2\sqrt{3}+1}$$

$$=\sqrt{\sqrt{3}^2+2\sqrt{3}+1}$$

$$=\sqrt{(\sqrt{3}+1)^2}$$

$$=\sqrt{3}+1$$

Example 7 shows that extracting a root can be done *without* using a calculator. While such an approach may be impractical, the problems test your algebraic skills for forming a perfect square under the radical sign.

Note that $\sqrt{a+b}\neq\sqrt{a}+\sqrt{b}$, so that the radical sign is itself a grouping symbol for addition. Thus what is under the radical must be summed before the root is found.

Exercises for Section 2.2

Simplify the following.

1. $\sqrt{144}$ **2.** $\sqrt{64}$ **3.** $-\sqrt{900}$

4. $-\sqrt{225}$ **5.** $\sqrt[3]{27}$ **6.** $\sqrt[3]{64}$

7. $\sqrt[3]{125}$ **8.** $\sqrt[5]{32}$ **9.** $\sqrt[5]{-243}$

10. $\sqrt[3]{-64}$ **11.** $\sqrt[4]{x^8}$ **12.** $\sqrt{a^{10}}$

13. $\sqrt[3]{4^9}$ **14.** $\sqrt[7]{5^{14}}$ **15.** $\sqrt[4]{3x^{12}}$

16. $\sqrt[4]{16z^{20}}$ **17.** $\sqrt{50}$ **18.** $\sqrt{63}$

19. $2\sqrt{72}$ **20.** $5\sqrt{243}$ **21.** $\sqrt[3]{\frac{32}{27}}$

22. $\sqrt{\frac{98}{121}}$ **23.** $\sqrt{\frac{x^7}{y^4}}$ **24.** $\sqrt[3]{\frac{c^{12}}{d^9}}$

25. $\sqrt{48x^3y}$ **26.** $\sqrt{63a^5b^8}$ **27.** $\sqrt[5]{8x^4y^5}$

28. $\sqrt[4]{2x^3y^2}$ **29.** $\sqrt{5}\sqrt{3}$ **30.** $\sqrt[6]{3}\sqrt[6]{7}$

31. $\sqrt[3]{2a}\sqrt[3]{4a^2}$ **32.** $\sqrt{5xy^2}\sqrt{10x^3y^4}$

33. $\sqrt{\frac{3}{7}}$ **34.** $\sqrt{\frac{5}{3}}$ **35.** $\sqrt[4]{\frac{2}{3}}$

36. $\sqrt[3]{\frac{5}{4}}$ **37.** $\sqrt[4]{\frac{x}{y}}$ **38.** $\sqrt[3]{\frac{a^4}{b^5}}$

39. $\frac{3}{\sqrt{2x}}$ **40.** $\frac{x}{\sqrt{x-3}}$ **41.** $\frac{X}{\sqrt{X-Z}}$

42. $\sqrt{x^{-1}+x}$ **43.** $\sqrt[3]{a^{-1}-b^{-1}}$ **44.** $\sqrt{2^{-3}+3^{-1}}$

45. $\sqrt{ab^{-1}+a^{-1}b-2}$ **46.** $\sqrt[3]{w^{-2}+z^{-2}}$

47. $\sqrt{x+y}\sqrt{x^2-y^2}$ **48.** $\sqrt[3]{\frac{x-y}{x^2+2xy+y^2}}$

49. $\sqrt{\frac{x+y}{x^3+3x^2y+3xy^2+y^3}}$

50. $\sqrt{(c-d)(c+d)}$

51. A student solving a calculus problem finds the expression $\sqrt[3]{ab^2}\sqrt[3]{a^5b}$. Write this expression without the radicals.

52. In calculating the thermal conductivity of a material a student encounters the expression $\sqrt{1+\delta t^{-2}}$. Write this in rationalized form.

53. A professor of mechanics writes the expression

$$\frac{\sqrt{t_1^2+t_2^2}}{\sqrt{t_1^{-2}+t_2^{-2}}}=t_1t_2$$

at the board. Is this a correct statement?

In Exercises 54–64, simplify the given radical by finding an "exact" square root. (*Hint for No. 61*: $7=6+1=3(\sqrt{2})^2+1$)

54. $\sqrt{6-2\sqrt{5}}$ **55.** $\sqrt{5+2\sqrt{6}}$

56. $\sqrt{37+20\sqrt{3}}$ **57.** $\sqrt{71-16\sqrt{7}}$

58. $\sqrt{8+2\sqrt{15}}$ **59.** $\sqrt{16+6\sqrt{7}}$

60. $\sqrt{19-8\sqrt{3}}$ **61.** $\sqrt[3]{7+5\sqrt{2}}$

62. $\sqrt[3]{26-15\sqrt{3}}$

63. Inattention to the "detail" of a principal root can lead to some interesting "logic." Tell what is wrong with the following argument, which "shows that all numbers are equal."

Consider any two numbers x and y with $x>y$. Let $m=(x+y)/2$. Then $2m=x+y$.

Multiplying both sides by the quantity $x-y$ gives

$$(x-y)(2m)=(x-y)(x+y)$$
$$2mx-2my=x^2-y^2$$
$$y^2-2my=x^2-2mx$$

Then, if m^2 is added to both sides,

$$y^2-2my+m^2=x^2-2mx+m^2$$
$$(y-m)^2=(x-m)^2$$
$$y-m=x-m$$
$$y=x$$

2.3 Addition and Subtraction of Radicals

Radicals that are of the same order and have the same radicand are said to be **similar**. Thus $2\sqrt{3}$, $-5\sqrt{3}$, and $x\sqrt{3}$ are similar, and $\sqrt{2}$, $\sqrt{3}$, $\sqrt[3]{3}$, and $\sqrt[4]{2}$ are dissimilar. Some dissimilar radicals can be made similar by extracting roots. For example, the radicals $\sqrt{2}$ and $\sqrt{8}$ are dissimilar but can be made similar because $\sqrt{8}=\sqrt{4\cdot 2}=2\sqrt{2}$.

Similar radicals are added or subtracted by adding or subtracting their coefficients. If radicals are not similar, their sum can only be indicated. For example, $\sqrt{2}+\sqrt{3}$ cannot be written as a single radical.

EXAMPLE 1

(a) $3\sqrt{5}+4\sqrt{5}-5\sqrt{5}=(3+4-5)\sqrt{5}=2\sqrt{5}$

(b) $\sqrt{17}-3\sqrt{17}+\sqrt{2}=(1-3)\sqrt{17}+\sqrt{2}=\sqrt{2}-2\sqrt{17}$

(c) $\sqrt{3}+\sqrt[4]{3}+5\sqrt{3}=(1+5)\sqrt{3}+\sqrt[4]{3}=6\sqrt{3}+\sqrt[4]{3}$

Calculator users must be aware of the important difference in the way the [$\sqrt{\ }$] *key operates for addition and for multiplication. Thus,* $\sqrt{2}+\sqrt{3}$ *is evaluated by the following sequence of keystrokes:*

2 [$\sqrt{\ }$] [+] 3 [$\sqrt{\ }$] = 3.14626

Since $\sqrt{ab}=\sqrt{a}\sqrt{b}$, *the product* $\sqrt{2}\sqrt{3}$ *can be evaluated by*

2 [$\sqrt{\ }$] [×] 3 [$\sqrt{\ }$] = 2.449

or equivalently,

2 [×] 3 [$\sqrt{\ }$] = 2.449

Notice that in the last sequence, the [$\sqrt{\ }$] *key is pushed only once.*

Sometimes, as Example 2 shows, you must express the radical in simplest form before combining like terms.

EXAMPLE 2 Simplify $\sqrt{20}+3\sqrt{5}-\sqrt{24}$.

SOLUTION Notice that $\sqrt{20}=\sqrt{4\cdot 5}=2\sqrt{5}$ and $\sqrt{24}=\sqrt{4\cdot 6}=2\sqrt{6}$. Therefore,

$$\begin{aligned}\sqrt{20}+3\sqrt{5}-\sqrt{24}&=2\sqrt{5}+3\sqrt{5}-2\sqrt{6}\\&=5\sqrt{5}-2\sqrt{6}\end{aligned}$$

EXAMPLE 3 (a) $\sqrt{6}-\sqrt{\frac{3}{2}}=\sqrt{6}-\sqrt{\frac{3}{2}}\cdot\frac{\sqrt{2}}{\sqrt{2}}=\sqrt{6}-\frac{\sqrt{6}}{2}=\frac{1}{2}\sqrt{6}$

(b) $\sqrt[3]{ab}+\sqrt[3]{\frac{a}{b^2}}=\sqrt[3]{ab}+\frac{\sqrt[3]{ab}}{b}=\left(1+\frac{1}{b}\right)\sqrt[3]{ab}=\frac{(b+1)\sqrt[3]{ab}}{b}$

EXAMPLE 4 Perform the multiplication $(\sqrt{x}+3)(\sqrt{x}+5)$.

SOLUTION

$$\begin{aligned}(\sqrt{x}+3)(\sqrt{x}+5)&=\sqrt{x}\sqrt{x}+5\sqrt{x}+3\sqrt{x}+15\\&=x+8\sqrt{x}+15\end{aligned}$$

EXAMPLE 5 Rationalize the expression $\dfrac{2}{\sqrt{3}-\sqrt{8}}$.

SOLUTION To rationalize the given fraction, use the knowledge that $(\sqrt{x})^2=x$ and the special product $(a+b)(a-b)=a^2-b^2$, to write

$$\frac{2}{\sqrt{3}-\sqrt{8}}=\frac{2}{\sqrt{3}-\sqrt{8}}\cdot\frac{\sqrt{3}+\sqrt{8}}{\sqrt{3}+\sqrt{8}}=\frac{2(\sqrt{3}+\sqrt{8})}{(\sqrt{3})^2-(\sqrt{8})^2}=\frac{2(\sqrt{3}+\sqrt{8})}{3-8}=-\frac{2(\sqrt{3}+\sqrt{8})}{5}$$

EXAMPLE 6 Rationalize the denominator in $\dfrac{x+y}{\sqrt[3]{x}+\sqrt[3]{y}}$.

SOLUTION We know (Section 1.7) that $x^3+y^3=(x+y)(x^2-xy+y^2)$. Hence, we multiply numerator and denominator by $\sqrt[3]{x^2}-\sqrt[3]{xy}+\sqrt[3]{y^2}$. By this operation, we get

$$\frac{x+y}{\sqrt[3]{x}+\sqrt[3]{y}}\cdot\frac{\sqrt[3]{x^2}-\sqrt[3]{xy}+\sqrt[3]{y^2}}{\sqrt[3]{x^2}-\sqrt[3]{xy}+\sqrt[3]{y^2}}=\frac{(x+y)(\sqrt[3]{x^2}-\sqrt[3]{xy}+\sqrt[3]{y^2})}{(\sqrt[3]{x})^3+(\sqrt[3]{y})^3}$$
$$=\frac{(x+y)(\sqrt[3]{x^2}-\sqrt[3]{xy}+\sqrt[3]{y^2})}{x+y}=\sqrt[3]{x^2}-\sqrt[3]{xy}+\sqrt[3]{y^2}$$

EXAMPLE 7 Write without negative exponents or radicals in the denominator

$$\frac{\sqrt[3]{x^{-2}}+\sqrt[3]{y^{-2}}}{\sqrt[3]{x^{-2}+y^{-2}}}$$

SOLUTION We first change the expression to one without negative exponents:

$$\frac{\sqrt[3]{1/x^2}+\sqrt[3]{1/y^2}}{\sqrt[3]{(1/x^2)+(1/y^2)}}$$

When the expression in the denominator is simplified, we have

$$\frac{\sqrt[3]{(1/x^2)}+\sqrt[3]{1/y^2}}{\sqrt[3]{(x^2+y^2)/(x^2y^2)}}$$

Multiplying numerator and denominator by $\sqrt[3]{x^2y^2}$ gives

$$\frac{(\sqrt[3]{1/x^2}+\sqrt[3]{1/y^2})\cdot\sqrt[3]{x^2y^2}}{\sqrt[3]{[(x^2+y^2)/x^2y^2]}\cdot\sqrt[3]{x^2y^2}}=\frac{\sqrt[3]{x^2y^2/x^2}+\sqrt[3]{x^2y^2/y^2}}{\sqrt[3]{x^2+y^2}}=\frac{\sqrt[3]{y^2}+\sqrt[3]{x^2}}{\sqrt[3]{x^2+y^2}}$$

To rationalize the denominator, multiply by $\sqrt[3]{(x^2+y^2)^2}$:

$$\frac{(\sqrt[3]{y^2}+\sqrt[3]{x^2})\sqrt[3]{(x^2+y^2)^2}}{x^2+y^2}$$

Exercises for Section 2.3

Simplify the expressions in Exercises 1–18.

1. $\sqrt{3}+5\sqrt{3}$
2. $\sqrt[3]{4}-3\sqrt[3]{4}+5\sqrt[3]{3}$
3. $11\sqrt[5]{a}-4\sqrt[5]{a}-5\sqrt[5]{a}$
4. $2\sqrt{7}+4\sqrt{3}-2\sqrt{7}$
5. $\sqrt{3}+\sqrt{48}$
6. $4\sqrt{2}-\sqrt{8}+\sqrt{50}$
7. $\sqrt{27}+5\sqrt{12}$
8. $\sqrt[3]{-16}+\sqrt[3]{128}$
9. $\sqrt{x^2y}-\sqrt{x^5y^3}$
10. $\sqrt{2x^3y^4}+\sqrt{2x}$
11. $\sqrt{\frac{1}{2}}+\sqrt{\frac{9}{2}}$
12. $\sqrt{\frac{3}{5}}+\sqrt{\frac{27}{5}}$
13. $\sqrt{5}+\sqrt{\frac{5}{2}}-\sqrt{\frac{1}{5}}$
14. $\sqrt{8}+3\sqrt{\frac{1}{2}}$
15. $\sqrt{xy}+\sqrt{x^3y^3}+\sqrt{x/y}$
16. $\sqrt{5}\sqrt{2}+\sqrt{40x^6}$
17. $\sqrt{\dfrac{x}{y}}+\sqrt{\dfrac{y}{x}}$
18. $\sqrt{\dfrac{ab}{c}}-\sqrt{\dfrac{c}{ab}}$

In Exercises 19–30 expand the indicated products and collect like terms.

19. $\sqrt{2}(\sqrt{3}+\sqrt{7})$
20. $\sqrt{3}(\sqrt{3}-\sqrt{27})$
21. $\sqrt{x}(3\sqrt{x}-5\sqrt{y})$
22. $(\sqrt{2}+\sqrt{8})(\sqrt{8}-\sqrt{2})$
23. $(\sqrt{7}-2\sqrt{3})(\sqrt{7}+2\sqrt{3})$
24. $(\sqrt{5}+\sqrt{x})^2$
25. $(\sqrt{a}-\sqrt{b})^2$

26. $(2+\sqrt{3})(5-\sqrt{7})$

27. $(1-\sqrt{2})(\sqrt{3}+\sqrt{2})$

28. $(\sqrt[3]{x}+2)(\sqrt[3]{x}-2)$

29. $(\sqrt{a}+\sqrt{b})(a+\sqrt{ab}+b)$

30. $(\sqrt{x+y}-3)(\sqrt{x+y}+3)$

Write Exercises 31–37 without negative exponents and with a rational denominator (if necessary).

31. $\dfrac{\sqrt{5+3}}{\sqrt{3}+\sqrt{5}}$

32. $\dfrac{2\sqrt{2}}{\sqrt{2}-\sqrt{7}}$

33. $\dfrac{\sqrt{x}-\sqrt{y}}{\sqrt{x}+\sqrt{y}}$

34. $\dfrac{\sqrt{x}+2\sqrt{y}}{\sqrt{x}-3\sqrt{y}}$

35. $\dfrac{\sqrt{a}+\sqrt{b}}{\sqrt{a}-\sqrt{b}}$

36. $\dfrac{\sqrt[3]{x^2y^2}}{\sqrt[3]{x}-\sqrt[3]{y}}$

37. $\sqrt[5]{x^{-12}}+\sqrt[5]{x^{-7}}$

38. A communications engineer derives a bandwidth equation that involves the expression $\frac{1}{2}(3\sqrt{f}-\sqrt{25f}+2\sqrt{3})$. Simplify this expression.

39. The expression

$\sqrt{t_1^2+t_2^2}=t_1+t_2$

appeared in a physics lab report. Is it correct?

40. Show by an example that $\sqrt{a+b}\neq\sqrt{a}+\sqrt{b}$.

2.4 Fractional Exponents

We now extend the definition of exponential notation to include fractional exponents. What meaning shall we give to $3^{1/2}$, $5^{1/3}$, and $(-7)^{1/3}$? We assume that the laws of exponents will continue to hold for fractional exponents. Thus

$$(3^{1/2})^2=3^{2/2}=3^1=3$$
$$(5^{1/3})^3=5^{3/3}=5^1=5$$
$$[(-7)^{1/3}]^3=(-7)^{3/3}=(-7)^1=-7$$

Consequently, we can interpret $3^{1/2}$ as a square root of 3, and $5^{1/3}$ as a third root of 5, and $(-7)^{1/3}$ as a cube root of -7. In general, since $(a^{1/n})^n=a^{n/n}=a^1=a$, we define $a^{1/n}$ as a number whose nth power is a. It follows that

$$a^{1/n}=\sqrt[n]{a}$$

Similarly, if the fraction p/q is in lowest terms, $a^{p/q}$ is interpreted as

$$a^{p/q}=(a^{1/q})^p=(a^p)^{1/q}$$

and hence $a^{p/q}$ is defined as

$$a^{p/q}=\sqrt[q]{a^p}=(\sqrt[q]{a})^p$$

This means in words that the numerator of a fractional exponent corresponds to the power to which the number is raised, and the denominator corresponds to the order of the root to be taken. Note that the sequence in which the root is taken is immaterial. Thus, $\sqrt[3]{x^2}=(\sqrt[3]{x})^2$.

EXAMPLE 1

(a) $\sqrt[5]{y^2}=(y^2)^{1/5}=y^{2/5}$

(b) $(x-1)^{3/4}=\sqrt[4]{(x-1)^3}$

(c) $2^{5/3}=\sqrt[3]{2^5}=\sqrt[3]{2^3\cdot 2^2}=2\sqrt[3]{4}$

EXAMPLE 2 Evaluate $(27)^{4/3}$.

SOLUTION We can write either $\sqrt[3]{(27)^4}$ or $(\sqrt[3]{27})^4$, since both will yield the correct answer. The second form is simpler, however. If we use the first form we get

$$\sqrt[3]{(27)^4}=\sqrt[3]{531{,}441}=81$$

By the second form,

$$(\sqrt[3]{27})^4=(3)^4=81$$

Since the rules of exponents apply here, we can also write

$$(3^3)^{4/3}=3^4=81$$

EXAMPLE 3 Simplify $(-8)^{2/3}\,(9)^{-3/2}$.

SOLUTION

$$(-8)^{2/3}\,(9)^{-3/2}=[(-8)^{1/3}]^2[(9^{1/2})]^{-3}=(-2)^2(3)^{-3}=\frac{(-2)^2}{3^3}=\frac{4}{27}$$

EXAMPLE 4 Simplify $\left[\dfrac{3x^{2/3}}{y^{5/2}}\right]^3\left[\dfrac{7y^{1/6}}{5x^{5/6}}\right]^{-1}$.

SOLUTION

$$\left[\frac{3x^{2/3}}{y^{5/2}}\right]^3\left[\frac{7y^{1/6}}{5x^{5/6}}\right]^{-1}=\frac{27x^2}{y^{15/2}}\cdot\frac{5x^{5/6}}{7y^{1/6}}=\frac{135x^{17/6}}{7y^{23/3}}$$

EXAMPLE 5

(a) $\sqrt[8]{81}=(3^4)^{1/8}=3^{1/2}=\sqrt{3}$

(b) $\sqrt[3]{x}\sqrt{x}=x^{1/3}x^{1/2}=x^{5/6}=\sqrt[6]{x^5}$

EXAMPLE 6 Simplify $(x^{1/2}-x^{-1/2})^{-2}$

SOLUTION

$$\begin{aligned}(x^{1/2}-x^{-1/2})^{-2}&=\left(x^{1/2}-\frac{1}{x^{1/2}}\right)^{-2}\\&=\left(\frac{x-1}{x^{1/2}}\right)^{-2}\\&=\frac{x}{(x-1)^2}\end{aligned}$$

EXAMPLE 7 Simplify (a) $\sqrt[3]{\sqrt{8}}$; (b) $\sqrt{x\sqrt{x}}$.

SOLUTION

(a) $\sqrt[3]{\sqrt{8}}=(8^{1/2})^{1/3}=[(2^3)^{1/2}]^{1/3}=(2^{3/2})^{1/3}=2^{1/2}=\sqrt{2}$

(b) $\sqrt{x\sqrt{x}}=(x\cdot x^{1/2})^{1/2}=(x^{3/2})^{1/2}=x^{3/4}=\sqrt[4]{x^3}$

The exponent concept can be further extended to include irrational numbers so that the rules of exponents hold for all real exponents.

Calculators that have a [y^x] *key can be used to give rational approximations to any real positive number taken to any real exponent. Notice that calculators evaluate powers of positive numbers only. Consequently, to evaluate an expression like* $(-5.75)^{3.1}$, *one must evaluate* $(5.75)^{3.1}$ *and prefix the correct numerical sign to the result.*

Example *Use a calculator to evaluate (a)* $\sqrt[4]{3.7}$; (b) $2^{-\sqrt{3}}$.

Solution (a) $\sqrt[4]{3.7}$ *is computed by observing that*

$$\sqrt[4]{3.7} = (3.7)^{1/4} = (3.7)^{0.25}$$

Thus, $\sqrt[4]{3.7}$ *is evaluated on a calculator using the following sequence of keystrokes:*

3.7 [y^x] 0.25 [=] 1.3869

Since $\sqrt[4]{3.7} = \sqrt{\sqrt{3.7}}$, *the same result follows from*

3.7 [$\sqrt{\ }$] [$\sqrt{\ }$] = 1.3869

(b) *To compute* $2^{-\sqrt{3}}$, *use the following keystrokes:*

2 [y^x] 3 [$\sqrt{\ }$] [+/−] [=] 0.30102

Exercises for Section 2.4

Write Exercises 1–8 in fractional exponent form.

1. $\sqrt[3]{x}$ **2.** $\sqrt{a}$ **3.** $\sqrt[4]{x/y}$

4. $\sqrt{x\sqrt{x\sqrt{x}}}$ **5.** $\sqrt[3]{x^2}$ **6.** $\sqrt{x}+\sqrt{y}$

7. $\sqrt[4]{x\sqrt{x\sqrt[3]{x}}}$ **8.** $\sqrt{5x\sqrt[3]{x}}$

Write Exercises 9–14 in radical form.

9. $x^{3/5}$ **10.** $x^{3/4}$ **11.** $(a^{3/2})^{1/4}$

12. $\dfrac{1}{x^{-1/2}}$ **13.** $(a+b)^{1/3}$ **14.** $x+y^{1/2}$

Simplify the expressions in Exercises 15–37. Write all answers with positive exponents when exponents are required.

15. $(64)^{1/2}$ **16.** $(64)^{2/3}$ **17.** $(81)^{3/4}$

18. $8^{-5/3}$ **19.** $(32)^{-3/5}$ **20.** $(-125)^{1/3}$

21. $x^{1/2}x^{2/3}$ **22.** $a^{1/4}a^{1/3}$ **23.** $y^{4/3}y^{-1/2}$

24. $x^{3/8}x^{1/4}$ **25.** $(125a^3b^9)^{1/3}$

26. $(9x^4y^{10})^{1/2}$ **27.** $(9x^{2/3})^{1/2}$

28. $(16a^{-8/9}b^{4/3})^{3/4}$ **29.** $2x^{-1/2}+3y^{-1/2}$

30. $x^{1/2}-x^{-1/2}$ **31.** $(a^{1/2}+b^{1/2})^2$

32. $(a^{1/2}+b^{1/2})(a^{1/2}-b^{1/2})$

33. $(y^{1/2}+y^{-1/2})^2$ **34.** $x^{1/2}y+x^{-1/2}$

35. $(x+3)(x-2)^{-1/2}+(x-2)^{1/2}$

36. $(x-1)(x+1)^{-1/2}+2(x+1)^{1/2}$

37. $\dfrac{(x-4)(x+5)^{-1/2}-(x+5)^{1/2}}{(x+5)^{-1/2}+(x+5)^{1/2}}$

Write the expressions in Exercises 38–41 using a single radical.

38. $\sqrt[3]{2}\sqrt{7}$ **39.** $\sqrt[3]{5}\sqrt[4]{4}$

40. $\sqrt[5]{x^2}\sqrt{y}$ **41.** $\sqrt{x}\sqrt[3]{(x+y)}$

With a calculator, approximate the numbers in Exercises 42–54.

42. $2^{\sqrt{3}}$ **43.** $8^{-\sqrt{8}}$ **44.** $\pi^{\sqrt{2}}$

45. π^{π} **46.** $(\sqrt{2})^{\pi}$ **47.** $3^{0.78}$

48. $(15.78)^{-3/2}$ **49.** $(0.95)^{2/5}$ **50.** $(102.67)^{3/8}$

51. $(226.4)^{-1.25}$

52. $\sqrt[3]{29.38}+41.6\sqrt[3]{29.38}$

53. $\dfrac{\sqrt[3]{0.115}}{\sqrt[3]{0.115}+\sqrt[3]{1.297}}$ **54.** $\dfrac{\sqrt[5]{16.79}+\sqrt[5]{394.6}}{\sqrt[5]{16.79}-3.01\sqrt[5]{394.6}}$

2.5 Common Errors

We have reviewed those skills necessary for any subsequent mathematics course. In addition to learning concepts, you must also develop manipulative skills. These skills depend not only on knowing the governing algebraic rule but also on such traits as precision, carefulness, and experience in "knowing what to do next."

Everyone has made careless errors in mathematics, such as writing a number illegibly, omitting signs, leaving out necessary parentheses, or writing the exponent of a number as a coefficient. Such "trivial" mistakes can be disastrous in some of the mathematically related disciplines. Computer programmers soon find out how a computer responds to the omission of a right parenthesis. The computer will unhesitatingly reject the entire program for such a "trivial" error. Likewise, failure to distinguish between $(3x)^2$ and $3x^2$ or to change the sign of every term when removing parentheses from expressions such as $-(x^2-4xy+y^2)$ might cause your bridge to collapse if you are a civil engineer. Perhaps these examples are dramatic, but carelessness and imprecision *can* be costly. You must know the laws of real numbers and be very careful when applying them. The following exercise set shows some trivial errors that have appeared in examination papers. See whether you can spot them.

Exercises for Section 2.5

Tell in precisely what way the mathematical statements in Exercises 1–21 are incorrect.

1. $2x^2=(2x)^2$

2. $(3x)^2=3x^2$

3. $-(2x-1)=-2x-1$

4. $-(x^2-5x-6)=-x^2-5x-6$

5. $\frac{1}{2}+\frac{3}{4}=\frac{4}{6}$

6. $\frac{2}{6}-\frac{1}{2}=\frac{1}{4}$

7. $\frac{x+2}{x}=x+\frac{2}{x}$

8. $(x+y)^2=x^2+y^2$

9. $(x^{-1}+y^{-1})^{-1}=x+y$

10. $\sqrt{x+y}=\sqrt{x}+\sqrt{y}$

11. $(x^3+x^2)=x^2(x+x^2)$

12. $(\sqrt{x}+\sqrt{y})^2=x+y$

13. $\frac{1}{x+y}=\frac{1}{x}+\frac{1}{y}$

14. $\sqrt{2x}=\sqrt{2}x$

15. $(-x)^2=-x^2$

16. $\frac{2}{x^2+2x+1}=\frac{1}{x^2+x+1}$

Test 1 for Chapter 2

Answer *True* or *False* to Exercises 1–10.

1. $\sqrt{-4}=\pm 2$

2. $\sqrt[3]{-8}=-2$

3. $a^{-2}=\sqrt{a}$

4. $(2xy^2)^3=8x^3y^6$

5. $\sqrt{x+y}=\sqrt{x}+\sqrt{y}$

6. $\sqrt{x\cdot y}=\sqrt{x}\cdot\sqrt{y}$

7. $\frac{1}{\sqrt[5]{(x-y)^3}}=(x-y)^{-3/5}$

8. $[(a^2b^5)^0]^3=a^6b^{15}$

9. $\sqrt{(-2)^2}=-2$

10. $(\sqrt{x}+\sqrt{y})^2=x+y$

11. Simplify (a) $[(25x^4y^{12})^{1/2}]^3$; (b) $\sqrt{\frac{a^4}{bc^2}}$.

12. Simplify $(xy^{-2}+1)^{-1}$

13. Simplify $\frac{x+y^{-1}}{2+x^{-2}+y^{-2}}$.

14. Simplify $\sqrt{2c^{-1}+d^{-2}}$.

15. Expand and simplify $(\sqrt{x}+\sqrt{y}-2)(\sqrt{x}-\sqrt{y}+5)$.

16. Simplify $\frac{\sqrt{a}-\sqrt{b}}{\sqrt{a}+\sqrt{b}}$.

Test 2 for Chapter 2

Simplify each of the expressions in 1-5.

1. $\sqrt[3]{128x^4y^{12}}$

2. $\dfrac{x+4}{\sqrt[3]{x+4}}$

3. $\left[\dfrac{x^3y^{-7}}{2x^5y^{-2}}\right]^{-3}$

4. $\sqrt[5]{a^4b^3}\,\sqrt[5]{ab^3}$

5. $\sqrt{48}+\sqrt{108}-\sqrt{50}$

6. Write in positive fractional exponent form.
(a) $\sqrt{\sqrt{\sqrt{x}}}$ (b) $\sqrt[3]{-x^{-1}}$

7. Rationalize (a) $\dfrac{\sqrt{2}}{5+\sqrt{2}}$ (b) $\dfrac{1}{x+\sqrt{y}}$.

8. Simplify $\dfrac{(xy)^{-2}}{x^{-2}+y^{-2}}+x^{-2}$.

9. Write $\sqrt[3]{2x\sqrt{y}}$ using a single radical.

10. Show that $x+y$ and $(x^{-1}+y^{-1})^{-1}$ are not identically equal.

3

Equations and Inequalities

3.1 Linear Equations

We begin the study of equations and inequalities by explaining some of the terminology used in the algebra of equations. Much of the same terminology is also used for inequalities.

An **equation** is a statement that two expressions are equal. An equation may do more than state an equality, however; it often raises a question. For instance, the equation $x+3=7$ states that the expressions "$x+3$" and "7" are equal; it also raises the question, what number x must be added to 3 to equal 7? We conclude that since $4+3=7$, then $x=4$ makes the equation a true statement; and we say that $x=4$ *satisfies* the equation. In general, a **solution**, or a **root**, of the equation is any real number in the domain of the variable of the equation that makes the equation a true statement. The set of all solutions of an equation is called the **solution set**. For the equation $x+3=7$, the solution set consists of the one element 4. The equation $x^2=4$ has both -2 and 2 in its solution set, since $(-2)^2=(2)^2=4$. The expression "to solve the equation" means to find the solution set.

Equations such as $x+3=7$ and $x^2=4$, which are true for some but not all of the permissible values of the variable, are called **conditional** equations.* If the statement is never true, the solution set of the equation is the empty set; that is, it contains no elements. For instance, $x^2=-1$ is considered a conditional equation even though there is no real number that equals -1 when squared. In some cases all permissible values of the variable make the statement true. Equations of this kind are called **identities**. The equation $x+3x=4x$ is an identity.

*The "permissible values" of the variable are the real numbers unless otherwise stated.

An equation that can be arranged in the form

$$ax + b = 0$$

where a and b are constants and $a \neq 0$, is called a **linear** equation in the variable x. The solution set of the linear equation $ax+b=0$ consists of one element, $-b/a$. Although we can frequently find the solution to a linear equation simply by examining the terms of the equation, we need a systematic method. You should be aware that the procedures we are about to develop for solving linear equations apply to more general equations.

Two equations are said to be **equivalent** if they have exactly the same solution set. Thus, the equations $x+3=7$ and $x+2=6$ are equivalent, since each has the one and only solution 4. The equations $x=2$ and $x^2=4$ are *not* equivalent, since the solution to the first one is 2 and to the second 2, -2. The idea in solving an equation is to manipulate the given equation to change it into an equivalent equation in which the solution is more or less obvious. Usually, the aim of the manipulation is to isolate the variable on one side of the equation. After a series of steps, the solution becomes obvious.

We say that a manipulation is **allowable** if it transforms an equation into an equivalent one. Two allowable manipulations are:

(1) Adding the same quantity to or subtracting the same quantity from both sides of a conditional equation.

(2) Multiplying or dividing both sides by any nonzero quantity.

Examples 1 and 2 show how to use these manipulations to solve linear equations in one unknown. The sequence in which the manipulations are made is not unique, but the solution set is unique.

EXAMPLE 1 Solve the equation $3x+12=48$.

SOLUTION Subtract 12 from each side, to get

$$(3x+12)-12=48-12$$

which can be simplified to the equivalent equation $3x=36$. Dividing both sides of this equation by 3 gives the explicit solution

$$x=12$$

EXAMPLE 2 Solve $2(x+3)=3x$.

SOLUTION Expanding the left-hand side, we get

$$2x+6=3x$$

Subtracting $3x+6$ from both sides gives

$$-x=-6$$

$$x=6$$

Some equations in which the variable is in the denominator of a term can be solved by the same methods, after multiplying each term in the equation by the lowest common denominator (LCD) to eliminate the fractions. If the resulting equation is linear, that is, of the form $ax+b=0$, it can be solved accordingly. A word of caution: *Any value of the variable that produces a zero denominator cannot be a solution*. Therefore, all apparent solutions must be checked for validity.

EXAMPLE 3 Solve $\frac{1}{x}+\frac{3}{2x}=\frac{3}{x-5}$.

SOLUTION Multiply each term by $2x(x-5)$ to get rid of fractions:

$$\frac{1}{x}\cdot 2x(x-5)+\frac{3}{2x}\cdot 2x(x-5)=\frac{3}{x-5}\cdot 2x(x-5)$$

Cancelling, we get

$$2(x-5)+3(x-5)=3(2x)$$

In this form, the equation is linear, so we proceed as before:

$$\begin{aligned} 2x-10+3x-15&=6x \\ 2x+3x-6x&=10+15 \\ -x&=25 \\ x&=-25 \end{aligned}$$

Check: Since none of the denominators is zero for $x=-25$, it is a valid solution.

EXAMPLE 4 Solve $\frac{2}{x+1}-\frac{4x}{x^2-1}=\frac{3}{x-1}$.

SOLUTION Multiply each term by $(x+1)(x-1)$, to get

$$2(x-1)-4x=3(x+1)$$

Expanding gives

$$2x-2-4x=3x+3$$

and if like terms are collected,

$$\begin{aligned} -5x&=5 \\ x&=-1 \end{aligned}$$

Check: Since $x=-1$ causes a zero in the denominator of the first and second terms, it is not a permissible solution. The value $x=-1$ is sometimes called an **extraneous solution**, since it appears to be a solution but is not. There is no solution to this equation.

EXAMPLE 5 Solve $\frac{x-2}{x+2} - \frac{2}{x} = 1$.

SOLUTION Multiply by $x(x+2)$ to get

$$x(x-2)-2(x+2)=x(x+2)$$

Expanding gives

$$x^2-2x-2x-4=x^2+2x$$

and

$$-6x=4$$

$$x=-\frac{2}{3}$$

Check: $x=-\frac{2}{3}$ is a solution since it does not cause a zero denominator in the original equation.

Algebraic expressions that involve two or more variables can often be solved explicitly for one of the variables. The following example is typical.

EXAMPLE 6 The expression $S=\frac{a-rL}{1-r}$ occurs in the study of geometric progressions. Solve for r.

SOLUTION Multiplying both sides of the equation by $1-r$, we get

$$S(1-r)=a-rL$$

Expanding the left-hand side yields

$$S-Sr=a-rL$$

Isolating the terms involving r on the left-hand side, we have

$$rL-Sr=a-S$$

and then factoring the value of r on the left-hand side, we get

$$r(L-S)=a-S$$

Finally, dividing both sides by the coefficient of r, we get

$$r=\frac{a-S}{L-S}$$

To solve equations that involve absolute values, we can often use the methods considered here. Notice that the absolute value expressions cause the equation to take on different forms.

EXAMPLE 7 Solve the equation $|x-1|+3x=6$.

SOLUTION The form of the given equation depends on the absolute value term. Since

$$|x-1|=\begin{cases} x-1 & \text{if } x\geqslant 1 \\ -(x-1) & \text{if } x<1 \end{cases}$$

we must solve the two equations

$$x-1+3x=6 \quad\text{and}\quad -(x-1)+3x=6$$

The first equation is valid for $x\geqslant 1$ and is equivalent to each of the following:

$$\begin{aligned} x-1+3x&=6 \\ x+3x&=7 \\ 4x&=7 \\ x&=\frac{7}{4} \end{aligned}$$

Since $7/4>1$, we have that 7/4 is a solution.

The second equation is valid for $x<1$ and is equivalent to each of the following:

$$\begin{aligned} -(x-1)+3x&=6 \\ -x+1+3x&=6 \\ -x+3x&=5 \\ 2x&=5 \\ x&=\frac{5}{2} \end{aligned}$$

But 5/2 is not less than 1. Thus 5/2 is not a solution. (Check it.)

Hence, the solution consists only of 7/4.

Exercises for Section 3.1

In Exercises 1–25, find the solution set. Check for extraneous solutions when there are fractional expressions.

1. $2x+5=2$

2. $6x-1=-x-3$

3. $3(2x-1)-x=4-(x+3)$

4. $3(x-2)+7(2x+4)=-x-1$

5. $3x+5=4(x-2)$

6. $3y+7(4-y)=3(5-2y)$

7. $\frac{4a}{3}-5a+2=\frac{a}{2}-1$

8. $\frac{3x}{5}+4=\frac{x}{2}+5$

9. $\frac{2x+1}{3}+16=3x$

10. $\frac{x+1}{3}+\frac{x+2}{7}=5$

11. $\frac{2}{c}+\frac{3}{c}=10$

12. $\frac{4}{y}-3=\frac{5}{2y}$

13. $\frac{3x}{x+5}-4=0$

14. $\frac{4}{x+2}+4=0$

15. $\frac{3}{6y+2}=\frac{4}{7y+3}$

16. $\frac{2}{x+3}=\frac{5}{2x-1}$

17. $\frac{2}{x-2}-\frac{3}{x+5}=\frac{10}{(x-2)(x+5)}$

18. $\frac{1}{2x+3}+\frac{1}{x-1}=\frac{1}{(2x+3)(x-1)}$

19. $\dfrac{5}{5m-11}=\dfrac{3}{m-5}-\dfrac{4}{2m-3}$

20. $\dfrac{2}{x+3}+\dfrac{3}{x-4}=\dfrac{5}{x+6}$

21. $(x+7)(x-1)=(x+9)^2$

22. $(x-3)(x+4)=x^2-1$

23. $2x^2+1=(2x-1)(x+3)$

24. $3x(x-1)=(3x-7)(x+1)$

25. $x(x^2+3)=5+x^3$

26. Solve for x: $\dfrac{1}{x-a}+\dfrac{1}{x+a}=\dfrac{2}{x-1}$

27. Solve for x: $\dfrac{p+q}{x+l}=\dfrac{p-q}{x-l}$

28. The expression $l=a+(n-1)d$ appears in the discussion of arithmetic progressions. Solve for d.

29. The formula $\dfrac{1}{R}=\dfrac{1}{R_1}+\dfrac{1}{R_2}$ occurs in electric circuit analysis. Solve for R.

30. Degrees Fahrenheit F and degrees Celsius C are related by the formula $F=\frac{9}{5}C+32$. Solve for C.

31. The expression for linear velocity v has the form $v=at+b$. Solve for t.

32. The simple interest formula is given by $A=P(1+in)$. Solve for n.

33. The area of a trapezoid with bases a and b and height h is given by $A=\frac{1}{2}h(a+b)$. Solve for b.

Solve the equations in Exercises 34–43 by considering the different forms of the equation.

34. $|x|+x=3$

35. $|x|+2=-2$

36. $2x-|x|=-3$

37. $3x-1=|x|$

38. $|x+3|=5$

39. $|2x-3|=-1$

40. $|x-2|+x=3x-4$

41. $|x+5|-2x=1+x$

42. $|2x+7|-3(x+5)=0$

43. $|5x-1|+2=3(x+13)$

3.2 Applications of Linear Equations

Mathematics helps us to describe and understand the physical world. In this section we show how to represent some simple physical problems with abstract algebraic language, which uses symbols to represent unknown quantities. Usually we let symbols, such as letters of the alphabet, stand for the unknowns, and we then try to express the desired relation by these symbols. You will need to know certain elementary relations, some of which are included in the following example.

EXAMPLE 1

(1) If we allow w to represent the width of a rectangle, l its length, and A its area, then the fact that "the area of the rectangle is the product of the length and the width" is written as $A=l\cdot w$.

(2) If you have to pay eight percent tax on the cost of a car, then by letting x represent the cost of the car, the tax you pay is written $0.08x$.

(3) The distance traveled in miles d equals the product of the speed in miles per hour, denoted by v, and the time in hours t. Symbolically, we have $d=vt$.

(4) The sum of two numbers is 12. If x is one of the numbers, $12-x$ is the other.

(5) By letting x be the units digit of a number and y the tens digit, we can write the value of the number as $10y+x$.

(6) If x represents a number, then $x+4$ represents the number that is larger than x by 4; $5x$ represents the number 5 times x; and $5x+4$ represents the number that is larger than 5 times x by 4.

(7) If it takes t hr to do a job, then $1/t$ is the part of the job that can be done in 1 hr.

The foregoing examples show how algebraic language can be used to represent an English expression in an abstract, symbolic form. In abstract form the given expression can be studied independently of the sentence itself. Further, in the abstract setting seemingly dissimilar problems often turn out to be mathematically identical.

To solve problems that arise from physical situations, you must be able to write equations that describe the stipulated conditions. There are no set "rules," but the following guidelines at least give you a systematic approach.

(1) **Read the problem.** Read the given problem carefully to determine precisely what is known and what you are trying to find. If at all possible, draw a picture of the situation. Drawing a picture frequently helps you to clarify your ideas.

(2) **Identify the variable quantities.** Assign letters to the quantities that you want to find, thereby denoting them as variables. In each new problem, your solution should begin with the phrase "Let $x=$" or "Let $n=$."

(3) **Form an equation.** The problem itself should tell you how the variables are related. Use the given data to establish an equation involving the variables.* You will need to know certain elementary relations, such as those mentioned in Example 1.

(4) **Solve the equation.** Solving the equation is a mathematical process that is independent of the physical conditions of the problem. Check your solution to make sure it is a sensible answer. For instance, -2 would not be realistic as the answer for one dimension of a rectangle.

EXAMPLE 2 A student pays \$278 for a calculator and a typewriter. If the calculator costs \$64 less than the typewriter, how much did each cost?

SOLUTION We let $x=$the cost of the typewriter. Then $x-64=$the cost of the calculator. Since the total cost of the two items is \$278, we have

$$\boxed{\text{Cost of the typewriter}} + \boxed{\text{Cost of the calculator}} = \boxed{\text{Total cost}}$$

Substituting the variable expressions for the cost of the typewriter and the cost of the

*An equation that represents a set of physical conditions is sometimes called a **mathematical model** of those conditions.

calculator, we get the desired equation:

$$x+(x-64)=278$$
$$2x=278+64$$
$$x=\frac{342}{2}=171$$

Hence, the typewriter costs \$171 and the calculator costs \$107.

EXAMPLE 3 A plane leaves New York for Los Angeles at 10:00 P.M., cruising at 400 mph. (See Figure 3.1.) A second plane leaves New York at 11:30 P.M. for the same destination, cruising at 550 mph. How long will it take for the second plane to overtake the first?

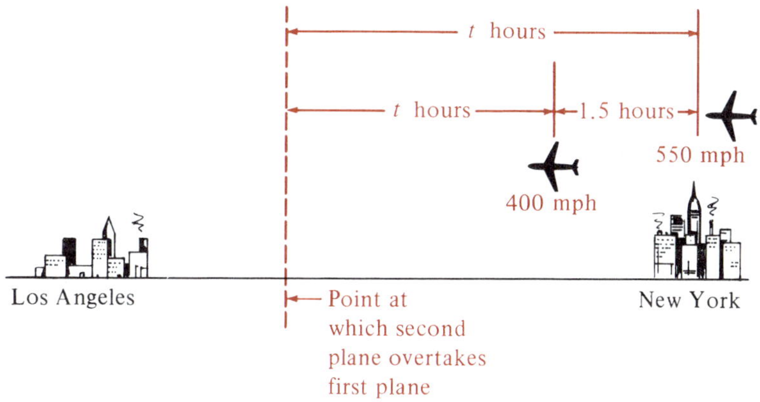

FIGURE 3.1

SOLUTION If we let t equal the time required for the second plane to overtake the first, then at the time of overtaking, the first plane will have been in the air $t+1.5$ hr (from 10:00 P.M. to 11:30 P.M.). Using the formula $d=vt$ (Example 1), we write

$400(t+1.5)=$ distance traveled by first plane and

$550t=$ distance traveled by second plane

Since the distance traveled by each plane is equal when the second plane overtakes the first, we have

Distance traveled by first plane	=	Distance traveled by second plane

Consequently, the desired equation is

$$400(t+1.5)=550t$$
$$150t=600$$
$$t=4 \text{ hr}$$

Since it requires almost 5.5 hr for the faster plane to make the trip (3000 mi), the second plane will overtake the first when almost three-fourths of the trip is completed.

EXAMPLE 4 A man has two types of solder; one is 55% tin and the other is 15% tin. How much of each must he mix to yield 75 lb of solder containing 40% tin?

SOLUTION Let n be the number of pounds of solder with 55% tin in the final mixture. Since the final mixture is to weigh 75 lb, then $75-n$ is the number of pounds of solder with 15% tin. The amount of tin in the two types of solder is then $0.55n$ and $0.15\,(75-n)$, and the total amount of tin in the 40% mixture is $0.40(75)$. Hence,

$$\boxed{\begin{array}{l}\text{Amount of tin}\\ \text{from 55\% mix}\end{array}}+\boxed{\begin{array}{l}\text{Amount of tin}\\ \text{from 15\% mix}\end{array}}=\boxed{\begin{array}{l}\text{Amount of tin}\\ \text{in 40\% mix}\end{array}}$$

$$\begin{aligned} 0.55n+0.15(75-n)&=0.40(75)\\ 0.55n+11.25-0.15n&=30.00\\ 0.40n&=18.75\\ n&=\frac{18.75}{0.40}=46.9 \text{ lb of solder with 55\% tin}\end{aligned}$$

and

$$75-n=75-46.9=28.1 \text{ lb of solder with 15\% tin.}$$

Note the implicit physical limitations on this solution. For example, if the man has only 40 lb of the solder with the 55% tin available, then there would be no solution.

EXAMPLE 5 The salt content of seawater is often reduced to the level at which it is usable for irrigation by combining it with "pure" drinking water. What percentage of seawater (3.8% salt) should be combined with drinking water (0.5% salt) to get irrigation water of 1% salt?

SOLUTION For simplicity, consider a 100-gal mixture. Let $x=$the number of gallons of seawater in the mixture; then $100-x$ is thus the number of gallons of drinking water. The amount of salt contributed from the seawater is $0.038x$, and from the drinking water, $0.005(100-x)$. The total amount of salt in the mixture is $0.01(100)$. Hence,

$$\boxed{\begin{array}{l}\text{Amount of salt}\\ \text{from seawater}\end{array}}+\boxed{\begin{array}{l}\text{Amount of salt}\\ \text{from drinking water}\end{array}}=\boxed{\begin{array}{l}\text{Amount of salt}\\ \text{in irrigation water}\end{array}}$$

Substituting the appropriate algebraic expressions yields

$$0.038x+0.005(100-x)=0.01(100)$$

By multiplying each side by 1000 (to clear the decimals), we get

$$38x+5(100-x)=10(100)$$

simplifying,

$$38x+500-5x=1000$$

and solving for x, we get

$$33x=500$$
$$x=15.15 \text{ gal}$$

On a percentage basis, there should be about 15% seawater and 85% drinking water in the mixture.

EXAMPLE 6 A swimming club manager needs to fill the pool in 8 hr, and she knows that the built-in water line will take 12 hr to fill the pool. What should be the minimum rating of an auxiliary hose, in "hours to fill a pool," to get the pool filled on time if both main line and auxiliary hose are used?

SOLUTION Let x = minimum rating of auxiliary hose. Then

$$\frac{1}{12}=\text{part of the pool filled by main line in 1 hr}$$

$$\frac{1}{x}=\text{part of the pool filled by auxiliary hose in 1 hr}$$

$$\frac{1}{8}=\text{part of the pool that must be filled in 1 hr by both hoses}$$

Since the part of the pool that can be filled in 1 hr is the sum of the capacities of the main and auxiliary lines, we have

[Amount filled by main line in 1 hr] + [Amount filled by auxiliary hose in 1 hr] = [Amount filled by both in 1 hr]

That is,

$$\frac{1}{12}+\frac{1}{x}=\frac{1}{8}$$

Multiply through by $24x$ to get

$$2x+24=3x$$

from which

$$x=24$$

Hence the auxiliary hose must be able to fill the pool (when operating alone) in 24 hr or less.

Exercises for Section 3.2

Express each of the sentences or phrases in Exercises 1–20 in algebraic symbols.

1. The area of a triangle equals one-half the product of the base times the height.

2. The area of a circle equals the product of π times the square of the radius.

3. The perimeter of a square is 4 times the side length.

4. The perimeter of a triangle is the sum of the lengths of the sides.

5. Acceleration equals velocity divided by time.

6. Voltage equals the current in the circuit times the resistance of the circuit.

7. A number that is less than another number by 8.

8. A number 3 times the square of another number.

9. A number 5 times the sum of another number plus 3.

10. A number 3 more than 5 times another number.

11. Twenty percent of the amount by which a number exceeds 12,000.

12. The sum of two numbers divided by 2.

13. The average of three numbers.

14. The portion of a job done in 1 day if it takes x days to do the entire job.

15. Interest equals principal times rate.

16. The product of the sum and difference of two numbers.

17. Twice the sum of two numbers.

18. The area of a triangle whose base is twice the height.

19. The surface area of a cube.

20. The volume of a rectangular box.

21. Two resistors in parallel carry a total current of 20 amperes (A). What is the current in each resistor if one resistor carries 2 A more than the other?

22. Two gears together weigh 17 lb. If one gear weighs 3 lb more than the other, what is the weight of each?

23. Find three consecutive even integers whose sum is 312.

24. The total monetary value of 73 nickels and dimes is $5.75. How many nickels are there?

25. A jogger who can run a 7 min mile starts one-tenth mile ahead of a jogger who can run a 6 min mile. How far does the faster jogger have to run to catch the slower jogger?

26. The average of the salaries of Jane, Jim, and Nancy is $15,000 a year. If Jane earns $3000 more than Nancy and Jim earns half as much as Jane, how much does each earn?

27. A car that averages 22 mi/gal of gasoline for city driving and 28 mi/gal for highway driving uses 20 gal of gasoline in 500 mi of driving. How much of the driving was in the city?

28. A woman wants to winterize her car and finds that her 10-qt radiator is 30% antifreeze. How much of the fluid should she drain and replace by pure antifreeze to double the strength of the mixture?

29. An airplane flew with the wind for 1 hr and returned the same distance against the wind in 2 hr. If the cruising (air) speed of the plane is 400 miles per hour, what is the velocity of the wind?

30. Suppose that in driving you average 70 mph outside the city limits and 30 mph within the city limits. If you take 5 hr to make a 270-mi trip, how much of the time did you spend inside the limits of cities and how much outside?

31. The distance around the Indianapolis speedway track is 2.5 mi. If one driver averages 170 mph and another averages 165 mph, in how much time will the first driver "lap" the second?

32. If you and a man together can build in 20 days a house that he would take 30 days to build alone, how long would it take you alone to build the house, assuming that you know how?

33. In football a team scored 44 points by scoring equal numbers of touchdowns, extra points, and field goals, and half as many safeties. How many of each scoring play did the team make?

34. An oil tanker is being filled by two pipelines, one of which could separately fill it in 12 hr and one in 16 hr. How long will it take to fill the tanker?

35. A girl weighing 100 lb sits on one end of a 20-ft teeterboard and a boy weighing 150 lb sits on the other. Where must the fulcrum be for the two to be balanced?

36. A man has a 20-lb mixture that is one-fourth cement and three-fourths sand. How much of a mixture that is half cement and half sand must be added to the 20-lb mixture to produce a mixture that is one-third cement?

37. A car leaves Cincinnati for Toledo at 55 mph. Thirty minutes later a truck leaves Toledo for Cincinnati at 45 mph. If the two cities are 206 mi apart, how many hours must the truck travel before it meets the car? Assume that both vehicles are using Interstate 75.

38. Jane can overhaul an engine in 20 hr and Fred can do the same job in 30 hr. If they both work together for a time and then Jane finishes the job by herself in 5 hr, how many hours did they work together?

3.3 Quadratic Equations

Suppose we want to construct a rectangle that has an area of 36 square inches and length 5 inches greater than its width. (See Figure 3.2.) We can write the conditions in equation form by letting x equal the width of the rectangle. The length is then $x+5$. Since the area of a rectangle equals the length times the width, we can write

$$x(x+5)=36$$

Expanding the left-hand side gives

$$x^2+5x=36$$

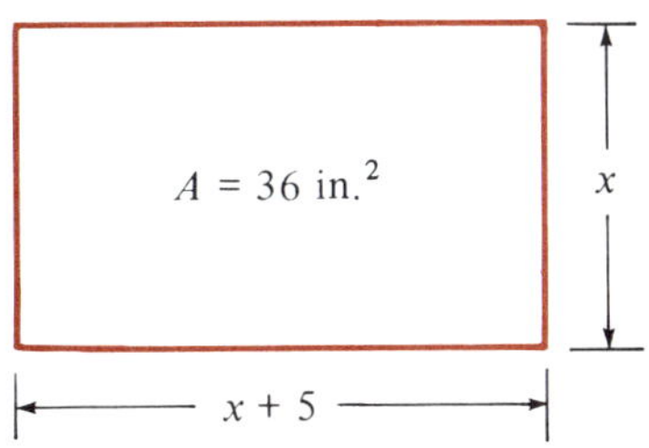

FIGURE 3.2

This equation, which is not a linear equation because of the x^2 term, typifies a whole class of equations called quadratic equations.

Equations that can be put into the form

$$ax^2+bx+c=0$$

where a, b, and c are constants and $a \neq 0$, are called **quadratic equations**. Quadratic equations must contain a square term, and they may contain a linear term and a constant term, depending on the values of b and c. Each of the following equations is a quadratic equation:

$$3t^2-5t-9=0$$

$$t^2=t+2$$

$$x^2+3x=0$$

$$3u^2-7=0$$

$$(y+2)(y-1)=0$$

The last of these equations is quadratic, since $(y+2)(y-1)=y^2+y-2$.

Two basic approaches to finding the solution set of a quadratic equation are: (1) Factoring the quadratic expression; and (2) completing the square. The second method generalizes to a formula in which the roots, or solutions, of the equation are given in terms of the coefficients a, b, and c.

Solution by Factoring

If a quadratic equation can be factored into the product of linear factors, it is said to be a **factorable** equation. The roots of a factorable equation can be found by taking advantage of the following fundamental property of real numbers:

If u and v are any two real numbers and

$$u \cdot v=0,$$

then either $u=0$ or $v=0$.

Thus, the roots of a factorable equation can be found by setting each factor equal to zero and solving the resulting linear equation. The procedure is summarized as follows:

(1) Arrange the given equation in the form $ax^2+bx+c=0$.

(2) Factor the quadratic expression into linear factors.

(3) Determine the value of x for which each factor will equal zero. These are the roots of the equation.

EXAMPLE 1 Solve the quadratic equation $x^2+5x=36$.

SOLUTION First, we write the equation in the form

$$x^2+5x-36=0$$

The quadratic expression factors into $(x+9)(x-4)$, so

$$(x+9)(x-4)=0$$

Since $x=-9$ will make the first factor equal to zero and $x=4$ will make the second factor zero, we conclude that the roots, are -9 and 4.

EXAMPLE 2 Solve the equation $t^2=3t$.

SOLUTION If $3t$ is subtracted from both sides, the given equation becomes

$$t^2-3t=0$$

Or if t is factored out of each term, then

$$t(t-3)=0$$

Since the first factor is zero for $t=0$ and the second is zero for $t=3$, the roots are 0 and 3.

EXAMPLE 3 Solve the equation $3m^2+14m-5=0$.

SOLUTION $3m^2+14m-5=0$

Factoring this, we get

$$(3m-1)(m+5)=0$$

The factor $3m-1$ will equal zero for $m=\frac{1}{3}$, and the factor $m+5$ will equal zero for $m=-5$. Hence, the roots of the equation are $\frac{1}{3}$ and -5.

Solution by the Quadratic Formula

The roots of $3x^2+9x+4=0$ are difficult to obtain by factoring. To find the roots of equations like this, we use the quadratic formula, which is obtained from the general equation $ax^2+bx+c=0$ by applying the method of completing the square.*

To complete the square on $ax^2+bx+c=0$, we first add $-c$ to both sides of the equation and then divide by the coefficient a, to get

$$x^2+\frac{b}{a}x=-\frac{c}{a}$$

Completing the square on the left-hand side by adding $(b/2a)^2$ to both sides yields

$$x^2+\frac{b}{a}x+\frac{b^2}{4a^2}=-\frac{c}{a}+\frac{b^2}{4a^2}$$

Simplifying gives

$$\left(x+\frac{b}{2a}\right)^2=\frac{b^2-4ac}{4a^2}$$

or

$$x+\frac{b}{2a}=\pm\sqrt{\frac{b^2-4ac}{4a^2}}=\frac{\pm\sqrt{b^2-4ac}}{2a}$$

*Completing the square was discussed in Section 1.7.

Subtracting $\frac{b}{2a}$ from both sides,

$$x=\frac{-b}{2a}\pm\frac{\sqrt{b^2-4ac}}{2a}$$

or

$$\textbf{(3.1)}\qquad x=\frac{-b\pm\sqrt{b^2-4ac}}{2a}$$

Equation (3.1) is the **quadratic formula**; we use it by matching the coefficient of x^2 with a, the coefficient of x with b, and the constant term with c in the formula, when the quadratic equation is written in the standard form $ax^2+bx+c=0$.
From Formula 3.1 we can see that the equation $ax^2+bx+c=0$ has two roots:

$$x=\frac{-b+\sqrt{b^2-4ac}}{2a} \quad \text{and} \quad x=\frac{-b-\sqrt{b^2-4ac}}{2a}$$

EXAMPLE 4 Use the quadratic formula to solve $3x^2=2x+5$.

SOLUTION To use the quadratic formula, write the given equation in the form $ax^2+bx+c=0$. Subtracting $2x+5$ from both sides of the given equation, we get

$$3x^2-2x-5=0$$

From this you can see that $a=3$, $b=-2$, and $c=-5$. Substituting these values in the quadratic formula, we have

$$x=\frac{-(-2)\pm\sqrt{(-2)^2-4(3)(-5)}}{2(3)}$$

$$=\frac{2\pm\sqrt{4+60}}{6}=\frac{2\pm\sqrt{64}}{6}$$

$$=\frac{2\pm 8}{6}$$

Therefore, the equation $3x^2-2x-5=0$ has the roots

$$x=\frac{2+8}{6}=\frac{10}{6}=\frac{5}{3}$$

and

$$x=\frac{2-8}{6}=\frac{-6}{6}=-1$$

With a calculator, the same effort is involved in solving quadratic equation with integer coefficients as in solving one with irrational coefficients. Begin by computing the quantity b^2-4ac inside the radical. Solve a few equations using the quadratic formula and a calculator. Notice any peculiarities of your particular calculator. For example, what kind of error message do you get if b^2-4ac is negative?

EXAMPLE 5 Solve the quadratic equation $\sqrt{2}x^2-3x-\pi=0$.

SOLUTION Using the quadratic formula with $a=\sqrt{2}$, $b=-3$, and $c=-\pi$, we have

$$x=\frac{-(-3)\pm\sqrt{(-3)^2+4\sqrt{2}\pi}}{2\sqrt{2}}$$

Using a calculator to approximate these values, we get

$$x=2.89 \qquad \text{and} \qquad x=-0.769$$

EXAMPLE 6 Solve the quadratic equation $x^2+2x+3=0$.

SOLUTION In this case, $a=1$, $b=2$, and $c=3$. Thus,

$$x=\frac{-2\pm\sqrt{4-12}}{2}=\frac{-2\pm\sqrt{-8}}{2}$$

The quantity inside the radical is negative, so we conclude that there are no real roots. Since $\sqrt{-8}=2\sqrt{2}i$, however, the solution to $x^2+2x+3=0$ can be written in terms of complex numbers as

$$x=\frac{-2\pm 2\sqrt{2}i}{2}=-1\pm\sqrt{2}i$$

One of the important uses of complex numbers is representing nonreal roots to equations. In solving quadratic equations, we choose to write the nonreal roots in terms of complex numbers. Thus, instead of saying that $x^2+4=0$ has no real roots, we shall say that its roots are $\pm 2i$. Aside from using complex numbers to represent nonreal roots of quadratic equations, however, we continue to restrict the permissible values to the set of real numbers.

EXAMPLE 7 Solve the quadratic equation $m^2+2m+2=0$. Represent the nonreal roots by complex numbers.

SOLUTION Using the quadratic formula with $a=1$, $b=2$, and $c=2$, we have

$$m=\frac{-2\pm\sqrt{2^2-4(1)(2)}}{2(1)}$$

$$=\frac{-2\pm\sqrt{-4}}{2}=\frac{-2\pm 2i}{2}=-1\pm i$$

The roots to this equation are a complex conjugate pair. This is typical of nonreal roots of quadratic equations with real coefficients: **Complex roots occur in conjugate pairs.**

The previous examples show that the nature of the roots of a quadratic equation depends on the expression b^2-4ac inside the radical.

The **discriminant** D of the quadratic equation $ax^2+bx+c=0$ is given by

$$D=b^2-4ac$$

By this notation, the roots of the quadratic can be expressed as

$$x=\frac{-b\pm\sqrt{D}}{2a}$$

If the coefficients of the quadratic are real, we can use the discriminant to determine the nature of the roots.

- If $D>0$, there are two distinct real roots of the form $(-b\pm\sqrt{D})/2a$.
- If $D=0$, there is one root of the form $-b/2a$. (*Note:* We sometimes say there are two equal real roots, or a **double** root.)
- If $D<0$, the roots are nonreal and can be represented by the complex conjugate pair $(-b\pm i\sqrt{|D|})/2a$.

EXAMPLE 8 Solve the quadratic equation $2z^2-3z-7=0$.

SOLUTION Since the discriminant is $(-3)^2-4(2)(-7)=9+56=65$, which is >0, this equation has two real roots, namely,

$$z=\frac{3\pm\sqrt{65}}{4}$$

EXAMPLE 9 Solve the quadratic equation $4x^2+20x+25=0$.

SOLUTION Since the discriminant is

$$(20)^2-4(4)(25)=400-400=0$$

this equation has two equal roots, namely,

$$\frac{-20}{2(4)}=\frac{-5}{2}$$

EXAMPLE 10 Does the quadratic equation $2x^2-5x+8=0$ have real roots?

SOLUTION The discriminant is $25-4(2)(8)=25-64<0$. Therefore the roots are nonreal, namely,

$$\frac{5\pm\sqrt{39}i}{4}$$

EXAMPLE 11 Determine k so that $kx^2+3kx+2k+1=0$ has equal roots.

SOLUTION In this case $a=k$, $b=3k$, and $c=2k+1$. Since for equal roots the discriminant must be 0,

$$b^2-4ac=(3k)^2-4k(2k+1)=0$$

Simplifying gives

$$9k^2-8k^2-4k=0$$
$$k(k-4)=0$$

Hence, $k=4$. Why is $k=0$ not a valid solution?

The two roots to a quadratic equation can be written separately as

$$x_1=\frac{-b+\sqrt{b^2-4ac}}{2a} \qquad x_2=\frac{-b-\sqrt{b^2-4ac}}{2a}$$

Then
$$x_1+x_2=\left(\frac{-b+\sqrt{b^2-4ac}}{2a}\right)+\left(\frac{-b-\sqrt{b^2-4ac}}{2a}\right)$$
$$=\frac{-b+\sqrt{b^2-4ac}-b-\sqrt{b^2-4ac}}{2a}$$
$$=\frac{-2b}{2a}=\frac{-b}{a}$$

Similarly
$$x_1x_2=\left(\frac{-b+\sqrt{b^2-4ac}}{2a}\right)\left(\frac{-b-\sqrt{b^2-4ac}}{2a}\right)$$
$$=\frac{(-b)^2-(\sqrt{b^2-4ac})^2}{4a^2}=\frac{b^2-(b^2-4ac)}{4a^2}$$
$$=\frac{c}{a}$$

Thus the sum and the product of the roots of a quadratic equation can be expressed in terms of the coefficients a, b, and c.

Consider, for example, the equation $2x^2+3x-1=0$. Without actually finding the roots, we know that their sum is $-b/a=-\frac{3}{2}$ and their product is $c/a=-\frac{1}{2}$. Since the roots can be found directly from the quadratic formula to be $(-3\pm\sqrt{17})/4$, this conclusion can be verified directly.

The coefficients can be found by the sum and product of roots formula *after* the roots have been specified.

EXAMPLE 12 Determine the values of b and c so that $2x^2+bx+c=0$ has roots 5 and -4.

SOLUTION Here we have $x_1=5$, $x_2=-4$, and $a=2$. Therefore,

$$x_1+x_2=5-4=\frac{-b}{2} \quad \text{or} \quad b=-2$$

Likewise,

$$x_1\cdot x_2=5(-4)=\frac{c}{2} \quad \text{or} \quad c=-40$$

Exercises for Section 3.3

Solve the equations in Exercises 1–18 by factoring.

1. $x^2-2x=0$ **2.** $3x^2+5x=0$

3. $x^2-x-2=0$ **4.** $x^2-x=12$

5. $x^2=x+6$ **6.** $x^2+7x+12=0$

7. $z^2+3z-10=0$ **8.** $x^2+10x=0$

9. $2x^2+5x-12=0$ **10.** $2x^2=x+6$

11. $3w^2-5w-2=0$ **12.** $4x^2+2=9x$

13. $10v^2-2v=0$ **14.** $6x^2-x-12=0$

15. $-t^2+3t+10=0$ **16.** $6y^2+7y=3$

17. $35x^2+2x-24=0$

18. $60y^2+y-10=0$

Solve Exercises 19–39 by using the quadratic formula. Express all nonreal roots in the form $a+bi$.

19. $3x^2+4x-4=0$ **20.** $2x^2+4x-5=0$

21. $5x^2+2x-3=0$ **22.** $8x^2+3x-5=0$

23. $2x^2-x-2=0$ **24.** $3x^2+x+9=0$

25. $x(2x-1)=4$ **26.** $x(x+1)=3$

27. $x^2-5x+5=0$ **28.** $12x^2+x-1=0$

29. $5x^2+12x+6=0$ **30.** $2x^2-x+5=0$

31. $3x^2-2x+2=0$ **32.** $6x(x-1)+1=0$

33. $2x^2-2x+5=0$ **34.** $\sqrt{3}x^2-2x-\pi=0$

35. $3x^2-\pi x+2=0$ **36.** $2x^2+4x-\sqrt{5}=0$

37. $\sqrt{6}x(x+1)=2.7$ **38.** $0.4x^2-1.3x+2.3=0$

39. $1.1x^2-0.2x+0.8=0$

Determine the value of k for which the roots of the equations in Exercises 40–49 satisfy the indicated condition.

40. $x^2+2x+k=0$; the product of the roots is -1.

41. $kx^2-3x+1=0$; the sum of the roots is π.

42. $2x^2-kx+7=0$; the sum of the roots is 10.

43. $kx^2+x+8=0$; both roots are real.

44. $x^2+4x+k=0$; both roots are complex.

45. $x^2-kx-5=0$; give two values of k for which the roots are rational.

46. $x^2+x-(k+3)=0$; one of the roots is zero.

47. $x^2-kx+k-1=0$; the two roots are equal.

48. $2x^2+x=-k$; one root is the reciprocal of the other.

49. $kx^2-2kx+5=0$; one root exceeds the other by 3.

3.4 Applied Problems That Lead to Quadatic Equations

EXAMPLE 1 Find the dimensions of a rectangle whose area is 36 cm² and whose length exceeds its width by 5 cm. See Figure 3.2.

SOLUTION Let x equal the length of the rectangle; then $x-5$ is its width. The area of a rectangle is the product of its length and its width, so

$$x(x-5)=36$$

or

$$x^2-5x-36=0$$

$$(x-9)(x+4)=0$$

Therefore, the two roots are $x=9$ and $x=-4$. Obviously, only $x=9$ has meaning, since a negative length is not possible. The width of the rectangle is $9-5$, or 4, and the desired dimensions are 9 by 4.

EXAMPLE 2 The height h, in feet, of a projectile t sec after it is fired from a gun is given by

$$h = kt - \tfrac{1}{2}gt^2$$

where g is the constant acceleration of gravity and k is a constant that depends on the angle of elevation of the gun and the initial velocity of the projectile. If we assume that $k = 100$ and $g = 32$, when is the projectile at a height of 50 ft ($h = 50$)?

SOLUTION Substituting these values into the equation gives

$$50 = 100t - \frac{1}{2}(32)t^2$$

or

$$-16t^2 + 100t - 50 = 0$$

Since this is a quadratic equation in the variable t, use the quadratic formula, to write

$$t = \frac{-100 \pm \sqrt{100^2 - 4(-16)(-50)}}{2(-16)}$$

$$= \frac{-100 \pm \sqrt{6800}}{-32}$$

$$\approx \frac{-100 \pm 82.5}{-32}$$

$$= \frac{-17.5}{-32} \quad \text{and} \quad \frac{-182.5}{-32}$$

$= 0.55$ and 5.7 sec (These two times correspond to the time for the projectile to go up and the time to come down.)

EXAMPLE 3 A physics student finds that the power P in watts (W) that is being consumed by an electrical circuit varies with elapsed time t, in seconds, according to the equation $p = 8t + 2t^2$. How long will it take the power consumption to reach 42 W (watts)?

SOLUTION Letting $p = 42$ in the given equation, we get

$$42 = 8t + 2t^2$$

or

$$2t^2 + 8t - 42 = 0$$

Dividing each term by 2,

$$t^2 + 4t - 21 = 0$$

and factoring, we get

$$(t + 7)(t - 3) = 0$$

The roots are then $t = -7$ and $t = 3$. Since a negative time does not make sense in the problem, we conclude that the power will equal 42 W when $t = 3$ sec.

EXAMPLE 4 Jack and Jim working together can complete a job in 6 days. Jack working alone can do it in 5 days less than the time Jim needs. In how much time does each do the job alone?

SOLUTION The desired equation has the form

Work done by Jack in 1 day	+	Work done by Jim in 1 day	=	Work done by both in 1 day

Let x be the time in days required by Jack to complete the job. Then Jim takes $x+5$ days. The fraction of the job done by Jack in 1 day is $1/x$; the fraction of the work done by Jim in 1 day is $1/(x+5)$. Since the work done by both in 1 day is $1/6$, we have

$$\frac{1}{x}+\frac{1}{x+5}=\frac{1}{6}$$

$$6(x+5)+6x=x(x+5)$$

$$6x+30+6x=x^2+5x$$

$$x^2-7x-30=0$$

$$(x-10)(x+3)=0$$

$$x=10,\ -3$$

The root $x=-3$ obviously has no meaning in the context of this problem. Hence, Jack could do the job in 10 days and Jim in 15 days.

Exercises for Section 3.4

1. In analyzing the motion of a certain spring-mass-damper system, the equation $D^2+6D+7=0$ must be solved. What values of D satisfy the equation?

2. The equation $R^2-3R+1=0$ arises in calculating the equivalent resistance of two electric circuits. What values of R satisfy this equation?

3. The bending moment of a certain simple beam is given by $M=20x-x^2$, where x is the distance in feet from one end. For what values of x is $M=60$?

4. The specific heat of a substance varies with temperature in accordance with $c=5.1T^2+0.3T$. For what values of T is $c=6.0$?

5. The ionization constant of an acid is $K=x^2/[5(1-x)]$. For what value of x is $K=3$?

6. The amount of propellant remaining in a certain rocket decreases with time according to $w=20-t-t^2$ lbs, where t is measured in seconds. How long after ignition will the amount of propellant remaining equal 8 lb?

7. A biologist finds that the number of water mites in a sample of river water depends on the ambient temperature in degrees Fahrenheit. If $N=100T-T^2$ describes this relation, at what temperature is $N=1100$?

8. The surface area of a right circular cylinder is given by $A=2\pi r^2+2\pi rh$, where r is the radius and h the height. Find the radius of the cylinder in the accompanying figure if the cylinder has a surface area of 50 in.^{2} and a height of 4 in.

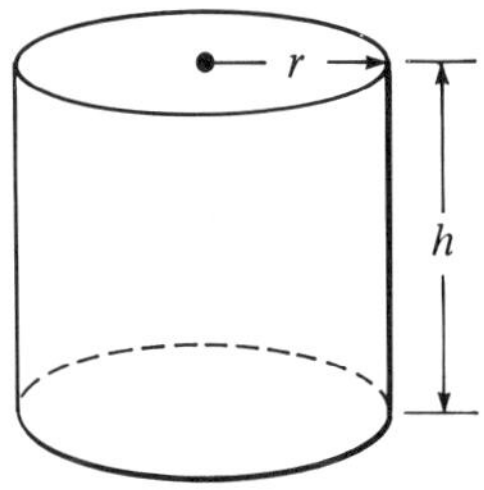

9. If the area of the L-shaped field shown in the accompanying figure is 10 km^2, what is the dimension x?

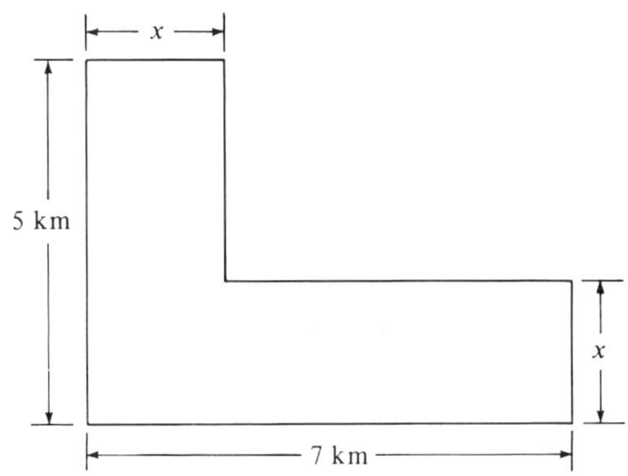

10. The area of a triangle is 14 ft^2. Determine what the base and the altitude of the triangle are if the altitude exceeds the base by 3 ft.

11. The hypotenuse of a right triangle is 13 m long. Find the lengths of the sides of the triangle, which has one side 7 m longer than the other.

12. Mary can audit a set of books in 7 hr less than Jane, and together they can do the work in 9 hr less than Mary alone. Find the time they take when they work together.

13. A rectangular piece of glass has an area of 1875 cm^2. If the length of the rectangle is 50 cm greater than the width, what are the dimensions of the piece of glass?

14. A man finds that a trip of 70 mi covered at his normal speed can be decreased by 15 min if he travels 5 mph faster. What is his normal speed?

15. A boy mows a strip 7 ft wide around a rectangular yard; he determines that the strip is 50% of the yard. Find the dimensions if the width of the yard is three-fourths its length.

16. A rectangular field 250 ft long and 180 ft wide has a concrete walkway of uniform width on its border. If there are 2616 ft^2 of concrete in the walkway, what is the width of the walkway?

17. A man can row 20 mi downstream and back in 7.5 hr. If his rate of rowing in still water is 6 mph, what is the rate of the stream?

18. The percentage of markup on the cost price of a pair of shoes is the same as the cost price in dollars. If the shoes sell for $24, what is the cost price of the shoes?

19. A new labor contract provides for an increase of $1 an hour and a reduction of 5 hr in the work week. A worker who receives $240 a week would receive $5 more a week under the new contract. How long was the work week before the new contract?

20. Jack and Jim working together can mow and manicure a lawn in 3 hr less than Jack alone. Jim alone takes 1 hr more than Jack. Determine the approximate time they take when they work together.

21. Find three consecutive integers having squares whose sum equals 434.

22. A 2-in. square is cut from each corner of a rectangular piece of cardboard whose length exceeds the width by 4 in. The sides are then turned up and a box is formed. Find the three dimensions of the box, which has the volume 64 $in.^3$.

23. The area of a triangular plate is 20 ft^2. What are its base and height if the base is 6 ft longer than the height?

24. A rectangular piece of metal is 10 cm long and 4 cm wide. Strips of equal width are to be added to one end and one side. Find what width of strip is necessary to double the area of the original rectangle.

25. Two fuel lines of different diameter can fill a tank in 5 hr. If the large pipe will fill the tank in 4 fewer hours than the smaller pipe, what time is needed for the large pipe to fill the tank?

26. A couple bought some steaks for their freezer for a total of $120. If each steak had cost $1 more, the couple would have got 6 fewer steaks for their $120. How many steaks did they receive?

27. The tens digit of a two-digit number is lower by 5 than the units digit. Twice the number is 1 more than the sum of the squares of the digits. What is the number?

28. The sum of the areas of the two inner circles in the accompanying figure is eight-tenths of the area of the outer circle. What is the radius of the smallest circle?

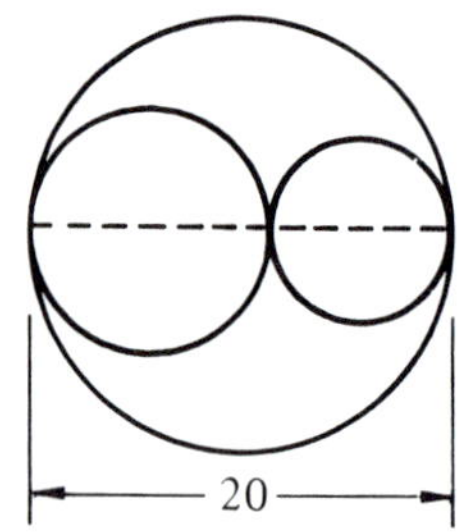

3.5 Equations in Quadratic Form

Some equations that are not strictly quadratic can be solved by the methods of the previous sections if an appropriate substitution is made first. For instance, by letting $z=\sqrt{x}$, we can transform the equation

$$3x-2\sqrt{x}-5=0$$

into the quadratic equation

$$3z^2-2z-5=0$$

Similarly, by letting $z=x^{1/3}$, we can write

$$2x^{2/3}+3x^{1/3}-2=0$$

as

$$2z^2+3z-2=0$$

EXAMPLE 1 Solve the equation $x^4+x^2-12=0$.

SOLUTION The solution is obtained by letting $z=x^2$ and writing $z^2+z-12=0$, which can be factored into

$$(z+4)(z-3)=0$$

From this we get $z=-4$ and $z=3$. To find the roots of the equation $x^4+x^2-12=0$, we note that $z=x^2$, and therefore,

$$x^2=-4 \quad \text{and} \quad x^2=3$$

The solution to $x^2=-4$ is $\pm 2i$ and to $x^2=3$ is $\pm\sqrt{3}$. Hence the solution set of the given equation is

$$\{\sqrt{3}, -\sqrt{3}, 2i, -2i\}$$

EXAMPLE 2 Solve $6/x^2+1/x=1$.

SOLUTION We can convert $6/x^2+1/x=1$ into a quadratic equation by making the transformation

$z=1/x$. Making this substitution, we get

$$6z^2+z=1$$

or

$$6z^2+z-1=0$$

Factoring yields

$$(3z-1)(2z+1)=0$$

from which $z=\frac{1}{3}$ and $z=-\frac{1}{2}$. Thus, $1/x=\frac{1}{3}$ and $1/x=-\frac{1}{2}$. Solving for x in both cases gives

$$x=-2 \quad \text{and} \quad x=3.$$

EXAMPLE 3 Solve $\dfrac{1}{x-2}+\dfrac{2}{x+3}=5$.

SOLUTION The given equation can be simplified by multiplying through by $(x-2)(x+3)$, to get

$$x+3+2(x-2)=5(x-2)(x+3)$$

Expanding and simplifying gives

$$3x-1=5x^2+5x-30$$
$$5x^2+2x-29=0$$

From the quadratic formula,

$$x=\frac{-2\pm\sqrt{4+580}}{10}=\frac{-2\pm\sqrt{584}}{10}$$
$$\approx -2.62,\ 2.22$$

Check: Since neither value produces a zero denominator in the original equation, both are valid solutions.

Equations that contain a square root term can frequently be solved by squaring both sides to eliminate the radical. The validity of each root must then be tested by substituting it in the given equation. Any root that does not satisfy the original equation is extraneous (that is, not valid for the original equation) and must be rejected.

EXAMPLE 4 Solve $\sqrt{5x-1}=x-3$.

SOLUTION Squaring both sides, we get

$$5x-1=x^2-6x+9$$

If like terms are collected,

$$x^2-11x+10=0$$
$$(x-1)(x-10)=0$$
$$x=1 \quad \text{and} \quad x=10$$

Only the root $x=10$ satisfies the original equation, since $\sqrt{50-1}=10-3$ and $\sqrt{5-1} \neq 1-3$. Consequently, the solution is $x=10$.

Solve $\sqrt{3x+1}-\sqrt{x-1}=2$.

Equations involving two radical terms are usually most easily solved by putting one radical on each side of the equation. Thus, we write the given equation as

$$\sqrt{3x+1}=\sqrt{x-1}+2$$

Squaring both sides, we get

$$3x+1=(x-1)+4\sqrt{x-1}+4$$

Collecting all terms without a radical on one side of the equation yields

$$2x-2=4\sqrt{x-1}$$

and if both sides are divided by 2, then $x-1=2\sqrt{x-1}$. Squaring both sides again gives

$$x^2-2x+1=4(x-1)=4x-4$$
$$x^2-6x+5=0$$
$$x=1 \quad \text{and} \quad x=5$$

Both values are roots, since $\sqrt{3+1} - \sqrt{1-1}=2$ and $\sqrt{15+1} - \sqrt{5-1}=2$.

Exercises for Section 3.5

Solve the equations in Exercises 1–18. (Find all real and complex roots.)

1. $x^4-x^2-12=0$ **2.** $4x^2-x^4=0$

3. $x^4-6x^2-16=0$ **4.** $3x^4=2x^2+1$

5. $x^4+9x^2+8=0$ **6.** $x^{-2}+x^{-1}-6=0$

7. $2x^{-2}-5x^{-1}-3=0$

8. $x^{-4}-9x^{-2}+14=0$

9. $3x^{-2}=4(x^{-1}+1)$

10. $(x-3)^2+5(x-3)+6=0$

11. $(x+2)^2-2(x+2)-8=0$

12. $(x^2+x)^2-18(x^2+x)+72=0$

13. $(x^2+2)^2+12(x^2+2)+11=0$

14. $(x+3)^4=4-3(x+3)^2$

15. $(x-2)^{-2}+35=(x-2)^{-1}$

16. $(x^2-1)^2-3(x^2-1)+2=0$

17. $4(x^2+1)^2-7(x^2+1)-2=0$

18. $(x^2+3)^2-7(x^2+3)-8=0$

Find the real solutions for the equations in Exercises 19–28.

19. $\dfrac{3}{x-1} - \dfrac{2}{x+3}=\dfrac{1}{2}$ **20.** $\dfrac{x}{x-4}=6 - \dfrac{x}{x+4}$

21. $\frac{1}{2}-\frac{1}{x-1}=\frac{6}{x^2-1}$

22. $\frac{x}{x-2}+\frac{x-1}{2}=x-1$

23. $\frac{2}{x-1}+\frac{5}{x-7}=\frac{4}{3x-1}$

24. $\frac{1}{x}+\frac{3}{x-1}-\frac{2}{x+1}=0$

25. $\frac{x}{x^2-4}+\frac{1}{x-2}=\frac{x+5}{x^2-3x+2}$

26. $\frac{2x+7}{x^2-5x+4}-\frac{x-8}{x^2-4x+3}=\frac{x}{x^2-7x+12}$

27. $\frac{x^2-7}{x^2+4}+\frac{3}{x^2-1}=\frac{x^2+3}{x^2}$

28. $\frac{2x^2+5}{x^2-9}+\frac{x^2-5}{3x^2-1}=3$

Find the real solutions for the radical equations in Exercises 29–46.

29. $\sqrt{x-6}=3$

30. $\sqrt{x+2}=4$

31. $\sqrt{2y+3}=5$

32. $\sqrt{3z-5}=1$

33. $\sqrt{x^2-16}=3$

34. $\sqrt{y^2-11}=5$

35. $\sqrt{z^2+5z+2}=4$

36. $2=\sqrt{x^2-3x}$

37. $y+1=\sqrt{3y+7}$

38. $z-2=\sqrt{z-2}$

39. $2x-1=\sqrt{2x+5}$

40. $\sqrt{1+9p}=p+1$

41. $\sqrt{x+5}=\sqrt{x}-1$

42. $2\sqrt{x}=x-3$

43. $\sqrt{3y-5}-\sqrt{y+7}=2$

44. $\sqrt{2x+3}-\sqrt{x+1}=1$

45. $\sqrt{5z+4}-\sqrt{z}=2$

46. $\sqrt{5x+6}-\sqrt{3x+7}=1$

47. One side of a right triangle is 12 in. long. If the other side is lengthened by 4 in., the hypotenuse is lengthened by 2 in. Find the perimeter of the triangle.

48. Find the dimensions of a rectangle that has an area of 192 in.^{2} and a diagonal of 20 in.

49. The surface area of a right circular cone is given by the formula

$$S=\pi r\sqrt{r^2+h^2}.$$

Solve this equation for h.

50. For an electric current in a circuit given by the equation $i=2t-\sqrt{2t+5}$, where t is time, at what time will $i=1$?

51. If the pressure in a waste disposal tank varies with time according to the equation $P=3+\sqrt{t+2}$ lb, at what time will $P=5$?

3.6 Linear Inequalities; Intervals

In addition to mathematical statements involving the equality of algebraic expressions, there is the corresponding problem of inequalities. The set of values for which an inequality is true is called its **solution set** and we say we have *solved* the inequality when we find its solution set. Two statements involving inequalities are **equivalent** if they have the same solution set.

Solving inequalities is similar to solving equations. We use manipulations that give a chain of equivalent inequalities to isolate the variable on one side of the inequality. The *allowable* manipulations look much like the allowable manipulations with equalities.

(1) Adding the same quantity to both sides of an inequality or subtracting therefrom results in an equivalent inequality; if $x<y$, for example, then $x+3<y+3$.

(2) Multiplying or dividing both sides of an inequality by the same positive quantity results in an equivalent inequality; if $x>y$, for example, then $2x>2y$.

(3) Multiplying or dividing both sides of an inequality by the same negative quantity results in an equivalent inequality **if the sense of the inequality is reversed**; if $x<y$, for example, then $-5x>-5y$.

EXAMPLE 1 Solve the inequality $x-2<3$.

SOLUTION Adding 2 to each side of the inequality, we have

$$x-2+2<3+2$$

or

$$x<5$$

Thus, the solution set consists of all real numbers less than 5.

EXAMPLE 2 Solve the inequality $x+1<3x-2$.

SOLUTION To get the terms involving x on one side and the constants on the other, add $-3x$ and -1 to both sides to get the equivalent inequality:

$$-2x<-3$$

Multiplying this result by $-\frac{1}{2}$ yields

$$x>3/2$$

Notice that the sense of the inequality changed when we multiplied by $-\frac{1}{2}$. The solution set is the set of all real numbers greater than $\frac{3}{2}$.

Solutions to inequalities can also be written in interval notation. An **interval** is the set of all real numbers between two given real numbers. Special notation is used to indicate intervals. The set of real numbers greater than a and less than b is called an **open interval**. An open interval is denoted by (a, b) and defined by

$$(a, b)=\{x|a<x<b\}$$

The notation $\{x|a<x<b\}$ is read "the set of all x such that x is greater than a and less than b." The notation (a, b) presumes that $a<b$. The set of real numbers greater than or equal to a and less than or equal to b is called a **closed interval**. A closed interval is denoted by $[a, b]$ and defined by

$$[a, b]=\{x|a\leqslant x\leqslant b\}$$

The points corresponding to a and b are called **endpoints** of the interval. Thus, the distinction between an open and a closed interval is the exclusion or inclusion of the endpoints. Graphically, an endpoint to be included in the interval is represented by a closed dot. Otherwise an open dot is used. See Figure 3.3.

FIGURE 3.3 Open Interval (a, b) Closed Interval $[a, b]$

If only one of the endpoints is included in an interval, the interval is said to be **half-open**. The half-open intervals $(a, b]$ and $[a, b)$ are depicted in Figure 3.4.

FIGURE 3.4

Finally, the set of all real numbers greater than a is represented in interval notation by (a, ∞). The symbol ∞ (infinity) means that there is no finite upper bound for the interval; it should not be thought of as a number. The symbol $-\infty$ is used to indicate that there is no lower bound for an interval; therefore, $(-\infty, a)$ denotes all real numbers less than a. Similarly, $(-\infty, a]$ is an interval that corresponds to $x \leqslant a$, and $[a, \infty)$ is an interval that corresponds to $x \geqslant a$. These intervals are shown in Figure 3.5.

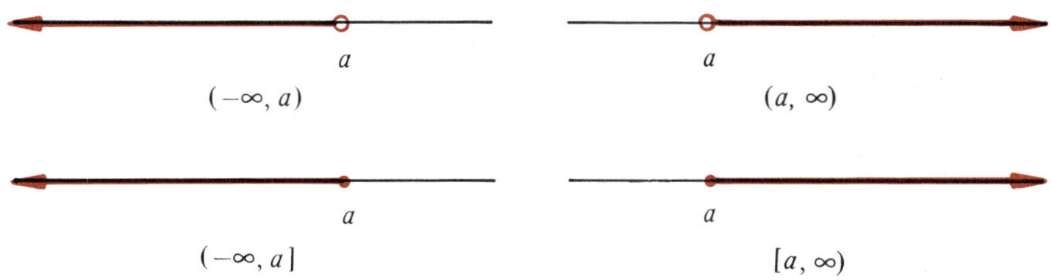

FIGURE 3.5

EXAMPLE 3 Solve the inequality $x^2-5x+3 \geqslant (x-2)^2$. Using interval notation, write the solution and sketch its graph.

SOLUTION Expanding the binomial on the right side of the given inequality, we have

$$x^2-5x+3 \geqslant x^2-4x+4$$

Adding $-x^2+4x-3$ to both sides, we get

$$-x \geqslant 1$$
$$x \leqslant -1$$

Notice that the sense of the inequality changed when both sides of the inequality were multiplied by -1. The desired solutions are given by $(-\infty, -1]$ and represented in Figure 3.6.

FIGURE 3.6

The solution set to an inequality of the form $q<p<r$ consists of all real numbers that are solutions to both of the inequalities, $q<p$ and $p<r$.

EXAMPLE 4 Solve the inequality $x-20<3x-8 \leqslant 4$ and graph the solution interval.

SOLUTION Any solution to the inequality $x-20<3x-8 \leqslant 4$ must also be a solution to both

$x-20<3x-8$ and $3x-8\leqslant 4$. Solving these two inequalities separately, we get the following results.

$x-20<3x-8$	$3x-8\leqslant 4$
$x-12<3x$	$3x\leqslant 12$
$-12<2x$	$x\leqslant 4$
$-6<x$	

Since the solution of the given inequality must satisfy both $-6<x$ and $x\leqslant 4$, we conclude that the solution is

$$-6<x\leqslant 4$$

The solution is all numbers in the half-open interval $(-6, 4]$. The graph of this interval is shown in Figure 3.7. ⊕

FIGURE 3.7

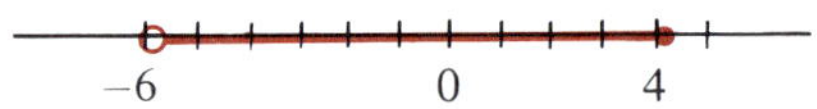

The points common to both $(-\infty, 4]$ and $(-6, \infty)$ are denoted $(-\infty, 4]\cap(-6, \infty)$, called the **intersection** of the intervals. Thus, the solution to this problem can be written

$$(-\infty, 4]\cap(-6, \infty)=(-6, 4].$$

Any consideration of linear inequalities would not be complete without mentioning absolute value inequalities. The absolute inequalities $|x|<b$ and $|x|>b$ both describe sets of real numbers that can be represented by intervals on the real line. From our knowledge of absolute value (Section 1.3), it follows that $|x|<b$ and $|x|>b$ have the following interpretations.

(3.2) $|x|<b$ **means that** $-b<x<b$
$|x|>b$ **means that** $x<-b$ **or** $x>b$

Equivalently, $|x|<b$ represents the open interval $(-b, b)$, and $|x|>b$ represents the union of $(-\infty, -b)$ and (b, ∞). The **union** of the intervals means that x can be either in $(-\infty, -b)$ or in (b, ∞) and is denoted $(-\infty, -b)\cup(b, \infty)$. Figure 3.8 pictures these two intervals.

FIGURE 3.8

We can extend the use of (3.2) to solving inequalities that involve the absolute value of linear expressions. For example, we can use (3.2) to solve $|t-3|<5$ by letting $x=t-3$ and $b=5$.

EXAMPLE 5 Solve the inequality $|t-3|<5$.

SOLUTION Using (3.2) as our guide with $x=t-3$ and $b=5$, we have the double inequality $-5<x<5$, or equivalently,

$$-5<t-3<5$$

Adding 3 to each member of this inequality, we get

$$-2<t<8$$

as the solution set. This is written in interval notation as $(-2, 8)$. See Figure 3.9.

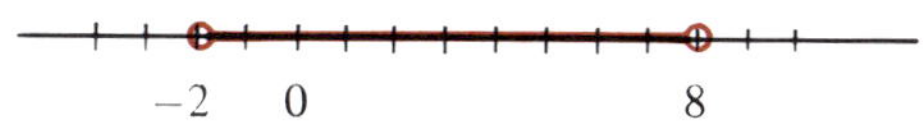

FIGURE 3.9

EXAMPLE 6 Solve the inequality $|2x+5|<7$.

SOLUTION The given absolute value inequality can be represented by the double inequality $-7<2x+5<7$. Subtracting -5 from all three terms gives

$$-12<2x<2$$

Dividing through by 2, we have the solution

$$-6<x<1$$

The graph of this solution set is shown in Figure 3.10.

FIGURE 3.10

EXAMPLE 7 Solve the inequality $|3x-1|\geqslant 8$.

SOLUTION The inequality $|3x-1|\geqslant 8$ is equivalent to the two inequalities

$$3x-1\leqslant -8 \quad \text{or} \quad 3x-1\geqslant 8$$

These two are independently solved to obtain

$$x\leqslant -\tfrac{7}{3} \quad \text{or} \quad x\geqslant 3$$

The solution is written in interval notation as $(-\infty, -7/3]\cup[3, \infty)$, where $\cup$ denotes the union of the two intervals. The graph is sketched in Figure 3.11.

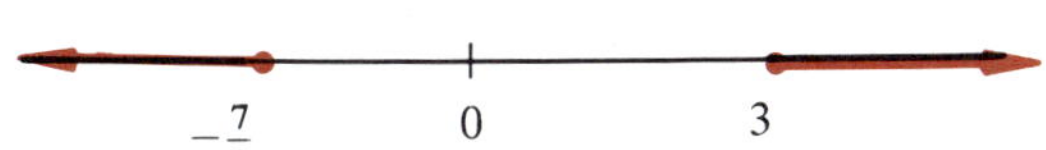

FIGURE 3.11

In the previous examples, we used interval notation to represent solutions to linear inequalities involving absolute values. Another important application of inequalities with absolute values is as an alternative way of describing intervals on the real line. Recall from Section 1.3 that the distance between two points a and b on the real line is given by

$$d(a, b)=|a-b|$$

The value $|x-2|$ denotes the distance from x to 2, and therefore the inequality $|x-2|<3$ represents the set of all points whose distance from 2 is less than 3. A sketch of this inequality is shown in Figure 3.12. It is clear from the figure that the interval within 3 units of 2 is $(-1, 5)$. Therefore, $|x-2|<3$ and $(-1, 5)$ describe the same interval.

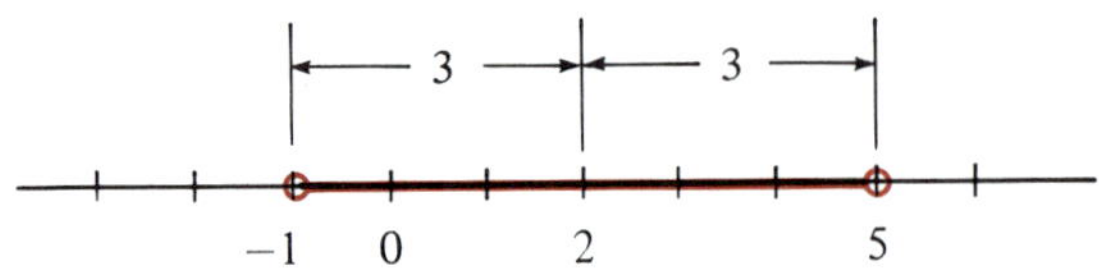

FIGURE 3.12

In general, $|x-p|<b$ represents all values of x within a distance of b units of p, and $|x-p|>b$ represents all values of x more than b units from p. The point p is the **midpoint** of the interval, and the endpoints of the interval are $p-b$ and $p+b$. See Figure 3.13.

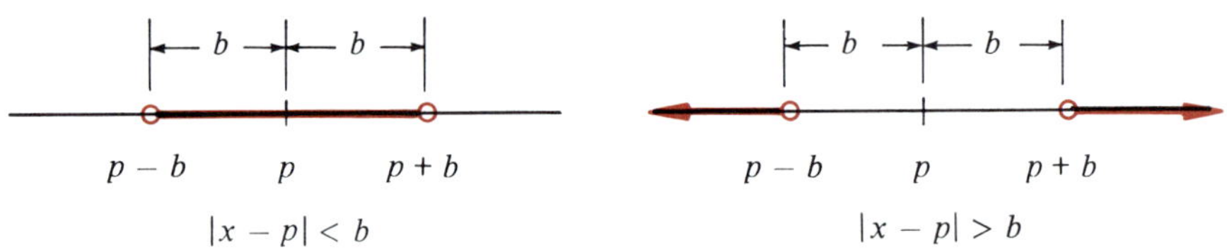

FIGURE 3.13

EXAMPLE 8 Using interval notation, write the interval on the real line that describes the inequality $|x-2|<5$. Also display graphically.

SOLUTION The point $x=2$ is the midpoint of the interval that has a total length of 10. The given inequality can be written

$$-5<x-2<5; \quad \text{that is,} \quad -3<x<7$$

or in interval notation, $(-3, 7)$. Figure 3.14 displays the interval.

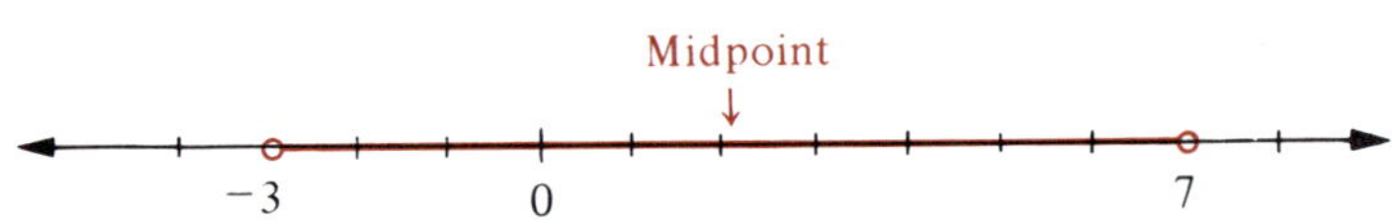

FIGURE 3.14

Sometimes we want to describe a given interval (a, b) using an absolute value with an inequality. To do this, compute the midpoint and half the length of the interval. The midpoint of (a, b) is given by

$$\textbf{Midpoint} = \frac{a+b}{2}$$

and half the length of the interval is given by

$$\textbf{Half-length} = \frac{|a-b|}{2}$$

Then, the inequality

$$|x-\textbf{midpoint}|<\textbf{half-length of interval}$$

is another representation of (a, b).

A similar analysis describes the union of the two intervals $(-\infty, a)$ and (b, ∞) with an absolute value inequality.

EXAMPLE 9 Describe the interval $(-4, 10)$ in terms of an absolute value inequality.

SOLUTION The midpoint and half-length of $(-4, 10)$ are given by

$$\text{Midpoint}=\frac{-4+10}{2}=\frac{6}{2}=3$$

and

$$\text{Half-length}=\frac{|-4-10|}{2}=\frac{14}{2}=7$$

Hence, the interval may be represented by $|x-3|<7$. The interval is shown in Figure 3.15.

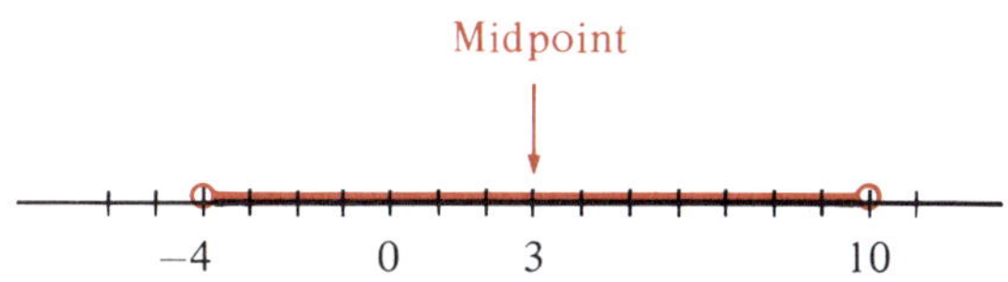

FIGURE 3.15

EXAMPLE 10 Describe the union of the two intervals $(-\infty, -8)$ and $(0, \infty)$ in terms of one absolute value inequality, as shown in Figure 3.16.

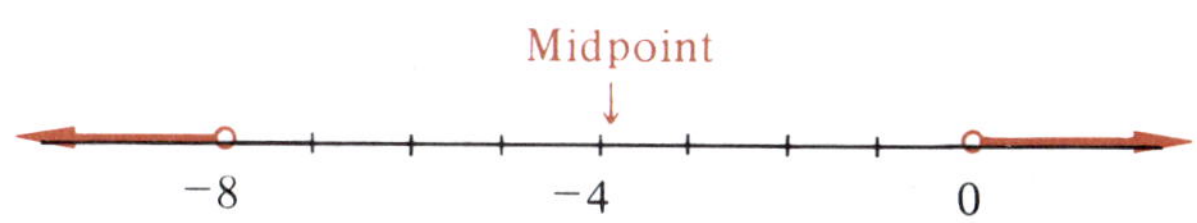

FIGURE 3.16

SOLUTION In this case, the midpoint of -8 and 0 is -4, and the half-length of the interval that we want to exclude is 4. Hence, the given interval can be described by the notation

$$|x-(-4)|>4; \quad \text{that is,} \quad |x+4|>4$$

The last example in this section is an inequality involving multiple absolute value expressions. As with linear equalities, these are best handled by considering the different forms of the inequality caused by the absolute value.

EXAMPLE 11 Solve the inequality $|2x+5|<|3x-7|$.

SOLUTION

$$|2x+5|=\begin{cases}(2x+5) & \text{for } x\geqslant -\dfrac{5}{2}\\ -(2x+5) & \text{for } x< -\dfrac{5}{2}\end{cases}$$

$$|3x-7|=\begin{cases}(3x-7) & \text{for } x\geqslant \dfrac{7}{3}\\ -(3x-7) & \text{for } x< \dfrac{7}{3}\end{cases}$$

For what interval are *both* $|2x+5|=2x+5$ and $|3x-7|=3x-7$ valid? Only for $x\geqslant 7/3$. (Although $|2x+5|=2x+5$ is true for $x\geqslant -5/2$, we know $|3x-7|=3x-7$ only when $x\geqslant 7/3$.) Therefore we write

$$2x+5<3x-7 \quad \text{for } x\geqslant 7/3$$

On the interval $[-5/2,\ 7/3)$ we know $|2x+5|=2x+5$ and $|3x-7|=-(3x-7)$. Therefore we can write

$$2x+5<-(3x-7) \quad \text{for } -5/2\leqslant x<7/3$$

Finally, $|2x+5|=-(2x+5)$ and $|3x-7|=-(3x-7)$ only when $x<-5/2$ so we write

$$-(2x+5)<-(3x-7) \quad \text{for } x<-5/2$$

The solutions of the three inequalities above are, in order,

$$x>12 \quad \text{for } x\geqslant\frac{7}{3}$$

$$x<\frac{2}{5} \quad \text{for } -\frac{5}{2}\leqslant x<\frac{7}{3}$$

$$x<12 \quad \text{for } x<-\frac{5}{2}$$

Thus, *considering the domain limitations for each of the solutions*, the solution set is the union of the intervals

$$\left(-\infty,\frac{2}{5}\right) \quad \text{and} \quad (12,\infty).$$

Exercises for Section 3.6

Sketch the graph of the intervals in Exercises 1–16.

1. $(-2, 3)$

2. $[-2, 3]$

3. $(-\infty, 0]$

4. $[1, \infty)$

5. $(-\infty, \infty)$

6. $(-3, 3]$

7. $[2, 3]$

8. $[\frac{1}{4}, \frac{1}{2})$

9. $(-1, 3)$

10. $[-3, -1]$

11. $[-5, 0)$

12. $(0, 1)\cup(\frac{1}{2}, 1)$

13. $(2, 3)\cap(3, \infty)$

14. $(-\infty, -3)\cup(3, \infty)$

15. $(-3, 5)\cap[4, 7)$

16. $(2, \infty)\cap(3, \infty)$

Solve the inequalities in Exercises 17–35 and sketch the graph of each solution set on the number line.

17. $3x+1>x-4$ **18.** $2-5x>-x$

19. $x+1\geqslant\sqrt{2}-x\sqrt{3}$ **20.** $x\leqslant x+1$

21. $\frac{2}{3}x+5<\frac{7}{5}x+\frac{1}{3}$ **22.** $3(2x-1)+4(x-2)>0$

23. $(x-3)(x+4)>(x-5)(x+5)$

24. $x^2-9\leqslant(x-3)(x+2)$

25. $x^2-1\geqslant(x+3)^2$

26. $2x+x(x-2)<(x-3)(3+x)$

27. $x(x-1)>x(x-3)$

28. $(2x-1)^2>(2x+3)^2$

29. $x^2-5x+4<x(x-2)$

30. $x(x^2-4)>x(x^2-9)+5$

31. $x+2<2x+5\leqslant 6x+7$

32. $x-1\leqslant 3x+5<2x+7$

33. $x<2x+3<3x-5$

34. $x^2-2x<x^2+2x-3<(x-3)(x-2)$

35. $x^2+1<(x-3)(x-4)\leqslant(x+3)(x+5)$

In Exercises 36–50, describe the given absolute value inequality by an interval or a union of intervals.

36. $|x|\leqslant 6$ **37.** $|x|<1.5$

38. $|x|>7$ **39.** $|x|\geqslant 4$

40. $|x|<-1$ **41.** $|x-2|<3$

42. $|3x+6|>9$ **43.** $|x+2|<5$

44. $|2x+7|<5$ **45.** $|x-5|>1$

46. $|x+3|>0$ **47.** $|\frac{1}{2}x-3|\leqslant 5$

48. $|3x|\geqslant 6$ **49.** $|2x-3|\leqslant 2$

50. $|x+1|\leqslant 1$

Describe the intervals in Exercises 51–65 using absolute value notation.

51. $(-5, 5)$ **52.** $(-10, 10)$

53. $(-\infty, -3)\cup(3, \infty)$

54. $(-\infty, -1]\cup[1, \infty)$

55. $(2, 6)$ **56.** $(-\infty, -2)\cup(0, \infty)$

57. $(-10, 16)$ **58.** $(-\infty, 4)\cup(12, \infty)$

59. $(-\infty, -2)\cup(2, \infty)$

60. $(-\infty, 2]\cup[6, \infty)$

61. $[2, 7]$ **62.** $(-12, -3)$

63. Complement* of the interval $(0, 5)$

64. Complement of the interval $[1, 7]$

65. $[0, 5]$

66. What is wrong with describing the set $\{x \mid |x|>a\}$ with the notation $-a>x>a$?

Solve the inequalities in Exercises 67–74 by considering the different expressions for the absolute value.

67. $|x+2x|>x-2x$ **68.** $|x+1|>x$

69. $|x-1|\leqslant|x+4|$ **70.** $x\leqslant|x|$

71. $|3x+1|\leqslant|x-5|$ **72.** $|x+3|<4-|x-1|$

73. $|x-1|+2|x+4|\leqslant|x|$

74. $|x+1|+|2x-1|\geqslant|x+5|+|x|$

3.7 Quadratic Inequalities

An inequality that contains a quadratic expression is called a *quadratic inequality*. A basic rule for solving nonlinear inequalities is to begin by gathering all terms to one side of the inequality. If the resulting expression is quadratic, factor it, and

*The complement of any interval consists of all points x such that x is *not* in the interval.

determine where the signs of the individual factors become positive and negative. We usually accompany this method with a diagram showing the signs of the linear factors; hence the method is often called the method of sign diagrams.

EXAMPLE 1 Solve the inequality $4x^2+2x+4<2x^2+13x+25$.

SOLUTION Gathering all the terms to the left side of the inequality and simplifying, we get

$$2x^2-11x-21<0$$

Factoring the quadratic expression gives

$$(2x+3)(x-7)<0$$

The linear factor $2x+3$ is positive for $x>-\frac{3}{2}$ and negative otherwise. The factor $x-7$ is positive for $x>7$. To display these facts, construct a sign diagram for the linear factors as shown in Figure 3.17. An algebraic product of like signs gives a positive number, whereas a product of unlike signs gives a negative number. The sign of the product of the two linear factors is shown in the figure. Since we want to find those values of x for which the product is negative, the diagram tells us immediately that the solution is $-\frac{3}{2}<x<7$.

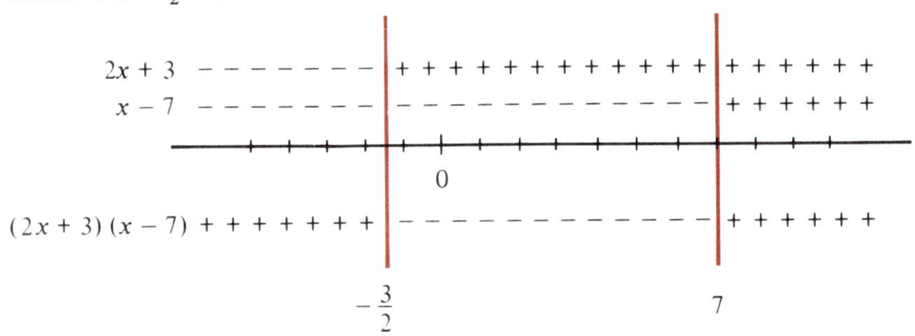

FIGURE 3.17

A mistake that beginners often make is to consider that the quadratic inequality $(2x+3)(x-7)<0$ is equivalent to the two linear inequalities $2x+3<0$ and $x-7<0$. Solutions to the two latter are $x<-\frac{3}{2}$ and $x<7$, and their union is $x<7$, which is clearly not the solution set to the given quadratic inequality. For example, the quadratic expression is positive when $x=-2$.

EXAMPLE 2 Solve the inequality $x^2<x+1$.

SOLUTION Bringing everything to the left-hand side of the inequality gives $x^2-x-1<0$. The quadratic expression x^2-x-1 cannot be factored by inspection, but the factors can be determined if the two roots r_1, r_2 of $x^2-x-1=0$ are known. The factors are $(x-r_1)$ and $(x-r_2)$. From the quadratic formula, the two roots are

$$\frac{1\pm\sqrt{1+4}}{2}=\frac{1\pm\sqrt{5}}{2}$$

Hence, we have the sign diagram of Figure 3.18, in which

$$r_1 = \frac{1+\sqrt{5}}{2} \quad \text{and} \quad r_2 = \frac{1-\sqrt{5}}{2}$$

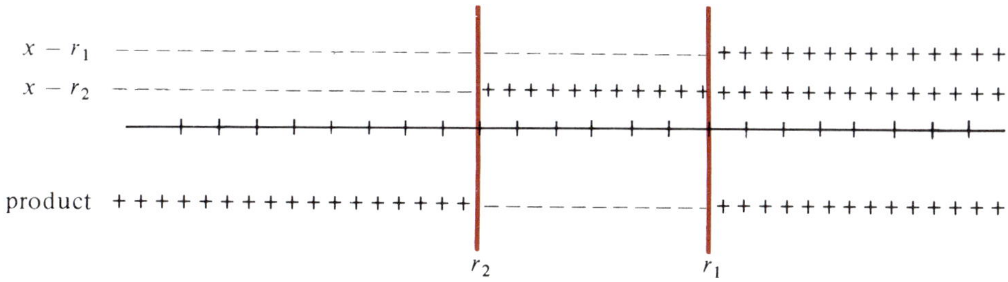

FIGURE 3.18

You see that $r_2 < r_1$. Therefore, since we want to know those numbers for which the product is negative, the solution is

$$\frac{1-\sqrt{5}}{2} < x < \frac{1+\sqrt{5}}{2}$$

EXAMPLE 3 Solve the inequality $x^2 < 2x - 5$.

SOLUTION The given inequality is equivalent to

$$x^2 - 2x + 5 < 0$$

The roots of $x^2 - 2x + 5$ are both nonreal because the discriminant D equals $(-2)^2 - 4 \cdot 5 = -16$. Hence the quadratic expression $x^2 - 2x + 5$ never changes sign. Since if $x = 0$, then $x^2 - 2x + 5$ is $+5$, the expression is *always* positive and the solution to the given inequality is the empty set. (We could also say, there is "no solution.")

Using sign diagrams is not restricted to solving quadratic inequalities; any time you can factor the inequality into linear factors, the sign diagram method is useful. Remember to bring all terms to one side of the inequality before simplifying and factoring.

EXAMPLE 4 Solve the inequality $\frac{2x+1}{x-2} \geqslant 1$.

SOLUTION Bring all nonzero terms to one side and solve the resulting inequality. Thus, if we subtract 1 from both sides of the given inequality, we have

$$\frac{2x+1}{x-2} - 1 \geqslant 0$$

and then simplifying,

$$\frac{2x+1-x+2}{x-2} \geqslant 0$$

$$\frac{x+3}{x-2} \geqslant 0$$

The sign diagram in Figure 3.19 associated with this quotient shows the signs of the individual factors $x+3$ and $x-2$ for any real number. The last line shows the sign of the quotient. From that line, read off the two intervals $(-\infty, -3)$ and $(2, \infty)$.

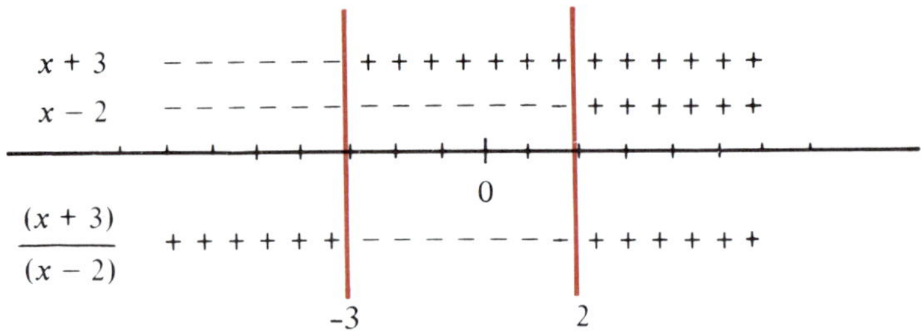

FIGURE 3.19

The endpoints -3 and 2 must be inspected separately. Inserting -3 in the original inequality, we find that the left-hand side becomes 1; hence -3 is a solution. Since 2 is not permissible (the denominator would become zero), it is not a solution. Hence the solution set is

$$(-\infty, -3] \cup (2, \infty)$$

EXAMPLE 5 Solve the inequality $\dfrac{x^2(3-x)(x+1)}{(x+10)(x-1)} \geqslant 0.$

SOLUTION Figure 3.20 shows the associated sign diagram. The factor x^2 is always nonnegative, so it affects the solution set only when it equals zero. From the sign diagram, we can determine the sign of the entire expression for all values of x. The solution set is $(-10, -1] \cup \{0\} \cup (1, 3]$.

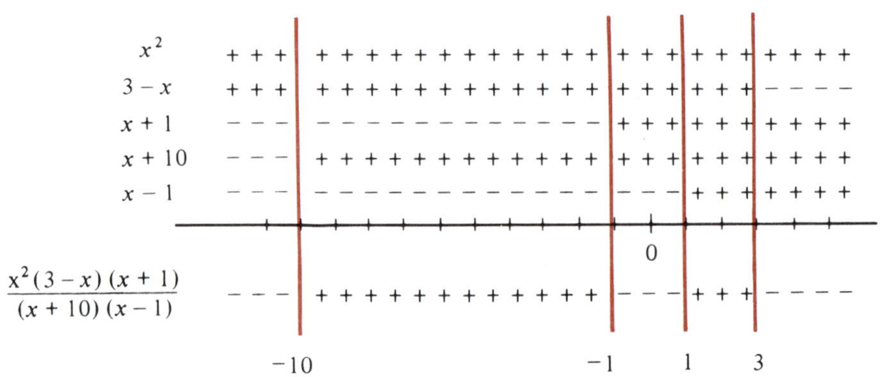

FIGURE 3.20

The values of x for which the numerator is zero (that is, -1, 0, 3) are included here because the inequality specifically asks for equality also. The values of x for which the denominator is zero (that is, -10 and 1) are never included, since the given expression is meaningless for those values.

Exercises for Section 3.7

Use sign diagrams to solve the quadratic inequalities in Exercises 1–10.

1. $x^2+2x\geqslant 3$ **2.** $x^2<x+2$

3. $5x<3x^2-2$

4. $(x+1)(x-3)>4x^2-10x+2$

5. $x(6-x)\geqslant 3x^2+2x-15$

6. $x^2+3x+2<11x^2+6x-16$

7. $x-1<x^2+1$ **8.** $3x^2-4<(x-4)^2$

9. $(x+4)<x^2-4$ **10.** $(x-3)^2>(2x+1)^2$

Solve the inequalities in Exercises 11–22 by sign diagrams. Be sure to bring everything to one side of the inequality before proceeding.

11. $\dfrac{(x+4)(x+5)}{x+1}<0$ **12.** $\dfrac{x+1}{x+2}<1$

13. $\dfrac{x+1}{2x+3}\leqslant 1$ **14.** $\dfrac{(x+5)(x+2)(x-1)}{x^2-4}\leqslant 0$

15. $\dfrac{(x^2-1)(x^2+1)}{x^2-4}\leqslant 0$

16. $\dfrac{(x+5)(x-3)(2-x)}{x^2-4}\geqslant 0$

17. $\dfrac{x^2+1}{x^2+4}\leqslant 1$ **18.** $\dfrac{x^2+x+1}{x+1}\geqslant 0$

19. $\dfrac{(x-2)}{x^2+x-1}\geqslant 0$ **20.** $\dfrac{x^2}{(x^2+4)(x-3)}\geqslant 0$

21. $\dfrac{x+2}{x^2-3}\leqslant 1$ **22.** $\dfrac{(x+1)(x^2-1)}{(x^2-9)(x+5)}\geqslant 0$

23. Determine what values of k would allow both roots of the equation $4x^2-kx+3=0$ to be real.

24. Determine the values of k for which the function $f(x)=2kx^2+7x+2k$ is always positive.

25. Determine the value or values of k for which the graph of the function $f(x)=kx^2+x+k$ touches but does not cross the x axis.

26. Determine the values of k that allow the function $f(x)=2x^2+kx+3$ to have real roots.

27. Suppose that the profit in a business varies in an 8-hr day according to the formula $P=15t-3t^2$, where t is the time in hours. For what values of t is $P>0$? (Assume that $t\geqslant 0$.)

28. The height of a certain projectile above ground level is given by $h(t)=32t-16t^2$, where t is the time in seconds. For what values of t $(t>0)$ is $h>4$ ft?

29. The tensile strength of a new plastic varies with temperature according to $S=500+600T-20T^2$ psi. For what temperature range is $S>4500$?

Test 1 for Chapter 3

Answer *True* or *False* to Exercises 1–10.

1. The equation $x+(1/x)=0$ is a linear equation.

2. The interval described by $|x-2|<2$ is 2 units long.

3. The inequality $|x-1|>2$ represents the same set as $2<x<-2$.

4. If a, b, and c are real numbers in the equation $ax^2+bx+c=0$, $a\neq 0$, then both roots are real.

5. If a quadratic equation has real roots, then it can be factored.

6. The equation $1-x^2=0$ is a quadratic equation.

7. The equation $\sqrt{x+4}=2-x$ has $x=5$ as a root.

8. The quadratic formula can be used to solve the equation $ax^3+bx^2+c=0$.

9. A linear equation can have two roots.

10. If the discriminant of $ax^2+bx+c=0$ is greater than zero, the equation has complex roots.

11. Solve the equation $3x+5=x-10$.

12. Represent the interval $(-10, 2)$ with an absolute value inequality.

13. Solve the equation $x^2+5x+2=0$.

14. For what values of k will the quadratic equation $2x^2+kx+2=0$ have real roots?

15. Solve the inequality $x(x-1)<2(x+2)(x+1)$.

16. Solve the inequality $x^2 \leqslant 11x-28$.

17. Jane has a 1-gal mixture of orange concentrate and water that is 20% concentrate. How much concentrate must she add to make a 30% mixture? (Express answer in gallons.)

Test 2 for Chapter 3

1. Solve $3(x+2)-5x=7$.

2. Solve $\dfrac{3}{x+2}-\dfrac{1}{x-2}=\dfrac{10}{x^2-4}$.

3. Solve $3+2x<5(x-2)$.

4. Solve $|x+4| \geqslant 8$ and graph the solution set.

5. Solve $2x(x+5)=12$ by factoring.

6. Solve $x^2-6x+2=0$ using the quadratic formula.

7. Find two real numbers such that $x^{2/3}-3x^{1/3}-4=0$.

8. Solve $\sqrt{5-y}=y+1$.

9. A girl paid \$35 for a tennis racket and a pair of shoes. If the racket cost \$7 more than the shoes, how much did each cost?

10. Two water lines can fill a swimming pool in 6 hr. The larger of the lines will fill the pool in 2 hr less than the smaller. What time will it take for the larger water line to fill the pool?

11. The formula for the area of a right circular cylinder is $A=2\pi r^2+2\pi rh$, where r is the radius and h is the height of the cylinder. If the height of a cylindrical can is 6 in., what must the radius be to give the area 14π in.^{2}?

12. Solve the inequality $\dfrac{2x-1}{x+1} \leqslant 1$.

4

The Idea of a Function

4.1 The Cartesian Coordinate System

You have learned that each point on a line can be associated with a real number, called the coordinate of the point, and that the entire line forms a one-to-one correspondence with the set of real numbers (Section 1.3). A one-to-one correspondence can also be established between points in a plane and ordered pairs of real numbers.

A pair of numbers a and b is said to be an **ordered pair** if the order in which the elements are written is significant. An ordered pair for which a is identified as the "first" element and b as the "second" is denoted by (a, b). Two ordered pairs are equal if they have the same first element and the same second element; otherwise they are not equal. The ordered pair (3, 2), for example, does not equal the ordered pair (2, 3).

The notation for ordered pairs of numbers is the same as what is used to designate an open interval on the real line. Whenever the possibility of confusion arises, we specify whether the notation (a, b) refers to an open interval or an ordered pair. In this chapter, the notation will always mean an ordered pair.

Just as we form a correspondence between real numbers and points on a line, so we also associate ordered pairs of numbers with points in a plane. Any one of several approaches could accomplish this association, but one is used the most. That is the **rectangular coordinate system**, or as it is just as often called, the **Cartesian coordinate system**.*

To construct such a system, first draw a pair of mutually perpendicular number lines to intersect at zero on each line, as shown in Figure 4.1. Normally the

*Named in honor of René Descartes (1596–1650), the first person to use ordered pairs of numbers systematically.

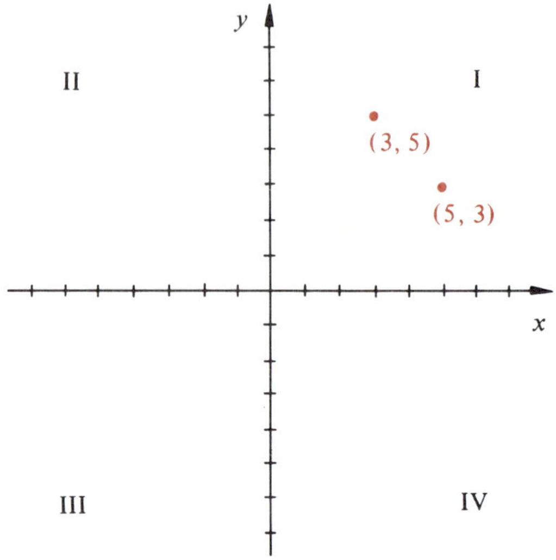

FIGURE 4.1

horizontal line is called the x-axis, the vertical line the y-axis, and their intersection the origin. The two axes considered together are called the coordinate axes. They divide the plane into four zones, or **quadrants**. The upper right quadrant is called the first quadrant and the others are numbered consecutively counterclockwise, as in Figure 4.1. The coordinate axes are not regarded as being in any quadrant.

To locate points in the plane, use the origin as a reference point and lay off a suitable scale on each of the coordinate axes. The displacement of a point in the plane to the right or the left of the y-axis is called the **x-coordinate**, or **abscissa**, of the point; it is denoted by x. Values of x measured to the right of the y-axis are considered *positive* and to the left, *negative*. The displacement of a point in the plane above or below the x-axis is called the **y-coordinate**, or the **ordinate**, of the point; it is denoted by y. Values of y above the x-axis are considered *positive* and below the x-axis, *negative*. The abscissa and the ordinate of a point taken together are called the **coordinates** of the point. The coordinates of a point are conventionally written in parentheses—the abscissa first, separated from the ordinate by a comma, that is, (x, y).

We see that a point (x, y) lies:

- In quadrant I if both coordinates are positive.
- In quadrant II if the x-coordinate is negative and the y-coordinate is positive.
- In quadrant III if both coordinates are negative.
- In quadrant IV if the x-coordinate is positive and the y-coordinate is negative.

Since the first number represents the horizontal displacement and the second the vertical displacement, we see the significance of order. The ordered pair (3, 5), for example, represents a point that is displaced 3 units to the right of the origin and 5 units above it; and the ordered pair (5, 3) represents a point that is 5 units to the right and 3 units up. The association of points in the plane with ordered pairs of real numbers is an obvious extension of a similar idea on the real line.

To be precise, we should always distinguish between the point and the ordered pair; however, it is common practice to blur the distinction and say "the point (x, y)" instead of "the point whose coordinates are (x, y)."

Using the rectangular coordinate system establishes a one-to-one correspoendence between the points in a plane and all possible ordered pairs of real numbers (x, y); that is, each point in the plane can be described by a unique ordered pair of numbers (x, y), and each ordered pair of numbers (x, y) can be represented by a unique point in the plane, called the **graph** of the ordered pair.

EXAMPLE 1 Locate the points $P(-1, 2)$, $Q(2, 3)$, $R(-3, -4)$, $S(3, -5)$, and $T(\pi, 0)$ in the plane.

SOLUTION $P(-1, 2)$ is in quadrant II, since the x-coordinate is negative and the y-coordinate is positive.

$Q(2, 3)$ is in quadrant I, since both coordinates are positive.

$R(-3, -4)$ is in quadrant III, since both coordinates are negative.

$S(3, -5)$ is in quadrant IV, since the x-coordinate is positive and the y-coordinate is negative.

$T(\pi, 0)$ is not in any quadrant but lies on the positive x-axis.

The points are plotted in Figure 4.2.

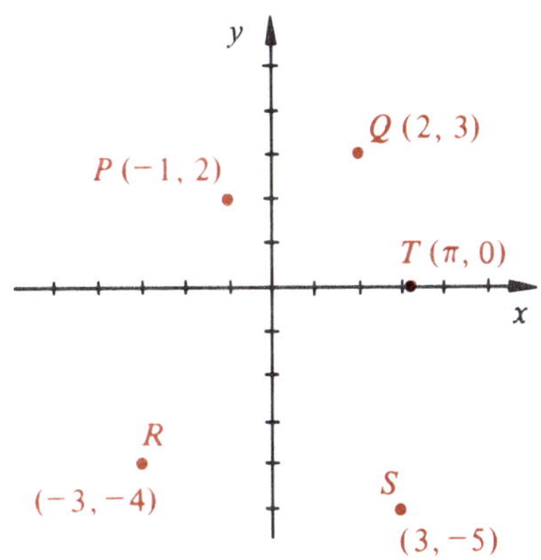

When an entire set of ordered pairs is plotted, the corresponding set of points in the plane represents the graph of the set. Sometimes the points in the graph form a recognizable pattern, such as a straight line or a circle.

EXAMPLE 2 Graph the set $\{(-4, -2), (-2, -1), (0, 0), (2, 1), (4, 7)\}$ and indicate the pattern of the points in the graph.

SOLUTION The graph is shown in Figure 4.3. The points in the graph fall in a straight line.

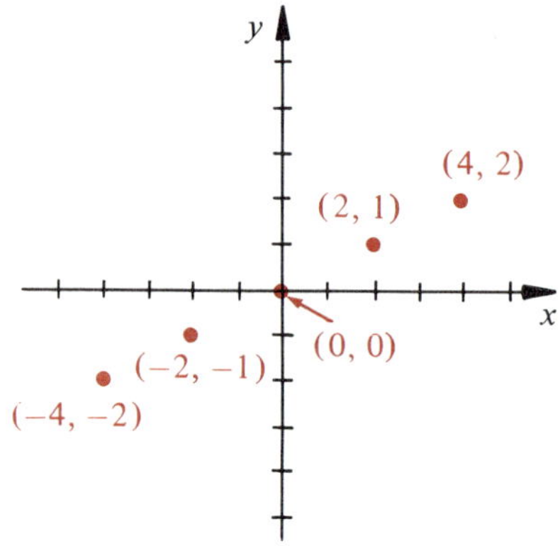

FIGURE 4.3

EXAMPLE 3 Graph the set of points in the plane whose ordinate, y, is 3.

SOLUTION This is the set defined by the equation $y=3$. The graph is shown in Figure 4.4.

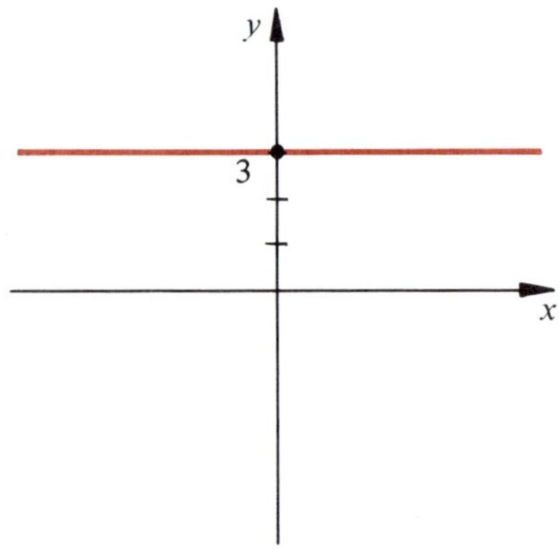

FIGURE 4.4

EXAMPLE 4 Graph the set of points whose abscissas are greater than -1 and whose ordinates are less than or equal to 4.

SOLUTION This set is described by the pair of inequalities $x>-1$, $y\leqslant 4$. The shaded region in Figure 4.5 is the graph of the set. The solid line is part of the region, whereas the broken line is not.

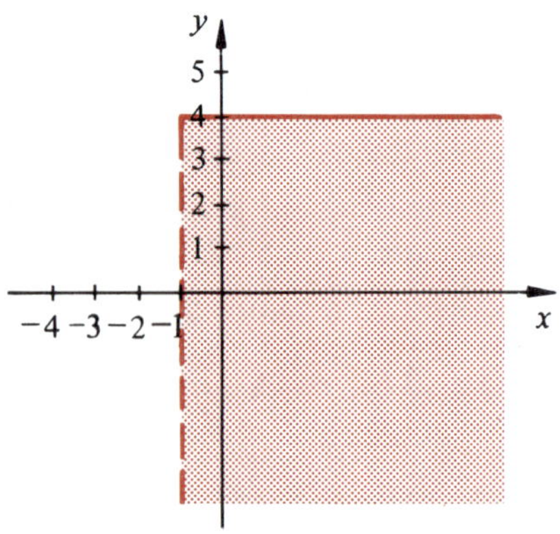

FIGURE 4.5

4.2 The Distance Between Two Points

Consider two points P_1 and P_2 on the x-axis as shown in Figure 4.6. The distance between these two points can be found simply by counting the number of units between them, or algebraically by subtraction. To be sure that the distance between P_1 and P_2 will be positive, we define it as the absolute value of the difference between P_1 and P_2. Thus

$$d(P_1, P_2) = |P_1 - P_2|$$

Computing the distance in Figure 4.6, we have

$$d(-2, 3) = |-2-3| = |-5| = 5$$

We do the same thing if the points lie on the y-axis.

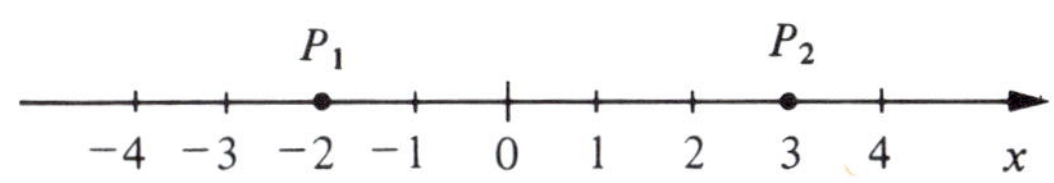

FIGURE 4.6

Now consider two points $P_1(x_1, y_1)$ and $P_2(x_2, y_2)$ that determine a slant line, as shown in Figure 4.7. Draw a line through P_1 parallel to the x-axis, and a line through P_2 parallel to the y-axis. These two lines intersect at the point $M(x_2, y_1)$. Hence, by the Pythagorean theorem,

(4.1) $$[d(P_1, P_2)]^2 = [d(P_1, M)]^2 + [d(M, P_2)]^2$$

We see from Figure 4.7 that $d(P_1, M)$ is the horizontal distance between P_1 and P_2.

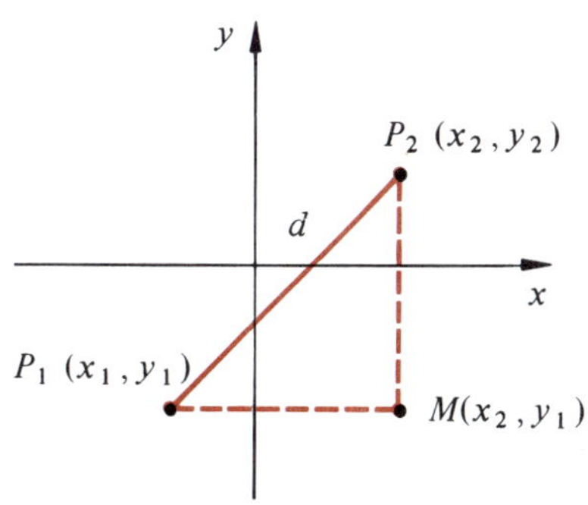

FIGURE 4.7

Therefore, the distance $d(P_1, M)$ is given by

$$d(P_1, M) = |x_2 - x_1|$$

Likewise, the vertical distance $d(M, P_2)$ is given by

$$d(M, P_2) = |y_2 - y_1|$$

Making these substitutions in Equation (4.1) and denoting $d(P_1, P_2)$ by d, we get

$$d^2 = (x_2 - x_1)^2 + (y_2 - y_1)^2$$

(4.2) $$d = \sqrt{(x_2 - x_1)^2 + (y_2 - y_1)^2}$$

Equation (4.2) is called the **distance formula**, and it is used to find the distance between two points in the plane directly from the coordinates of the points. The order in which the two points are labeled is immaterial, since

$$(x_2-x_1)^2=(x_1-x_2)^2 \quad \text{and} \quad (y_2-y_1)^2=(y_1-y_2)^2$$

EXAMPLE 1 Find the distance between $(-3, -6)$ and $(5, -2)$. (See Figure 4.8.)

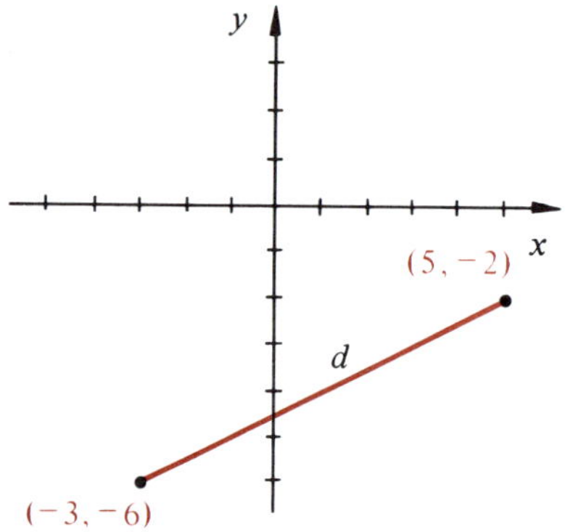

FIGURE 4.8

SOLUTION Let $(x_1, y_1)=(-3, -6)$ and $(x_2, y_2)=(5, -2)$. Substituting these values in the distance formula, we get

$$\begin{aligned} d&=\sqrt{(x_2-x_1)^2+(y_2-y_1)^2} \\ &=\sqrt{[5-(-3)]^2+[-2-(-6)]^2} \\ &=\sqrt{64+16}=\sqrt{80}=4\sqrt{5} \end{aligned}$$

When you are substituting values in the distance formula, remember to include the signs of the coordinates.

EXAMPLE 2 Find the distance between $(2, 5)$ and $(2, -1)$.

SOLUTION The two given points lie on a vertical line, since they have the same abscissa. (See Figure 4.9.) The distance between the two points can therefore be found directly:

$$d=|5-(-1)|=|5+1|=6 \text{ units}$$

The distance can also be found from the distance formula. Letting $(x_1, y_1)=(2, 5)$ and $(x_2, y_2)=(2, -1)$, we have

$$d=\sqrt{(2-2)^2+(-1-5)^2}=\sqrt{36}=6 \text{ units}$$

One of the important uses of the distance formula is in determining expressions that relate the x and y coordinates of a point as it moves along a curve. Such an expression is called the **equation of the curve**. We show the technique here as it applies to a circle. Equations of other curves can be derived by the same approach (see Chapter 11).

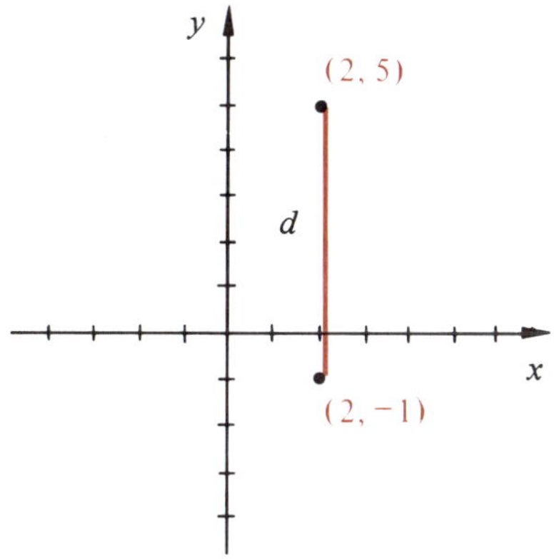

FIGURE 4.9

DEFINITION

> A circle is a curve consisting of a set of points in a plane that are equidistant from a fixed point.

The fixed point is called the **center** and the given distance is called the **radius** of the circle. To get the equation of the circle, suppose that the center has the coordinates (h, k) and a general point on the circle has coordinates (x, y) as shown in Figure 4.10.

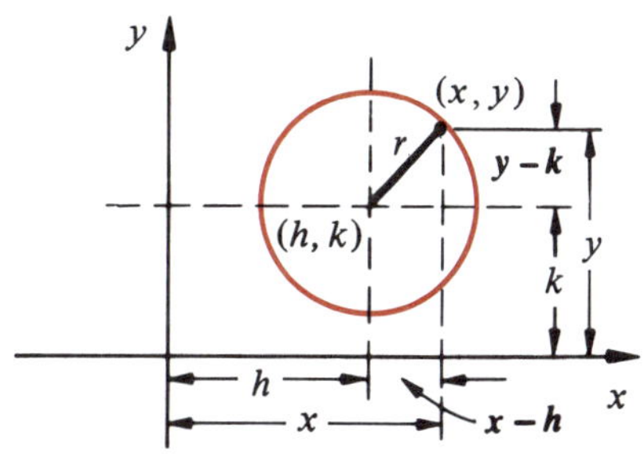

FIGURE 4.10

From the distance formula, since the distance from (h, k) to (x, y) is r, we have

$$(x-h)^2+(y-k)^2=r^2$$

This is called the standard form of the equation of a circle with center at (h, k) and radius r. If the center is the origin, then $h=0$ and $k=0$, and the equation becomes

$$x^2+y^2=r^2$$

This is called the standard form of a circle with center at the origin and radius r. For example, the equation of the circle centered at the origin with radius 4 is

$$x^2+y^2=16$$

By expanding the standard form of a circle with center at (h, k), we get $x^2-2hx+y^2-2ky+h^2+k^2-r^2=0$. If we let $-2h=C$, $-2k=D$, and $h^2+k^2-r^2=E$, the equation of a circle becomes

$$x^2+y^2+Cx+Dy+E=0$$

In this form, neither the center nor the radius is obvious; but if we complete the square, first on the x terms and then on the y terms, these values can be determined. Example 3 shows how.

EXAMPLE 3 Find the center and radius of the circle whose equation is

$$x^2+y^2+3x-8y=0$$

SOLUTION Rewriting this equation, we get

$$x^2+3x+(\quad)+y^2-8y+(\quad)=0$$

To complete the square on the x term, add $(\frac{3}{2})^2$ to both sides of the equation. To complete the square on the y term, add 4^2 to both sides. Hence,

$$x^2+3x+(\tfrac{3}{2})^2+y^2-8y+4^2=\tfrac{9}{4}+16$$

or

$$(x+\tfrac{3}{2})^2+(y-4)^2=\tfrac{73}{4}$$

The form of the given equation shows that the circle is centered at $(-\frac{3}{2}, 4)$ and has the radius $\sqrt{73}/2$. (See Figure 4.11.)

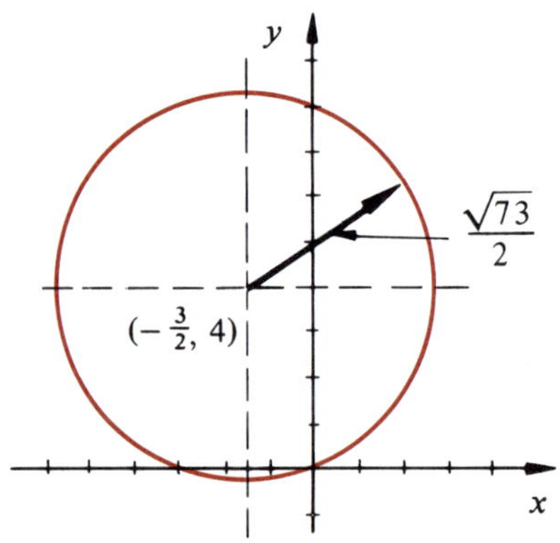

FIGURE 4.11

Exercises for Sections 4.1 and 4.2

Plot the ordered pairs in Exercises 1–6.

1. $(3, 2)$
2. $(4, 6)$
3. $(-2, \frac{1}{2})$
4. $(-6, -5)$
5. $(\frac{1}{4}, \frac{1}{2})$
6. $(-2.5, 1.7)$
7. In which two quadrants do the points have positive abscissas?
8. In which two quadrants do the points have negative ordinates?
9. In which quadrant are both the abscissa and ordinate negative?
10. In which quadrants is the ratio y/x negative?
11. What is the ordinate of a point on the x-axis?

Graph the given set in Exercises 12–21.

12. $\{(0, 0), (1, 1), (2, 2), (3, 3)\}$
13. $\{(0, 5), (1, 4), (2, 3), (3, 2), (4, 1), (5, 0)\}$
14. $\{(-2, 7), (0, 7), (3, 7), (7, 7)\}$
15. $\{(\sqrt{2}, 5), (\sqrt{2}, 2), (\sqrt{2}, 0), (\sqrt{2}, -\sqrt{3})\}$
16. $\{(-2, -1), (-1, -1), (0, -1), (1, \frac{1}{2}), (2, 1), (3, \frac{3}{2})\}$

17. $\{(0, 0), (1, 1), (2, 2), (3, 2), (4, 2), (5, 2)\}$

18. $\{(-5, 0), (-4, 3), (-3, 4), (0, 5), (3, 4), (4, 3), (5, 0)\}$

19. $\{(0, 4), (1, \sqrt{15}), (2, \sqrt{12}), (3, \sqrt{7}), (4, 0)\}$

20. $\{(1, 1), (3, 1), (1, 3), (3, 3)\}$

21. $\{(-3, 0), (-2, \sqrt{5}), (-1, \sqrt{8}), (0, 3), (1, \sqrt{8}), (2, \sqrt{5}), (3, 0)\}$

Plot the pairs of points in Exercises 22–34 and find the distance between the points.

22. $(1, 2), (5, 4)$ 23. $(0, 4), (-1, 3)$

24. $(-1, 5), (-1, -6)$ 25. $(\frac{1}{2}, \frac{1}{2}), (\frac{1}{2}, -\frac{3}{4})$

26. $(-5, 3), (2, -1)$ 27. $(0.5, 1.6), (6.2, 7.5)$

28. $(-3, 4), (0, 4)$ 29. $(2, -6), (-\sqrt{3}, -3)$

30. The point $(x, 3)$ is 4 units from $(5, 1)$. Find x.

31. Find the distance between the points $(\sqrt{x}, \sqrt{y})$ and $(-\sqrt{x}, -\sqrt{y})$.

32. Find the distance between the points (x, y) and $(-x, y)$.

33. Find the distance between the points (x, y) and $(x, -y)$.

34. Find the point on the x-axis that is equidistant from $(0, -1)$ and $(3, 2)$.

Graph the set of points defined by the given equations or inequalities in Exercises 35–45.

35. $y \geq 0$ 36. $x=0$ 37. $x=-2$

38. $x \geq 0$ 39. $y=2$ 40. $y=-1$

41. $x>-1$ and $y \geq 0$

42. $y=x$ 43. $y=-x$

44. The points common to $x \geq 0, y \geq 0$ and $x<1, y<1$.

45. The points common to $x>-1$, $y<-1$, and $x \leq -1, y \geq -1$.

Find the center and the radius of each of the circles in Exercises 46–53. Sketch.

46. $x^2+y^2-y=0$ 47. $2x^2+2y^2-5x+y=3$

48. $x^2+y^2+2x-6y=15$

49. $3x^2+3y^2+7x-y-9=0$

50. $x^2+y^2+4x-4y=17$

51. $x^2+y^2+8x-10y-8=0$

52. $x^2+y^2-6x+7=0$ 53. $x^2+y^2-5y=0$

Find the equation of each of the circles in Exercises 54–62.

54. With center at the origin and radius 8.

55. With center at the origin and passing through $(1, 5)$.

56. With center at $(1, 5)$ and passing through the origin.

57. With ends of diameter at $(-6, 1)$ and $(4, 3)$.

58. With center at $(1, 3)$ and touching the x-axis.

59. With center at $(-2, 5)$ and radius $\sqrt{3}$.

60. With center at $(-1, -2)$ and radius 5.

61. With center at $(2, 1)$ and passing through $(5, 3)$.

62. With center at $(0, 4)$ and passing through $(1, -1)$.

4.3 Variation

Most applications of mathematics concern the relation of one quantity to another, for example, how the acceleration of an object is related to the force causing the acceleration, or how the profit of a company is related to the work force. Mathematics can be used to describe such relations.

The relation between two physical quantities can often be expressed by a mathematical formula. For instance, the area of a circle is related to its radius by the formula $A=\pi r^2$, where π is an irrational number approximated by 3.14159. Since the radius can be any positive real number, we call r the **independent variable**. The area A is called the **dependent variable**, since it depends on the value of r.

EXAMPLE 1 When temperature is converted from degrees Celsius to degrees Fahrenheit, the formula $F=\frac{9}{5}C+32$ is used. In this form, C is the independent variable and F the dependent variable.

The ratio of a to b is the number obtained by dividing a by b, that is a/b. For example the ratio of 3 to 5 is $\frac{3}{5}$, and the ratio of 12 to 4 is $\frac{12}{4}=3$. Example 2 shows some physical quantities that are expressible as ratios.

EXAMPLE 2 (a) Velocity is the ratio of displacement to elapsed time; that is,

$$\text{Velocity}=\frac{\text{Displacement}}{\text{Time}}$$

(b) Electric current is the ratio of voltage to resistance; that is,

$$\text{Current}=\frac{\text{Voltage}}{\text{Resistance}}$$

(c) Slope of an incline is the ratio of vertical rise to horizontal run; that is,

$$\text{Slope}=\frac{\text{Rise}}{\text{Run}}$$

The equality of two ratios is called a **proportion**. Many applied problems are solved by writing an appropriate proportion.

EXAMPLE 3 A model airplane for use in wind tunnel tests is built to a scale of $\frac{1}{4}$ in. to 1 ft. What is the wing span of the model if the actual airplane has a wing span of 50 ft?

SOLUTION Establish a proportion using S_M, S_A, W_M, and W_A to represent the values of the model scale, the actual scale, the model wing dimension, and the actual wing dimension. Thus,

$$\frac{S_M}{S_A}=\frac{W_M}{W_A}$$

When you supply the known values $S_M=\frac{1}{4}''$, $S_A=1'$, and $W_A=50'$, you must use the appropriate units.

$$\frac{\frac{1}{4}''}{1'}=\frac{W_M}{50'}$$

$$W_M(1')=\tfrac{1}{4}''(50')$$

$$W_M=12.5'' \quad \text{(wing span of wind tunnel model)}$$

This example used the simple idea of a ratio, or a proportion, to establish a desired relation. If the ratio of two variables is always a constant, then the two variables are said to **vary directly**.

DEFINITION

The variable y is said to vary directly with x if $y/x=k$, or equivalently, if $y=kx$, where k is the *constant of direct proportionality*.

Given that two variables are related by direct variation, then if you know the value of the variables in only one case, you will have the formula of y in terms of x for all cases.

EXAMPLE 4 Newton's second law of motion says that the acceleration a of an object varies directly with the force F applied to it. Suppose that an object is accelerated 10 m/sec^2 by a force of 5 newtons (N). What acceleration will the same object experience if a force of 20 N is applied?

SOLUTION Symbolically, Newton's law is written $a=k\cdot F$. Since $a=10$ when $F=5$, we have that $10=k\cdot 5$, or $k=2$. Hence, the general formula is $a=2F$. In particular, if $F=20$, then $a=2\cdot 20=40$ m/sec^2.

Another important kind of variation, called **inverse variation**, occurs when the *product* of the variables is a constant.

DEFINITION

The variable y is said to vary inversely with x if $xy=k$, or equivalently, if $y=k/x$, where k is the *constant of inverse proportionality*.

In direct variation the two variables increase (or decrease) together. But in inverse variation, an increase in one variable means a decrease in the other.

By considering the values of x and y as determining ordered pairs, we can draw a picture of direct and inverse variation on a Cartesian coordinate system. Figure 4.12 shows typical cases of direct variation ($y=kx$) and inverse variation ($y=k/x$).

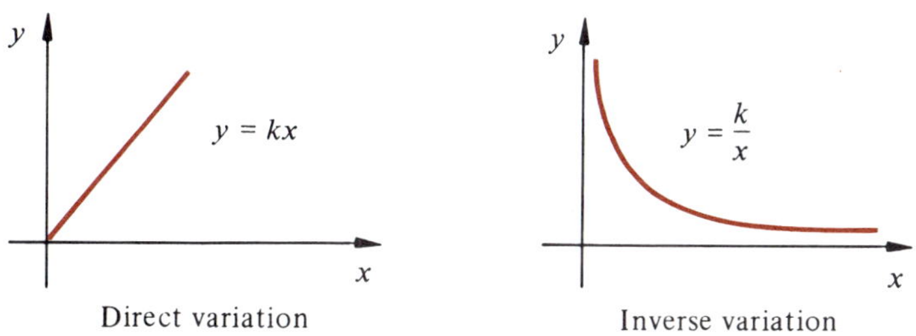

FIGURE 4.12

EXAMPLE 5 Boyle's law states: *If the temperature of a gas is constant, then the product of the volume V and pressure P is constant.* Suppose that the volume of a gas is 100 in.^{3} when the pressure is 20 psi. What is the volume when the pressure is 14.7 psi?

SOLUTION Boyle's law states that $VP=k$, which can be expressed as $V=k(1/P)$. Hence, if $V=100$ when $P=20$, we have

$$100=k(\tfrac{1}{20})$$

or

$$k=20(100)=2000$$

Boyle's law for this gas is then

$$V=(2000)\frac{1}{P}$$

Substituting $P=14.7$ psi into this formula yields

$$V=\frac{2000}{14.7}=136.1 \text{ in.}^3$$

The preceding examples show variation in which one quantity depends on another. In some cases, the variation is in terms of either the product or the quotient of two other quantities. The words *directly* and *inversely* describe the relations between the variables, as Example 6 makes clear.

EXAMPLE 6 The ACE Light Bulb Company finds that the price P of each light bulb it produces varies directly with the cost C of material and inversely with the square root of the number produced n. Materials to produce 40,000 light bulbs priced at \$0.30 cost \$2000. How much should the company charge for a new bulb if \$1500-worth of material will make 10,000 bulbs?

SOLUTION The relation among P, C, and n can be written

$$P=k\frac{C}{\sqrt{n}}$$

The condition $P=0.30$, $C=2000$, and $n=40{,}000$ yields

$$0.30=k\frac{2000}{\sqrt{40{,}000}}=\frac{2000k}{200}=10k$$

Solving for k, we have $k=0.03$. Thus, the price is given by

$$P=\frac{0.03C}{\sqrt{n}}$$

Substituting $C=1500$ and $n=10{,}000$ gives

$$P=\frac{0.03(1500)}{\sqrt{10{,}000}}=0.45$$

Hence, the company should charge \$0.45 for each of these bulbs.

Exercises for Section 4.3

• Write the indicated variation in Exercises 1–10 in equation form.

1. v varies directly with t.
2. a varies directly with k.
3. r varies inversely with $\sqrt{p}$.
4. q varies inversely with $(t-2)$.
5. m varies directly with the square of x and the cube of y.
6. z varies directly with t and inversely with the square of v.
7. y varies directly with the fourth power of h and inversely with the square of g.
8. x varies directly with the square root of t and inversely with the cube of c.
9. n varies inversely with the product of r times t.
10. w varies inversely with the product of p, q, and t.

In the remaining exercises solve the indicated problem.

11. A road map uses a scale of 1 in. = 20 mi. Determine how far Cincinnati is from Cleveland if the map distance measures 12.6 in.
12. In a blueprint scale, 1 in. = 5 ft. What are the dimensions of a room that measures 9 in. by 7 in. on the blueprint?
13. The ratio of sodium to chlorine in common table salt is approximately $\frac{5}{3}$. Find the approximate amount of each element in a salt compound weighing 100 lb.
14. A sample revealed that on a certain day, for every 1000 baseballs manufactured, 20 were defective. If 15,250 were manufactured on that day, how many were defective?
15. A piece of wire 8 ft long is to be cut so that the parts are in the ratio of 5 to 4. How long should each part be?
16. A saloon keeper advertises 18-oz steins of beer for \$0.40, while the neighborhood grocer advertises a case of twenty-four 12-oz cans for \$6.10. Which is the better buy and why?
17. A punch press applies a force of 20,000 lb to a piece of metal with an area of 50 in.^{2}. What is the ratio of force to area?
18. The *acceleration* of an object equals the ratio of the applied force f to the mass m of the object. What is the acceleration if $f=12$ and $m=5$?
19. The *slope* of an inclined plane is the ratio of the vertical rise to the corresponding horizontal run. What is the slope of an incline having a vertical rise of 15 ft for a horizontal run of 100 ft?
20. The *density* of an object is the ratio of the object's weight to its volume. What is the density of an object having a volume of 25 in.^{3} and a weight of 75 lb?
21. If y varies directly with x and $y=5$ when $x=10$, what is the value of y when $x=12$?
22. If r varies directly with the square of t and $r=3$ when $t=4$, what is the value of r when $t=5$?
23. If z varies inversely with t and $z=4$ when $t=\frac{1}{2}$, what is the value of z when $t=2$?
24. If y varies inversely with the square root of x and $y=2$ when $x=9$, what is the value of y when $x=4$?
25. When an object falls from rest, its velocity varies directly with the time of fall. Determine what the velocity of an object is 2 sec after release if it has a velocity of 96 ft/sec at the end of 3 sec.
26. Hooke's law states that the force required to stretch a spring is directly proportional to the elongation of the spring. If a 5-lb force stretches the spring 1.5 in., how much force is needed to stretch it 2.0 in.?
27. The unit cost to produce a monthly magazine varies inversely with the square root of the circulation. If it costs \$0.40 a unit when the circulation is 200,000, what will it cost if the production jumps to 400,000?
28. The distance an object falls from rest varies directly with the square of the elapsed time. If an object fell 64 ft in 2 sec, how far would it have fallen at the end of $\frac{1}{2}$ sec?
29. The intensity of light varies inversely with the square of the distance from the source. If the intensity 5 ft from a light is 4 foot-candles, what is the intensity 10 ft from the light?

30. The law of gravitational attraction states that the gravitational force between two objects varies inversely with the square of the distance between them. If the gravitation force between two objects is 150 lb when they are 200 mi apart, what is the gravitational force between them when they are 50 mi apart?

31. The power in a resistive circuit varies directly with the resistance and the square of the current. If the power in a 50-ohm resistor is 50 watts when the current is 0.5 amp, what is the power when the current is 1.0 amp?

32. The cost of labor varies directly with the number of workers and the number of days they work. If three men work 5 days for $1500, how many days must 5 men work to earn $1500?

33. Ohm's law states that the electric current in a circuit varies directly with the voltage and inversely with the resistance. If the current in a 20-ohm resistor is 0.5 A when a voltage of 6 V is applied, what is the current in the resistor when the applied voltage is 12 V?

34. The volume of a gas varies directly with the temperature and inversely with the pressure. Find a formula relating volume, temperature, and pressure. Suppose that the volume is 100 in.^{3} when $T = 200°$C and the pressure is 50 lb/in.^{2}. Find the volume for a temperature of 300°C and a pressure of 100 lb/in.^{2}.

35. The cost of publishing a pamphlet is $2000 plus an amount varying directly with the number of copies published. If 1000 copies of the pamphlet cost $2200, how much will it cost to publish 1500 copies?

36. The number of gallons of gasoline used by an automobile traveling at a constant speed is directly proportional to the number of miles traveled, and the number of miles traveled is in turn directly proportional to the number of hours spent in driving. Can you correctly conclude that the number of gallons used is directly proportional to the time driven?

37. If y is directly proportional to t and if x is directly proportional to t, what can you say about (a) xy; (b) x/y; (c) $x+y$?

38. Kepler's third law states that the time in which a planet revolves about the sun varies directly with the 3/2 power of the maximum radius of its orbit. Using 93 million miles as the maximum radius of Earth's orbit and 142 million miles as the maximum radius of Mars' orbit, in how many days will Mars make one revolution about the sun?

4.4 The Functional Relation

Perhaps the most fundamental concept in all mathematics is pairing numbers by establishing a *rule of correspondence* between the elements of the pairing. You have learned (Section 4.3) that direct variation means that a number kx is paired with each number x by the rule of correspondence $y = kx$; direct variation establishes a unique value of y for each value of x. The pairings given in the accompanying table were generated using $y = 3x$ as the rule of correspondence.

x	-2	-1	0	1	2	3
y	-6	-3	0	3	6	9

Rules of correspondence that assign a unique value of the dependent variable to each value of the independent variable are particularly interesting to us, and the following definition formalizes this concept.

DEFINITION

> If the relation between two variables x and y is such that there is exactly one value of y for each value of x, then, we say, **y is a function of x**.

Functions are frequently defined by some formula or expression that involves x and y. The following remarks relate the definition of a function to using formulas.

- A formula such as $y = \pm\sqrt{x}$ does not give a functional relation, since two values of y correspond to each value of x. However, each of the formulas $y = \sqrt{x}$ and $y = -\sqrt{x}$ taken separately *does* define a function.
- The definition does not require the value of the dependent variable to change when the independent variable changes; it only requires the dependent variable to have a unique value corresponding to each value of the independent variable. Thus, $y = 5$ is a function, since y has the unique value 5 for any value of the independent variable. However, $x = 5$ is not a function, since many (in fact, infinitely many) values of y correspond to the value $x = 5$.
- We are not saying that expressions such as $y = \pm\sqrt{x}$ and $x = 5$ do not occur or that they are unimportant. These formulas simply do not describe functions; they describe a broader concept, called a *relation*.

The definition of a function requires that for each x there correspond exactly one value of y. This idea, in conjunction with the idea of an ordered pair, gives rise to another interpretation of the word *function* in terms of sets.

DEFINITION

> A function is a set of ordered pairs $\{(x, y)\}$ in which no two distinct pairs have the same first element.

The set $\{(2, 3), (-1, 4), (0, -5), (3, 4)\}$, for example, is a function, but the set $\{(-3, 4), (2, 5), (2, -6), (9, 7)\}$ is not; in the second set, the ordered pairs (2, 5) and (2, -6) have the same first element.

EXAMPLE 1 The expression $y = 3x^2 + 5$ defines a function. The accompanying table lists some of the corresponding ordered pairs. Note that only one value of y is obtained for each value of x.

x	0	-1	$+1$	-2	$+2$	-3	$+3$
y	5	8	8	17	17	32	32

EXAMPLE 2 The expression "y is greater than x" (in symbols, $y > x$) does not define a function, since there are many values of y for each x.

Although the functional relation between two variables is usually given by a formula, it can be given in other ways. For instance, let A denote the set of letters $\{a, b, c\}$ and let B denote the set of numbers $\{5, 2, 4, \sqrt{6}, 12\}$. Suppose a rule of correspondence is given that assigns a to 5, and b to $\sqrt{6}$, and c to 4. Figure 4.13 diagrams this situation. Then the rule, together with the sets A and B, is said to be a function from A to B. The set A is called the *domain* of the function. The set of all elements of B that are assigned by the rule to some member of A is called the *range* of the function, that is, the set $\{5, 4, \sqrt{6}\}$. This terminology applies to all functions.

DEFINITIONS

The **domain** of a function is the set of all values that the independent variable can assume.

The **range** of a function is the corresponding set of values that the dependent variable assumes.

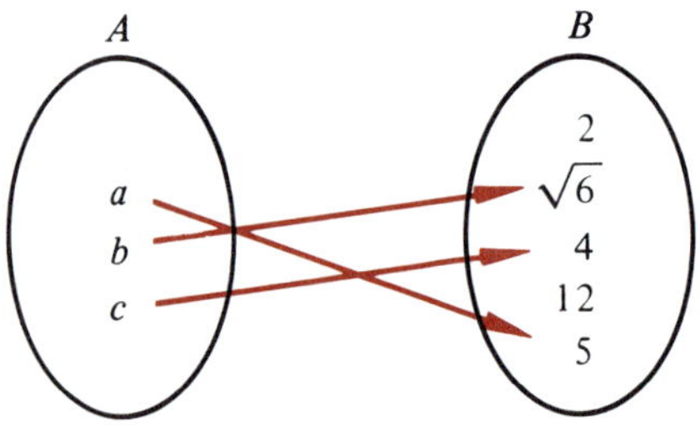

FIGURE 4.13

The domain and the range of a function are usually sets of numbers; but as Figure 4.13 shows, the domain or the range, or both, may be sets that contain no numbers.

Although a functional relation can be given in a variety of forms, it must never assign more than one range element to the same domain element. Figure 4.14 does not describe a function, since c is assigned to the two different elements 5 and 6.

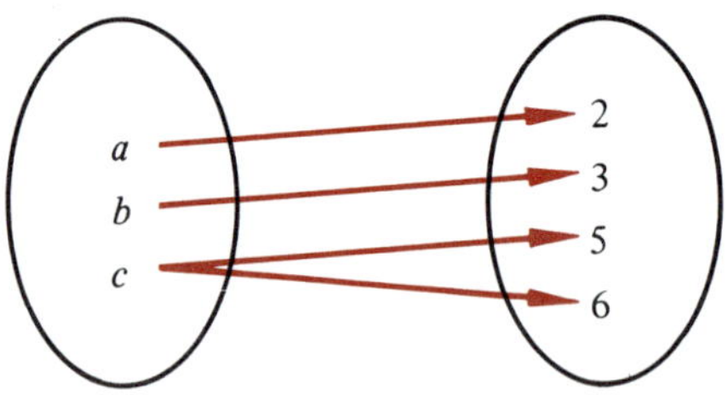

FIGURE 4.14

Single letters or symbols customarily denote functions; examples are f, g, h, ϕ. If x is an element of the domain of a function f, the corresponding element of the range is denoted by $y=f(x)$. Thus, from Figure 4.13, we see that $f(a)=5$, and $f(b)=\sqrt{6}$, and $f(c)=4$. The notation $f(x)$ reads "f of x" or "f at x," meaning the range *value* of the function for a particular x. Note that $f(x)$ does *not* mean the product of f times x.

In a useful analogy, we can compare a function f to a machine that has an input and an output. (See Figure 4.15.) When an element x enters the machine, it is formed by the machine (function) into a new element, $f(x)$. The set of elements that can

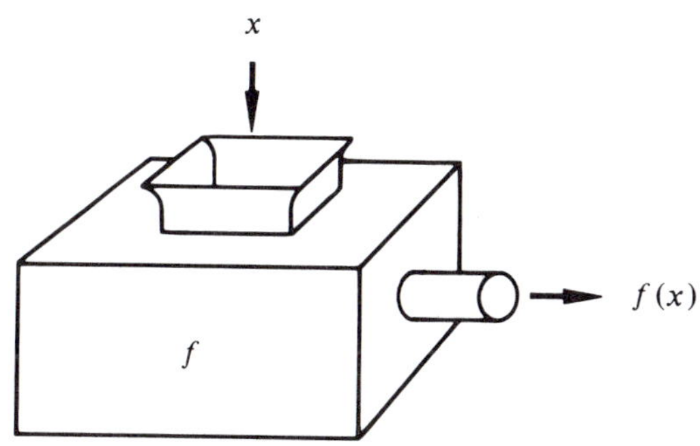

FIGURE 4.15

be put into the machine represents the domain of the function and the output represents the range.

EXAMPLE 3 An ice-vending machine will dispense a 5-lb bag of crushed ice for \$0.50, an 8-lb bag for \$0.75, and a 10-lb bag for \$0.90. This machine represents a function, since a unique output exists for each input (Figure 4.16). This function can be described by the set $\{(50, 5), (75, 8), (90, 10)\}$, where the first number gives the cost and the second the weight.

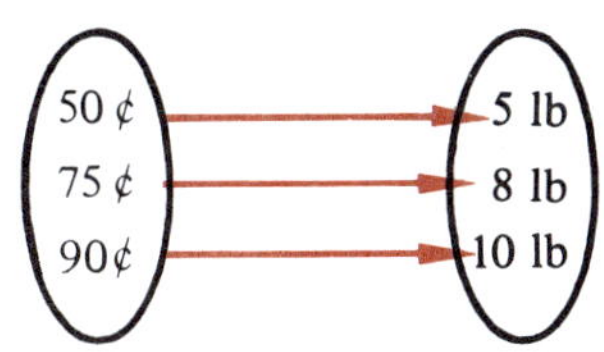

FIGURE 4.16

While there is a tendency to use the letter f for naming a function, any other letter or symbol will do. In practice, it is often necessary to use different functional names to avoid confusion. For instance, $v_1=f(t)$ and $v_2=g(t)$ might indicate that the atomic velocities of two particles p_1 and p_2 are different functions of time. Had either f or g symbolized both functional names, the distinction between the velocity functions would not be obvious. The worth of function notation lies in its flexibility.

The function f may or may not have a definite mathematical expression describing the rule of correspondence. If, however, such a specific expression is available, functional notation is convenient for indicating numbers in the range. To find a number in the range that is associated with any number in the domain, merely replace the letter x wherever it appears in the expression for $f(x)$ by the number from the domain.

EXAMPLE 4 Find the value of the function $f(x)=x^2+3x$ at $x=2$.

SOLUTION We denote the value of $f(x)$ at $x=2$ by $f(2)$. To find $f(2)$, substitute 2 for x in x^2+3x; that is,

$$f(2)=(2)^2+3(2)=10$$

It is sometimes helpful to think of x as a blank, so that the functional notation tells us what to put in the blank. The function in Example 4 could be written

$$f(\quad)=(\quad)^2+3(\quad)$$

Then, for example,

$$f(a^2)=(a^2)^2+3(a^2)=a^4+3a^2$$

Calculators are indispensable tools for evaluating functions, since they take the drudgery out of evaluating complex functions for specified values of the domain. Thus, with a calculator, it is just as easy to evaluate $f(x)=\pi x^3+\sqrt{x+2}$ *for* $x=3.105$ *as for* $x=2$.

EXAMPLE 5 Calculus frequently calls for calculating the "difference quotient" of a function; this is defined by

$$\Delta f=\frac{f(a+h)-f(a)}{h}$$

What is the difference quotient if $f(x)=1/x$?

SOLUTION Here $f(a+h)=1/(a+h)$ and $f(a)=1/a$. Therefore,

$$\Delta f=\frac{\dfrac{1}{a+h}-\dfrac{1}{a}}{h}=\frac{1}{h}\left(\frac{a-(a+h)}{a(a+h)}\right)=\frac{-1}{a(a+h)}$$

When the rule of correspondence is given by a formula, the domain and the range are often not specifically mentioned. If they are not otherwise stated, think of the domain as including all those real numbers for which the expression itself is real. The range is then the corresponding set of functional values. For example, the function whose rule is given by $f(x)=\sqrt{x}$ has a domain of all nonnegative real numbers, since $\sqrt{x}$ is an imaginary number if x is negative. The range also comprises the nonnegative real numbers, since $\sqrt{x}$ always yields a nonnegative number. On the other hand, the function $f(x)=x^2$ has a domain of all real numbers (the value of x has no limitations), but all the range values are nonnegative reals, since any number squared cannot be negative. Determining the domain and range is generally left to you. Typically, the domain values are limited by whether the variable is inside a radical or in a denominator. While domain limitations are often clear, the corresponding limitations on the range may require a little work.

EXAMPLE 6 Find the domain and range of the function defined by $y=(2-x)/(2x+3)$.

SOLUTION Since $2x+3$ is not permitted to be zero, $x\neq-\frac{3}{2}$, and the domain consists of all real values except this number. To find the range, solve for x as follows:

$$
\begin{aligned}
y(2x+3) &= 2-x \\
2xy+3y &= 2-x \\
2xy+x &= 2-3y \\
x(2y+1) &= 2-3y \\
x &= \frac{2-3y}{2y+1}
\end{aligned}
$$

From this equation, we see that since $2y+1$ may not equal zero, the range consists of all real numbers except $-\frac{1}{2}$.

Inferring that a rule exists for a function from a finite set of data can be "dangerous." Suppose, for example, that you know that $f(1)=1, f(2)=4, f(3)=9, f(4)=16$, and $f(5)=25$. In such a case, the rule for $f(x)$ *could* be

$$f(x)=x^2$$

or $$f(x)=x^2+(x-1)(x-2)(x-3)(x-4)(x-5)$$

or $$f(x)=x^2+(x^2-1)(x^2-4)(x^2-9)(x^2-16)(x^2-25)$$

Lastly we note that the dependent variable may be a function of two or more independent variables. For example, the formula for the area of a triangle is $A=\frac{1}{2}bh$, where b is the length of the base and h is the altitude. Thus, the area of a triangle is a function of *both* the base and the altitude of the triangle. This idea could be noted in general by using functional notation and writing $A=F(b, h)$. In the same way, if we want to indicate that the acceleration a of a body is a function of force f and mass m, we could write

$$a=G(f, m)$$

Functions of the type $y=f(x)$ are called functions of *one* variable; functions of the type $z=f(x, y)$ are called functions of *two* variables.

Exercises for Section 4.4

Which of the expressions in Exercises 1–6 define functions? (Assume that y depends on x.)

1. $y=2x+5$ **2.** $y=x^2$ **3.** $y=10$

4. $y<3x$ **5.** $y=\sqrt[3]{x}$ **6.** $y^2=x^3$

Which of the sets in Exercises 7–12 define functions?

7. $\{(1, 2), (1, 3), (2, 4), (3, 7)\}$

8. $\{(2, 3), (7, 9), (8, 3)\}$

9. $\{(x, y) \mid y=5x+1\}$ **10.** $\{(s, t) \mid s=1/t\}$

11. $\{(x, y) \mid y=\pm\sqrt{x}\}$ **12.** $\{(x, y) \mid y=x^{1/2}\}$

13. Which of the diagrams define functions?

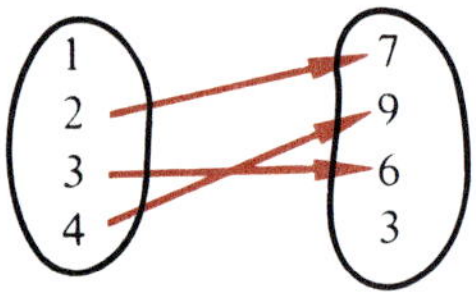

(a)

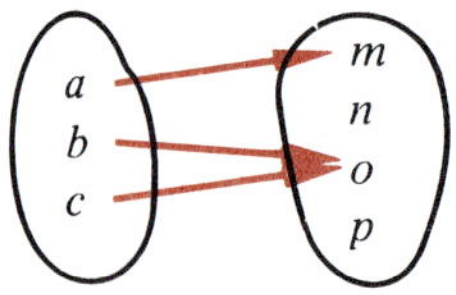

(b)

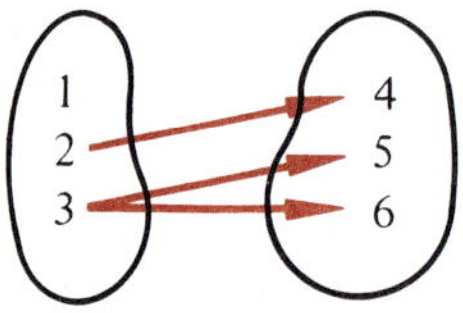

(c)

Find the domain and the range of each of the real-valued functions in Exercises 14–28.

14. $f(x)=2x$ **15.** $g(x)=4-3x$ **16.** $h(t)=3t^2+5$

17. $F(x)=\sqrt{x}$ **18.** $G(x)=1/x^2$ **19.** $y=x^{1/3}$

20. $y=\dfrac{1}{x}$ **21.** $y=\dfrac{4}{x+3}$ **22.** $f(x)=\dfrac{5-x}{x-2}$

23. $g(t)=\dfrac{2+t}{4t+1}$

24. $\{(1, 2), (3, 5), (7, 1), (12, -2)\}$

25. $f(x)=\sqrt{-x}$ **26.** $f(x)=\sqrt{-x^2}$

27. $f(x)=\sqrt{x-1}$ **28.** $f(x)=\sqrt{1-x}$

29. Suppose that the function f has the rule $f(x)=3x+1$. Compute the following:

(a) $f(3)$ (b) $f(\pi)$ (c) $f(z)$ (d) $f(x-h)$
(e) The element in the domain that maps onto 10.
(f) The range of the function.
(g) $f(f(3))$ (h) $f(f(f(0)))$
(i) The difference quotient of f.

30. Suppose that the function G has the rule $G(t)=t^2-2t+1$. Compute the following:

(a) $G(2)$ (b) $G(-1)$ (c) $G(x^2)$
(d) $G(x-h)$ (e) $G(\sqrt{t})$ (f) $G(G(0))$
(g) The difference quotient of G.

31. Suppose that the function H is defined by the set $\{(2, 3), (5, 9), (7, -3)\}$. Compute the following:

(a) $H(2)$ (b) $H(7)$ (c) $H(3)$
(d) The range of H. (e) The domain of H.
(f) Compare $H(2+5)$ to $H(2)+H(5)$.

Compute the difference quotient for the functions in Exercises 32–39.

32. $f(x)=x$ **33.** $f(x)=x^2$ **34.** $f(x)=2x^2$

35. $f(x)=x^2-2x$ **36.** $f(x)=x^3$

37. $f(x)=x^3-1$ **38.** $f(x)=\dfrac{1}{x^2}$

39. $f(x)=\dfrac{1}{x+1}$

40. Suppose that y is directly proportional to x, that is, that $y=f(x)=kx$.

(a) Compute $f(x_1)/f(x_2)$ for any two numbers x_1 and x_2 $(f(x_2)\neq 0)$.
(b) Compare $f(1/x)$ with $1/f(x)$.
(c) Compare $f(x^2)$ with $[f(x)]^2$.
(d) Find the difference quotient.
(e) Compare $f(x)+1$ with $f(x+1)$.
(f) Compare $f(x_1+x_2)$ with $f(x_1)+f(x_2)$.
(g) Compare $af(x)$ with $f(ax)$, where a is a constant.

41. Suppose that y is inversely proportional to x, that is, that $y=f(x)=k/x$. Answer the questions in Exercise 40 for this function.

42. Let $f(x)=mx+b$, where m and b are constants. (a) Compare $f(ax)$ with $af(x)$. (b) Compare $f(x)+1$ with $f(x+1)$.

43. What can you say about the following?

(a) $f(x_1+x_2)$ and $f(x_1)+f(x_2)$
(b) $f(ax)$ and $af(x)$
(c) $f(x_1x_2)$ and $f(x_1)f(x_2)$
(d) $f(\sqrt{x})$ and $\sqrt{f(x)}$ (e) $f\left(\dfrac{1}{x}\right)$ and $\dfrac{1}{f(x)}$

Use a calculator to find the indicated range value.

44. Given $f(x)=2x^3+5.73x-125.65$, find $f(3.75)$.

45. Given $f(s)=3.101s^2+29.46s$, find $f(-0.415)$.

46. Given $g(t)=4356t(3.7+t)$, find $g(287.7)$.

47. The current i in a resistor is found to vary as time according to

$i(t) = 1.196t^2 + 0.076t,$

where i is in amperes and t is in seconds. Find $i(1.375)$.

48. The equation of motion of a rocket is given by

$h(t) = 150.6t - 16.1t^2$

where h is altitude in feet when t is measured in seconds. Find $h(0.45)$.

4.5 The Graph of a Function

Most functions used in science, engineering, and business are "real" in the sense that both the domain and the range of the functions are sets of real numbers. A function f defined by the set of ordered real number pairs $\{(x, f(x))\}$ can be plotted as points in the Cartesian plane. We call the resulting points the *graph of the function*. Traditionally, the first element of the ordered pair is plotted along the horizontal axis, and the second element is plotted along the vertical axis.

EXAMPLE 1 The graph of the function defined by $f = \{(1, 3), (-2, 2), (5, -1)\}$ is shown in Figure 4.17.

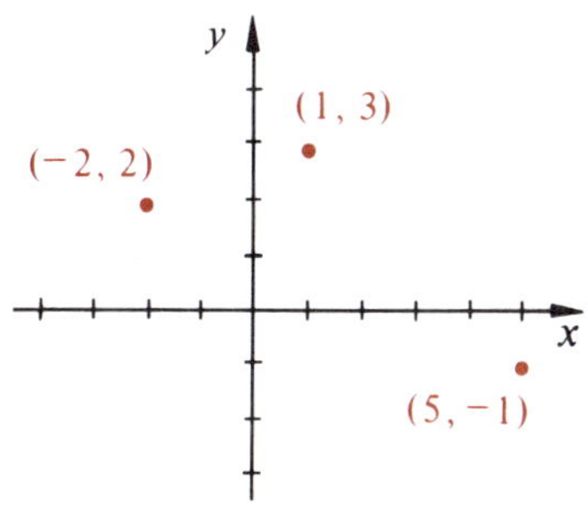

FIGURE 4.17

EXAMPLE 2 Draw the graph of the function g for the conditions that the domain of g is $\{0, 1, 4, 9\}$ and the rule of correspondence is $g(x) = -\sqrt{x}$.

SOLUTION From the given conditions, we have

$$g = \{(0, 0), (1, -1), (4, -2), (9, -3)\}$$

The graph of g is shown in Figure 4.18.

The domain of a function is usually an interval of real numbers. The graphs of such functions can be constructed by plotting all points in the plane whose coordinates satisfy the function. This approach is impractical, however, since there is no limit to the number of such points that can be plotted. (Remember the makeup of the real line.) In practice, the graph of such a function is constructed by plotting a few selected points and then connecting these points with a smooth curve. Consider, as an illustration of this technique, the function defined by $y = x^2$, whose domain is the

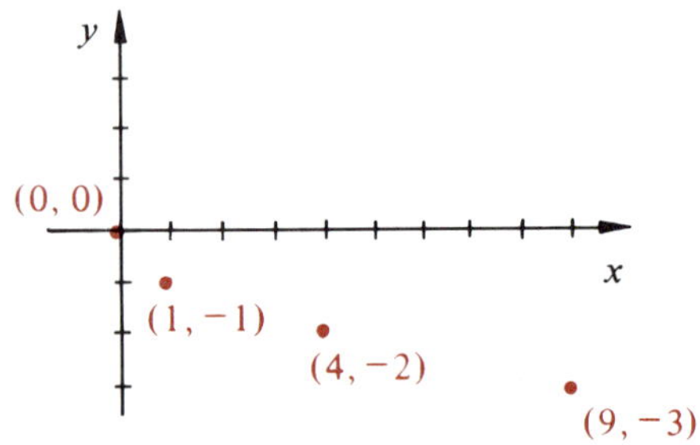

FIGURE 4.18

entire set of real numbers. By assigning values to x, we get the set of ordered pairs in the accompanying table.

x	−3	−2	−1	0	1	2	3
y	9	4	1	0	1	4	9

Now plot the set of ordered pairs in the Cartesian plane, as shown in Figure 4.19(a). The graph of $y=x^2$ can then be obtained by connecting the points with a smooth curve, as in Figure 4.19(b). This graph only approximates the actual graph of the function; its accuracy depends on the number of points plotted and the care taken in drawing the smooth curve connecting the points.

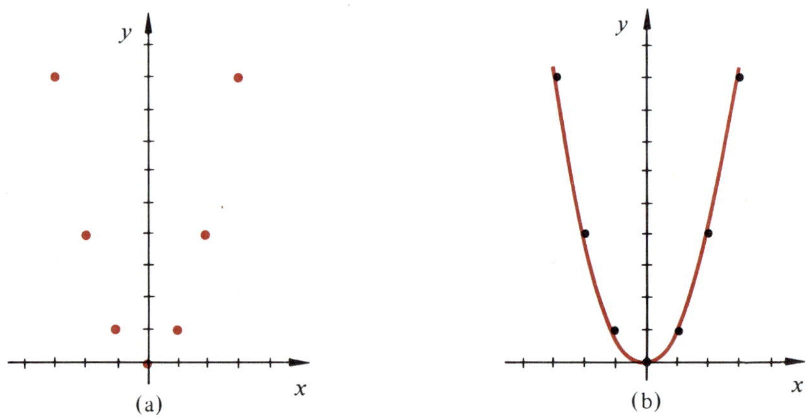

FIGURE 4.19

Usually only a segment of the graph near the origin is plotted, even though the graph may extend much farther. In Figure 4.19, for instance, only the portion of the curve from $x=-3$ to $x=3$ is plotted, although the domain of the function is the set of all real numbers.

It is always helpful in graphing a function if the domain is known beforehand, since the domain determines the extent of the graph with respect to the x-axis.

EXAMPLE 3 Graph the function $y=\sqrt{4-x}$.

SOLUTION The domain consists of all real numbers x for which $4-x \geqslant 0$. (Otherwise the range values would not be real numbers.) Hence the domain is the set of all real numbers x such that $x \leqslant 4$. This means that the graph must be to the left of the vertical line $x=4$, except for the point (4, 0), which *is* part of the graph, as you can see in Figure 4.20.

$y=\sqrt{4-x}$

x	y
4	0
2	$\sqrt{2}$
0	2
-2	$\sqrt{6}$

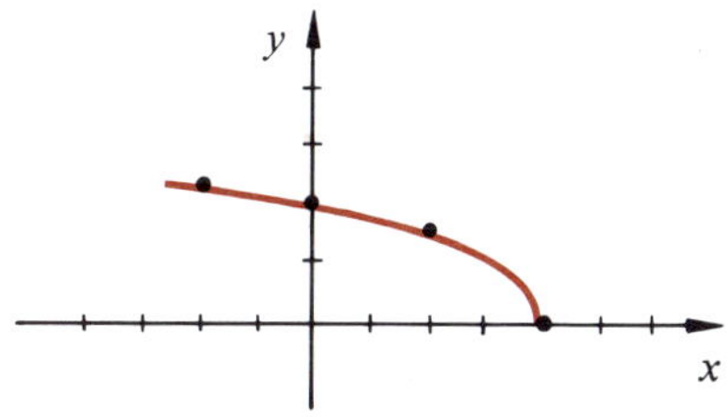

FIGURE 4.20

EXAMPLE 4 A numerically controlled milling machine produces a die whose cross-sectional height is described by the function $y=\sqrt{16-4x^2}$. Let y represent the height of the piece above the bed of the milling machine and let x represent the horizontal distance from the center line of the piece to the outer extreme. Draw the shape of the die.

SOLUTION The domain of this function is $\{x \mid -2 \leqslant x \leqslant 2\}$, since $16-4x^2 \geqslant 0$ for $-2 \leqslant x \leqslant 2$. The values in the table are computed from the given function and then plotted in Figure 4.21.

$y=\sqrt{16-4x^2}$

x	y
-2	0
-1	$\sqrt{12}$
0	4
1	$\sqrt{12}$
2	0

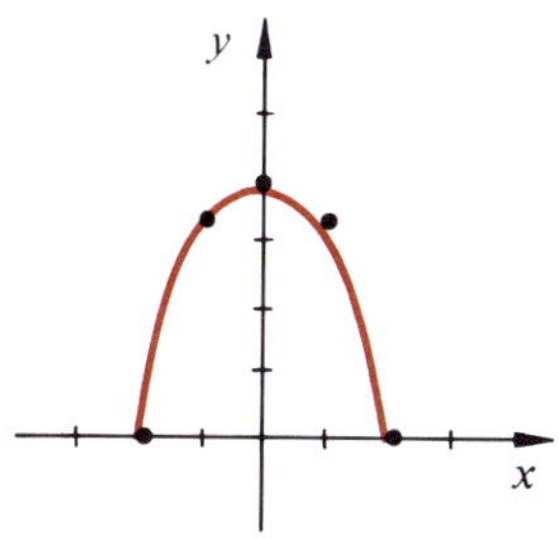

FIGURE 4.21

In many physical situations the nature of the problem restricts the domain of the function. For instance, the motion of a ball that is thrown upward with a velocity of 32 ft/sec is described by the function $h=32t-16t^2$, where t is the elapsed time and h is the vertical height. An analysis of this problem reveals that the ball will strike the ground 2 sec after it is thrown upward. The domain of the indicated function is therefore $\{t \mid 0 \leqslant t \leqslant 2\}$, since the equation has no meaning before the ball is thrown or after it hits the ground.

EXAMPLE 5 Draw the graph of $h=32t-16t^2$ on $0 \leqslant t \leqslant 2$.

SOLUTION The table and graph in Figure 4.22 result from computing some convenient values of h in the interval $0 \leqslant t \leqslant 2$. The graph reveals that the maximum height to which the ball rises is 16 ft.

$h=32t-16t^2$

t	h
0	0
.5	12
1.0	16
1.5	12
2.0	0

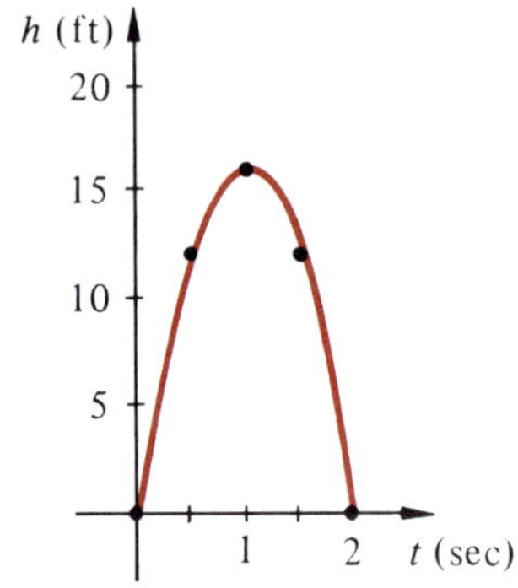

FIGURE 4.22

Example 5 clearly shows how the graph of a function can reveal important features of the function. Indeed, a graph almost always provides the best way of understanding a function.

Using a Graph to Define a Function

A set of ordered pairs of real numbers is represented graphically by points in the Cartesian plane. Likewise, a graph determines a set of ordered pairs of real numbers corresponding to the coordinates of the points on the graph. The set of ordered pairs determined by a graph may or may not define a function. Remember, to define a function the set of ordered pairs must obey the rule that *no two ordered pairs may have the same first element*. Graphically this means that a vertical line will not intersect the graph of a function at more than one point. Thus, the graphs in Figures 4.23(a) and (b) define functions, and the graphs in 4.23(c) and (d) do not.

The Zeros of a Function

The elements of the domain of a function f for which the range element is zero are called the **zeros** of the function. Thus, if $f(C)=0$, where C is a number in the domain of f, then C is called a zero of f. The concept of a zero of a function has a simple graphical analogue: *The zeros of a function correspond to points at which the graph of the function crosses the x-axis*. These points are called ***x*-intercepts**. In Figure 4.24(a), the values x_1, x_2, and x_3 are zeros of the function. Figure 4.24(b) shows a function that has no zeros. Be sure to distinguish carefully between the zeros of a function and the value of

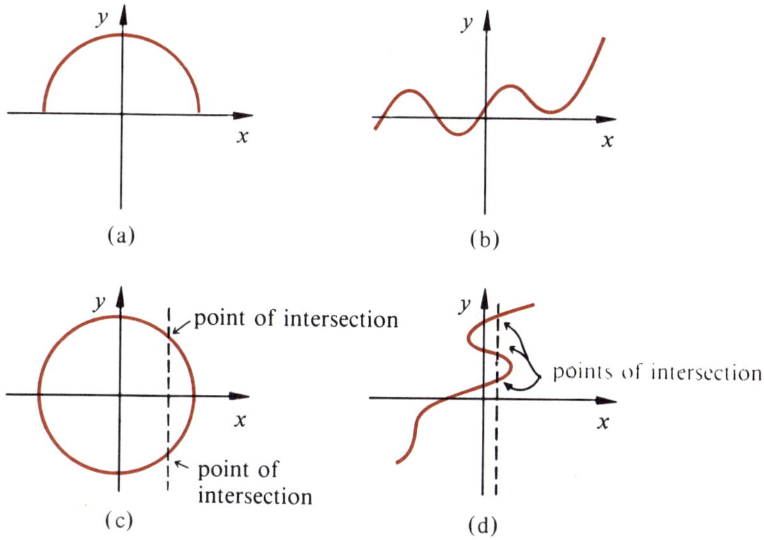

FIGURE 4.23

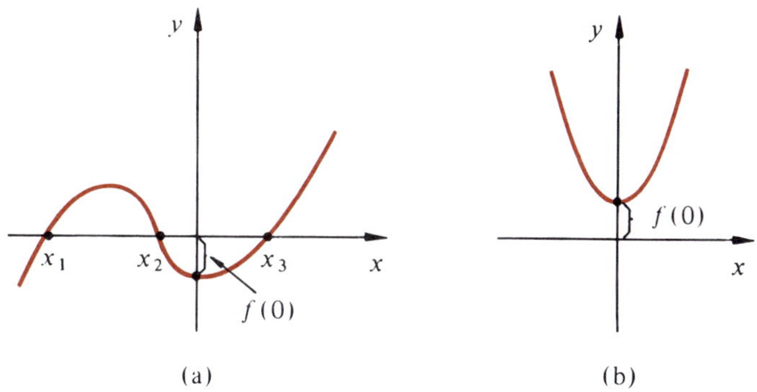

FIGURE 4.24

the function when $x=0$. The x-intercepts give the zeros, and the value of the y-intercept is $f(0)$. The zeros should always be indicated on the graph, at least approximately.

EXAMPLE 6 Draw the graph of $f(x)=x^2-3x+1$ and indicate the zeros.

SOLUTION Computing $f(x)$ for some convenient values of x, we get the graph shown in Figure 4.25. Here the graph crosses the x-axis at approximately $x=0.4$ and $x=2.6$. Therefore, these points are the zeros of the function. By a graphical method, we can only estimate the zeros. Analytically, the zeros are obtained by setting $f(x)=0$. In this case, that results in the quadratic equation

$$x^2-3x+1=0$$

whose roots are found, from the quadratic formula, to be

$$x=\frac{3\pm\sqrt{5}}{2}$$

These two roots of the quadratic equation give the zeros of the function precisely.

$f(x)=x^2-3x+1$

x	$f(x)$
-1	5
0	1
1	-1
2	-1
3	1
4	5

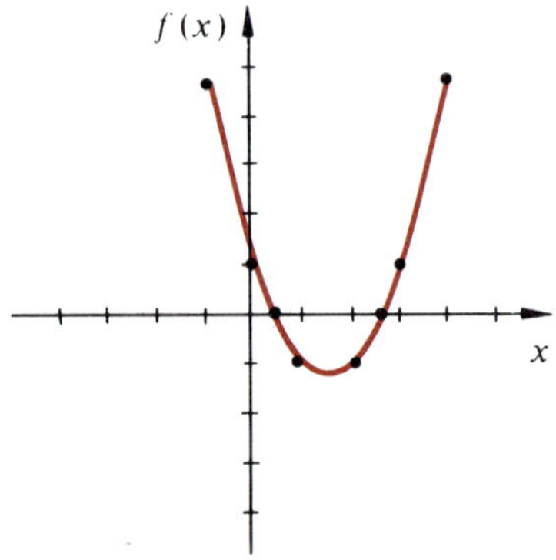

FIGURE 4.25

Exercises for Section 4.5

In Exercises 1–6, indicate whether the given graph determines a function.

1. 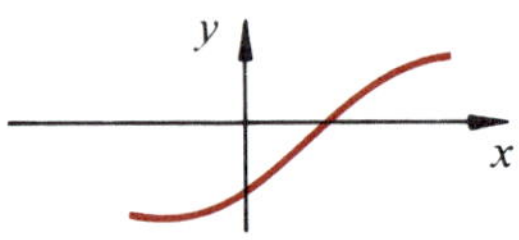

2. 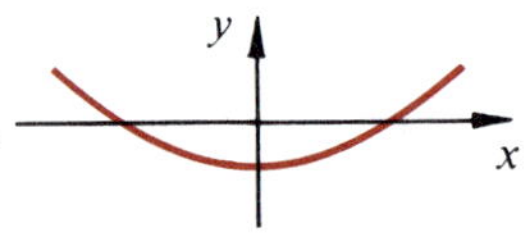

3. 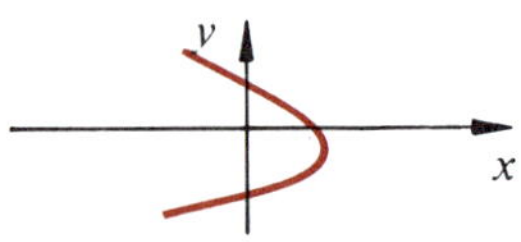

4. 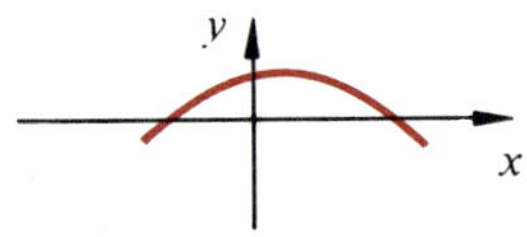

5. 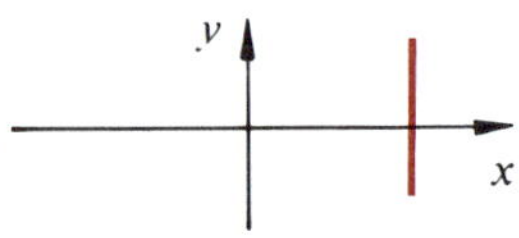

6. 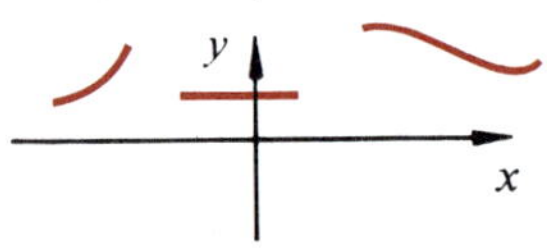

Graph the functions in Exercises 7–28 and indicate the zeros where appropriate. If possible, find the exact value of the zeros.

7. $\{(-2, 3), (0, 4), (2, 5)\}$

8. $\{(-1, 0), (0, 0), (1, 2), (3, -4)\}$

9. $f(x)=3x+5$

10. $y=2-\frac{1}{2}x$

11. $y=x^3$

12. $f(x)=-x^2$

13. $z=t^2+4$

14. $i=r-r^2$

15. $y(x)=\sqrt{x}$

16. $\phi=\dfrac{w^2}{2}$

17. $p=z^2-z-6$

18. $v=10+2t$

19. $y=\sqrt{16-4x^2}$

20. $y=\sqrt{25-x^2}$

21. $\alpha=\theta^{1/3}$

22. $z=\sqrt[3]{t^2}$

23. $y(x)=\sqrt{x-2}$

24. $y=\sqrt[3]{2-x}$

25. $v=s^3-4s^2$

26. $a=\frac{1}{4}b^4$

27. $y=-x^3$

28. $\beta=-\alpha^{1/2}$

29. The work w done in moving an object varies with the distance s according to $w=\sqrt[3]{2s}$. Graph this function.

30. The path of a certain projectile is described by the function $h=100x-2x^2$, where h is the vertical height in feet and x is the horizontal displacement in feet. Draw the path of the projectile.

31. An office machine is supposed to be serviced once a month. If it is not serviced, the cost of repairs is \$20 plus 5 times the square of the number of months the machine goes unserviced. Express the cost of repairs as a function of the number of months the machine goes unserviced and draw the graph.

32. An archway is to be built in the form of a semi-ellipse. Draw the shape of the archway, which is described analytically by the function $y=\sqrt{64-4x^2}$, where y is the vertical height of the arch and x is the horizontal distance from the center line of the arch. Both x and y are measured in feet.

33. Torricelli's law states that the velocity of a stream of liquid issuing from an orifice is given by $v=\sqrt{2gy}$, where g is the acceleration of gravity and y is the depth of the orifice below the surface of the liquid. Show the relation between v and y if $g=32$ ft/sec^2.

34. The following table describes the functional relation between the safe speed at which a car can round a curve and the degree of the curve. Draw the graph of safe speed as a function of the degree of the curve.

Degrees of curve	5	10	15	20	25	30	35
Safe speed (mph)	70	68	66	63	58	50	34

35. A thermocouple generates a voltage when its two ends are kept at different temperatures. In laboratory experiments, the cold end of the thermocouple is usually kept at 0°C, while the temperature of the other end varies. The accompanying table represents the results of an experiment in which the output voltage of the thermocouple was recorded for various thermocouple temperatures. Draw the graph of voltage versus temperature.

T (°C)	0	50	100	150	200	250
V (mV)	0	2.1	4.1	5.9	7.0	7.8

36. The accompanying table shows men's height and corresponding normal weight in pounds. Draw the graph of weight vs height.

Ht	5′2″	5′4″	5′6″	5′8″	5′10″	6′0″	6′2″	6′4″
Wt	130	136	144	152	161	170	184	196

4.6 Inverse Functions and Composition

You have now experienced numerous instances of being given a function $y=f(x)$ and being asked to determine the value of y corresponding to a given x. Sometimes, instead of computing values of the function for various values of the independent variable, we are interested in computing the value of the independent variable corresponding to a given functional value. Suppose, for example, that the velocity of a car (ft/sec) varies with time (sec) according to $v=10+4t$ and we should like to know the amount of time that elapses until the car reaches a velocity of 70 ft/sec. Inspection of the given function reveals that the answer is $t=15$ sec. Underlying this example is an important mathematical concept—the idea of getting a domain element from a given range element.

Under certain conditions, a new function can be obtained from a given function f merely by interchanging the numbers in the functional pairings $(x, f(x))$. A function obtained in this way is called the **inverse** of the given function, and it is denoted by f^{-1}. (The $^{-1}$ notation does not represent a negative exponent; it is simply a symbol for the inverse function.)

Suppose that we are given the function

$$f = \{(2, 1), (-3, 2), (0, 5)\}$$

Then the inverse function is given by

$$f^{-1} = \{(1, 2), (2, -3), (5, 0)\}$$

You should verify that f^{-1} is truly a function by noting that no two first elements are the same.

Not every function has an inverse function. Suppose, for instance, that f is given by the set

$$f = \{(2, 7), (4, 6), (5, 7), (6, 9)\}$$

The set g, obtained by interchanging the first and second elements in each of the ordered pairs of f is

$$g = \{(7, 2), (6, 4), (7, 5), (9, 6)\}$$

The set g does not define a function, since 7 appears as the first element in the ordered pairs (7, 2) and (7, 5).

If some functions have inverse functions and others do not, as the preceding examples show, is it possible to determine this in advance? The answer to this query is, Yes, it is. Specifically, a function has an inverse function if its rule of correspondence does not assign the same range element to any two different domain elements. Functions that have this special property are called **one-to-one functions**, and they can be defined as follows.

DEFINITION

A function f is a one-to-one function if for a and b in the domain of f, the relation $f(a) = f(b)$ implies that $a = b$.

Notice that the function

$$f = \{(2, 7), (4, 6), (5, 7), (6, 9)\}$$

is not a one-to-one function, since $f(2) = 7 = f(5)$, but $2 \neq 5$.

Graphically, a one-to-one function is a function for which *both* horizontal and vertical lines intersect the graph in, at most, one point. In Figure 4.26, the first function is one-to-one; the second is not.

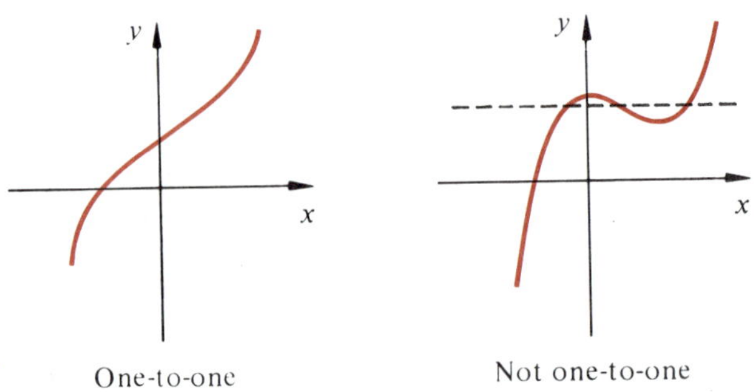

FIGURE 4.26

Suppose the rule of correspondence of a one-to-one function is given by a formula. Then simply by interchanging the x and y variables we get the inverse relation; such an equation will not give y explicitly in terms of x, however. Sometimes finding such an explicit expression for y means merely solving for y after this interchange of variables. If so, this is the inverse function; that is, $y=f^{-1}(x)$.

We summarize the algebraic procedure for finding the inverse function of $y=f(x)$ as follows:

(1) Interchange the x and y variables.

(2) Solve the new equation for the y variable (the resulting expression is $f^{-1}(x)$).

EXAMPLE 1 What is $f^{-1}(x)$ if $f(x)=2x+4$?

SOLUTION Let $y=2x+4$ and then interchange x and y, to get

$$x=2y+4$$

Solving this equation for y, we have

$$2y=x-4$$
$$y=\tfrac{1}{2}x-2$$

Substituting $f^{-1}(x)$ for y, the inverse is

$$f^{-1}(x)=\tfrac{1}{2}x-2$$

EXAMPLE 2 Let $f(x)=(2x+1)/(x+3)$. Find the inverse of f.

SOLUTION Let $y=(2x+1)/(x+3)$. Then interchanging the x and y variables gives

$$x=\frac{2y+1}{y+3}$$

Solving for y, we get

$$xy+3x=2y+1$$
$$xy-2y=1-3x$$
$$y=\frac{1-3x}{x-2}$$

Hence, the inverse of f is

$$f^{-1}(x)=\frac{1-3x}{x-2}$$

The domain of f^{-1} is the set of all real numbers except $x=2$. Further, the domain of f is the range of f^{-1}, and the range of f is the domain of f^{-1}. Thus, the function and its inverse are found by interchanging the domain and range and keeping the same correspondence.

Some calculators have an inv *button, but the operation of this button applies to only a few selected preprogrammed functions and will not perform the inverse operation in general.*

The graphs of f and f^{-1} have an interesting relation. As Examples 3 and 4 show, the graphs of f and f^{-1} are mirror reflections in the line that bisects the first and third quadrants.

EXAMPLE 3 Draw the graphs of $f=\{(2, 1), (-3, 2), (0, 5)\}$ and $f^{-1}=\{(1, 2), (2, -3), (5, 0)\}$ on the same coordinate axes.

SOLUTION By our definition of an inverse function, clearly the second set of ordered pairs is the inverse of the first. The graphs of f and f^{-1} are shown in Figure 4.27. The points in the graphs are mirror reflections in the solid line.

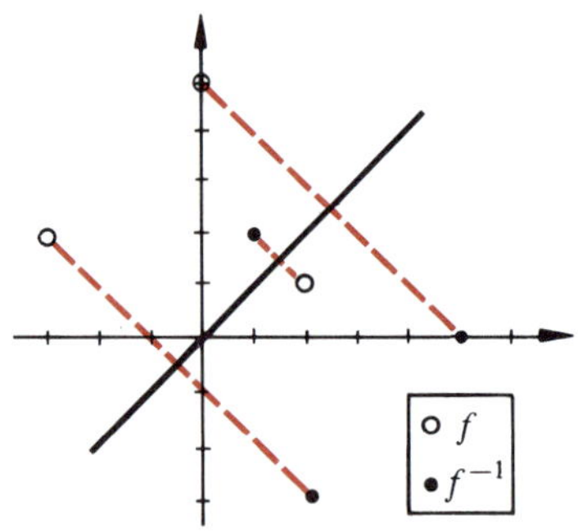

FIGURE 4.27

EXAMPLE 4 Draw the graphs of $y=2x+4$ and $y=\frac{1}{2}x-2$ on the same coordinate axes.

SOLUTION Several solution pairs for each equation are given in the respective tables and the graphs plotted in Figure 4.28. Notice that each graph is the mirrored reflection of the other relative to the dashed line. The inverse nature of these functions was established in Example 1.

$y=2x+4$

x	y
-6	-8
-4	-4
-2	0
0	4
2	8

$y=\frac{1}{2}x-2$

x	y
-8	-6
-4	-4
0	-2
4	0
8	2

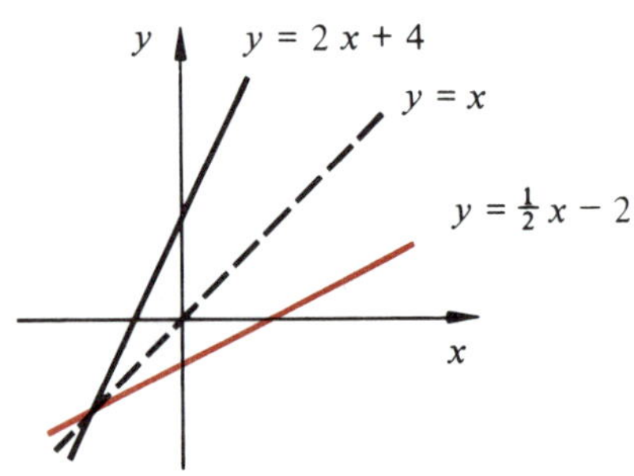

FIGURE 4.28

Composite Functions

An inverse function can be characterized in another way by the concept of composition of functions. A composite function is formed by the successive actions of other functions, according to the following definition.

DEFINITION

If f and g are two functions for which the range of g is contained in the domain of f, the **composite function of f with g** is denoted by $f\,g$ and defined by

$$f \circ g(x) = f(g(x))$$

for every x in the domain of f. The notation $f(g(x))$ is read "f of g of x."

Note the important assumption that $g(x)$ is in the domain of f; otherwise, $f(g(x))$ does not make sense. The process of composition is represented in diagrammatic form in Figure 4.29. The diagram shows us that to get the composition $f(g(x))$, we first apply the rule for $g(x)$ and then the rule for $f(x)$.

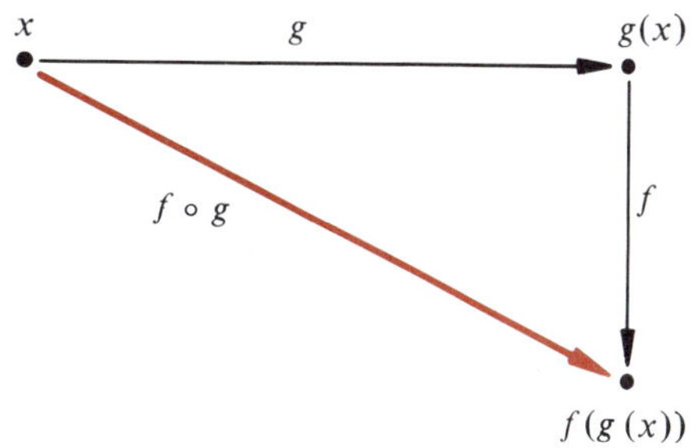

FIGURE 4.29

EXAMPLE 5 What is the composition $f(g(x))$ and $g(f(x))$ if $f(x)=3x+1$ and $g(x)=x-4$?

SOLUTION $g(f(x))=g(3x+1)=(3x+1)-4=3x-3$

$f(g(x))=f(x-4)=3(x-4)+1=3x-11$

Notice that $f(g(x))\neq g(f(x))$ here. This is typical of the composition of functions; so we say that composition does not have the commutative property.

EXAMPLE 6 Let $f(x)=x+1$ and $g(x)=\sqrt{x}$. What is the domain of the composite $g(f(x))$?

SOLUTION The composite function is $g(f(x))=g(x+1)=\sqrt{x+1}$. Hence, the domain of the composite is $x\geqslant -1$, since these values yield real numbers for the range.

The formation of the composition of two functions is straightforward. However, one of the most important applications of compositions in calculus is recognizing a complicated functional expression as a composite of more elementary functions.

EXAMPLE 7 (a) The function $(x+1)^2$ can be regarded as the composition $f(g(x))$, where $f(x)=x^2$ and $g(x)=x+1$.

(b) The function $(x^5+1)^{1/3}$ can be regarded as $f(g(x))$, where $f(x)=x^{1/3}$ and $g(x)=x^5+1$.

The composition of two functions can be used to give a test for determining when two functions are inverses of one another.

> Suppose f is a one-to-one function and f^{-1} its inverse.
>
> - For all x in the domain of f, $f^{-1}(f(x))=x$.
> - For all x in the domain of f^{-1}, $f(f^{-1}(x))=x$.

This fact can be used to show that two functions are respective inverses; it cannot be used, however, to find the inverse of a given function. By this composition property, the function of Example 3, $f=\{2, 1), (-3, 2), (0, 5)\}$, can be shown to have the inverse $\{(1, 2), (2, -3), (5, 0)\}$. The verification is:

$$f^{-1}(f(2))=f^{-1}(1)=2$$
$$f^{-1}(f(-3))=f^{-1}(2)=-3$$
$$f^{-1}(f(0))=f^{-1}(5)=0$$

with similar verifications for $f(f^{-1}(x))$:

$$f(f^{-1}(1))=f(2)=1$$
$$f(f^{-1}(2))=f(-3)=2$$
$$f(f^{-1}(5))=f(0)=5$$

EXAMPLE 8 Show that $f(x)=2x+1$ and $g(x)=(x-1)/2$ are inverse functions.

SOLUTION

$$f(g(x))=f\left(\frac{x-1}{2}\right)=2\left(\frac{x-1}{2}\right)+1=x$$

$$g(f(x))=g(2x+1)=\frac{(2x+1)-1}{2}=x$$

Therefore, f and g are inverses.

Exercises for Section 4.6

1. If $f(x)=2x+3$, for which x is $f(x)=6$?

2. If $f(x)=4/x$, for which x is $f(x)=3$?

3. If $v=10+4t$, when is $v=20$?

4. If $f=\{(3, 0), (2, 6), (1, 5)\}$, for which x is $f(x)=5$? For which x is $f(x)=0$?

5. If $f=\{(0, 1), (2, 5), (3, 8), (5, 6)\}$, for which x is $f(x)=8$? For which x is $f(x)=3$?

Determine the inverse function in Exercises 6–13 *if such exists.*

6. $\{(2, 6), (3, 5), (0, 4)\}$ **7.** $\{(3, 7), (5, 9), (7, 3), (9, 5)\}$

8. $\{(-1, 2), (2, 3), (6, -2)\}$

9. $\{(1, 2), (2, 2)\}$ **10.** $\{(3, 2), (5, 4), (7, 2), (9, 8)\}$

11. $\{(-2, 3), (-1, 4), (0, 0)\}$

12. $\{(0, 1), (1, 0)\}$ **13.** $\{(0, 3), (1, 5), (2, 3), (6, 7)\}$

Determine graphically which of the pairs of functions in Exercises 14–21 are inverses of one another.

14. $y=3x,\ y=\frac{1}{3}x$ **15.** $y=-2x,\ y=-\frac{1}{2}x$

16. $y=6-3x,\ y=-\frac{1}{3}x+2$

17. $y=x+1,\ y=x-2$ **18.** $y=\frac{1}{2}x+1,\ y=2x-1$

19. $y=5(x+1),\ y=\frac{1}{5}x-1$

20. $y=x^2,\ y=\sqrt{x}$ **21.** $y=x^3,\ y=\sqrt[3]{x}$

22. Suppose that f and f^{-1} are defined by identical formulas. What can you say about the graph of f?

23. Which of the functions whose graphs are shown in the accompanying figures have inverses? Sketch the graph of the inverse for those that do.

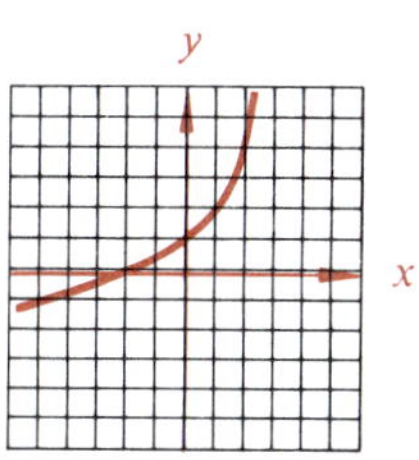

(a)

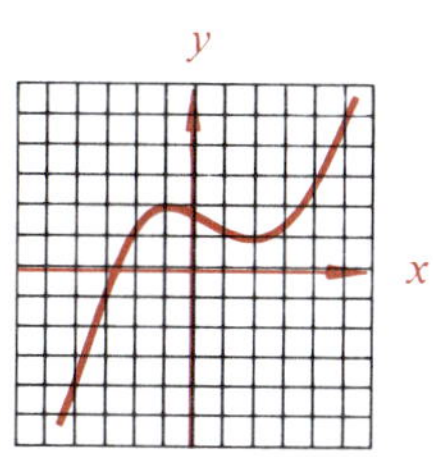

(b)

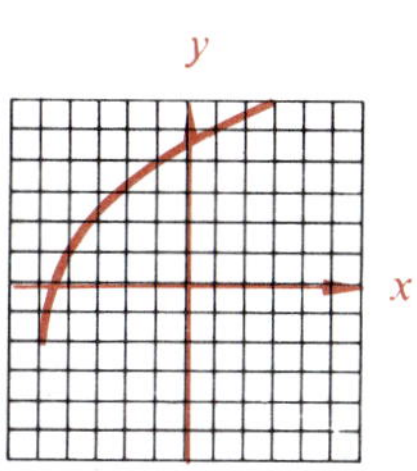

(c)

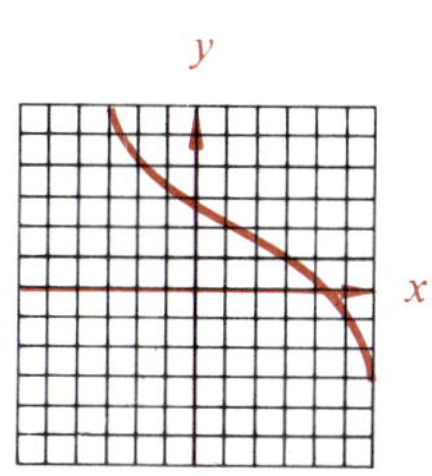

(d)

In the statements in Exercises 24–30, find $f(g(x))$ and $g(f(x))$.

24. $f(x)=2x+3,\ g(x)=5x-1$

25. $f(x)=3x+4,\ g(x)=2x$

26. $f(x)=\dfrac{1}{x+1},\ g(x)=x+1$

27. $f(x)=x,\ g(x)=x^2$

28. $f(x)=\sqrt{x},\ g(x)=x^2$ **29.** $f(x)=x^3-1,\ g(x)=x^{1/3}$

30. $f(x)=1/(x^2+1),\ g(x)=1/x$

In each of the statements in Exercises 31–35, use composition to show that f and g are inverses.

31. $f(x)=-2x,\ g(x)=-x/2$

32. $f(x)=3x,\ g(x)=x/3$

33. $f(x)=6-3x,\ g(x)=-\frac{1}{3}x+2$

34. $f(x)=\dfrac{1}{2x+1},\ g(x)=\dfrac{1-x}{2x}$

35. $f(x)=5(x+1),\ g(x)=\frac{1}{5}x-1$

In each of the statements in Exercises 36–41, find the inverse function of f.

36. $f(x)=3x+2$ **37.** $f(x)=x-3$

38. $f(x)=\frac{1}{2}x+5$ **39.** $f(x)=\dfrac{1}{x+1}$

40. $f(x)=\dfrac{2x-1}{3x+5}$ **41.** $f(x)=\dfrac{x-1}{x+1}$

Test 1 for Chapter 4

Answer *True* or *False* to Exercises 1–10.

1. If $a>0$ and $b<0$, then the point (a, b) is in quadrant IV.
2. The distance from a to b is the same as the distance from b to a.
3. For any function f, $f(x+y)=f(x)+f(y)$.
4. $\{(2, 1)\}$ is a function.
5. $\{(x, y)|x^2+y^2=1\}$ represents a function.
6. $\{(x, y)|x+y=1\}$ represents a function.
7. For a formula to represent a function, there must be a real value of $f(x)$ for every real x.
8. Only one-to-one functions have inverse functions.
9. If y varies directly with x and x varies directly with t, then y varies directly with t.
10. The range of a real valued function is a subset of the real numbers.
11. The amount of money a man earns is directly proportional to the number of hours he works. If he makes \$18.80 in an 8-hour day, what formula could describe the amount of money he makes as a function of the hours he works?
12. Let $f(x)=x^2+5x+2$. Find and simplify $[f(2+h)-f(2)]/h$.
13. Sketch the function $f(x)=x^2-3x$ on the interval $-1 \leqslant x \leqslant 4$.
14. Find the domain and range of each of the following real valued functions:

    (a) $f(x)=|x|$ (b) $f(x)=1/x$ (c) $f(x)=\sqrt{x}$
15. Find the distance from $(2, \sqrt{2})$ to $(-2, 0)$.
16. Graph the set of points $\{(x, y)\,|\,x=1\}$. Is this a function?

Test 2 for Chapter 4

1. If s varies directly with t and if $s=2$ when $t=5$, what is s when $t=7$?
2. Find the distance between the points $(-2, -2)$ and $(-5, 3)$.
3. Graph $\{(2, -1), (3, 0), (3, -1), (2, 0)\}$.
4. Which of the following define functions?

    (a) $\{(2, 1), (3, 5), (-2, 0)\}$ (b) $\{(-1, 3), (0, 3)\}$

    (c) $y<3x+2$ (d) $y=3-x^2$
5. Find the domain and range for the following functions

    (a) $\{(-1, 0), (0, 5), (5, 3)\}$

    (b) $y=5x-2$

    (c) $y=\sqrt{x-3}$
6. Given $f(t)=3t^2-4t+6$, find $f(-2)$ and $f(x^2)$.
7. Draw the graph of $y=\sqrt{x-5}$.
8. Draw the graph of $y=6+x-x^2$. What are the zeros?
9. Determine the inverse function for

    (a) $\{(-1, 2), (7, 6), (12, -3)\}$ (b) $y=5x+3$
10. Given $f(x)=4/x^2$ and $g(x)=3x$, find $f(g(x))$ and $g(f(x))$.

5

Elementary Functions

5.1 Linear Functions

Several elementary functions are extremely important in applied work, and you should know as much as you can about them. To "know a function" means to be able to determine:

(1) The domain of the function.

(2) The range of the function.

(3) The rule of correspondence (usually given by some formula.)

(4) The graph of the function.

We spend much time studying graphs of functions, since they provide the easiest description of how a function behaves.

By a **linear function** we mean a function expressible in the form

(5.1a) $f(x) = ax + b$

where a and b are constants. If we replace $f(x)$ with y, the linear function can be written

(5.1b) $y = ax + b$

In this form, it is called a **linear equation in the two variables x and y**. More generally, any equation in the variables x and y that can be arranged in the form $y = ax + b$ is a linear equation. Thus,

(5.2) $ax + by = c$

is a linear equation, since if $b \neq 0$, the equation can be written $y = (-a/b)x + (c/b)$. If

$b=0$, we get $ax=c$, which is not a linear function but is still considered a linear equation. Equation (5.2) is called the *general form* of a linear equation.

The phrases *linear function* and *linear equation in two variables* often mean the same thing. In fact, the words *function* and *equation* are used interchangeably whenever an equation can be solved for y in terms of x. We shall use *linear function* if we are stressing the functional concept of pairing x and y; otherwise, we shall use *linear equation*.

EXAMPLE 1 (a) The equations

$$y=x, \qquad y=2x-5, \qquad 3x+2y=7, \qquad y=0, \qquad x=3$$

are examples of linear equations. Notice that all but $x=3$ are also linear functions.

(b) Equations that are not linear are said to be nonlinear. Examples of nonlinear equations are

$$y=\sqrt{x}, \qquad y=x^2-3x, \qquad y=\frac{1}{x}$$

Any pair of numbers x and y for which the equation $ax+by=c$ is satisfied is called a **solution** to the equation. For example, $x=3$ and $y=-2$ represent a solution to $4x+y=10$, since $4(3)+(-2)=10$. Also, $x=0$, $y=10$, and $x=1$, $y=6$, are solutions. (Check each pair for yourself.) An infinite number of such pairs satisfy the equation.

EXAMPLE 2 List some of the solutions to the linear equation $2x+y=2$.

SOLUTION The usual way of finding solutions is to assign values to x and find the corresponding values of y. Thus, when $x=0$, then $y=2$; when $x=1$, then $y=0$; and so on. Some of the solutions are shown in the table. You see that we list only a few of the solutions. To list them all is not possible, since the solution set consists of infinitely many ordered pairs.

x	0	1	-1	2	-2
y	2	0	4	-2	6

A remark here about the domain and range of $y=ax+b$. If a is not zero, then the domain and range are both the set of all real numbers. If a equals zero, then $y=b$. In this case, the domain is still the set of all real numbers (that is, the independent variable is not restrained in any way), whereas the range is the one-element set $y=b$.

A set of ordered pairs is represented graphically by points in the Cartesian plane. Therefore, to graph a linear function or a linear equation, you must get some of the solutions, plot the corresponding points, and then, since the domain consists of all real numbers, connect these points with a smooth curve.

EXAMPLE 3 Graph the linear equation $2x+y=2$. (This is the equation of the previous example.)

SOLUTION See Figure 5.1.

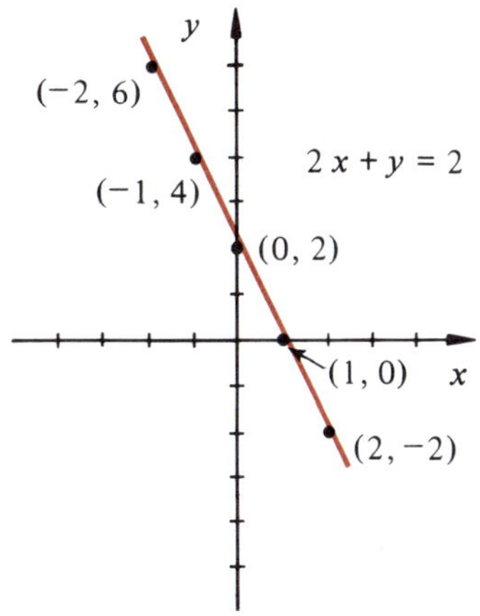

FIGURE 5.1

Figure 5.1 suggests that the graph of the linear equation $2x+y=2$ (*and* of the linear function $y=2-2x$) is a straight line. This is indeed true, and more generally, the **graph of every linear equation is a straight line**. We shall not prove this. Suffice it to say that the terms *linear function* and *linear equation* are chosen to suggest the geometric nature of their graphs.

Since the graph of every linear equation is a straight line, the work of graphing one can be considerably shortened. Recall that a straight line is determined by two distinct points. Thus, any two ordered pairs (x_1, y_1) and (x_2, y_2) that satisfy a given linear equation can be used to determine its graph. Although any two solutions can be used, we prefer the most convenient. The two solutions usually chosen are the ones that correspond to the **intercepts** of the line with the x- and y-axes; these are the ***x*-intercept** and the ***y*-intercept.**

- To find the x-intercept of a graph, let $y=0$ in the equation and solve for x.*
- To find the y-intercept of a graph, let $x=0$ in the equation and solve for y.

EXAMPLE 4 Graph the equation $4x-2y=8$.

SOLUTION Letting $x=0$ gives $y=-4$, and letting $y=0$ gives $x=2$. Thus the intercepts are $(0, -4)$ on the y-axis and $(2, 0)$ on the x-axis. The graph of this equation is shown in Figure 5.2.

Suppose you were asked to describe in words the graph in Figure 5.2. Obviously, you would say that the graph is a straight line that crosses the x-axis at $x=2$ and the y-axis at $y=-4$. You might also say that the line rises from left to right. Another bit of information that would be useful is the steepness of the rise. Thus, you

*Recall from Section 4.5 that an x-intercept is the geometric counterpart of a zero of a function.

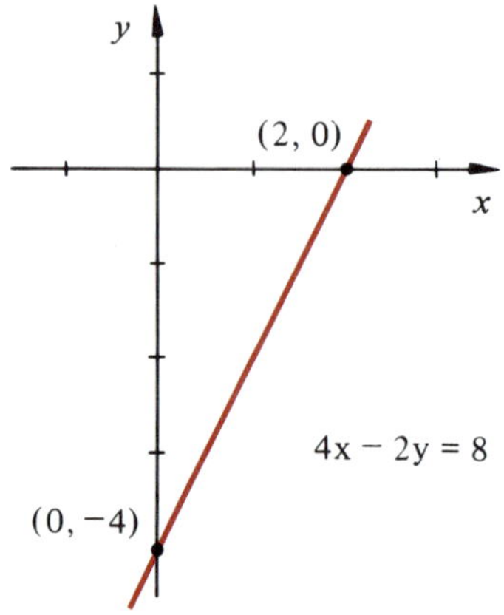

FIGURE 5.2

might observe that the line rises two units for every one unit moved to the right. Implicit in this description of steepness is an idea called the slope of the line.

The slope of a straight line is defined as the ratio of the vertical rise of the line to the corresponding horizontal run; that is,

$$\text{Slope} = \frac{\text{vertical rise}}{\text{horizontal run}}$$

If this definition is applied to the line segment $\overline{P_1P_2}$ in Figure 5.3, the slope m is given by the equation

$$m = \frac{y_2 - y_1}{x_2 - x_1} \qquad (5.3)$$

where $y_2 - y_1$ is the vertical distance between the given points and $x_2 - x_1$ the horizontal distance between the points.

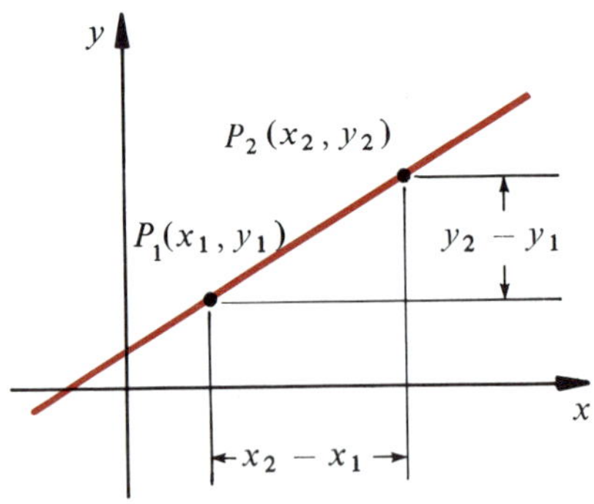

FIGURE 5.3

EXAMPLE 5 Find the slope of the straight line passing through the points $(-5, 1)$ and $(2, -3)$. (See Figure 5.4.)

SOLUTION Letting $(x_1, y_1) = (-5, 1)$ and $(x_2, y_2) = (2, -3)$ and using Equation (5.3), we can find

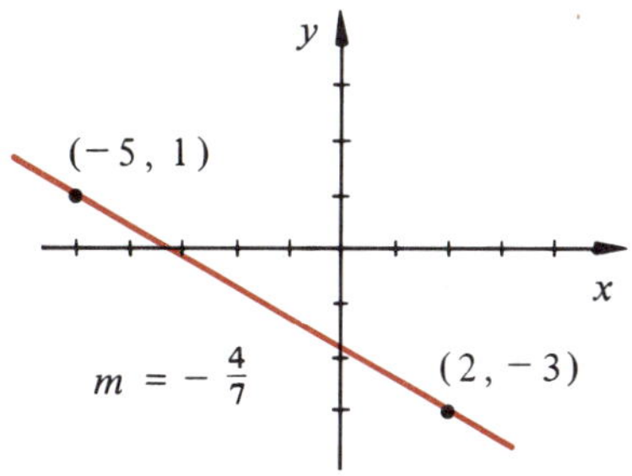

FIGURE 5.4

the desired slope:

$$m=\frac{y_2-y_1}{x_2-x_1}=\frac{-3-1}{2-(-5)}=\frac{-4}{7}=-\frac{4}{7}$$

If we interchange the labels of the given points and let $(x_1, y_1)=(2, -3)$ and $(x_2, y_2)=(-5, 1)$, the result is

$$m=\frac{y_2-y_1}{x_2-x_1}=\frac{1-(-3)}{-5-2}=\frac{4}{-7}=-\frac{4}{7}$$

Hence, the order in labeling the given points is immaterial. We interpret a slope of $-4/7$ to mean that for every 7 units moved to the right in the direction of the x-axis, the straight line moves down 4 units. Using similar triangles, you can show that the value of the slope is independent of which two distinct points on the line are used to compute m.

Example 6 shows how we can sketch the graph of a line if we know *one* point and the slope of the line.

EXAMPLE 6 Draw the straight line passing through the point (2, 1) with a slope $m=2$.

SOLUTION The slope $m=2$ can be written as $m=2/1$. To lay off the indicated slope, start at the point (2, 1) and move 1 unit to the right and 2 units upward. This brings us to the point (3, 3). The desired straight line is then the line that passes through the points (2, 1) and (3, 3). (See Figure 5.5.)

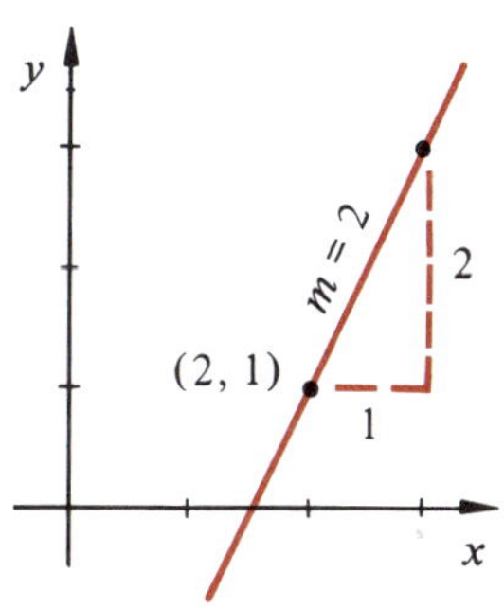

FIGURE 5.5

The following facts about slopes are useful in working with straight lines. These properties can be verified by referring to the definition of slope.

- The slope of a straight line parallel to the x-axis is zero.
- The slope of a straight line parallel to the y-axis is undefined.
- Parallel straight lines have equal slopes.
- Perpendicular straight lines have slopes that are negative reciprocals; that is, $m_1 = -1/m_2$. (It is easy to verify this property using similar triangles).
- A positive slope means the line is rising from left to right and a negative slope means the line is falling from left to right.

EXAMPLE 7 Find the slope of a line drawn perpendicular to a line passing through (2, 1) and (−4, 6).

SOLUTION The slope m of the given line is

$$m = \frac{y_2 - y_1}{x_2 - x_1} = \frac{6-1}{-4-2} = -\frac{5}{6}$$

The slope m' of a line perpendicular to the given line is then

$$m' = -\frac{1}{m} = -\frac{1}{-\frac{5}{6}} = \frac{6}{5}$$

EXAMPLE 8 Draw the graph of the equation $y = 3$. What is its slope?

SOLUTION The graph is a straight line parallel to the x axis and passing through (0, 3). The slope is zero. (See Figure 5.6.)

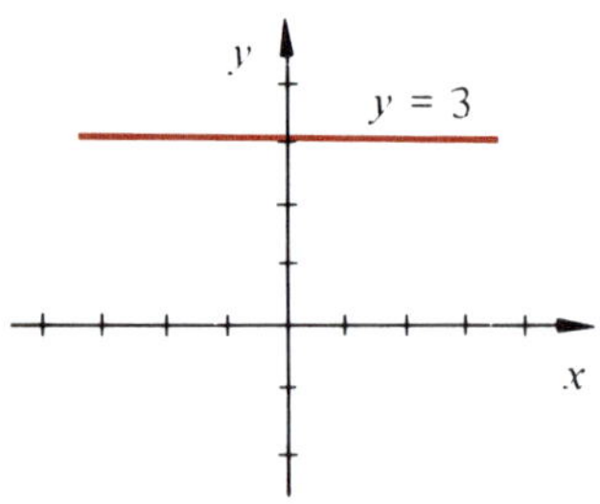

FIGURE 5.6

The concept of slope is defined as the ratio of the difference of two quantities. Such ratios exist in many areas of mathematics and have other interpretations besides the slope of a line. Since the slope expresses the ratio of a change in y to the corresponding change in x, we also refer to this ratio as the *rate of change* of y with respect to x.

EXAMPLE 9 The average velocity of an object is defined as the distance traveled divided by the corresponding elapsed time. If a car covers 100 miles in 2.5 hr, what is the average velocity of the car?

SOLUTION Since the distance traveled is 100 miles and the elapsed time is 2.5 hr, the average velocity of the car is

$$v_{\text{av}} = \frac{100}{2.5} = 40 \text{ mph}$$

Exercises for Section 5.1

Draw the graph of the linear equations in Exercises 1-12.

1. $x+y=1$ **2.** $x-y=1$ **3.** $x=5$

4. $y=-2$ **5.** $2x-y=5$ **6.** $3x+2y=5$

7. $4x-y+1=0$ **8.** $x=7-y$ **9.** $3y-4x=4$

10. $y=\frac{1}{2}x-3$ **11.** $y=\dfrac{x-2}{3}$ **12.** $15(x+y)=10$

Sketch the straight line through the pairs of points in Exercises 13-20 and compute the slope.

13. $(1, 2), (5, 4)$ **14.** $(-5, 2), (3, -7)$

15. $(-1, -1), (3, -6)$ **16.** $(7, 3), (0, 5)$

17. $(-2, -3), (-5, -7)$ **18.** $(3, -2), (7, 6)$

19. $(-2, 3), (5, 3)$ **20.** $(\frac{1}{2}, \frac{1}{2}), (\frac{1}{2}, -\frac{2}{5})$

21. Draw the straight line passing through $(2, 1)$ with a slope of $\frac{1}{3}$.

22. Draw the straight line passing through $(-3, 4)$ with a slope of 3.

23. Draw the straight line passing through $(2, -5)$ with a slope of $-\frac{2}{5}$.

24. Draw the straight line passing through $(-1, -2)$ with a slope of -2.

25. Find the slope of a 25-ft ladder that is leaning against a building and has its foot 12 ft from the building.

26. Find the slope of a line drawn perpendicular to the line through $(1, 2)$ and $(-2, -5)$.

27. Find the slope of a line drawn perpendicular to the line through $(-3, 0)$ and $(5, -4)$.

28. The acceleration of a particle is defined as the change in velocity divided by the corresponding change in time. If the initial velocity is 25 ft/sec and the velocity is 40 ft/sec at the end of 10 sec, what is the acceleration? What units does acceleration have?

29. The equation $S=25{,}000N+200{,}000$ gives the sales of a company for its first five years of operation; here N is the number of years since the company started and S is sales in dollars. Plot the sales curve of the company for the first five years.

30. A surveyor finds that the elevation h of a roadway varies with the distance x from a reference point according to $h=0.3x+1.5$. Draw the elevation of the roadway as a function of x.

31. Using the definition of the slope of a line and properties of similar triangles, show that the product of the slopes of perpendicular lines is -1.

.32. Show that the composition of a linear function with a linear function is linear.

5.2 Methods of Describing a Line

We have just examined some basic characteristics of straight lines in the Cartesian plane. The following example shows that a straight line can be represented by a linear equation; we know that, conversely, the graph of every linear equation is a straight line (Section 5.1).

EXAMPLE 1 Show that the function describing the straight line passing through (2, 1) with a slope of $\frac{1}{3}$ is linear. (See Figure 5.7.)

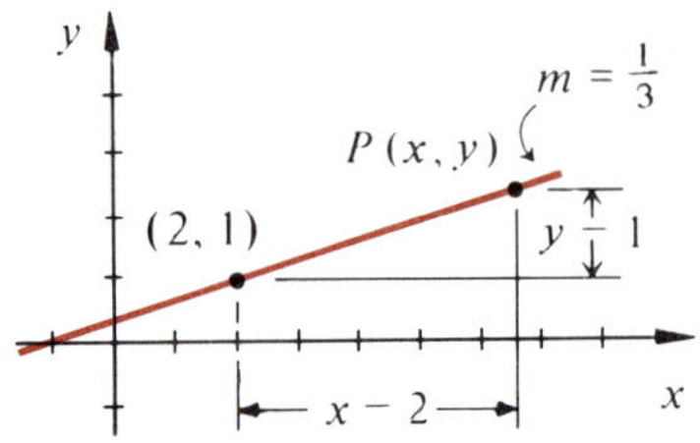

FIGURE 5.7

SOLUTION Let $P(x, y)$ represent any point on the straight line except (2, 1). Note that $P(x, y)$ is a general point and can be placed anywhere on the straight line. Then, by definition, the slope of the given line can be written $(y-1)/(x-2)$. This quantity must equal the given slope, so that

$$\frac{y-1}{x-2}=\frac{1}{3}$$

On simplification,

$$3y-3=x-2$$
$$3y-x=1$$
$$y=\frac{1}{3}x+\frac{1}{3}$$

which describes a linear equation. This equation expresses the relation that exists between the abscissa and the ordinate of *any* point on the given straight line.

Whereas every straight line can be represented by a linear equation, sometimes certain *forms* of representation are important. Two fundamental methods exist for describing a line with a linear equation: the point-slope form and the slope-intercept form. Each form shows two properties of a line that can be determined by inspection.

The **point-slope form** is a generalization of Example 1. With this form, assume that we are given one point $P(x_1, y_1)$ on the line and that the line has a slope m, as shown in Figure 5.8. Choose another arbitrary point $P(x, y)$ on this straight line. Then, by the same reasoning as in Example 1,

$$\frac{y-y_1}{x-x_1}=m$$

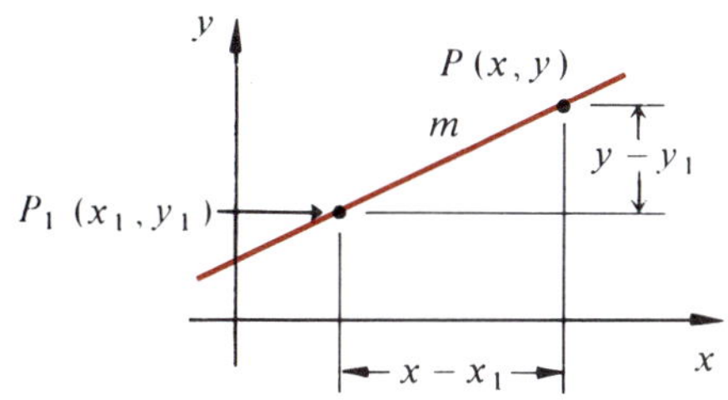

FIGURE 5.8

so that

(5.4) $\boldsymbol{y-y_1=m(x-x_1)}$

This is the point-slope form. Note that since *any* fixed point (x_1, y_1) can be used, the equation will *appear* different for various choices of the point, although each equation will be equivalent.

EXAMPLE 2 The point-slope form of the equation representing the line that passes through (2, 1) with slope $\frac{1}{3}$ is

$$y-1=\frac{1}{3}(x-2)$$

The two points $(0, \frac{1}{3})$ and $(-1, 0)$, however, are also on the line, so that two other point-slope forms of the same line are

$$y-\frac{1}{3}=\frac{1}{3}(x-0) \quad \text{and} \quad y=\frac{1}{3}(x+1)$$

Infinitely many other point-slope forms are possible.

If the point used in the point-slope form is the y-intercept $(0, b)$, a special case, called the **slope-intercept form**, arises. By substituting $x_1=0$ and $y_1=b$ into the point-slope form, we get

$$\frac{y-b}{x}=m$$

from which

(5.5) $\boldsymbol{y=mx+b}$

When an equation is in slope-intercept form, the constant on the right-hand side is the y-intercept and the coefficient of x is the slope. For any linear equation, a few simple manipulations can represent the line in any desired form.

EXAMPLE 3 Draw the graph of the linear equation $3x+2y=-5$ and rearrange the representation into slope-intercept form.

SOLUTION To rearrange into slope intercept form, solve for y.

$$2y = -3x - 5$$

$$y = \underbrace{-\frac{3}{2}}_{m}x + \underbrace{\left(-\frac{5}{2}\right)}_{b}$$

from which we recognize the slope as $-\frac{3}{2}$ and the y-intercept as $(0, -\frac{5}{2})$. (See Figure 5.9.)

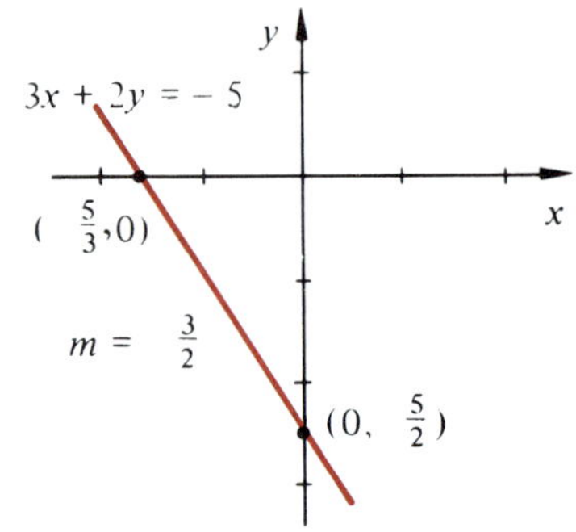

FIGURE 5.9

EXAMPLE 4 Write the equation of the line perpendicular to $x - 3y = 7$ at the point $(4, -1)$.

SOLUTION First, we rewrite the given equation in slope-intercept form, that is,

$$y = \frac{1}{3}x - \frac{7}{3}$$

From this form of the given equation, we see that the slope of its graph is $m = \frac{1}{3}$. Denoting the slope of the perpendicular line by m', we have

$$m' = -\frac{1}{m} = -\frac{1}{1/3} = -3$$

Finally, from the point-slope form of a straight line, the equation of the perpendicular line at $(4, -1)$ is

$$y - (-1) = -3(x - 4)$$
$$y + 1 = -3x + 12$$
$$y = -3x + 11$$

The general relation between equations of lines that are perpendicular is given in Exercise 33 of the exercise set for this section. Before looking at Exercise 33, see whether you can find any relation between the given equation and the equation of the perpendicular line.

Exercises for Section 5.2

For Exercises 1–4, draw the line passing through the given point with the given slope and then determine its equation.

1. $(2, 5), m=\frac{1}{2}$ **2.** $(-1, -3), m=3$

3. $(5, -2), m=-7$ **4.** $(3, 4), m=-\frac{2}{5}$

For each of Exercises 5–10, draw the line through the given points and then find its equation.

5. $(1, 3), (6, 2)$ **6.** $(2, 5), (-3, -7)$

7. $(-1, -1), (1, 2)$ **8.** $(0, 0), (3, -2)$

9. $(0, 2), (-5, 0)$ **10.** $(\frac{1}{2}, \frac{1}{3}), (-\frac{1}{2}, \frac{1}{3})$

Find the slope and the y-intercept of the equations in Exercises 11–22 and then draw the line.

11. $2x-3y=5$ **12.** $3x+4y=0$

13. $x+y=2$ **14.** $5x+2y=-3$

15. $4y-2x+8=0$ **16.** $-5x-y-2=0$

17. $2y=x+5$ **18.** $3x+6=2y$

19. $y=5$ **20.** $x+y=1$

21. $x-5y+7=0$ **22.** $y=1-x$

In Exercises 23–32, write the equation of the line perpendicular to the given line at the indicated point.

23. $3x+2y=7$ at $(1, 2)$ **24.** $x+3y=11$ at $(2, 3)$

25. $x-y=2$ at $(5, 3)$ **26.** $2x-5y=2$ at $(-4, -2)$

27. $2y-3x=1$ at $(1, 2)$ **28.** $y=x$ at $(0, 0)$

29. $5x-7y=0$ at $(0, 0)$ **30.** $5x-7y=3$ at $(2, 1)$

31. $-x+3y=5$ at $(-2, 1)$

32. $-2y-x=3$ at $(-1, -1)$

33. Show that the equation of the line perpendicular to $Ax+By=C$ at (x_1, y_1) is $Bx-Ay=Bx_1-Ay_1$.

5.3 Mathematical Models

Representing the relation between two or more physical variables by a mathematical equation is called *mathematical modeling*. If the relation is represented by a linear equation, the model is called a **linear model**. Three examples of linear models follow.

EXAMPLE 1 The amount that a spring stretches is directly proportional to the applied force. What equation would relate the length d of a spring to applied force f if the length of the spring is 5 in. when a 6-lb weight is applied and 7 in. when a 14-lb weight is applied?

SOLUTION Since spring length is directly proportional to applied force, the desired equation is linear. Figure 5.10 shows this graphically. The slope of the straight line between these two points is then given by

$$m=\frac{d_2-d_1}{f_2-f_1}=\frac{7-5}{14-6}=\frac{1}{4}$$

The point-slope form of this straight line can be written in the form

$$d-d_1=m(f-f_1)$$

Now letting $(f_1, d_1)=(6, 5)$ and $m=\frac{1}{4}$, we get

$$d-5=\frac{1}{4}(f-6)$$

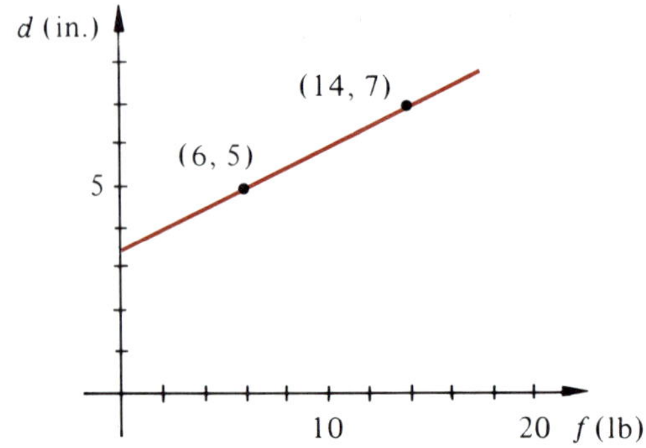

FIGURE 5.10

or

$$d = \frac{1}{4} f + \frac{7}{2}$$

In physics, the slope of the graph is called the **spring constant**. Also, the **free length** of the spring equals the vertical intercept of the graph. In the language of the physicist, the spring in this problem has a spring constant of $\frac{1}{4}$ and a free length of 3.5 in.

In economics, the total cost C in dollars of producing x units of a product is represented by a linear cost-unit model of the form

$$C = mx + b$$

where m and b are constants to be described. A typical graph of this model is shown in Figure 5.11.

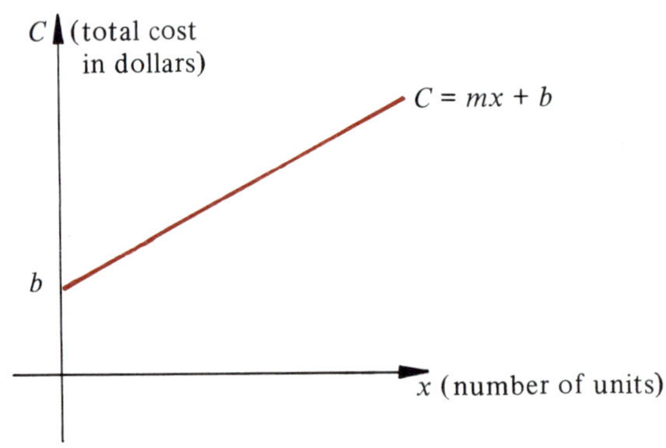

FIGURE 5.11

The vertical intercept b corresponds to the cost of operation when zero units are produced, that is, cost comprising utilities, insurance, and mortgage interest, and the like. Cost of this type is called **fixed cost**. Cost such as labor and materials is called **variable cost**.

The ratio of variable cost to the number of units made is called **marginal cost**. Thus, marginal cost corresponds to the slope of the graph of the linear function $C = mx + b$. Marginal cost is constant when we are using a linear model.

Total cost divided by the number of units produced is referred to as the **average cost** of a unit. Example 2 shows that average cost declines as the number of units produced increases when a typical linear model is used.

EXAMPLE 2 The manufacturing cost C when x units are produced by a company is given by $C = 0.2x + 5000$ dollars.

(a) Sketch the graph of total cost versus units produced.

(b) What is the marginal cost?

(c) Compute the average cost for the 100th unit; for the 200th unit.

SOLUTION (a) The graph is shown in Figure 5.12.

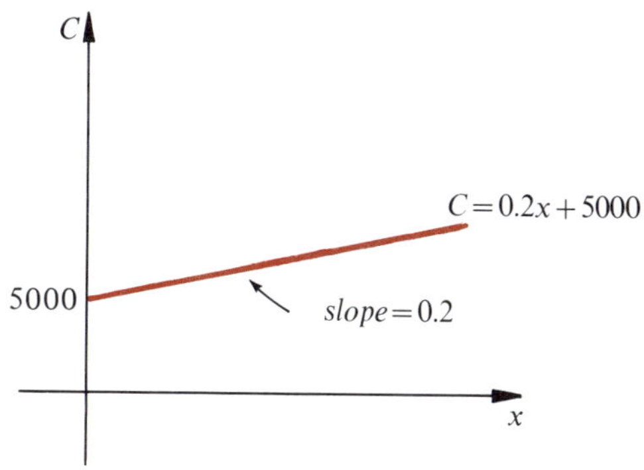

FIGURE 5.12

(b) The marginal cost is \$0.20. Marginal cost is numerically equal to the slope of the cost-unit equation.

(c) The average cost of the 100th unit is

$$\frac{5020}{100} = \$50.20$$

the average cost of the 200th unit is

$$\frac{5040}{200} = \$25.20$$

Cost is also described by a linear **cost-sales model** of the form $C = mx + b$, for which the variable x is the total sales volume, m the variable cost per dollar of sales volume, and b the fixed cost. The graph is called the cost-sales line. Profit can be determined from this linear model through what is called **break-even analysis**. Since Profit = total sales − total cost, we have

$$\text{Profit} = x - C = x - (mx + b)$$
$$P = (1 - m)x - b$$

The value of x for which $P = 0$ is called the **break-even point**.

EXAMPLE 3 A grocery store manager finds that for monthly sales of \$40,200, the fixed costs are \$10,800 and the variable costs are \$22,500.

(a) What is the variable cost per dollar of sales?

(b) Write the cost-sales equation.

(c) What is the profit for the month?

(d) What is the break-even point?

SOLUTION (a) The variable cost per dollar of sales is $m = \$22{,}500/\$40{,}200 = \$0.55970$.

(b) The cost-sales equation is

$$C = 0.55970x + 10{,}800$$

(c) The monthly profit is

$$\begin{aligned} P &= \$40{,}200 - \$10{,}800 - \$22{,}500 \\ &= \$6{,}900 \end{aligned}$$

(d) The profit equation is $P = (1-m)x - b$. Since the break-even point occurs for $P = 0$, it is determined from the equation $0 = (1 - 0.55970)x + 10{,}800$. Solving for x we get $x = \$24{,}529$.

The graph of the cost-sales equation is shown in Figure 5.13.

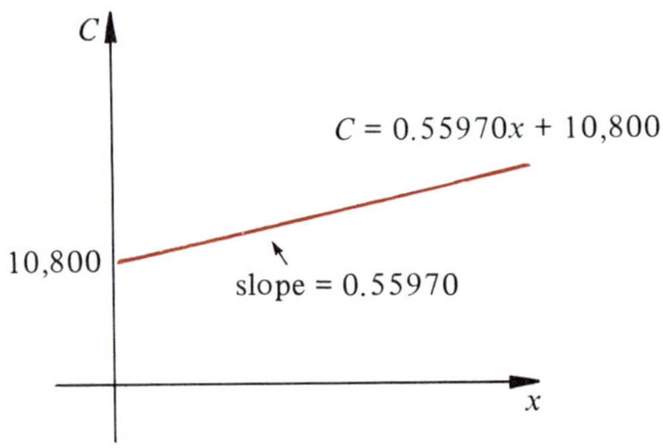

FIGURE 5.13

Exercises for Section 5.3

1. The current I in a resistor is a linear function of the applied voltage V. What equation will relate the current to the voltage if the current is $\frac{1}{2}$ amp when the voltage is 6 volts and $\frac{2}{3}$ amp when the voltage is 8 volts?

2. In an experiment to determine the coefficient of friction, it is found that a 10-lb block has a frictional force of 3 lb and a 25-lb block has a frictional force of 7.5 lb. What equation will relate frictional force to weight if the frictional force is a linear function of weight?

3. *Linear depreciation* is one of several methods approved by the Internal Revenue Service for depreciating business property. If the original cost of the property is C dollars and if it is depreciated linearly over N yr, its value V remaining at the end of n yr is

$$V = C - \frac{C}{N}n.$$

Find the value after 5 yr of a typewriter whose initial cost of \$300 is to be depreciated over 20 yr.

4. If you borrow P dollars at the simple interest rate i, the annual interest is Pi. Hence, the amount owed at the end of n years is given by

$$A = P(1 + i \cdot n)$$

Find the amount A you owe after 5 yr if you borrow \$2000 at 8% simple interest.

5. Mr. Smith wants to borrow \$4000 to buy a new car. He wants to pay off the loan with monthly payments stretching over 3 yr. If he is charged 12% simple interest and if he computes the monthly payments by dividing the total amount due in 3 yr by 36, how much will he have to pay each month?

6. A manufacturer of fountain pens can expect to sell 21,000 felt-tip pens if he charges \$0.80 per pen but only 10,000 if he raises the price to \$1.00. Assuming that the relation is linear, find the equation of the line relating the number of pens to their price. How many felt-tip pens can he sell if he charges \$0.90?

7. The error E of a radar varies with slant range R, according to the equation $5E - R = 3$. Graph the relation between radar error and slant range.

8. The pressure at a point below the surface of a body of water is given by $P = 0.04D + 14.7$, where P is the pressure in psi and D is the depth in inches. Graph the relation.

9. If the total factory cost C of making x units of a product is $C = 5x + 400$, what are the values for the following?
   (a) The cost of making 300 units.
   (b) The average cost per unit of making 300 units.
   (c) The average cost per unit of making 500 units.
   (d) The marginal cost per unit.
   (e) The number of units that can be made for a cash outlay of \$2000.

10. For a marginal cost of producing a unit that is constant at \$0.50, and a fixed cost of \$500:
   (a) Write the cost-unit equation.
   (b) What is the marginal cost of producing the 100th unit?
   (c) What is the cost of producing 100 units?
   (d) What is the average cost per unit of producing 100 units?

11. If the cost-unit equation is $C = 2x + 300$:
   (a) What is the marginal cost of the 100th unit?
   (b) What is the variable cost for 100 units?
   (c) How does the fixed cost change from the 100th to the 101st unit?
   (d) How many units can be made for a cash outlay of \$100? of \$500? of \$1000?

12. If the cost-unit equation is $C = 0.5x + 175$:
   (a) What is the variable cost for 300 units?
   (b) What is the fixed cost?
   (c) What is the average cost per unit to produce 100 units?
   (d) What is the average cost per unit to produce 200 units?

13. A small shoe store finds that it has fixed annual costs of \$28,700. For 1978, its sales were \$112,400 and its variable costs, \$71,600.
   (a) What was the profit for 1978?
   (b) What was the variable cost per dollar of sales?
   (c) Write the cost-sales equation for 1978.
   (d) What was the break-even point?

14. Suppose the cost-sales equation is $C = 0.25x + 1200$.
   (a) Determine the profit or loss on sales of \$1000; of \$2000.
   (b) What is the break-even point?

15. For a total cost related to sales by $C = 0.37x + 18{,}000$, find:
   (a) Variable cost on sales of \$50,000.
   (b) Profit or loss on sales of \$50,000.
   (c) The break-even point.

5.4 Quadratic Functions

Functions of the form

$$f(x) = ax^2 + bx + c \tag{5.6}$$

where a, b, and c are real constants and $a \neq 0$, are called **quadratic functions**. Quadratic functions arise naturally in describing physical variations such as the path of a projectile, the shape of a radar antenna, and the area of a circle. We see from the rule of correspondence that there is no restriction on the values that x can assume; that is, *the domain of a quadratic function is the set of all real numbers.* Determining the range of a quadratic function is usually more difficult, but sometimes it too can be found by inspection.

EXAMPLE 1 Determine the range of the following quadratic functions:

(a) $f(x)=x^2$ (b) $g(x)=x^2+1$ (c) $h(x)=x^2+4x+4$

SOLUTION

(a) Since x^2 is always greater than or equal to 0, it follows that the range of $f(x)$ is the set of all nonnegative real numbers.

(b) Similarly, x^2+1 is always greater than or equal to 1, so the range of $g(x)$ is the set of all real numbers greater than or equal to 1.

(c) The quadratic x^2+4x+4 can be written $(x+2)^2$, so the range of $h(x)$ is the set of all nonnegative real numbers.

Example 2 shows how to find the range of a quadratic function by a systematic approach. The technique is based on the process of completing the square (Section 1.7).

EXAMPLE 2 Determine the range of the quadratic function

$$f(x)=2x^2-x-4$$

SOLUTION First factor out the coefficient of x^2, to get

$$f(x)=2[x^2-\tfrac{1}{2}x-2]$$

Then "complete the square" by adding and subtracting $\frac{1}{16}$ to the expression inside the brackets:

$$\begin{aligned} f(x) &= 2\left[x^2-\frac{1}{2}x+\frac{1}{16}-2-\frac{1}{16}\right] \\ &= 2\left[\left(x-\frac{1}{4}\right)^2-\frac{33}{16}\right] \end{aligned}$$

The minimum value of the term in brackets is $-\frac{33}{16}$. This occurs when $x=\frac{1}{4}$. Hence $f(x)$ has a minimum value of $-\frac{33}{16}\cdot 2$, or $-\frac{33}{8}$, and the range is $f(x) \geqslant -\frac{33}{8}$.

The procedure of Example 2 can be generalized. Beginning with $f(x)=ax^2+bx+c$, we factor the coefficient of x^2 to get

$$f(x)=a\left(x^2+\frac{b}{a}x+\frac{c}{a}\right)$$

Adding and subtracting $[(1/2)(b/a)]^2$, to complete the square, gives

$$f(x)=a\left(x^2+\frac{b}{a}x+\frac{b^2}{4a^2}+\frac{c}{a}-\frac{b^2}{4a^2}\right)$$

and simplifying yields

$$f(x)=a\left[\left(x+\frac{b}{2a}\right)^2+\frac{4ac-b^2}{4a^2}\right]$$

$$=a\left[\left(x+\frac{b}{2a}\right)^2-\left(\frac{b^2-4ac}{4a^2}\right)\right]$$

The term in brackets has a minimum value of $-(b^2-4ac)/4a^2$ when $x=-b/2a$. Hence, the range is $f(x)\geqslant-(b^2-4ac)/4a$ if $a>0$, and $f(x)\leqslant-(b^2-4ac)/4a$ if $a<0$.

EXAMPLE 3 Find the range of $F(x)=-3x^2+4x+2$.

SOLUTION Since $a=-3$, and $b=4$, and $c=2$, the range of $F(x)$ is

$$F(x)\leqslant\frac{-(b^2-4ac)}{4a}=\frac{-(4^2-4(-3)(2))}{4(-3)}=\frac{10}{3}$$

Thus, the range of $F(x)$ is the set of all real numbers less than or equal to $\frac{10}{3}$.

The graph of a quadratic function is a U-shaped curve called a **parabola**. The parabola is the graph of the quadratic function just as the straight line is the graph of the linear function. The graphs of $y=x^2$ and $y=-x^2$, which are shown in Figure 5.14, typify the parabolic shape. The sign of the coefficient a determines how the parabola opens. If $a>0$, the parabola opens upward, as in Figure 5.14(a); if $a<0$, the parabola opens downward, as in Figure 5.14(b). The low (or high) point of the parabola is called the **vertex**. The parabola is symmetric about a vertical line through the vertex.

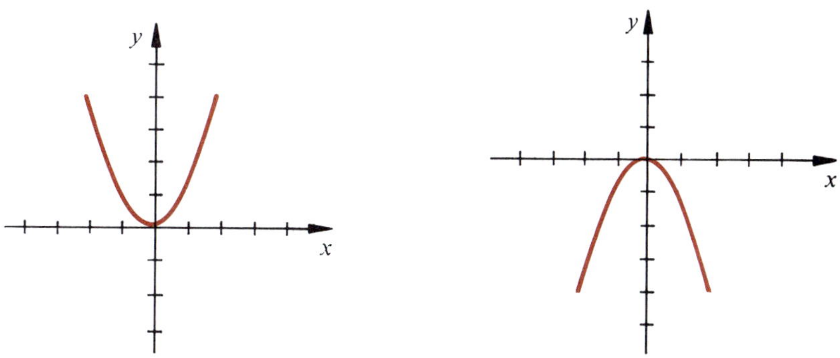

FIGURE 5.14 (a) $y=x^2$ (b) $y=-x^2$

The coefficients a, b, and c of the general quadratic affect both the shape and the location of the parabola. We have already seen how the sign of the coefficient of x^2

determines in which direction the parabola opens. Figure 5.15 shows the graphs of $y=x^2$, and of $y=4x^2$, and of $y=9x^2$. It is apparent that *the larger the coefficient of* x^2, *the smaller the opening of the parabola.* In Figure 5.15, the opening of each parabola is measured with a horizontal line segment passing through (0, 1). (Any other fixed point on the y-axis will do.) Using this point, you can determine the length of the openings; they will be 2, 1, and $\frac{2}{3}$.

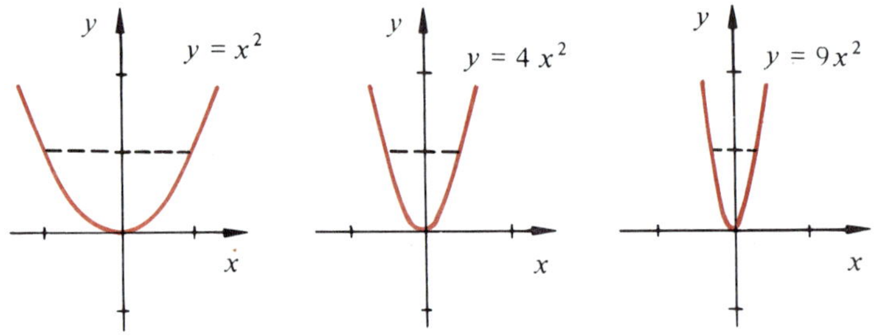

FIGURE 5.15

The constant b in the quadratic function has no fundamental effect on the shape of the parabola but causes the parabola to move to right or left. Figure 5.16 shows the graphs of three functions: $y=x^2-3x+2$, and $y=x^2+0x+2$, and $y=x^2+3x+2$. As you can see, the basic parabola stays the same but is *translated to the left* as the coefficient of x increases. Note that the y-intercept is the same for each curve.

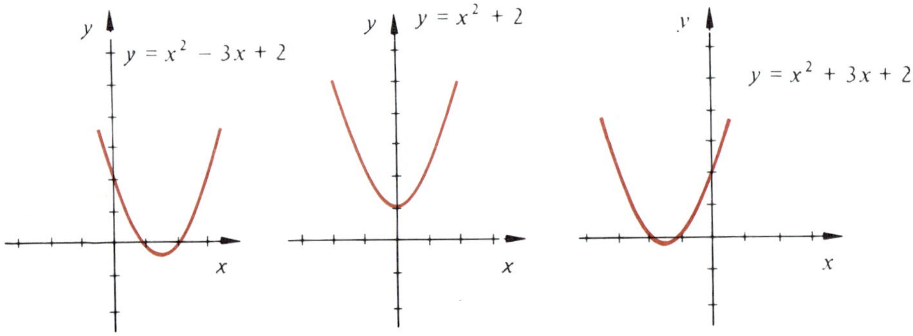

FIGURE 5.16

The constant c causes vertical translation. Figure 5.17 shows the cases $y=x^2-x-6$, $y=x^2-x$, and $y=x^2-x+6$ and demonstrates that each case produces the same kind of parabola, although it is shifted vertically by the difference in the value of the constant term. Clearly, *the constant term is the y-intercept.*

Our examination of parabolic graphs has not been exhaustive; it is intended only to help you understand how the constants a, b, and c affect the shape and location of the graph of a quadratic function. One obvious conclusion from Figures 5.16 and 5.17 is that the vertex of a parabola is not always at the origin; the vertex is at the origin only if both b and c equal zero. It follows from our earlier discussion of the range of the quadratic function $f(x)=ax^2+bx+c$ that **the coordinates of the vertex of a parabola are $(-b/2a, f(-b/2a))$.** The vertex of the graph of $y=2x^2-6x$, for example, is located

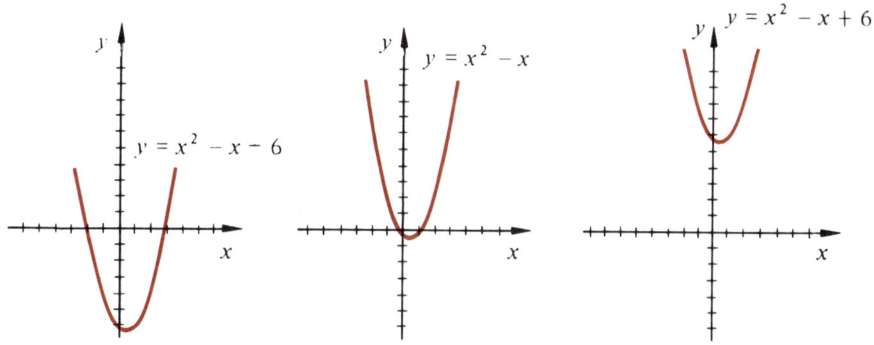

FIGURE 5.17

at $x = \frac{-(-6)}{2(2)} = \frac{3}{2}$ $\quad y = 2\left(\frac{3}{2}\right)^2 - 6\left(\frac{3}{2}\right) = -\frac{9}{2}$

We recommend the following procedure for making an accurate sketch of the graph of a quadratic function.

Procedure for Sketching the Graph of $f(x) = ax^2 + bx + c$

(1) Determine the sign of the x^2 term. If $a > 0$, the parabola opens upward; if $a < 0$, it opens downward.

(2) Determine the y-intercept by letting $x = 0$; that is, $f(0) = c$ is the y-intercept.

(3) Determine the zeros of the function by letting $f(x) = 0$. The zeros are given by

$$x = \frac{-b \pm \sqrt{b^2 - 4ac}}{2a}$$

If $b^2 - 4ac > 0$, the function has two real zeros; if $b^2 - 4ac = 0$, the function has one zero, $x = -b/2a$; if $b^2 - 4ac < 0$, the function has no real zeros and therefore does not cross the x-axis. *Note*: The quadratic formula can always be used to find the zeros of $f(x)$; if $f(x)$ is factorable, however, it is sometimes easier to find the zeros from the factors.

(4) Determine the location of the vertex by computing $x = -b/2a$ and $y = f(-b/2a)$.

If the given quadratic function does not have two real zeros, locating the parabola with precision is more difficult. A few additional points should then be plotted.

EXAMPLE 4 Sketch the graph of $f(x) = x^2 - x - 6$.

SOLUTION First we note that $a=1$, $b=-1$, and $c=-6$.

(1) The parabola opens upward, since $a>0$.

(2) The y-intercept is $f(0)=-6$.

(3) The zeros are found by letting $f(x)=0$. Thus,

$$x^2-x-6=0$$

Since the quadratic is factorable, we have

$$(x-3)(x+2)=0$$

Therefore, the zeros are $x=3$ and $x=-2$.

(4) To locate the vertex, we note that

$$x=-\frac{b}{2a}=-\frac{-1}{2(1)}=\frac{1}{2}$$

and

$$f(-b/2a)=\left(\frac{1}{2}\right)^2-\frac{1}{2}-6=-\frac{25}{4}$$

Hence, the vertex is located at $(\frac{1}{2}, -\frac{25}{4})$.

The graph of $f(x)=x^2-x-6$ is shown in Figure 5.18.

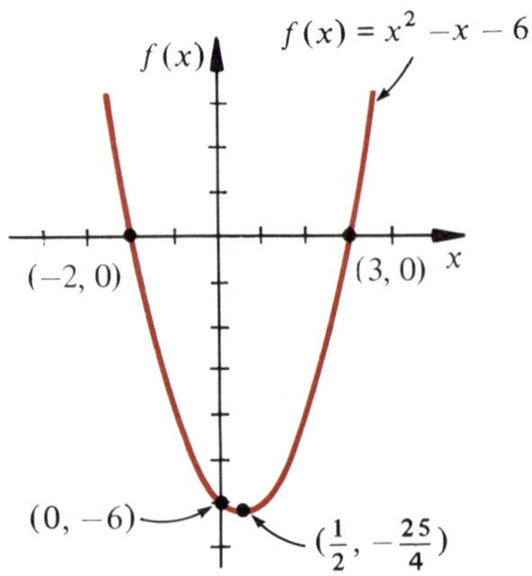

FIGURE 5.18

EXAMPLE 5 Sketch the graph of $y=x^2+3x$.

SOLUTION Here, $a=1$, $b=3$, and $c=0$.

(1) The parabola opens upward, since $a>0$.

(2) The y intercept is $y(0)=0$.

(3) The zeros are obtained from $x^2+3x=0$. Since x^2+3x factors into $x(x+3)$, we conclude that the zeros are $x=0$ and $x=-3$.

(4) The vertex is at

$$x = -\frac{3}{2(1)} = -\frac{3}{2},$$

and

$$y\left(-\frac{3}{2}\right) = \left(-\frac{3}{2}\right)^2 + 3\left(-\frac{3}{2}\right) = \frac{9}{4} - \frac{9}{2} = -\frac{9}{4}$$

These points and our knowledge of the parabolic shape yield the graph in Figure 5.19.

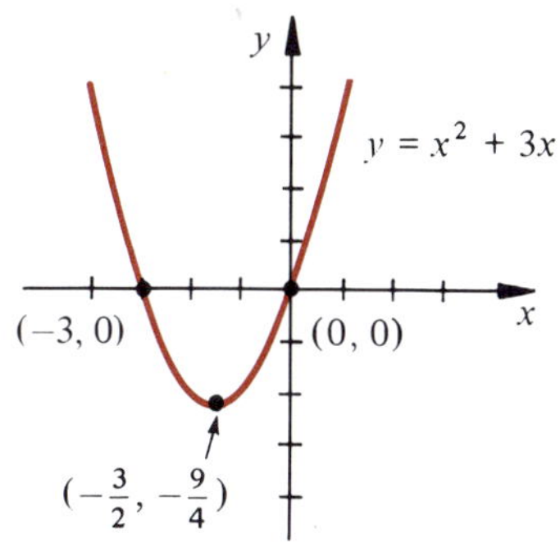

FIGURE 5.19

EXAMPLE 6 Sketch the graph of $G(t) = -t^2 + 2t - 2$.

SOLUTION

(1) Since $a < 0$, the parabola opens downward.

(2) The y-intercept is $G(0) = -2$.

(3) There are no real zeros, since the discriminant

$$b^2 - 4ac = (2)^2 - 4(-1)(-2) = -4$$

which is less than 0. This means that the graph cannot cross the t-axis.

(4) The vertex is located at $(1, G(1))$, where

$$G(1) = -(1)^2 + 2(1) - 2 = -1$$

Since there are no real zeros, we evaluate a couple of additional solution points. The graph is shown in Figure 5.20.

t	$G(t)$
-1	-5
2	-2

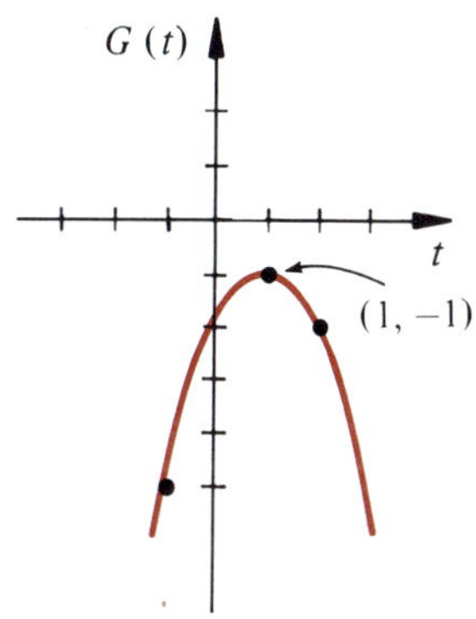

FIGURE 5.20

Sometimes we want to construct a quadratic function having certain zeros. Suppose the zeros are a and b. This implies that the function has linear factors $(x-a)$ and $(x-b)$; hence, the quadratic function has the form $f(x)=k(x-a)(x-b)$, where k is a constant. The value of k can be determined if an additional point on the parabola is known.

EXAMPLE 7 Determine the form of the quadratic function whose graph is shown in Figure 5.21.

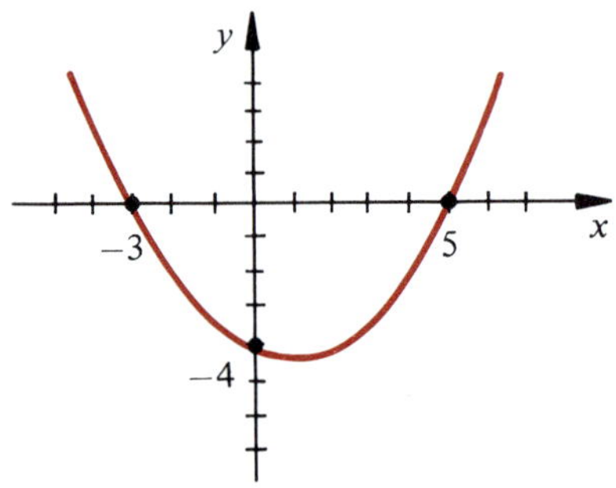

FIGURE 5.21

SOLUTION

(a) The quadratic function has zeros of 5 and -3. (Why?) Hence the form of the function is

$$y=k(x-5)(x+3)$$

(b) Since the y-intercept is -4, substitute $x=0$ and $y=-4$ into the functional expression to get

$$-4=k(-5)(3)$$

or

$$k=\frac{4}{15}$$

The desired function is then

$$y=\frac{4}{15}(x-5)(x+3)$$

$$=\frac{4}{15}x^2-\frac{8}{15}x-4$$

Exercises for Section 5.4

Sketch the graph of the given quadratic function in each of Exercises 1–20. Give the domain and range and the zeros of each function. Be sure to indicate the intercepts and the vertex of each.

1. $f(x)=x^2-4$
2. $y=2x^2-6$
3. $y=x^2+1$
4. $g(t)=t^2+5$
5. $f(x)=9-x^2$
6. $y=x^2+2x$
7. $x(t)=t^2-3t$
8. $s=3t-16t^2$
9. $y=x^2-5x+6$
10. $y=x^2+7x+12$
11. $y=3x^2+2x+1$
12. $y=x^2+7x+13$
13. $y=2-x^2+x$
14. $y=x-x^2-1$
15. $f(x)=2x^2-3x-5$
16. $f(t)=-2t^2-4t+5$
17. $y=3+x-x^2$
18. $g(x)=3+x+x^2$
19. $G(x)=1+x+x^2$
20. $f(x)=\frac{1}{2}x^2-x-2$

In Exercises 21–30, determine the quadratic function whose graph is given.

21.

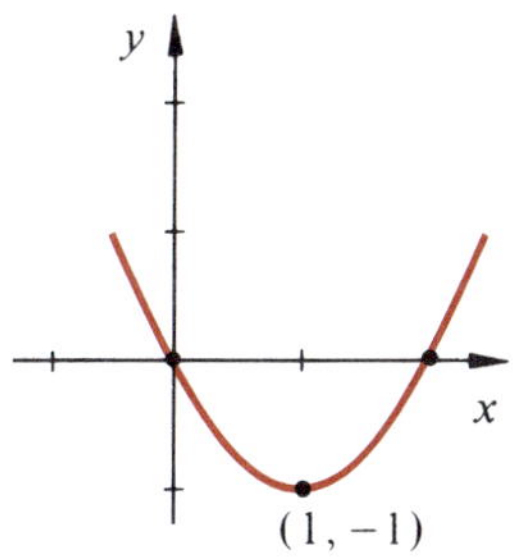

22.

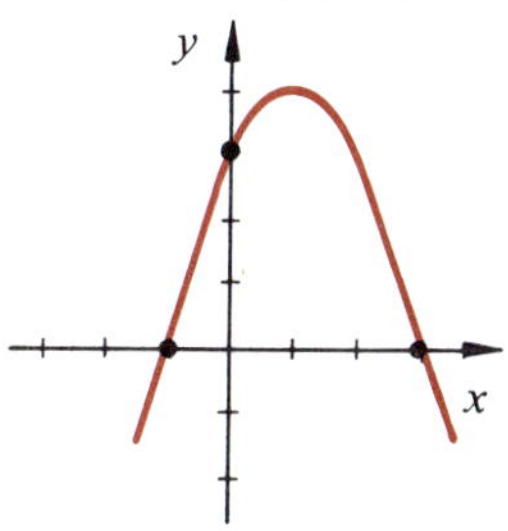

23.

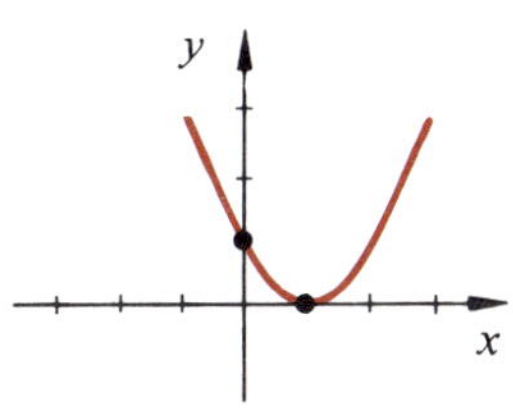

24.

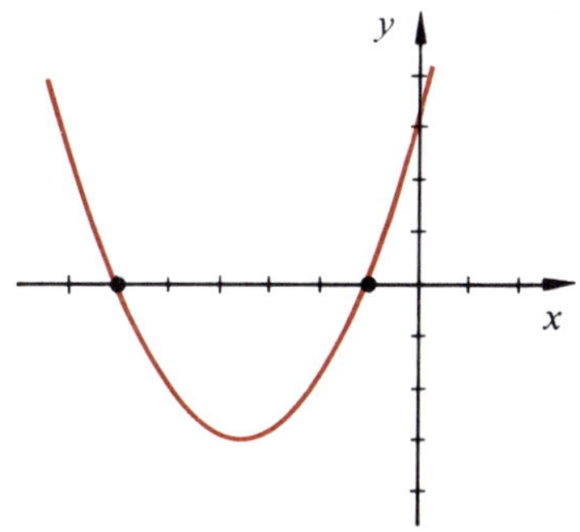

25.

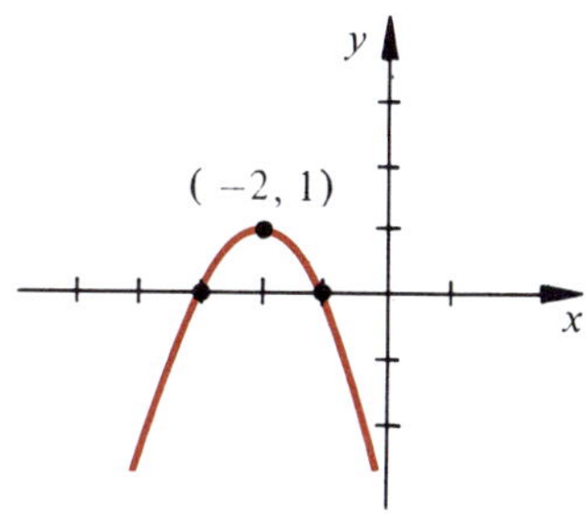

26.

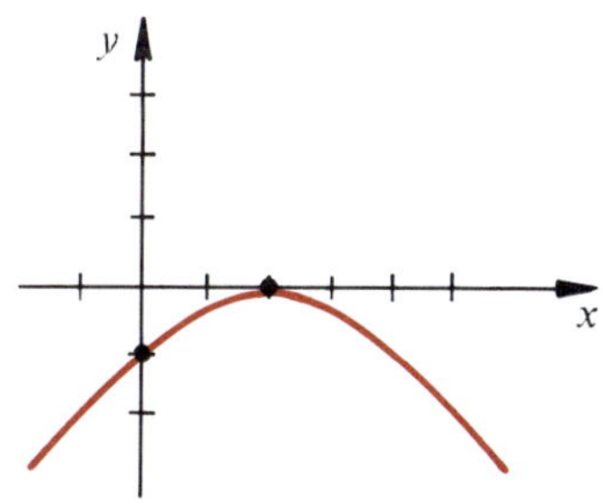

27.

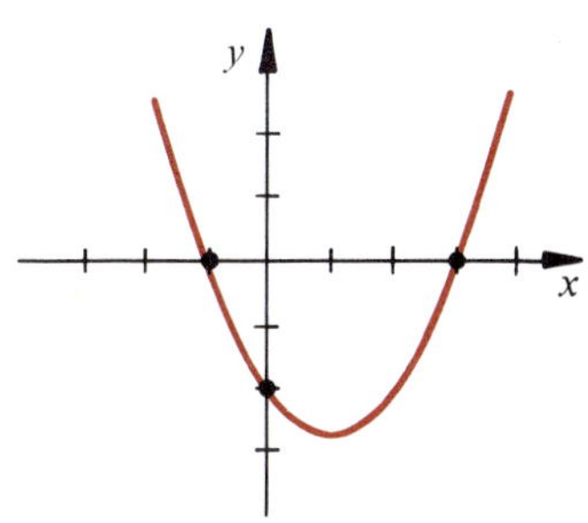

28.

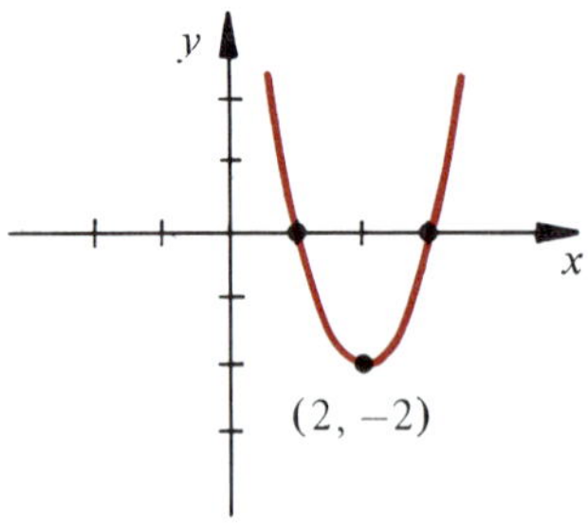

29.

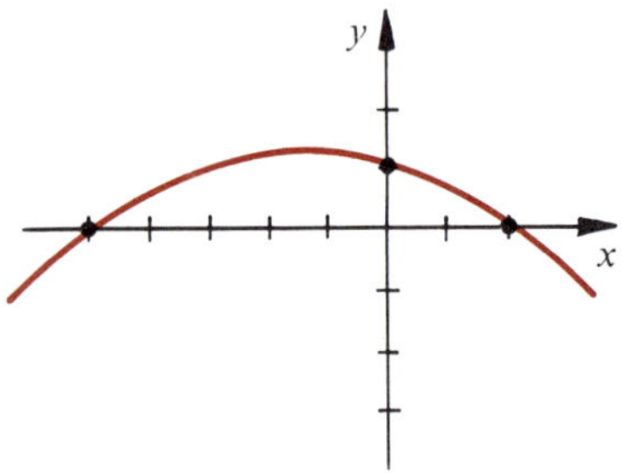

30.

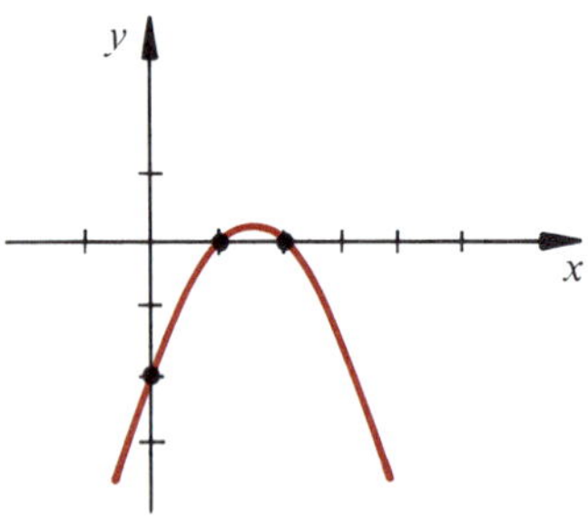

31. If $f(x)=x^2$, how are the graphs of $f(x)$ and $-f(x)$ related?

32. If $f(x)=x^2$, how are the graphs of $f(x)$ and $f(-x)$ related?

33. A radar antenna is to be built with a cross section that is described by the function $y=0.05x^2$, where x and y are in feet. Draw the cross section of the antenna, which has a diameter of 15 ft.

34. When an object is thrown into the air, its height above the ground is a quadratic function of time. Sketch the graph of height versus time from $t=0$ to $t=3$ for the functional relation $h=48t-16t^2$. What is the maximum height of the object?

35. Sketch the graph of electric current versus time for a current that varies according to $i=t^2-5t+6$ amps. When is the current zero? Assume that the domain is all $t \geqslant 0$.

36. The load being applied to a cantilever beam varies with the distance from the fixed end according to $F=32-2x^2$ lb. Sketch the graph of this quadratic function from $x=0$ to $x=4$. Where is the maximum load?

37. The velocity of a certain rocket varies with time according to $v=100+200t-25t^2$, where v is in ft/sec when t is in seconds. Sketch the graph of the velocity function for $t \geqslant 0$. What is the maximum velocity of the rocket? At what time does the velocity equal zero?

38. Using the graph from Exercise 37, estimate the time it takes for the velocity of the rocket to equal its initial velocity.

39. Determine whether or not the composition of a quadratic function with a quadratic function is quadratic.

5.5 Polynomial Functions

Linear and quadratic functions are special cases of a class of functions called **polynomial functions**. A function defined by the rule

$$f(x)=a_nx^n+a_{n-1}x^{n-1}+\cdots+a_1x+a_0 \qquad a_n \neq 0$$

is generally a *real* polynomial function of degree n if n is a positive integer and a_n, $a_{n-1}, \ldots, a_0$ are arbitrary but constant real numbers. Using this definition, you can see that a quadratic function is a polynomial function of degree 2 and a linear function is a polynomial function of degree 1.

The sum of two or more terms is not necessarily a polynomial. For instance, functions such as

$$f(x)=x+\frac{1}{x} \quad \text{or} \quad g(x)=\sqrt{x}+1$$

are not polynomial functions.

The discussion of the graph of the general polynomial function is best left until you have available the tools of calculus. An elementary study of the graph of some single-term or factored polynomials is possible at this time.

Single-Term Polynomials

A single-term polynomial is a polynomial of the form $y=x^n$ for n, a nonnegative integer. You already know what the graphs of $y=x$ and $y=x^2$ look like. The graphs of $y=x^3$ and $y=x^4$ are shown in Figure 5.22. These graphs show the characteristic shape of the simple cubic and quartic.

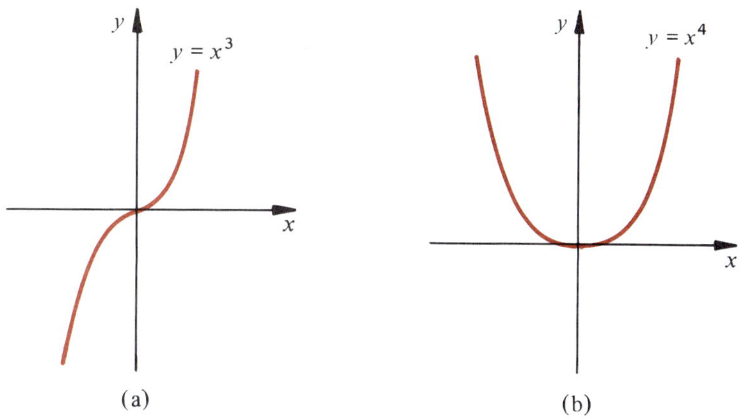

FIGURE 5.22

Figure 5.23(a) shows the graphs of $y=x^n$, $n=1, 2, 3, 4$, on the interval $0<x<5$. For $x>1$, the higher-degree function always appears above the lower-degree function; and the difference in the values of any two powers increases as x becomes large. We say that the higher-degree function **dominates** the lower-degree functions as x becomes large.

In Figure 5.23(a), the relative position of each of the graphs is barely distinguishable on $0 \leqslant x \leqslant 1$. So that we can see what happens to the graphs on this interval, the four functions are redrawn in Figure 5.23(b) on a more appropriate scale. Clearly the relative position of the graphs are reversed from their positions for $x>1$; that is, x dominates x^2, which dominates x^3, and so forth.

Understanding that one function dominates another can be useful in making a quick sketch of a simple polynomial. If $f(x)=x^4+x$, for example, we should consider the x term the most important for, say, $x<0.8$, and the term x^4 dominant for, say,

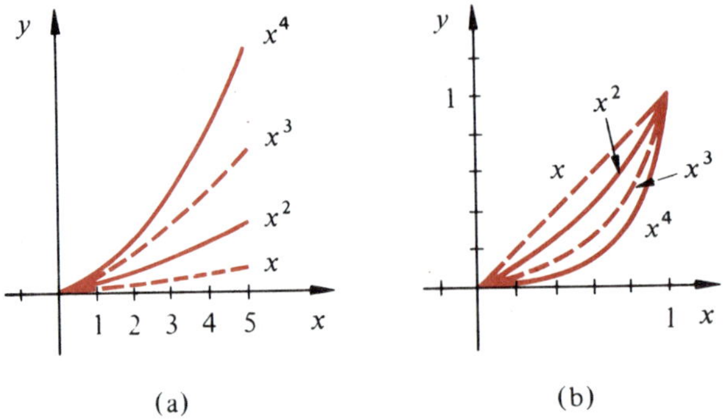

FIGURE 5.23

$x > 2$. In between, the values of both terms would contribute significantly to the sum. Figure 5.24 shows the graph of $f(x) = x^4 + x$ from $x = 0$ to $x = 3$.

Once you know how the graph of the basic power function looks, the extension to some simple linear combinations is straightforward.

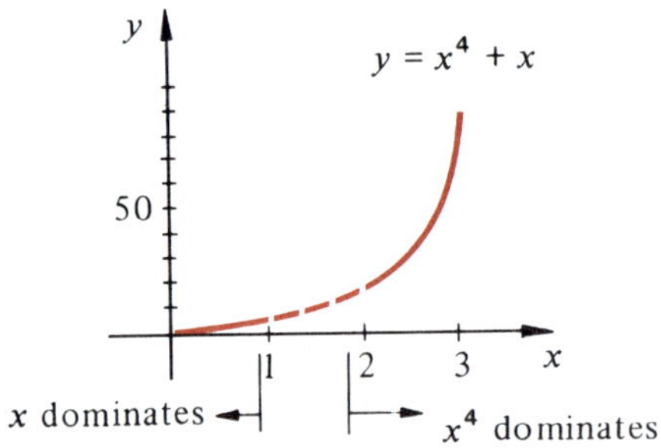

FIGURE 5.24

EXAMPLE 1 Sketch the graph of $f(x) = -x^3 + 1$.

SOLUTION Figure 5.25(a) presents the graph of $-x^3$. Adding 1 to each ordinate value gives the graph of $y = -x^3 + 1$, as shown in Figure 5.25(b).

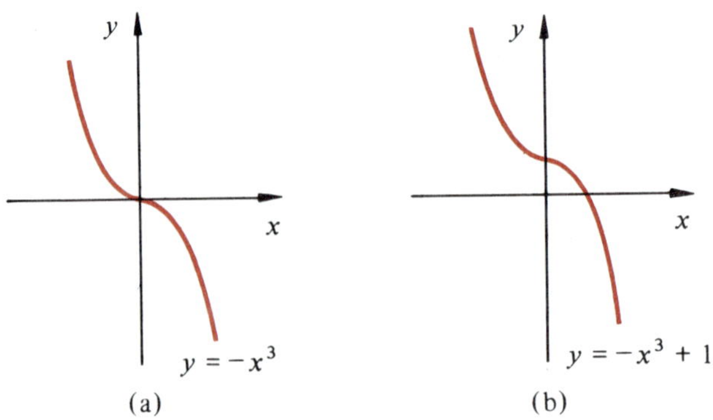

FIGURE 5.25

Factored Polynomials

It is easy to sketch polynomials in factored form since the zeros of each factor give the value of the intercepts on the x-axis. To determine where above or below the axis the graph is, we use a sign diagram as in solving an inequality.

EXAMPLE 2 Sketch the graph of $y=(x-1)(2x+1)(x-3)$.

SOLUTION Since $x=1$, $-\frac{1}{2}$, and 3 are zeros of this polynomial, they are the x-intercepts. By letting $x=0$, we get $y=3$ as the y-intercept. The value of y is positive for $x>-\frac{1}{2}$ and then alternates in sign as the curve passes through each of its intercepts. Figure 5.26 shows not only the curve but some additional values of y for x between the intercepts. The graph has the distinctive cubic character.

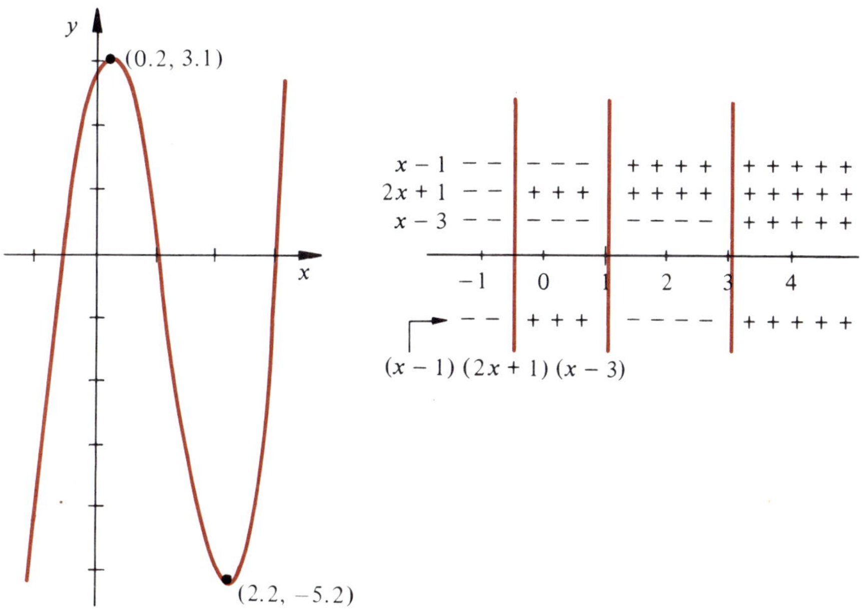

FIGURE 5.26

EXAMPLE 3 Sketch the graph of $y=x^2(x+1)(x+2)$.

SOLUTION Figure 5.27 shows the graph of this quartic. The points on the x-axis for $x=0$, -1, and -2 are easily obtained. The graph does not cross the x-axis at $x=0$, since x^2 is always nonnegative in the vicinity of $x=0$.

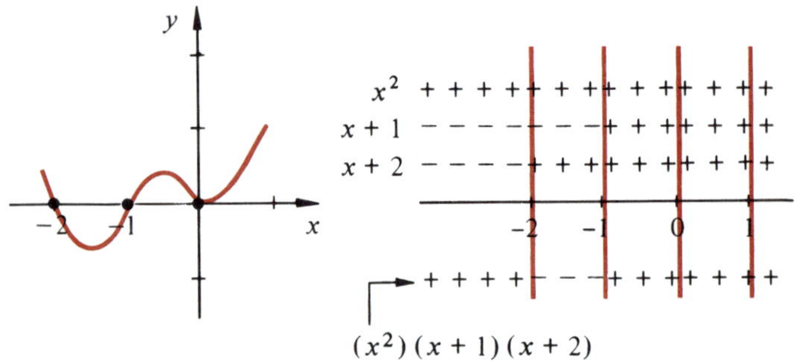

FIGURE 5.27

Exercises for Section 5.5

Sketch the graph of the polynomials in Exercises 1–20.

1. $y = x^3 + 1$
2. $y = x^4 + x$
3. $y = 1 - x^4$
4. $y = x^4 + x^2$
5. $y = x^4 - x^2$
6. $y = x^3 + x$
7. $y = x^3 - x$
8. $y = x^3 - x^2$
9. $y = x^5 + 1$
10. $y = x^6 - 1$
11. $y = (x)(x-1)(x+3)$
12. $y = (x-1)(x+4)(x-10)$
13. $y = (x+1)^2(x)(x-1)$
14. $y = (x^2 - 3x + 2)(x^2 - 9)$
15. $y = (x^2 - 9)(x^2 + 4)$
16. $y = (x^2 - 1)(x+1)^2$
17. $y = (x+2)(x-3)(x+5)$
18. $y = (x+5)(x+2)(x-1)x$
19. $y = (x^2 - 6x + 5)(x^2 + 4x + 4)$
20. $y = x^2(x^2 - 1)(x+2)$

5.6 Rational Functions

Functions that are the ratio of two polynomials are called **rational functions**. Specifically, if $N(x)$ and $D(x)$ are polynomials without common factors, then

$$f(x) = \frac{N(x)}{D(x)}$$

is a rational function of x. Notice the similarity between rational functions and rational numbers. Rational numbers are quotients of integers, whereas rational functions are ratios of polynomials. Some examples of rational functions are

$$\frac{3x^5 + 5}{20x^3 + x - 16}, \qquad \frac{4}{x^2 - 4}, \qquad \frac{3x^2 + 5x - 6}{(x-5)(x+3)}$$

If the degree of N is less than the degree of D, then N/D is a *proper fraction*; otherwise, N/D is an *improper fraction*. By division, any improper fraction can be expressed as a polynomial quotient and a remainder that is a proper fraction.

In studying the graphs of rational functions, let us begin by considering an arbitrary rational function

$$f(x)=\frac{N(x)}{D(x)}$$

The zeros of a rational function are the same as the numerator polynomial, $N(x)$. Graphically, the zeros of the function are the x-intercepts.

The zeros of the denominator are important, too. They are the values of x for which $f(x)$ does not exist, since division by zero is undefined. Let us see how the function behaves in the neighborhood of a value x_0 for which $D(x_0)=0$.

Since N/D is well defined for all values except x_0, we can consider the value of the function at real numbers as close to x_0 as we please. This consideration is usually described by saying that x "approaches" arbitrarily close to x_0. As x approaches x_0, the value $D(x)$ approaches zero; since $N(x)$ and $D(x)$ have no common factors, the value of $f(x)$ becomes larger and larger as x approaches x_0. Graphically, think of the curve moving farther and farther from the x-axis as x approaches x_0. In this case, the vertical line through (x_0) is called a **vertical asymptote** of the graph of the rational function.

To understand the meaning of a vertical asymptote, consider the rational function

$$f(x)=\frac{1}{x}$$

The function does not exist at $x=0$; it is, however, defined for all other values of x. Here the y-axis is a vertical asymptote. To determine the behavior of the graph, note that as x approaches 0 from the right, $f(x)$ is positive and increases indefinitely. We say that $f(x)$ **increases without bound**. Similarly, as x approaches 0 from the left, $f(x)$ is negative and decreases indefinitely. Using this knowledge and computing some points on either side of $x=0$ yield the graph shown in Figure 5.28.

As x increases, the graph comes closer to the x-axis; and similarly, as x decreases, the graph comes closer to the x-axis. In fact, as x increases without bound, the value of $1/x$ approaches zero. (But no matter how large x becomes, $1/x$ will not equal zero.) The same observation holds as x decreases without bound. The line $y=0$ is called a **horizontal asymptote** of the graph.

From the foregoing discussion you can see that the asymptotes of a rational function are very helpful for sketching its graph. Use the following rule to determine the vertical asymptotes of a rational function:

If $D(x)$ represents the denominator function and $D(k)=0$, then $x=k$ is a vertical asymptote.

A rational function may have more than one vertical asymptote, determined by the zeros of the denominator function. Thus, the rational function $(2x^2+5)/(x^2-9)$ has two vertical asymptotes, the lines $x=3$ and $x=-3$.

The graph of a rational function can have more than one vertical asymptote but at most only one horizontal asymptote. A convenient method for finding the horizontal asymptote of $N(x)/D(x)$ is to divide $N(x)$ and $D(x)$ by the highest power of x that occurs in either. If the resulting expression approaches a constant c as x becomes

x	y
-3	$-\frac{1}{3}$
-2	$-\frac{1}{2}$
-1	-1
$-\frac{1}{2}$	-2
$-\frac{1}{4}$	-4
$\frac{1}{4}$	4
$\frac{1}{2}$	2
1	1
2	$\frac{1}{2}$
3	$\frac{1}{3}$

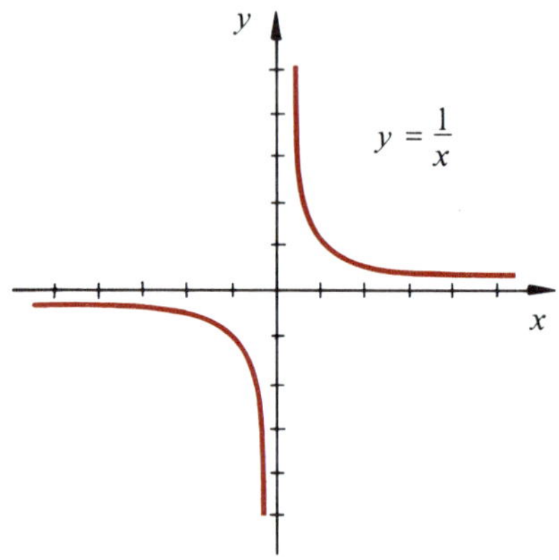

FIGURE 5.28

large, then the line $y=c$ is a horizontal asymptote of the function. If the expression becomes large as x becomes large, there is no horizontal asymptote.

For an example of the technique, consider the function $y=(2x^2+5)/(x^2-9)$. Dividing the numerator and the denominator of this function by x^2 gives

$$y=\frac{2+5/x^2}{1-9/x^2}$$

As x becomes large, both $5/x^2$ and $9/x^2$ approach zero. Hence, the value of the function approaches 2 as x becomes large, and we conclude that $y=2$ is a horizontal asymptote. An example of a rational function that does not have a horizontal asymptote is

$$y=\frac{x^3-6}{5x+1}$$

Dividing the numerator and the denominator by x^3 gives

$$y=\frac{1-6/x^3}{5/x^2+1/x^3}$$

As x becomes large, both terms in the denominator approach zero, and therefore y cannot approach a finite number.

The horizontal asymptote of a proper rational function is the x-axis. This condition follows from the fact that in a proper rational function, the numerator is of a lower degree than the denominator.

EXAMPLE 1 Sketch the graph of $y=3x/(x-2)$.

SOLUTION *Intercepts*: $x=0$, $y=0$.

Vertical Asymptote: The denominator of the function is zero for $x=2$. Since the numerator and the denominator have no common factors, the line $x=2$ is a vertical asymptote.

Horizontal Asymptote: To find the horizontal asymptote, divide the numerator and the denominator by x, to get

$$y=\frac{3}{1-2/x}$$

As x becomes large, the fraction $2/x$ approaches zero, and therefore the value of y approaches 3. The line $y=3$ is the horizontal asymptote.

Finally, determine some additional points in the vicinity of the vertical asymptote, as shown in the table. The graph is now readily determined. (See Figure 5.29.)

$$y=\frac{3x}{x-2}$$

x	y
-2	$\frac{3}{2}$
-1	1
0	0
1	-3
2	undef.
3	9
4	6

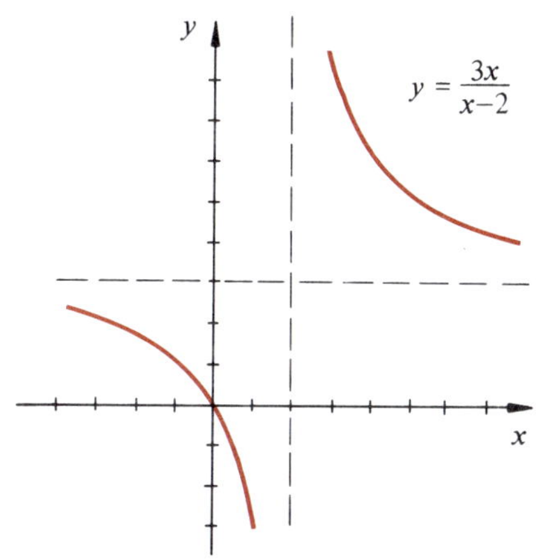

FIGURE 5.29

EXAMPLE 2 Sketch the graph of $y=\dfrac{2x}{x^2-4}$.

SOLUTION *Vertical Asymptotes*: The graph of this function has vertical asymptotes at $x=2$ and $x=-2$, since these are the zeros of the denominator.

Horizontal Asymptote: Divide the numerator and denominator by x^2 to get

$$y=\frac{2/x}{1-4/x^2}$$

As x gets very large, the numerator approaches zero and the denominator approaches 1. Therefore, $y=0$ is the horizontal asymptote. The graph is then determined by plotting a few additional points, as shown in the table. See Figure 5.30.

x	y
-4	$-\frac{2}{3}$
-3	$-\frac{6}{5}$
-1	$\frac{2}{3}$
0	0
1	$-\frac{2}{3}$
3	$\frac{6}{5}$
4	$\frac{2}{3}$

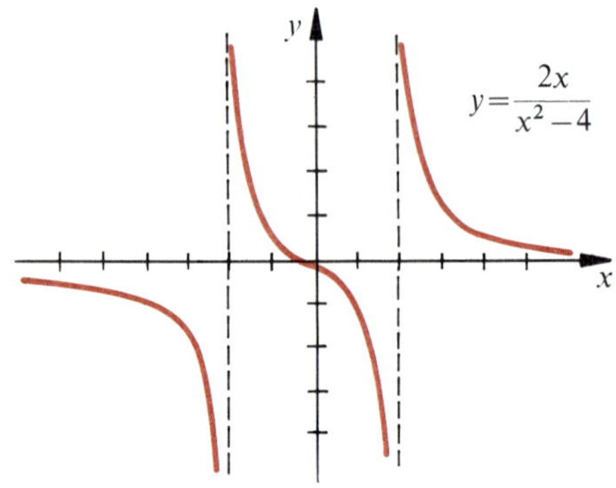

FIGURE 5.30

EXAMPLE 3 The heat generated by a certain chemical process varies with time according to

$$H=\frac{t}{t^2+4}$$

where H is the heat in calories and t the time in seconds. Draw the graph of this function and find the maximum heat output.

SOLUTION This rational function has no vertical asymptotes, since there is no real value of t for which $t^2+4=0$. The t-axis is a horizontal asymptote, however. Since a negative time has no physical significance, we consider only positive values of t. Plotting a few additional points and drawing the curve so that it approaches zero as t increases, we get the graph in Figure 5.31. The graph illustrates that the heat output increases rapidly at first, reaching a peak of 0.25 cal at about 2.0 sec. From this point, the heat output decreases gradually toward zero.

$$H=\frac{t}{t^2+4}$$

t	H
0	0
2	.25
4	.20
6	.15
8	.12
10	.09

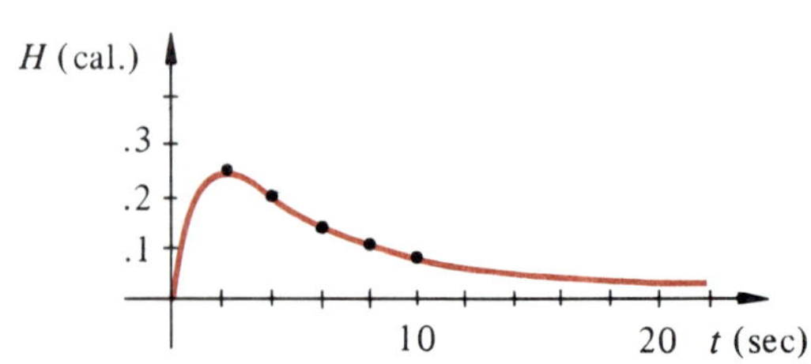

FIGURE 5.31

Exercises for Section 5.6

Compute intercepts and locate asymptotes (if any); then graph the rational functions in Exercises 1–18.

1. $y=-\dfrac{1}{x}$ **2.** $y=\dfrac{5x+2}{2-x}$ **3.** $y=\dfrac{3x-2}{x-4}$

4. $z=\dfrac{2}{t-1}$ **5.** $p=\dfrac{1}{w+1}$ **6.** $v=\dfrac{t^2-1}{t^2+1}$

7. $s=\dfrac{5w^2}{4-w^2}$ **8.** $m=\dfrac{u+1}{u^2-1}$ **9.** $y=\dfrac{x^2}{x^2-9}$

10. $y=\dfrac{-2x}{x^2-4}$ **11.** $s=\dfrac{(t+2)^2}{t^2+2t}$ **12.** $w=\dfrac{3x^2}{x^2-3x}$

13. $y=\dfrac{3}{w^2-9w}$ **14.** $r=\dfrac{5z+10}{3z-z^2}$

15. $y=\dfrac{1}{p^2-3p-4}$ **16.** $n=\dfrac{1+s^2}{s^2-s-2}$

17. $y=\dfrac{5x+3}{x^2+x}$ **18.** $y=\dfrac{x^2+3x+2}{x^2+x}$

19. The force between two unit masses varies with the distance between them according to the formula $F=1/d^2$. The force is in dynes when the distance is measured in centimeters. Graph this function.

20. Due to leakage, the pressure in a hydraulic system varies with time according to $P=10/(t^2+1)$ psi, where t is the time in seconds. Graph this pressure variation.

21. The coefficient of friction μ of a plastic block sliding on an aluminum table varies according to $\mu=(v^2+3)/(4v^2+5)$, where v is the velocity of the block in cm/sec. Graph this function.

5.7 Multipart Functions

Many practical situations are described by functions whose graphs contain discontinuities in the form of "jumps", or "gaps." When you mail a letter, for example, you pay \$0.15 for the first ounce and \$0.13 for each extra ounce. The table in Figure 5.32 gives the function describing this situation for weights up to 5 ounces. Plotting the graph of this function produces a series of line segments that are separated by discrete "jumps," as shown in Figure 5.32.

x	$f(x)$
Weight (oz)	Postage (cts)
$0\leqslant x<1$	15
$1\leqslant x<2$	28
$2\leqslant x<3$	41
$3\leqslant x<4$	54
$4\leqslant x<5$	67

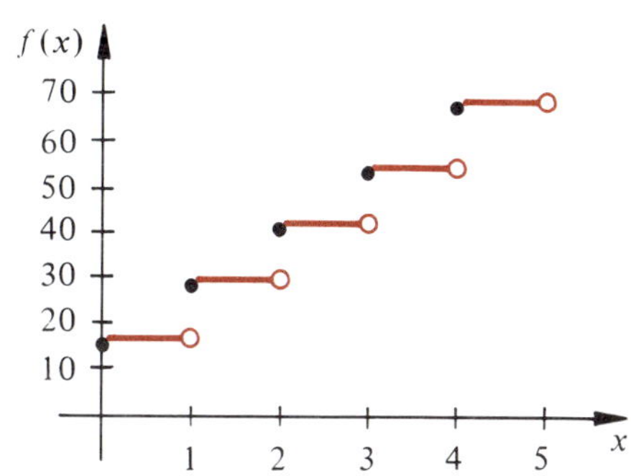

FIGURE 5.32

The cost of postage is a true function, since a vertical line at any x of the domain intersects the graph in only one place. The following examples show some additional functions that are defined by multipart rules. In each case, the rule changes over different subsets of the domain.

EXAMPLE 1 Graph f if the function is given by

$$f(x)=\begin{cases} 4, & x<0 \\ -2x+4, & 0\leqslant x\leqslant 3 \\ -2, & x>3 \end{cases}$$

SOLUTION The analytic description given describes *one* function, but to use the rule of correspondence, we need *three* parts. From Section 5.1, we know that each of the individual rules describes a straight line; these lines must be restricted to their intervals of definition, however. Figure 5.33 shows the graph of the function. The graphs of $y=4$, $y=-2x+4$, and $y=-2$ are first made as dashed curves and then the portions relevant to this function are made solid, as is shown in Figures 5.33(a) and 5.33(b).

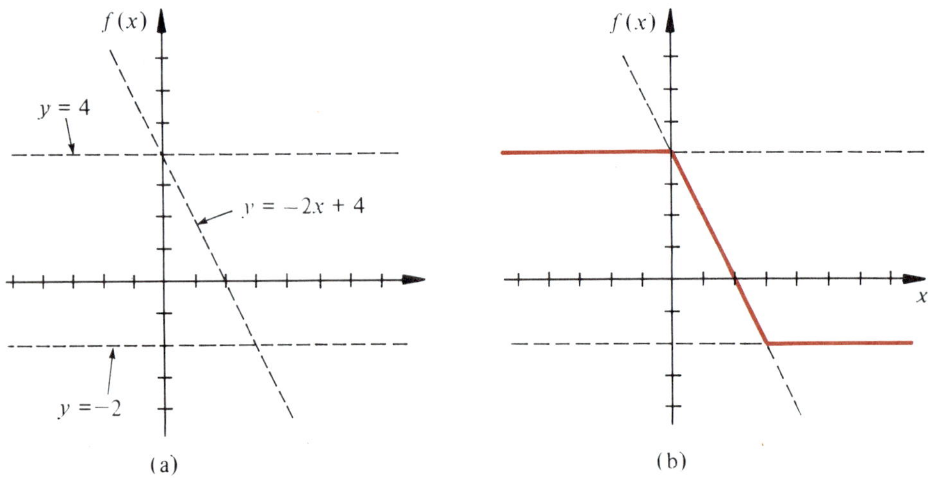

FIGURE 5.33

EXAMPLE 2 Sketch the graph of the function whose definition is

$$f(x)=\begin{cases} 1, & x<0 \\ x^2+1, & 0\leqslant x<2 \\ -x+7, & 2\leqslant x. \end{cases}$$

SOLUTION The graphs of $y=1$, $y=x^2+1$ and of $y=-x+7$ are drawn in dashed lines in Figure 5.34. The portions of these curves that make up the given function are shown as a solid line.

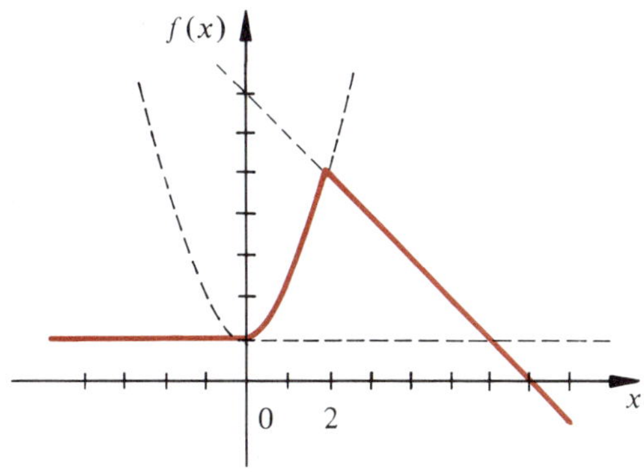

FIGURE 5.34

The kind of function defined in Examples 1 and 2 has more than just academic interest. Multipart functions can be used to describe physical situations, as Examples 3 and 4 show.

EXAMPLE 3 In static-testing an airplane wing for fatigue, a load is to be applied that simulates the actual loads experienced in a 5-hr flight. For a load function L given by

$$L(t)=\begin{cases} 20{,}000 \text{ lb}, & 0\leqslant t<1 \\ 15{,}000 \text{ lb}, & 1\leqslant t<2 \\ 25{,}000 \text{ lb}, & 2\leqslant t<4 \\ 20{,}000 \text{ lb}, & 4\leqslant t\leqslant 5 \end{cases}$$

plot the graph of the load applied to the wing as a function of time.

SOLUTION See Figure 5.35. Notice that the physical situation restricts the domain of the load function to the set $\{t|0\leqslant t\leqslant 5\}$.

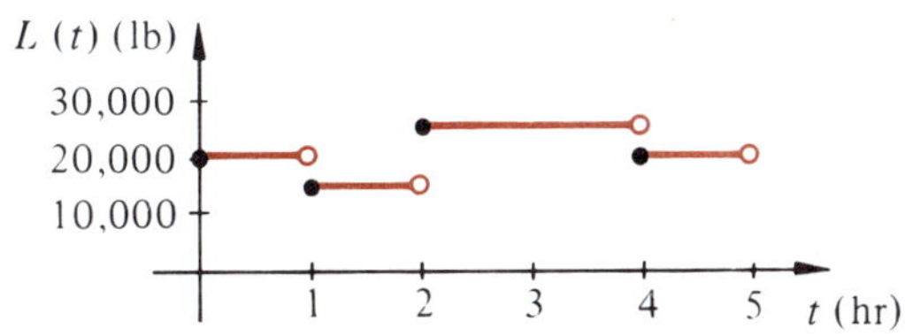

FIGURE 5.35

EXAMPLE 4 It is expected that yearly maintenance cost for a water purification system will vary with time according to

$$C(t)=\begin{cases} 20t, & 0\leqslant t<10 \\ 30t-100, & 10\leqslant t\leqslant 25 \end{cases}$$

Graph this function for C in dollars and t in years.

SOLUTION The graph of $C=20t$ is a straight line through the origin, but only the portion for $0\leqslant t\leqslant 10$ is a part of the graph of the given function. The equation $C=30t-100$ is another straight line and represents the function for $10 \leqslant t \leqslant 25$. The given function is undefined for $t<0$ and for $t>25$. Its graph is shown in Figure 5.36.

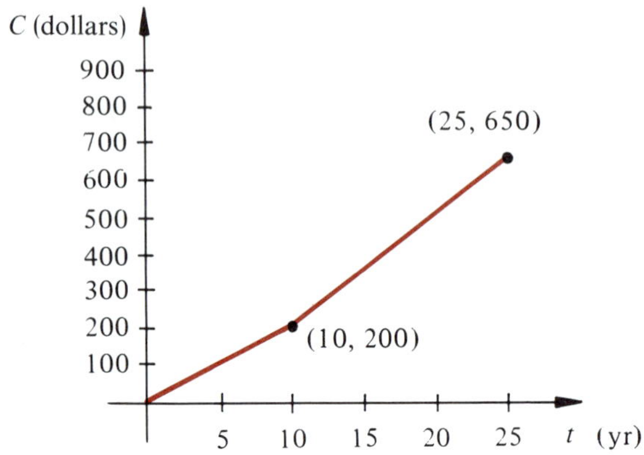

FIGURE 5.36

Exercises for Section 5.7

Graph the function specified in each of Exercises 1–20.

1. $f(x)=\begin{cases} 2, & x<0 \\ 3, & x \geqslant 0 \end{cases}$

2. $g(x)=\begin{cases} -1, & x<0 \\ 1, & x \geqslant 0 \end{cases}$

3. $F(t)=\begin{cases} 0, & t \leqslant 0 \\ t, & t>0 \end{cases}$

4. $f(x)=\begin{cases} x, & x<0 \\ -2, & x \geqslant 0 \end{cases}$

5. $h(x)=\begin{cases} 0, & x<2 \\ 2x-4, & x \geqslant 2 \end{cases}$

6. $H(t)=\begin{cases} -2, & t<0 \\ t-2, & t \geqslant 0 \end{cases}$

7. $G(x)=\begin{cases} 0, & x<0 \\ 1, & 0<x<2 \\ 0, & x \geqslant 2 \end{cases}$

8. $f(t)=\begin{cases} 0, & t<0 \\ 2t, & 0 \leqslant t<3 \\ 6, & t \geqslant 3 \end{cases}$

9. $y(x)=\begin{cases} 0, & x<-2 \\ x^2+2, & -2<x<0 \\ 0, & x>0 \end{cases}$

10. $z(x)=\begin{cases} x^2, & x<0 \\ -x^2, & x \geqslant 0 \end{cases}$

11. $f(x)=\begin{cases} x, & x<1 \\ -x, & x \geqslant 1 \end{cases}$

12. $f(x)=\begin{cases} 0, & x \text{ an integer} \\ 1, & x \text{ not an integer} \end{cases}$

13. $f(x)=\begin{cases} x+2, & x<-2 \\ x^2, & -2 \leqslant x<3 \\ x-2, & x \geqslant 3 \end{cases}$

14. $g(x)=\begin{cases} x, & x \leqslant -2 \\ x^2-4, & -2<x<1 \\ x, & x \geqslant 1 \end{cases}$

15. $f(x)=\begin{cases} (x-3)(x+2), & x<-2 \\ x^2-4, & x \geqslant -2 \end{cases}$

16. $f(x)=\begin{cases} x+3, & x \leqslant -3 \\ (x+3)(x-2), & x>-3 \end{cases}$

17. $h(x)=\begin{cases} x^2+4x+4 & x<0 \\ 4x-4-x^2, & x \geqslant 0 \end{cases}$

18. $f(x)=\begin{cases} x^2+5x+4, & x<0 \\ 5x-4-x^2, & x \geqslant 0 \end{cases}$

19. $f(x)=\begin{cases} 2x+3, & x<-2 \\ x^2-1, & -2 \leqslant x \leqslant 0 \\ 1-x^2, & 0<x \leqslant 1 \\ 2x-3, & x>1 \end{cases}$

20. $f(x)=\begin{cases} 2x^2-x-1, & x<-1 \\ x^2-2x-3, & -1 \leqslant x \leqslant 0 \\ 3, & 0<x<1 \\ x+2, & x \geqslant 1 \end{cases}$

Exercises 21–32 refer to the greatest integer function. The **greatest integer function**, denoted by $[\![x]\!]$, is defined such that for a particular value of x, the value of

$[\![x]\!]$ is the largest integer that is less than or equal to x. For example, $[\![2.1]\!]=2$, $[\![-3.1]\!]=-4$, $[\![-1]\!]=-1$, and so on. More specifically,

$$[\![x]\!]=\begin{cases} \vdots & \\ -3, & -3\leqslant x<-2 \\ -2, & -2\leqslant x<-1 \\ -1, & -1\leqslant x<0 \\ 0, & 0\leqslant x<1 \\ 1, & 1\leqslant x<2 \\ 2, & 2\leqslant x<3 \\ \vdots & \end{cases}$$

21. Sketch the graph of the greatest integer function.

In Exercises 22–32, sketch the graph of the given function.

22. $f(x)=2[\![x]\!]$

23. $f(x)=[\![2x]\!]$

24. $f(x)=2[\![x]\!]-[\![x]\!]$

25. $f(x)=[\![x^2]\!]$

26. $f(x)=[\![x^2+1]\!]$

27. $f(x)=[\![x]\!]-3[\![x]\!]$

28. $f(x)=[\![x^2]\!]+1$

29. $f(x)=2[\![x]\!]-3[\![x]\!]+1$

30. $f(x)=[\![x-1]\!]$

31. $f(x)=[\![x]\!]^2$

32. $f(x)=[\![x^2]\!]-[\![x]\!]^2$

33. Sketch the load-time curve for a load, in kilograms, applied to a tensile specimen, that varied with time according to

$$F(t)=\begin{cases} 1000t, & 0\leqslant t\leqslant 5 \text{ min} \\ 5000, & 5<t\leqslant 60 \text{ min} \end{cases}$$

34. Sketch the voltage-time curve for a voltage applied to a given circuit that varied with time according to

$$v(t)=\begin{cases} 0 \text{ volts}, & t<2 \\ 6 \text{ volts}, & t\geqslant 2 \end{cases}$$

35. The cost of concrete to a contractor varies with the number of yards delivered according to

$$c(n)=\begin{cases} 60 \text{ dollars}, & 0\leqslant n\leqslant 3 \\ 20n \text{ dollars}, & 3<n\leqslant 10 \end{cases}$$

Sketch the graph.

36. A chemical supply house charges \$1.50 a gallon for muriatic acid on orders up to 10 gal. For orders from 10 to 20 gal, the price is \$1.25 a gallon; and for orders from 20 to 30 gal, the price is \$1.10 a gallon. Orders in excess of 30 gal cost \$1.00 a gallon. Write the rule for the function that relates the cost per gallon to the size of the order. Sketch the graph.

5.8 The Absolute Value Function

One of the most useful multipart functions is the **absolute value function** $f(x)=|x|$, where $|x|$ is the "absolute value of x," defined as follows.

DEFINITION

> The absolute value of x, denoted by $|x|$, is given by
>
> $$|x|=\begin{cases} x, & x\geqslant 0 \\ -x, & x<0 \end{cases}$$

When we consider $|x|$, we sometimes say that we are "taking the absolute value of x." This obviously means considering the numerical value of x independent of the positive or negative quality. Thus, since the domain of the absolute value of x is the set of all real numbers, the range is the set of nonnegative real numbers.

The graph of $f(x)=|x|$ consists of a combination of the graphs of $y=-x$ (for $x<0$) and $y=x$ (for $x\geqslant 0$). These two graphs are shown in Figure 5.37; the portion relevant to the given function is drawn as solid lines.

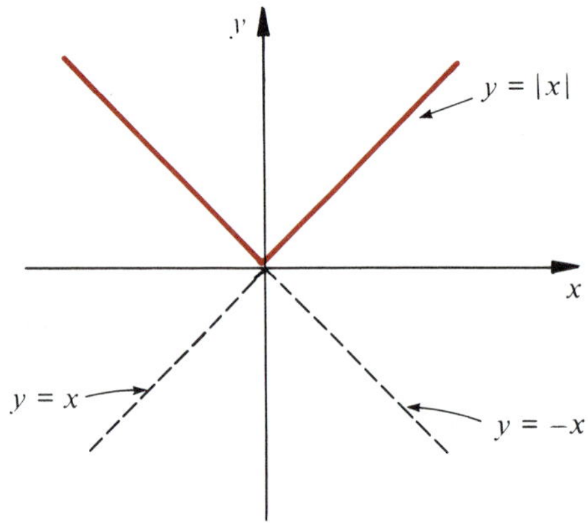

FIGURE 5.37

Often we want to take the absolute value of entire algebraic quantities. The absolute value of the quantity $u(x)$ is

$$|u(x)| = \begin{cases} u(x), & u(x) \geqslant 0 \\ -u(x), & u(x) < 0 \end{cases}$$

EXAMPLE 1 Sketch the graph of $f(x) = |x-2|$.

SOLUTION This function is given by the two-part rule

$$f(x) = \begin{cases} x-2, & x-2 \geqslant 0 \text{ (i.e., for } x \geqslant 2) \\ -(x-2), & x-2 < 0 \text{ (i.e., for } x < 2) \end{cases}$$

Thus the graph consists of a part of $y = x-2$ and a part of $y = -x+2$. See Figure 5.38.

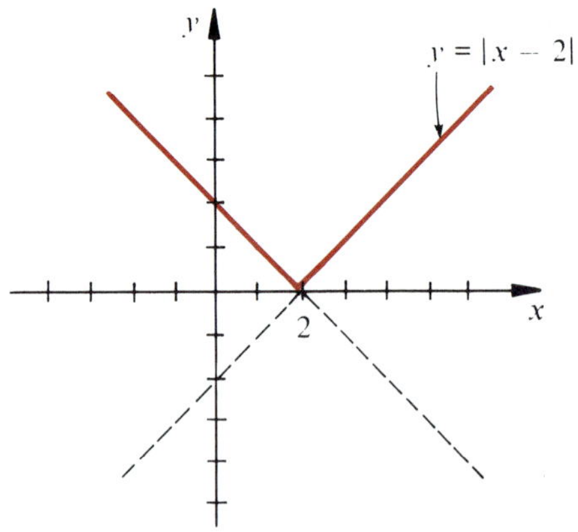

FIGURE 5.38

EXAMPLE 2 Sketch the graph of $f(x) = |x^2 + x - 6|$.

SOLUTION In this case

$$f(x)=\begin{cases} x^2+x-6 & \text{if } x^2+x-6\geqslant 0 \\ -x^2-x+6 & \text{if } x^2+x-6<0 \end{cases}$$

It should be easy for you to show that $f(x)$ has the equivalent definition:

$$f(x)=\begin{cases} (x+3)(x-2) & \text{for } x\leqslant -3 \quad \text{and} \quad x\geqslant 2 \\ -(x+3)(x-2) & \text{for } -3<x<2 \end{cases}$$

Effectively, you are asked to sketch the parabola $y=x^2+x-6$ and reflect any negative portion in the x-axis. The graph is shown in Figure 5.39.

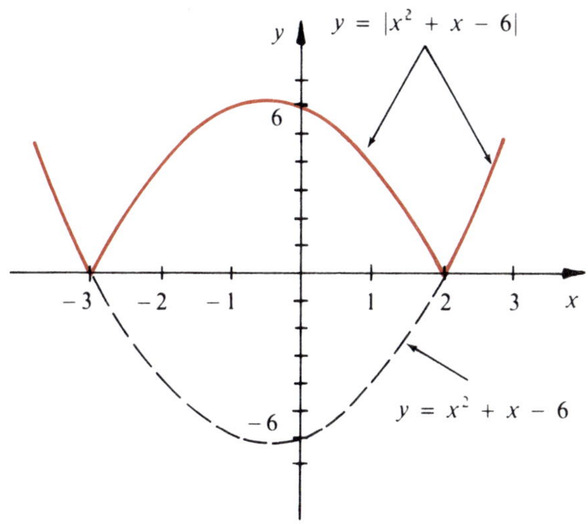

FIGURE 5.39

To solve equations that involve absolute values, the different forms of the absolute value expressions must be considered. The equation itself will take on different forms caused by the absolute value expressions.

EXAMPLE 3 Solve the equation $|x-1|+|2x+5|=6$.

SOLUTION Since

$$|x-1|=\begin{cases} x-1 & \text{for } x\geqslant 1, \\ -(x-1) & \text{for } x<1 \end{cases}$$

and

$$|2x+5|=\begin{cases} 2x+5 & \text{for } x\geqslant -\frac{5}{2} \\ -(2x+5) & \text{for } x<-\frac{5}{2} \end{cases}$$

there are three intervals on the real line to consider:

$$\left(-\infty, -\frac{5}{2}\right), \left[-\frac{5}{2}, 1\right), \text{ and } [1, \infty)$$

The given equation has the following three forms:

$$(x-1)+(2x+5)=6 \quad \text{on } [1, \infty)$$

$$-(x-1)+(2x+5)=6 \quad \text{on } \left[-\frac{5}{2}, 1\right)$$

$$-(x-1)-(2x+5)=6 \quad \text{on } \left(-\infty, -\frac{5}{2}\right)$$

Each of these equations is linear. The solutions are, in order,

$$x=\frac{2}{3}, \quad x=0, \quad \text{and} \quad x=-\frac{10}{3}$$

The "solution" $x=\frac{2}{3}$ is not in the allowable domain of values of that particular equation and hence must be rejected. Thus, the solution set for the given equation is

$$\left\{0, -\frac{10}{3}\right\}$$

Equalities in two variables that involve absolute value expressions reduce to several different linear equations, each valid on some subset of the plane.

EXAMPLE 4 Sketch the graph of the equation $|2x|+|y|=1$.

SOLUTION Since

$$|2x|=\begin{cases} 2x & \text{for } x\geqslant 0 \\ -2x & \text{for } x<0 \end{cases} \quad \text{and} \quad |y|=\begin{cases} y & \text{for } y\geqslant 0 \\ -y & \text{for } y<0 \end{cases}$$

the given equation reduces to

$$2x+y=1 \text{ in quadrant I}$$
$$-2x+y=1 \text{ in quadrant II}$$
$$-2x-y=1 \text{ in quadrant III}$$
$$2x-y=1 \text{ in quadrant IV}$$

The graph is shown in Figure 5.40.

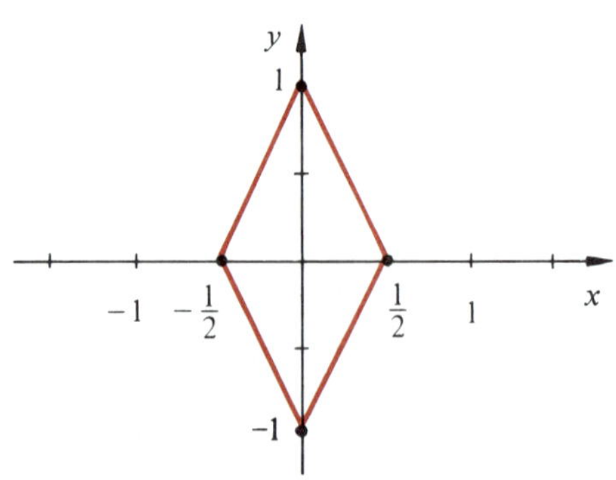

FIGURE 5.40

Exercises for Section 5.8

In Exercises 1–20, sketch the graph of the given function.

1. $f(x)=|x+1|$
2. $y=|x-4|$
3. $H(x)=|2x|$
4. $f(x)=2|x|$
5. $f(x)=|x|-3$
6. $G(t)=|t-3|$
7. $y=|x|+3$
8. $f(x)=|x|-5$
9. $f(t)=|t^2|$
10. $f(x)=|x|^2$
11. $f(x)=x^2-|x|+1$
12. $f(x)=|x^2-x+1|$
13. $g(x)=|2x^2+x-1|$
14. $y=|x^2+x+1|$
15. $y=x^2+3|x|+2$
16. $f(x)=|x^2+3x+2|$
17. $f(x)=(x-1)|x+1|$
18. $F(x)=(x-3)|x+2|$
19. $f(x)=x|x-4|$
20. $f(x)=(x-5)|x+1|$
21. Give an example that shows that $f(x+y)\neq f(x)+f(y)$ if $f(x)=|x|$.
22. Compare $f(x)=|x|$ and $g(x)=|x-2|$ by sketching their graphs on the same set of coordinates. How do the two graphs differ?
23. Compare $f(x)=|x|$ and $g(x)=|x|+2$ by sketching their graphs on the same set of coordinates. How do the two graphs differ?
24. Compare $f(x)=|x|$, $g(x)=|2x|$, $h(x)=2|x|$ by sketching their graphs on the same set of coordinates. How do the graphs differ?

Solve the equations of Exercises 25–40 by considering the different forms that the equation takes on different intervals.

25. $|x|+1=3$
26. $|x|+2=3$
27. $|x+1|=5$
28. $|x+2|=4$
29. $|x|+2=3x$
30. $2|x|+1=|x|$
31. $|x|+x=3$
32. $|x|+x=-2$
33. $|x-1|+x=|x|$
34. $|x+4|=|x-5|$
35. $|3x+1|=|x-5|$
36. $|x+3|+|x-1|=2$
37. $|x-1|+2|x+4|=x$
38. $|x-1|+2|x+4|=|x|$
39. $|x-4|+|x+1|=|2x+5|$
40. $|3x+4|+|2x-1|=|x+6|+|x|$

Sketch the graph of the equations in Exercises 41–50, considering the different expressions caused by the absolute value function.

41. $|x|+|y|=1$
42. $|x|-|y|=1$
43. $|x-1|=|y+2|$
44. $|2x+5|+|y+3|=5$
45. $|y|=2|x-1|$
46. $|y|=|x|$
47. $|x+1|+y=|y|$
48. $|y|=-|2x+3|$
49. $|x|+|x-1|+|y|=2$
50. $|y|=2|x|-1$

5.9 Tips on Curve Sketching

Much of our study of the elementary functions has concerned graphs of functions. We conclude the chapter with a discussion of graphs of more general relations such as $x^2-y^2=16$ and $y^2(x-1)=x$. The graphs of these relations can be sketched by plotting points and then drawing a smooth curve through the plotted points. Plotting points is tedious, however, and usually not very illuminating about the nature of the graph, so we look for ways of minimizing this effort. It is generally beneficial to analyze the relation for certain basic properties before plotting any points. The basic properties used in curve plotting are *intercepts*, *symmetry*, *excluded values*, *behavior for large values of the variables*, and *factorability of the equation*.

Intercepts

The intercepts of a curve are the points at which the curve crosses the x- and y-axes. To find the x-intercepts, let $y=0$ and solve for x. To find the y-intercepts, let $x=0$ and solve for y. The curve $x^2-y^2=4$ for example, has x-intercepts $x=\pm 2$ and no y-intercepts.

Symmetry

Curves can be symmetric relative to many different points and lines. However, we are interested in symmetry with respect only to the origin or the two axes, since it is easy to check this kind of symmetry. (See Figure 5.41.)

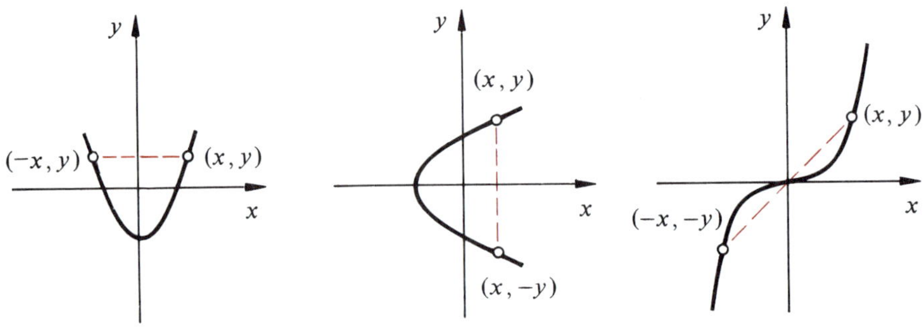

FIGURE 5.41 Symmetry about the y–axis Symmetry about the x–axis Symmetry about the origin

(1) A curve is symmetric with respect to the y-axis if both the points (x, y) and $(-x, y)$ are on the curve. Hence, the simple test for symmetry with respect to the y-axis is: *If the equation remains unchanged when x is changed to $-x$, then the curve is symmetric with respect to the y-axis.* For example, the curve $y+x^2-2=0$ is symmetric with respect to the y-axis since $y+(-x)^2-2=y+x^2-2$.

(2) A curve is symmetric with respect to the x-axis if both (x, y) and $(x, -y)$ are on the curve. Hence, *if the equation remains unchanged when y is changed to $-y$, then the curve is symmetric with respect to the x-axis.* For example, the graph of $x^2+2x+y^2=7$ is symmetric with respect to the x-axis, since $x^2+2x+(-y)^2=x^2+2x+y^2$.

(3) A curve is symmetric with respect to the origin if both (x, y) and $(-x, -y)$ are on the curve. *A simple test for this kind of symmetry is to change x to $-x$ and y to $-y$. If the equation remains unchanged, the curve is symmetric with respect to the origin.* The equation $y=x^3+4x$ is symmetric with respect to the origin, since $-y=(-x)^3+4(-x)$ simplifies to the original equation.

Excluded Values

By noticing which values of the variable are not permitted, the curve may be restricted to certain portions of the rectangular plane. Values of x and y must be excluded for which the other variable would be nonreal. For example, the equation $x^2+2y^2=1$ implies that $x^2=1-2y^2$. Hence, since $1-2y^2$ must be greater than or equal to 0, it follows that y is limited by the values $-1/\sqrt{2} \leq y \leq 1/\sqrt{2}$. Likewise, since $y^2=\frac{1}{2}(1-x^2)$, x is limited to the values $-1 \leq x \leq 1$.

For many functions, portions of the xy plane can be excluded merely by noticing where the value of y is positive and negative.

EXAMPLE 1 Determine where the graph of $y=(x-1)/(x+2)$ is above and below the x-axis.

SOLUTION To find where the graph is above the x-axis, solve the inequality $(x-1)/(x+2)>0$. Note that $x-1$ is positive for $x>1$ and negative for $x<1$. The factor $x+2$ is positive for $x>-2$ and negative for $x<-2$. The accompanying schematic represents the sign of the factors. Hence, the graph is above the x-axis for $x<-2$ and for $x>1$, since the quotient is positive there. The graph is below the x-axis between -2 and 1, since the quotient is negative for this interval.

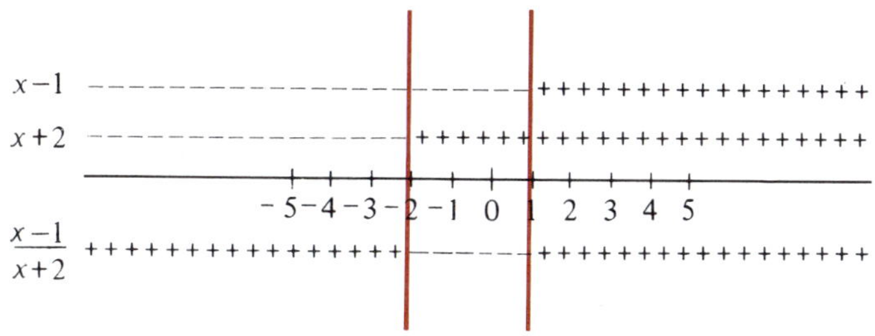

Behaviour for Large Values of the Variables

Sometimes when x and y become large, it is possible to approximate the given equation by a far simpler equation. The equation $y_1=x^2+(1/x)$, for example, tends to behave like $y_2=x^2$ as x becomes large. The curves representing y_1 and y_2 are said to be *asymptotic*. In Section 5.6, we introduced a particular kind of asymptotic behavior, in which y (or x) approaches a finite value as x (or y) becomes large. Such asymptotes are either horizontal or vertical lines. A more general analysis of asymptotic behavior is particularly useful for sketching some curves.

EXAMPLE 2 Sketch the graph of $y=x^2+(1/x)$.

SOLUTION For large positive x, the graph approaches $y=x^2$ from above, and for large negative x, the graph approaches $y=x^2$ from below. Thus, the desired graph is asymptotic to

$y=x^2$ for large $|x|$. As x approaches zero through positive values, the value of y becomes arbitrarily large. Similarly, the value of y becomes arbitrarily large but negative as x approaches zero through negative values. Hence, the line $x=0$ is a vertical asymptote. The point $(-1, 0)$ is obviously the only intercept. Figure 5.42 shows the graph using this information.

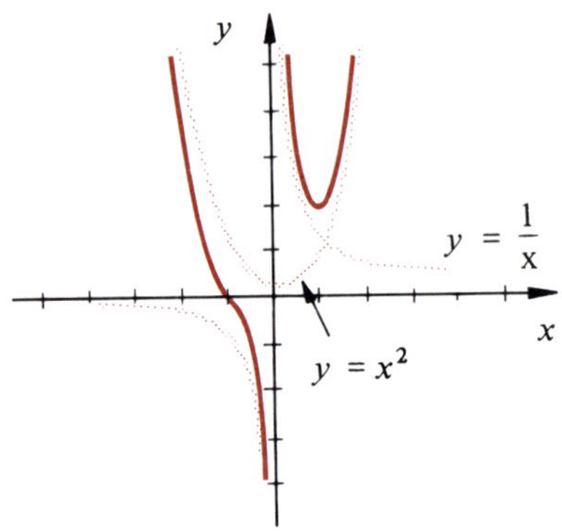

FIGURE 5.42

Asymptotic behavior is usually not obvious without some preliminary algebraic operation.

EXAMPLE 3 Locate the vertical and horizontal asymptotes for the curve whose equation is $xy-y-2x+3=0$. Sketch the graph.

SOLUTION Solving the given equation for y, we get

$$y=\frac{2x-3}{x-1}$$

We see by this form that as x approaches 1, the value of y becomes arbitrarily large. Hence, $x=1$ is a vertical asymptote. Solving the original equation for x gives

$$x=\frac{y-3}{y-2}$$

By this we see that there is a horizontal asymptote at $y=2$. With just a few more facts, we can easily draw the graph of the equation. From the first equation, the y-intercept is $y=3$; and from the second equation, the x-intercept is $x=\frac{3}{2}$.

Finally, the first equation shows that y is positive for $x>\frac{3}{2}$ and $x<1$ and negative for $1<x<\frac{3}{2}$. Therefore, the curve is below the x-axis for $1<x<\frac{3}{2}$ and above otherwise. Figure 5.43 shows the sketch of the curve.

Factorable Equations

If an equation in x and y can be factored into two or more factors, then the values of (x, y) that satisfy either one of the resulting equations will satisfy the total equation

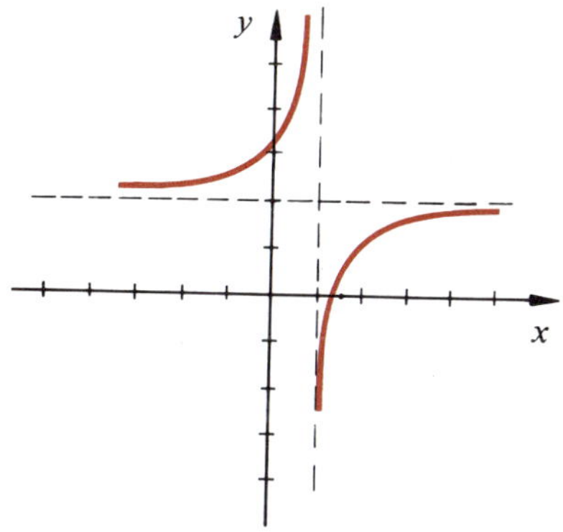

FIGURE 5.43

and hence determine a point on the curve. Thus, the total graph is made up of the graphs of the two separate factors.

EXAMPLE 4 Graph $x^2y-xy^2-x+y=0$.

SOLUTION The expression x^2y-xy^2-x+y can be factored into $(xy-1)(x-y)$. Therefore, the required graph is the union of the graph of $xy-1=0$ and the graph of $x-y=0$. Figure 5.44 displays the desired graph.

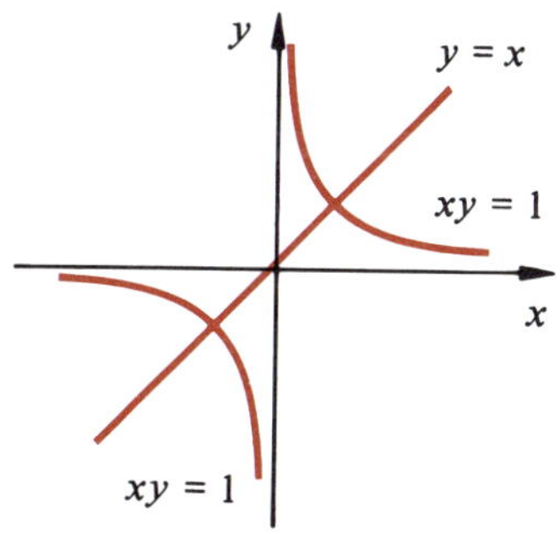

FIGURE 5.44

Exercises for Section 5.9

Discuss and draw the graphs of the equations in Exercises 1–25. In each case, determine intercepts, excluded regions, and asymptotic behavior.

1. $x^2-3y=9$
2. $y^2-x^2=16$
3. $y^2-x^2=0$
4. $y^3=x+4y$
5. $y^2=x(x-4)^2$
6. $y=x^3-4x$
7. $y^2-4y-x+3=0$
8. $xy=0$
9. $x^2y^2=2xy$
10. $x^2+y^2=1-2xy$
11. $y=\dfrac{5}{x-1}$
12. $y=\dfrac{x-1}{3-x}$
13. $y=\dfrac{x^2-2x}{x^2-2x-3}$
14. $y=\dfrac{x^2-1}{x^2+1}$

15. $y^2(x-1)=x$

16. $x^2+2xy+y^2+1=0$

17. $y^2=x^2(x-3)$

18. $x^2y^2=4-x^2$

19. $x^2y+xy^2-x-y=0$ (*Hint*: It factors.)

20. $y(1-x)=x+4$

21. $y^2(1-x)=x+4$

22. $y=x+\frac{1}{x}$

23. $y=x-\frac{1}{x}$

24. $y=x^2-\frac{1}{x}$

25. $y=x^2-\frac{1}{x^2}$

Test 1 for Chapter 5

For the statements in Exercises 1–10, answer *True* or *False*.

1. $x+1/x=0$ is a linear equation.
2. The slope of a straight line is a constant.
3. If a line passes through the origin, one of its forms is $x+y=1$.
4. If $f(x)=|x|$, then $f(x^2)=x^2$.
5. The slope of a line parallel to the line $x=4$ is 1.
6. The graph of a quadratic function is a parabola.
7. The graph of a quadratic function with real coefficients always intersects the y-axis.
8. If a quadratic function has real zeros, the value of the x-coordinate of the vertex is the average of these zeros.
9. The function $f(x)=(x-1)(1+x)$ is a quadratic.
10. Any quadratic function with real zeros can be written as a product of real linear factors.
11. Write the equation of the line that is perpendicular to $x-3y=7$ at the point $(1, -2)$.
12. Write the equation of the line that passes through $(1, 3)$ and $(-1, -2)$. Express in slope-intercept form and sketch the line.
13. Sketch the function

$$f(x)=\begin{cases} x^2, & x\leqslant 0 \\ 2, & 0<x<3 \\ -x, & 3\leqslant x \end{cases}$$

14. Sketch the graph of $y=\dfrac{2}{x^2-2x}$
15. Sketch the graph of $y=3+x-x^2$ after locating the x-intercepts, if any, the y-intercept, and the vertex.
16. Sketch the graph of

$$f(x)=|(x-3)(2+x)|.$$

Test 2 for Chapter 5

1. Draw the graph of $2x+3y=-6$.
2. Find the equation of the line through $(-1, 3)$ and $(0, -5)$.
3. Draw the graph of $y+|x|=2$.
4. Draw the graph of $y=x^2-x-12$, indicating intercepts and vertex.
5. Solve the equation $|x|+2|x-1|=3$.
6. Draw the graph of $f(x)=\begin{cases} 2, & x<-1 \\ x, & -1\leqslant x<2 \\ 2-x, & x\geqslant 2 \end{cases}$
7. Write the equation of the line perpendicular to $5x-2y=4$ at $(2, 3)$.
8. Write the equation of the quadratic function whose graph has x-intercepts $(-3, 0)$ and $(1, 0)$ and y-intercept $(0, 2)$.
9. Let $f(x)=|x-2|$. Give the domain and range of the function and graph it.
10. Sketch the graph of $y=\dfrac{x+2}{x-3}$.

6

Exponential and Logarithmic Functions

6.1 Exponential Functions

The initial use of an exponent is as an indicator of the *power* of a given number. Thus, the notation a^n means

$$\underbrace{a \cdot a \cdot a \cdot a \cdot \cdots \cdot a}_{n \text{ factors}}$$

From this definition come several important properties, or rules, called the *rules for exponents*.

Rule 1 $\qquad a^m a^n = a^{m+n}$

Rule 2 $\qquad (a^m)^n = a^{mn}$

Rule 3 $\qquad a^m \div a^n = a^{m-n}$

The idea of an exponent can be extended to the entire set of integers by the two definitions $a^{-n} = 1/a^n$ and $a^0 = 1$, $a \neq 0$. These two definitions are consistent in that the choice of a^0 to be 1 is at least a necessity if we want the same rules for exponents to hold for integers as for natural numbers. Thus,

$$\begin{aligned} a^0 &= a^{n-n} \\ &= (a^n)(a^{-n}) && \text{by Rule 1} \\ &= (a^n) \div (a^n) && \text{by definition of a negative exponent} \\ &= 1 && \text{by the law of cancellation} \end{aligned}$$

The rules can be extended to rational numbers by defining $y=a^{1/q}$ to mean $y^q=a$. Then, as before,

$$a^{p/q}=(a^{1/q})^p$$

if it is assumed that the rules for exponents continue to hold; this thereby allows any fractional exponent.

For $a<0$ and q a positive even integer, $a^{1/q}$ is not real. Hence, the consideration of exponents is limited to the case in which $a>0$. When the idea of the exponent is extended from the natural numbers to the integers and to the rational numbers, the three basic laws of exponents continue to hold. There is nothing mysterious about this—it is deliberately planned that way!

EXAMPLE 1 Use the laws of exponents to solve the equation

$$4^{x+1}=8.$$

SOLUTION Since $4=2^2$ and $8=2^3$, the given equation can be written in the equivalent form

$$(2^2)^{x+1}=2^3$$

or

$$2^{2(x+1)}=2^3$$

For these two expressions to be equal, the exponents must be equal. Therefore, we write

$$2(x+1)=3$$
$$2x+2=3$$
$$x=\tfrac{1}{2}$$

Now we want to extend the definition of an exponent to include all real numbers. Although detailed discussion of irrational exponents belongs more properly to a course in calculus, we can get some idea of the meaning attached to the quantity, say 2^π, by the following argument. Since π has a decimal expansion 3.14159 . . . , we can consider the numbers 3, 3.1, 3.14, 3.141, and so forth to be successively closer rational approximations to the irrational number π. Each obtained power of 2 from these approximations of π is in some sense an approximation of 2^π, with the accuracy increasing as more of the digits of π are used. In this way, we see that irrational exponents have a definite interpretation to which the three basic exponential laws apply. Accepting that a^x is defined for all real x if $a>0$ we make the following definition.

DEFINITION

If a is any fixed constant >0, $a\neq 1$, then the exponential function with base a is defined by

$$f(x)=a^x$$

The domain of the exponential function is all real numbers and the range is the positive real numbers.

Example 2 shows the characteristic shape of the graph of an exponential function.

EXAMPLE 2 Sketch the graph of 2^x and of $(\frac{1}{2})^x$.

SOLUTION Figure 6.1 shows a table of values for each of the functions along with the graph.

x	$(\frac{1}{2})^x$
-2	4
-1	2
0	1
1	$\frac{1}{2}$
2	$\frac{1}{4}$

x	2^x
-2	$\frac{1}{4}$
-1	$\frac{1}{2}$
0	1
1	2
2	4

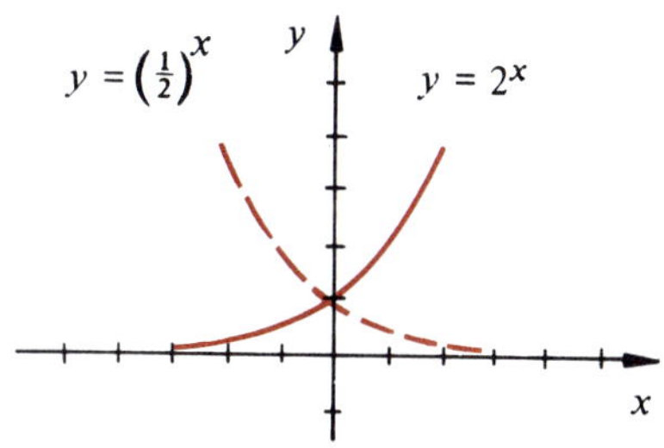

FIGURE 6.1

If we let $f(x)=2^x$, then $f(-x)=2^{-x}=(\frac{1}{2})^x$. Thus, the graphs of 2^x and $(\frac{1}{2})^x$ are "reflections" in the y-axis.

In general, we can make the following observations about the function $f(x)=a^x$ for $a>1$:

(1) $a^x>0$ for all values of x.

(2) When $x=0$, then $y=a^0=1$.

(3) When $x=1$, then $y=a^1=a$.

(4) As x increases without bound, a^x increases without bound.

(5) As x decreases without bound, a^x approaches 0.

(6) If $x_1<x_2$, then $a^{x_1}<a^{x_2}$.

Figure 6.1 is typical of the shape of *exponential curves*, while Properties (1) through (6) are the corresponding *analytic* or *functional properties* of exponential functions. If $0<a<1$, then properties (4) through (6) become:

(4′) As x increases without bound, a^x approaches 0.

(5′) As x decreases without bound, a^x increases without bound.

(6′) If $x_1<x_2$, then $a^{x_1}>a^{x_2}$.

Property (6) is described by saying that if $a>1$, then a^x is an *increasing function*, whereas Property (6′) says that if $0<a<1$, then a^x is a *decreasing function*. An increasing function has a graph that rises from left to right, while a decreasing function has a graph that falls from left to right.

A different exponential function is obtained for each value of a, although the shape of the graph remains the same and the same kind of functional properties as (1)–(6) continue to hold.

Figure 6.2 shows the graph of a^x for $a=\frac{1}{4}, \frac{1}{2}$, 1, 2, and 4. The value of $a=1$ is excluded from the class of exponential functions, since not only is the graph of 1^x trivial, but it is not of exponential shape. Further, $f(x)=1^x$ does not obey the functional properties (1) through (6), which are ordinarily associated with the exponential functions. Henceforth, the only acceptable bases will be $a>0$, $a\neq 1$.

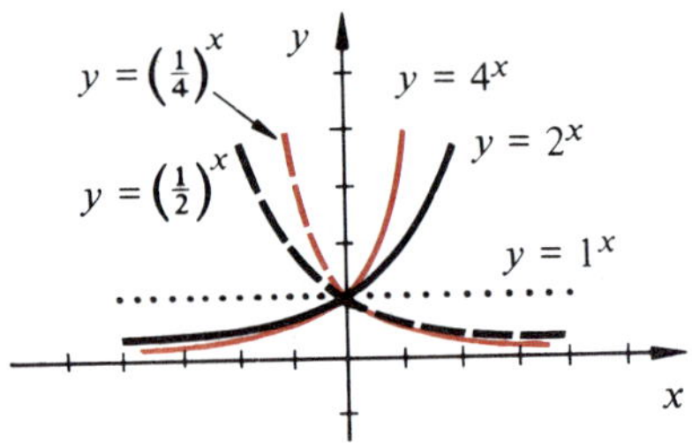

FIGURE 6.2

Example 3 shows that when f is the exponential function, $f(x)$ and $f(y)$ are related to $f(x+y)$ in a unique way.

EXAMPLE 3 Given that $f(t)=a^t$, show that $f(x+y)=f(x)\cdot f(y)$

SOLUTION If $f(t)=a^t$, then

$$f(x+y)=a^{x+y}=a^x a^y=f(x)\cdot f(y)$$

The formula $f(x+y)=f(x)\cdot f(y)$ is called the *addition formula* for the exponential function. This formula is a distinctive functional property of the exponential function in that no other function has an addition formula quite like this one.

Be aware of the fact that -3^x and $(-3)^x$ have different meanings (for example, $-3^2=-9$, but $(-3)^2=9$). The latter is an "unacceptable" exponential function since $-3<0$, while -3^x is merely the negative of 3^x. (See Figure 6.3.)

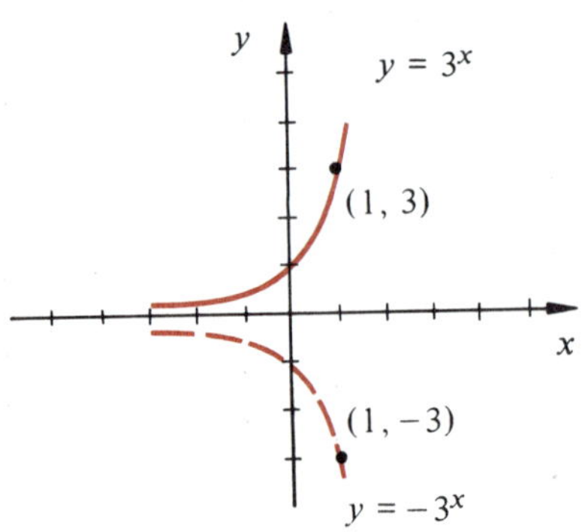

FIGURE 6.3

In some applications, we work with functions defined by

$$f(x) = a^{g(x)}$$

where $g(x)$ might be an algebraic expression such as x^2 or $\sqrt[3]{x}$. To sketch such functions, we use the basic functional properties of an exponential function supplemented by a few selected points.

EXAMPLE 4 Sketch $y = 2^{-x^2}$.

SOLUTION When $x = 0$, then $y = 1$. When $x = \pm 1$, then $y = \frac{1}{2}$. And as x increases positively without bound, y approaches 0. Also, for $x < 0$, the value y is greater than zero, approaching zero as x decreases. The graph is shown in Figure 6.4.

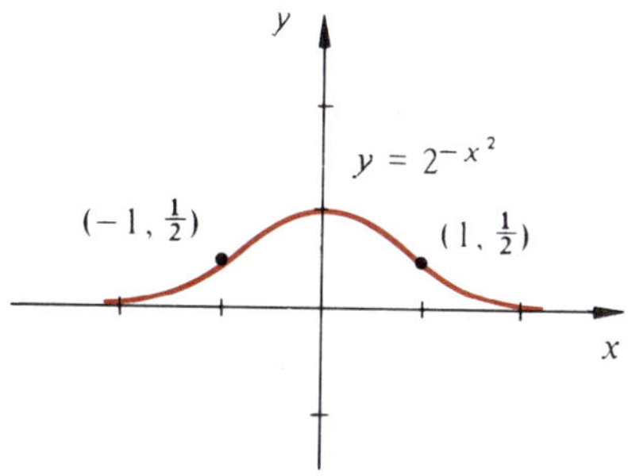

FIGURE 6.4

As you learn more about physical quantities that "vary exponentially," you will find that by far the most important base is the irrational number e. A decimal approximation of e is 2.718. Clearly, it is difficult to evaluate e^x without a calculator or access to a computer. Tables have been prepared using computers, and Table A in the appendix gives values of e^x and e^{-x}.

As an indication of how important the exponential function is, most scientific calculators have an [e^x] *button. All you have to do is to enter x and push the* [e^x] *button. Thus,* $e^{3.1} = 22.1979$ *is obtained by entering 3.1 and pushing the* [e^x] *button.*

EXAMPLE 5 Graph $y = 2 + e^x$.

SOLUTION The graph, which was constructed using values of e^x from Table A, is shown in Figure 6.5. As x decreases without bound, $2 + e^x$ approaches 2. This is shown in the figure by the dashed line.

6.2 Applications of Exponential Functions

The exponential function is one of the most useful functions for describing physical and social phenomena, of which the following are just a few:

x	e^x	$2+e^x$
-3	0.05	2.05
-2	0.14	2.14
-1	0.37	2.37
0	1.00	3.00
1	2.72	4.72
2	7.39	9.39

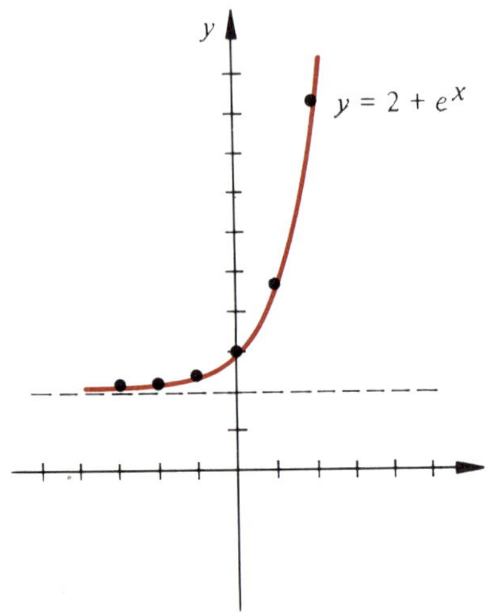

FIGURE 6.5

- Electric current
- Atmospheric pressure
- Decomposition of uranium
- Population growth
- Learning
- Compound interest

If a physical quantity obeys a law that is described by an exponential function and the quantity is increasing, it is said to **increase exponentially**. If it is exponential but decreasing, it is said to **decay exponentially.**

EXAMPLE 1 A company finds that the net sales of its product double every year. Write this fact as an exponential function.

SOLUTION Let $s(t)$ represent sales at any time t, and let s_0 be the initial sales; then we can write

$$s(0)=s_0$$
$$s(1)=2s_0$$
$$s(2)=2s(1)=2^2s_0$$
$$s(3)=2s(2)=2^3s_0$$

We could continue this process but we should not gain any new information. It is obvious from what we have done that $s(t)=s_0 2^t$ is the desired equation.

EXAMPLE 2 A psychologist finds that the number of mazes a laboratory animal can learn to run is given by $R=5(1-3^{-t})$, where t is time in days. Graph this function for $t \geqslant 0$, assuming R is a function of real numbers.

SOLUTION The graph in Figure 6.6 shows that the ability of the animals to learn new mazes levels off at 5 as t becomes large. This curve is called a *learning curve*.

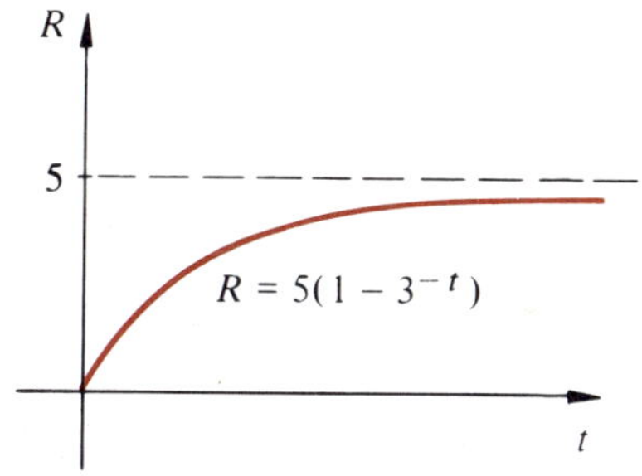

FIGURE 6.6

EXAMPLE 3 An approximate rule for atmospheric pressure at altitudes less than 50 mi is the following: *Standard atmospheric pressure, 14.7 pounds per square inch, is halved for each 3.25 mi of vertical ascent.*

(a) Write an exponential function to express this rule.

(b) Compute the atmospheric pressure for an altitude of 19.5 miles.

SOLUTION (a) Letting P denote the atmospheric pressure at altitudes less than 50 miles and h the altitude in miles, we get

$$P = 14.7\left(\frac{1}{2}\right)^{h/3.25}$$

(b) Using the expression for P and letting $h = 19.5$ gives finally

$$\begin{aligned} P &= 14.7\left(\frac{1}{2}\right)^{19.5/3.25} \\ &= 14.7\left(\frac{1}{2}\right)^{6} \\ &= \frac{14.7}{64} \\ &= 0.23 \text{ lb/in.}^2 \end{aligned}$$

EXAMPLE 4 An unmanned satellite has a radioisotope power supply whose power output in watts is given by

$$P = 50e^{-t/260}$$

where t is the time in days that the battery has been in operation. How much power will be available at the end of a year?

SOLUTION Applying the given formula with $t = 365$ yields

$$P = 50e^{-365/260} = 50e^{-1.4}$$

From Table A, $e^{-1.4}=0.2466$. Hence,

$$P=(50)(0.2466)$$
$$=12.33 \text{ watts}$$

EXAMPLE 5 Population growth is described by an exponential function of the form $P=P_0e^{kt}$, where P_0 is the initial population size, t is the elapsed time, and k is a constant. Estimate what the population of the asteroid Malthus will be in 2025 and 2075 if the population in 1975 is 500. Assume that $k=0.032$.

SOLUTION The growth equation for this population is

$$P=500e^{0.032t}$$

Using this formula with $t=50$, we have

$$P=500e^{1.6}=500(4.9530)=2476$$

If $t=100$, the population is

$$P=500e^{3.2}=500(24.533)=12{,}266$$

Thus, the population of the asteroid in 2025 will be 2476, or almost five times what it was in 1975. By 2075, the population will have increased to 12,266, which is about 25 times the population of 1975.

Exercises for Sections 6.1 and 6.2

Solve for x in Exercises 1–10.

1. $x^{-3/2}=\frac{27}{8}$

2. $27^{-2/3}=x$

3. $27^{x+2}=9^{2-x}$

4. $2^{x+4}=8$

5. $(-2)^{x+2}=-8$

6. $(-x)^{-3}=\frac{27}{64}$

7. $[(3)^x]^x=1$

8. $(2^x)^{x-1}=64$

9. $3^{x^2+4x}=\frac{1}{81}$

10. $4^{x^2-3x}=256$

Sketch the graph of the exponential functions in Exercises 11–26.

11. $y=5^x$

12. $h=10^x$

13. $y=3^{-x}$

14. $p=10^{-t}$

15. $s=3(2^x)$

16. $y=2+2^x$

17. $y=2-2^x$

18. $y=(\sqrt{2})^x$

19. $m(t)=3^{3-t}$

20. $y=1+e^x$

21. $y=e^{x/2}$

22. $y=10+e^{3x/2}$

23. $f(t)=e^{1-t}$

24. $z=2e^T$

25. $y=e^2e^x$

26. $g(x)=\dfrac{e}{e^x}$

27. Sketch the graph of $y=2^x$. Using this graph, approximate the values of (a) $2^{\sqrt{2}}$; (b) 2^{π}; (c) $2^{3/2}$; (d) $2^{-3/2}$; (e) $2^{1/3}$.

28. Why is the base of exponential functions restricted to positive numbers?

29. Discuss the exponential function with base 1.

In Exercises 30–35, say how the graph of the given exponential compares with the graph of $y=2^x$. In each case, make a sketch to show the comparison.

30. 2^{-x}

31. -2^x

32. $2^{|x|}$

33. $|2^x|$

34. $3+2^x$

35. 2^{x+3}

36. The population of water spiders on a pond is estimated at 1000. Using the population growth equation, estimate the population for $t=25$ if $k=0.02$.

37. The mass m of a certain radioactive substance decays with time according to $m=100e^{-0.1t}$ kilograms (kg), where t is the time in days. Use Table A to draw an accurate sketch of mass versus time for $t>0$. Using this graph, estimate the half-life of the substance. *Hint:* The half-life is the time in which a substance decays to one-half its original value.

38. A company has found that a person new to the assembly line produces items according to the function

$N=50-50e^{-0.5t}$

where t is the number of days he has worked on the line. How many items will the new worker produce on the fifth day? Graph this function for $t>0$, assuming it to be a function of real numbers, not just the positive integers. This graph is called the *learning curve*.

39. If P dollars is deposited in a bank paying i interest compounded annually for t years, the amount of money A in the account is given by the formula $A=P(1+i)^t$. Find the amount of money in an account after 5 yr for an initial investment of $1000 at an interest rate of 5 percent.

40. With no promotional activity, the sales of a product decreases at a rate described by the exponential function $A(t)=A_0e^{-kt}$, where A_0 is the amount of the sales when $t=0$ and k is the *sales decay constant*. What will the sales be after 1 yr if $A_0=10{,}000$ for a sales decay constant of 0.5?

Use a calculator to evaluate Exercises 41–48.

41. Find the velocity of a rocket for $t=2.75$ sec, and $v=10.772e^{2.032t}$ m/sec.

42. If the mass of a radioactive material varies with time according to $m=558.76e^{-0.01749t}$ grams (g), what is the mass when (a) $t=3.95$? (b) $t=5.50$?

43. What is the temperature of an object for $t=7.5$ min if the temperature T varies according to $T=198.6+95.3e^{-0.147t}$?

44. Given $F(x)=16.79x^2e^{-3.11x}$, find $F(1.665)$.

45. Given $f(t)=\sqrt{t+3.346}\,e^{0.0178t}$, find $f(30.542)$.

46. Given $g(t)=t^3e^{t^2}+156.79$, find $g(2.069)$.

47. Given $H(x)=(e^{0.15x}+e^{-0.15x})e^{3.19-x}$, find $H(0.7683)$.

48. Given $r(n)=0.00156(8013.6-13.607e^{1.116n})/n$, find $r(9.0075)$.

6.3 The Logarithm Function

The graph of the exponential function $y=a^x$ shows that any horizontal line will intersect the graph only once.* Hence, the function is one-to-one; that is, for each value of x there is at most one value of y, and for each value of y, there is a unique x. Using this property, we can define another important function.

DEFINITION

> Let $a>0$ and $a\neq 1$. Then
>
> $y=\log_a x$ means $x=a^y$
>
> The function $y=f(x)$ is called the **logarithm function** of x to the base a.

Thus, the value of $y=\log_a x$ is nothing more than the exponent to which a must be raised to give the value x. Another way of saying the same thing is, *Every logarithm is an exponent of the base*. Example 1 shows how the logarithm of a number is found directly from the definition.

*Section 4.6 should be reviewed before you read this section.

EXAMPLE 1 (a) $\log_2 16 = 4$ because $2^4 = 16$.

(b) $\log_2 (\frac{1}{8}) = -3$ because $2^{-3} = \frac{1}{8}$.

(c) $\log_{10} (\frac{1}{1000}) = -3$ because $10^{-3} = \frac{1}{1000}$.

Note that the logarithm is an exponent in each of the cases.

EXAMPLE 2 Find $\log_2 32$.

SOLUTION Let $y = \log_2 32$. Then $32 = 2^y$, or $y = 5$.

EXAMPLE 3 Find b, given $\log_b 4 = -\frac{1}{2}$.

SOLUTION Writing this in exponential form, we have $4 = b^{-1/2}$. Solving $4 = 1/(b^{1/2})$ for b, we find

$$16 = \frac{1}{b}, \quad \text{from which} \quad b = \frac{1}{16}$$

EXAMPLE 4 Solve for x in $3 = \log_4 (x^2 + 2)$.

SOLUTION By the definition of the logarithm,

$$3 = \log_4 (x^2 + 2)$$
$$4^3 = x^2 + 2$$

which means that $x^2 = 62$. Hence,

$$x = \pm\sqrt{62}$$

Because of the definition, the domain and the range of the logarithmic function are the same as the range and the domain of the exponential function, respectively. Thus, the domain of $y = \log_a x$ is all positive reals (sometimes we say that the logarithm of a nonpositive number is undefined) and the range is all real numbers. You will recall (Section 4.6) that a pair of functions, such as the exponential and logarithmic functions with their domains and ranges interchanged and with the same correspondence, are **inverses** of one another. In practice, many problems that involve logarithms call for little more than shifting from the exponential relation to the logarithmic relation or vice versa.

EXAMPLE 5 The intensity (energy) level β of a sound wave with intensity I is defined to be

$$\beta = 10 \log_{10} \frac{I}{I_0} \text{ decibels}$$

where I_0 is the minimum intensity detectable by the human ear. When two sounds differ in intensity by a factor of 10, they differ in loudness by 1 bel; a difference of 100 means a loudness difference of 2 bels. In practice, the unit used is the decibel, one-tenth of a bel.

EXAMPLE 6 The velocity of a space vehicle at launch when its propellant is burned to depletion is expressed by the equation $v = c \log_e R$, where v is the velocity gained during the burn, c is the exhaust velocity of the burned fuel, and R is the mass ratio, defined by takeoff weight/burn-out weight.

EXAMPLE 7 The intensity of a beam of radiation after passing through a material is given by the equation $I = I_0 10^{-kt}$, where I_0 is the original intensity, t the thickness in centimeters, and k an absorption coefficient. If a beam of gamma radiation is reduced from 1 million electron volts (eV) to 100,000 eV while passing through a sheet of material with $k = 0.08$, what is the thickness of the material?

SOLUTION From the given values in the equation,

$$10^5 = (10^6)10^{-0.08t}$$

from which $0.1 = 10^{-0.08t}$. This exponential relation can be expressed in logarithms to base 10 as

$$\log_{10} 0.1 = -0.08^t$$

Hence,

$$t = \frac{-\log_{10} 0.1}{0.08} = \frac{-(-1)}{0.08} = 12.5 \text{ cm}$$

Although the base of a logarithm can theoretically be any positive number except 1, in practice we seldom use any bases except 10 and e. Logarithms with base 10 are called **common logarithms**. By convention, $\log_{10} x$ is written **log *x***, with the subscript 10 understood. This convention is far from universal, however, so use it with care. Logarithms with base e, called **natural logarithms**, are most often denoted with the alternative notation **ln *x***. That is,

$\log_e x$ and $\ln x$ mean the same thing.

Most scientific calculators have both a [log x] *and an* [ln x] *button, or if not, they have the capacity for making these computations. Consult your particular owner's manual for specific directions.*

Since $y = \log_a x$ means that $x = a^y$, we can sketch the graph of the logarithmic function by first constructing a table of values. The tables for $y = \log_2 x$ and $y = \log x$ are shown in Figure 6.7 along with the corresponding graphs.

Each of the curves of Figure 6.7 is characteristic of what is called **logarithmic shape**. This figure clearly demonstrates the following functional characteristics of the logarithm function for $a > 1$. Any function that obeys these properties is said to *behave logarithmically*.

x	$\frac{1}{8}$	$\frac{1}{4}$	$\frac{1}{2}$	1	2	4	8
$\log_2 x$	-3	-2	-1	0	1	2	3

x	$\frac{1}{1000}$	$\frac{1}{100}$	$\frac{1}{10}$	1	10	100
$\log x$	-3	-2	-1	0	1	2

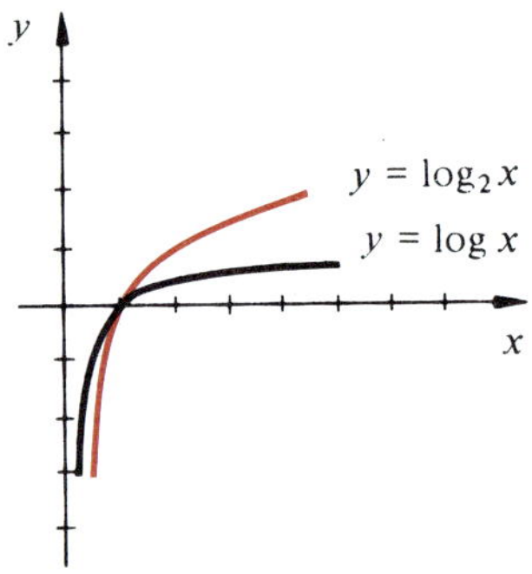

FIGURE 6.7

(1) $\text{Log}_a\, x$ is not defined for $x \leqslant 0$.

(2) $\text{Log}_a\, 1 = 0$.

(3) $\text{Log}_a\, a = 1$.

(4) $\text{Log}_a\, x$ is negative for $0 < x < 1$ and positive for $x > 1$.

(5) As x approaches 0, the value of $\log_a x$ decreases without bound.

(6) As x increases without bound, $\log_a x$ increases without bound.

Generally, in applications we must work with functions of the type $y = \log [g(x)]$, where $g(x)$ is some function of x.

EXAMPLE 8 Sketch the graph of the function $y = \log_2 (-x)$.

SOLUTION At first glance, you might conclude that this function is undefined, since the logarithm of a negative number is undefined. However, $-x$ is positive when $x < 0$, so $\log_2 (-x)$ has a domain of $x < 0$. Its graph is shown in Figure 6.8.

EXAMPLE 9 Sketch the graph of $y = \ln x$.

SOLUTION Recalling that $\ln x = \log_e x$, we note that $\ln 1 = 0$ and $\ln e = 1$. These facts and our knowledge of the characteristic logarithmic shape yield the graph in Figure 6.9.

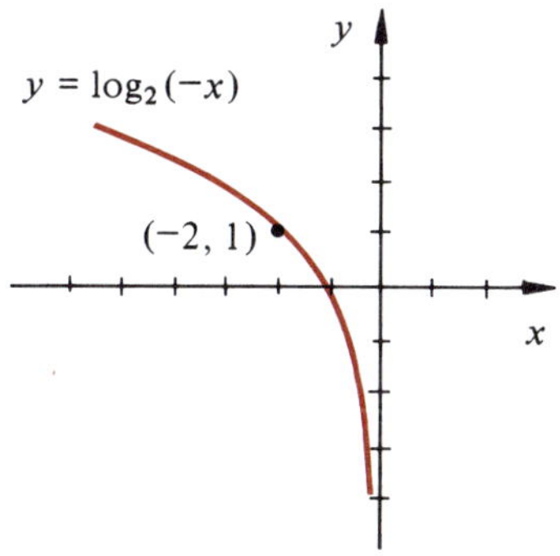

FIGURE 6.8

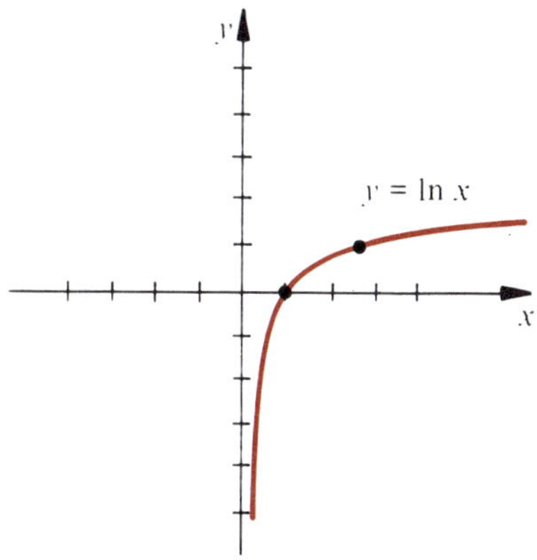

FIGURE 6.9

EXAMPLE 10 Sketch the graph of the $y = \log_5 (2x - 1)$.

SOLUTION Here we note that the function is *not* defined for values of $x \leqslant \frac{1}{2}$. Furthermore, when $x = 1$, $y = \log_5 1 = 0$, and when $x = 3$, $y = \log_5 5 = 1$. These facts are summarized in the table and the graph drawn in Figure 6.10.

$y = \log_5 (2x - 1)$

x	y
$\frac{1}{2}$	undef.
1	0
3	1
inc. w/o bound	inc. w/o bound

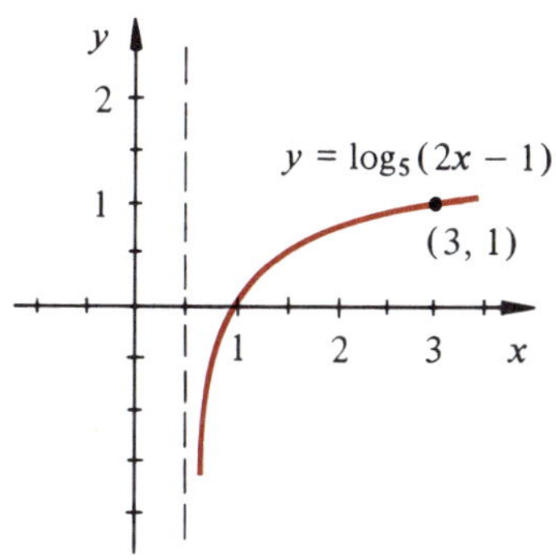

FIGURE 6.10

Exercises for Section 6.3

Write a logarithmic expression equivalent to the given exponential expression.

1. $x=2^3$ **2.** $y=3^8$ **3.** $M=5^{-3}$

4. $N=10^{-2}$ **5.** $L=7^2$

Write an exponential expression equivalent to the given logarithmic expression. In each case, find the base of the logarithm function.

6. $\log_b 8=3$ **7.** $\log_b 4=2$ **8.** $\log_b 0.25=-2$

9. $\log_b 100=2$ **10.** $\log_b b=1$

Solve the equations in Exercises 11–30 for the unknown.

11. $\log_{10} x=4$ **12.** $\log_5 N=2$

13. $\log_x 10=1$ **14.** $\log_x 25=2$

15. $\log_x 64=3$ **16.** $\log_{16} x=2$

17. $\log_{27} x=\frac{2}{3}$ **18.** $\log_2 \frac{1}{8}=x$

19. $\log_3 9=x$ **20.** $\log_{10} 10^7=x$

21. $\log_b b^a=x$ **22.** $\log_b x=b$

23. $\log_x 2=\frac{1}{3}$ **24.** $\log_x 0.0001=-2$

25. $\log_x 6=\frac{1}{2}$ **26.** $\log_b x=0$

27. $6^{\log_6 x}=6$ **28.** $x^{\log_x x}=3$

29. $\log_2 (x+3)=-1$ **30.** $\log_2 (x-1)=3$

In Exercises 31–34, solve the given logarithmic inequalities.

31. $\log_3 (x+1)<2$ **32.** $\log_2 (2-x)>-1$

33. $2\leqslant\log_2 x\leqslant 3$ **34.** $0<\log x<1$

Sketch a graph of the functions in Exercises 35–43. (Remember that $\log x$ means $\log_{10} x$.)

35. $f(x)=\ln x$ **36.** $f(x)=-\log x$

37. $f(x)=\log |x|$ **38.** $f(x)=\log (x-1)$

39. $f(x)=\log 2x$ **40.** $g(x)=\log_3 (-x)$

41. $f(x)=\log_2 (1-x)$ **42.** $f(x)=\log_2 (1+x)$

43. $f(x)=\log_2 |1+x|$

In Exercises 44–49, say how the graph of the given logarithmic function is related to the graph of $y=\log x$. In each case make a sketch to show the comparison.

44. $\log (-x)$ **45.** $-\log x$ **46.** $\log |x|$

47. $|\log x|$ **48.** $3+\log x$ **49.** $\log (x+3)$

Find the base of the logarithmic function $y=\log_a x$ whose graphs contain the following points:

50. (100, 2) **51.** (4, 2) **52.** (64, 4)

53. (0.1, −1) **54.** (0.5, −1)

55. Let $f_b(x)=\log_b x$. For any acceptable base, the graph of $f_b(x)$ passes through a point independent of b. What is that point?

56. Let $f(x)=\log_3 x$. Find $f(9)$, $f(\frac{1}{27})$, and $f(81)$.

57. Let $f(x)=\log_2 x$. By example show that
(a) $f(x+y)\neq f(x)+f(y)$ (b) $f(ax)\neq af(x)$

58. On the same set of axes, sketch $\log_2 x$ and $\log_3 x$. Then:
(a) Solve the equation $\log_2 x=\log_3 x$.
(b) Solve the inequality $\log_2 x>\log_3 x$.

59. How are the graphs of $y=\log_2 x$ and $y=2^x$ related?

In Exercises 60–67, use a calculator to give the approximate value of the logarithm.

60. $\log 2.67$ **61.** $\log 0.5$ **62.** $\log 5000$

63. $\log 5+\log 2$ **64.** $\ln 10$

65. $\ln e^{-1}$ **66.** $\ln 1000$ **67.** $\ln 0.5$

68. A power supply has a power output in watts approximated by the equation $P=64e^{-3t}$, where t is in days. Solve this expression for t.

69. A certain radioactive material decays exponentially by the equation $A(t)=A_0e^{-t/5}$. Find the half-life of the material.

6.4 Basic Properties of Logarithms

There are three rules for simplifying logarithmic expressions that you must know. These three rules or properties of logarithms correspond precisely to the three fundamental rules for exponents and are necessary consequences of them.

RULE 1 $\log_a MN = \log_a M + \log_a N$

PROOF Let $u = \log_a M$ and $v = \log_a N$. Then,

$$a^u = M \quad \text{and} \quad a^v = N$$

from which

$$MN = a^u a^v = a^{u+v}$$

Putting the expression in logarithms again, we get

$$\log_a MN = u + v = \log_a M + \log_a N$$

RULE 2 $\log_a M^c = c \log_a M$, where c is any real number.

PROOF Let $u = \log_a M$. Then,

$$a^u = M \quad \text{and} \quad (a^u)^c = a^{uc} = M^c$$

In logarithms, this can be expressed as

$$\log_a M^c = uc = c \log_a M$$

RULE 3 $\log_a (M/N) = \log_a M - \log_a N$

PROOF $M/N = M(N)^{-1}$. Now apply the previous two rules.

In words, Rule 1 states that the logarithm of a product equals the sum of the logarithms of the individual terms; Rule 2 states that the logarithm of a number to a power is the power times the logarithm of the number; and Rule 3 states that the logarithm of a quotient is the difference of the logarithms of the individual terms. Examine these rules carefully and notice where they apply as well as where they do *not* apply. For example, there is *no* rule for simplifying expressions of the form $\log_a (x+y)$ or $\log_a (x-y)$.

EXAMPLE 1

(a) $\log_2 (8)(64) = \log_2 8 + \log_2 64 = 3 + 6 = 9$

(b) $\log_3 \sqrt{243} = \log_3 243^{1/2} = \frac{1}{2} \log_3 243 = \frac{1}{2}(5) = 2.5$

(c) $\log_2 (\frac{3}{5}) = \log_2 3 - \log_2 5$

(d) $\log 4 \cdot 29/5 = \log 4 + \log 29 - \log 5$

EXAMPLE 2 Write the expression $\log x - 2 \log x + 3 \log (x+1) - \log (x^2-1)$ as a single term.

SOLUTION Proceed as follows:

$$\log x - 2 \log x + 3 \log (x+1) - \log (x^2-1)$$
$$= \log x - \log x^2 + \log (x+1)^3 - \log (x^2-1)$$
$$= \log \frac{x(x+1)^3}{x^2(x^2-1)} = \log \frac{(x+1)^2}{x(x-1)}$$

EXAMPLE 3 Given $\log_a x = 3$, find $\log_a (1/x)$.

SOLUTION By definition, $\log_a x = 3$ means $x = a^3$. Also, letting $\log_a (1/x) = y$, we can write $1/x = a^y$ or $x = a^{-y}$. Since a^3 and a^{-y} both equal x, we get

$$a^3 = a^{-y}$$

Therefore, $y = \log_a (1/x) = -3$.

Exercises for Section 6.4

Evaluate the logarithms in Exercises 1–8.

1. $\log_2 32 \cdot 16$ **2.** $\log_2 16^5$ **3.** $\log_5 25^{1/4}$

4. $\log_3 27$ **5.** $\log_3 27 \cdot 9 \cdot 3$

6. $\log_2 64 \cdot 32 \cdot 8$ **7.** $\log_2 (8 \cdot 32)^3$

8. $\log_3 (9 \cdot 81)^8$

Given that $\log 2 = 0.3010$, $\log 3 = 0.4771$, and $\log 7 = 0.8451$, find the logarithms in Exercises 9–20.

9. $\log \frac{3}{2}$ **10.** $\log 4$ **11.** $\log 12$

12. $\log 30$ **13.** $\log 90$ **14.** $\log \sqrt{2}$

15. $\log \sqrt{5}$ **16.** $\log 21^{1/3}$ **17.** $\log 2400$

18. $\log 0.00018$ **19.** $\log 0.0014$ **20.** $\log 42{,}000$

In Exercises 21–28, write the given expression as a single logarithmic term.

21. $\log_2 x^2 - \log_2 x$

22. $\log_2 (x^2-1) - \log_2 (x-1)$

23. $\log x + \log \frac{1}{x}$

24. $\log 3x + 3 \log (x+2) - \log (x^2-4)$

25. $\log 5t + 2 \log (t^2-4) - \frac{1}{2} \log (t+3)$

26. $\log z - 3 \log 3z - \log (2z-9)$

27. $3 \log u - 2 \log (u+1) - 5 \log (u-1)$

28. $\log t + 7 \log (2t-8)$

29. Let $\ln I = (-R/L)t + \ln I_0$. Show that $I = I_0 e^{-Rt/L}$.

30. If y is directly proportional to x^p, what relation exists between $\log y$ and $\log x$?

31. Compare the functions $f(x) = \log x^2$ and $g(x) = 2 \log x$. In what way are they the same? In what way different?

32. Let $f(x) = \log_a x$ and $g(x) = \log_{1/a} x$. Show that $g(x) = f(1/x)$.

33. If $\log_a x = 2$, what are $\log_{1/a} x$ and $\log_a (1/x)$?

34. Compare the graphs of the functions $f(x) = \log_2 2x$, $g(x) = \log_2 x$, $h(x) = \log_2 \sqrt{x}$, and $m(x) = \log_2 x^2$.

35. Given the graph of $y=\log x$, how can you conveniently get the following graphs?
(a) $\log x^p$ (b) $\log px$
(c) $\log (x+p)$ (d) $\log (x/p)$

36. If $f(x)=\log_b x$, is $f(x+y)=f(x)+f(y)$?

In Exercises 37–47, use a calculator to verify the given statement.

37. $\ln 6=\ln 2+\ln 3$

38. $\ln 100=\ln 50+\ln 2$

39. $\ln 9=2\ln 3$

40. $\ln 16\sqrt{2}=2\ln 4+\frac{1}{2}\ln 2$

41. $\ln (62.3\cdot 28.6)=\ln 62.3+\ln 28.6$

42. $\ln e^{\pi}=\pi$

43. $\ln 12=\ln 36-\ln 3$

44. $\log (25\cdot 34)=\log 25+\log 34$

45. $\log (16\cdot \pi)=\log 16+\log \pi$

46. $\log \frac{125}{73}=\log 125-\log 73$

47. $\log \frac{17}{35}=\log 17-\log 35$

6.5 Common Logarithms

Logarithms to the base 10 are called common logarithms. Historically, common logarithms assumed the most practical importance, especially in simplifying certain numerical computations. With the advent of the digital computer, however, the importance of logarithms as a computational tool has diminished. Nonetheless, any engineer or scientist must know something about how to use logarithms in computation.

Table B in the appendix is a listing of common logarithms of numbers between 1 and 10 in steps of 0.01. This is a "four-place" table, which means that it gives an approximation accurate to four decimal places. To find the logarithm of a number between 1 and 10, locate the first two digits of the number in the left column. The third digit of the number heads the column. Thus, to find the logarithm of 5.31, look down the left-hand column to 5.3. Then move over to the column headed by 1 to find that $\log 5.31=0.7251$.

To find the number whose logarithm is given, use Table B in reverse, a procedure often called finding **antilogarithms**. Thus, since $\log 2=0.3010$, antilog $0.3010=2$.

Table B lists logarithms for numbers m such that $1\leqslant m\leqslant 10$. The corresponding range values are $0\leqslant \log m\leqslant 1$; that is, all the values in the table are fractions between 0 and 1.

To find the logarithm of any number, recall that any positive number M can be written as the product of a number m (where m is between 1 and 10) and 10^c (where c is an integer). That is,

$$M=m\cdot 10^c, \qquad 1\leqslant m<10$$

A number so written is in *scientific form*. Several examples follow.

EXAMPLE 1

$$53.1=5.31\times 10^1$$
$$0.00531=5.31\times 10^{-3}$$
$$5{,}310{,}000=5.31\times 10^6$$

Thus, for any number M,

$$\log M = \log (m \cdot 10^c)$$

From Rule 1 for logarithms,

$$\log M = \log 10^c + \log m$$

From Rule 2,

$$\log M = c \log 10 + \log m$$
$$= c + \log m$$

This is the **standard**, or uniform, way of expressing logarithms—as the sum of an integer and a positive number between 0 and 1. The quantity $\log m$, called the **mantissa** of $\log M$, is always a number between 0 and 1. The integer c, called the **characteristic** of $\log M$, is always an integer.

EXAMPLE 2 Given that $\log 5.31 = 0.7251$, find (a) $\log 53.1$; (b) $\log 5{,}310{,}000$; (c) $\log 0.00531$.

SOLUTION

(a) Since $53.1 = 5.31 \times 10^1$,

$$\log 53.1 = 1 + \log 5.31 = 1 + 0.7251 = 1.7251$$

(b) Since $5{,}310{,}000 = 5.31 \times 10^6$,

$$\log 5{,}310{,}000 = 6 + \log 5.31 = 6 + 0.7251 = 6.7251$$

(c) Since $0.00531 = 5.31 \times 10^{-3}$,

$$\log 0.00531 = -3 + \log 5.31 = -3 + 0.7251$$

In (c), the negative characteristic is not combined with the mantissa, since its combination would obscure it and cause difficulty in recovering it when finding the antilog. Usually, we express a negative characteristic as a positive number minus 10. Thus, since $-3 = 7 - 10$, we prefer to write

$$\log 0.00531 = 7.7521 - 10$$

Separating the characteristic and the mantissa is convenient for computations using tables of logarithms, but this separation is unnecessary for calculator logic. Calculators (and computers) carry the characteristic and mantissa in combined form. For numbers greater than 1, the two forms are the same, but for numbers less than 1 the forms are different. For example, log 35.7 *is displayed to 4 decimal places as* 1.5527, *which is the same value one would get from a table. However,* log 0.00357 *is displayed as* -2.4473, *which does not have the same appearance as the table value of* 7.5527 − 10. *To convert the table value form to the calculator form, simply subtract as indicated. To convert the calculator form to the table form, write*

$$-2.4473 = -2.4473 + 10 - 10 = 7.5527 - 10$$

EXAMPLE 3 Given that antilog $(0.4099) = 2.57$, find (a) antilog (2.4099); (b) antilog $(7.4099 - 10)$; (c) antilog (-6.5901).

SOLUTION

(a) antilog $(2.4099) = 2.57 \times 10^2 = 257$

(b) antilog $(7.4099 - 10) = 2.57 \times 10^{-3} = 0.00257$

(c) antilog $(-6.5901) =$ antilog $(3.4099 - 10) = 2.57 \times 10^{-7}$

In part (c), write the number -6.5901 as the sum of a number between 0 and 1 and an integer before consulting a table of logarithms. Note, for example, that -6.5901 is *not* equal to $-6 + 0.5901$, but rather it is equal to $-7 + 0.4099 = 3.4099 - 10$.

There are two popular schemes for getting antilogs with calculators. Some models of calculators use an [inv] *key with the* [log] *or* [ln] *key to compute antilogs. Under this scheme, if* $\log N = 3.7971$, *the value of* N *is given by the following keystrokes:*

$N = 3.7971$ [inv] [log] $= 6267.58$

Other models use a [10^x] *key or a* [e^x] *key for antilogs. The use of these keys follows from the fact that* $\log_b N = x$ *means* $N = b^x$. *Thus, if* $\log N = 3.7971$, *the value of* N *is given by*

$N = 3.7971$ [10^x] $= 6267.58$

Similarly if $\ln M = 1.0136$, the value of M is given by

$M = 1.0136$ [e^x] $= 2.7555$

Expressions like $\log N = 7.0915 - 10$ *must be written in the form* $\log N = -2.9085$ *before a calculator can be used to give* $N = 0.0012345$.

EXAMPLE 4 Find the value of $10^{-4.0969}$.

SOLUTION Let $N = 10^{-4.0969}$. Then

$$\log N = -4.0969 = -5 + 0.9031 = 5.9031 - 10$$

From Table B, in the appendix, antilog $0.9031 = 8$. Therefore

$$N = 8 \times 10^{-5} = 0.00008$$

The logarithm of a number to any base can be found from a table of common logarithms. For example, to find the *natural* logarithm of M, denoted $\ln M$, let $y = \ln M$. Then, by definition,

$$M = e^y$$

Taking the common logarithm of both sides, we get

$$\log M = y \log e$$

Solving for y and recalling that $y = \ln M$, we have

(6.1) $$\ln M = \frac{\log M}{\log e}$$

Tables of natural logarithms are often available; but if they are not, the above formula will be sufficient along with knowing that

$$\log e = 0.4343$$

EXAMPLE 5 Solve for x in $e^x = 2640$.

SOLUTION In terms of natural logarithms, this equation becomes $x = \ln 2640$. From Formula (6.1),

$$x = \frac{\log 2640}{\log e}$$

Since $\log 2640 = \log (2.64 \times 10^3) = 3 + \log 2.64 = 3.4216$, we have

$$x = \frac{3.4216}{0.4343} = 7.88$$

Taking the common logarithm of each side of the given equation and solving for x gives the same result.

EXAMPLE 6 Solve for x in $2^x = 0.0075$.

SOLUTION By the definition of a logarithm, $x = \log_2 0.0075$. Converting this to common logarithm form gives

$$x = \frac{\log 0.0075}{\log 2}$$

Since $\log 0.0075 = -3 + 0.8751 = -2.1249$, we get

$$x = \frac{-2.1249}{0.3010} = -7.059$$

For each example we have worked so far, using Table B in the Appendix was sufficient for finding the mantissas directly. In the next section we discuss a method of extending the use of this table to numbers having four significant digits.

6.6 Interpolation

Table B does not give the logarithm of *every* number between 1 and 10; it gives only the numbers that are in steps of 0.01. For mantissas with more than two decimal places, you can use the tabulated values to approximate or interpolate the logarithm. Obviously, interpolation to find the logarithm of a mantissa with three decimal placed, like 4.286, is unnecessary if a calculator or a larger table is used. The material in this section is presented in the spirit of showing how to use tables in general to approximate values not specifically tabulated. Hence for this material, you should not use your [log] key except to compare answers.

The interpolation technique used most frequently assumes that small changes in a number bring about proportional changes in the tabulated values—in this case, the values of the logarithm. Thus, if the number M is in the interval $M_1 < M < M_2$, then

$$\frac{\log M - \log M_1}{M - M_1} = \frac{\log M_2 - \log M_1}{M_2 - M_1}$$

Solving for log M, we get

$$\log M = \log M_1 + \frac{M - M_1}{M_2 - M_1}(\log M_2 - \log M_1)$$

This formula is called interpolation by proportional parts, or more commonly, **linear interpolation**.

From a graphical standpoint, linear interpolation assumes that the graph of $y = \log x$ is essentially linear for small changes in the value of x. Figure 6.11 shows an exploded view of a small portion of the logarithm graph between 4.72 and 4.73, along with the corresponding straight line connecting the points (4.72, log 4.72) and (4.73, log 4.73). The curvature of the logarithmic curve is purposely exaggerated to show how it differs from the interpolated line.

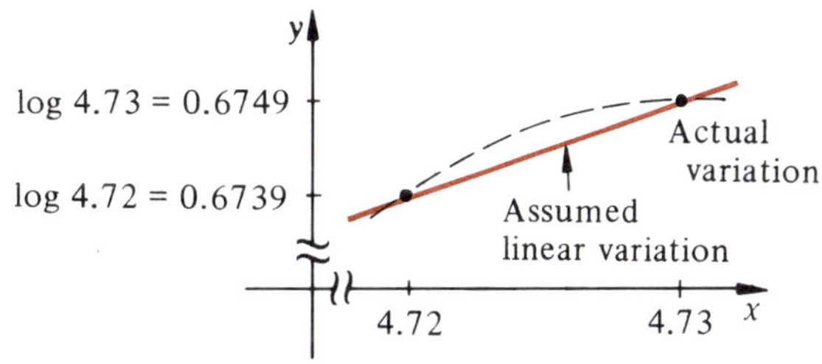

FIGURE 6.11

Linear interpolation is often used in mathematics. It seems to be a very natural way to approximate, but do not use it indiscriminately. Results may be inaccurate if the interval of interest is not restricted to a relatively small size.

The main idea of linear interpolation in finding, say log 2.573, is to note that 2.573 is three-tenths of the way from 2.57 to 2.58, and hence log 2.573 is three-tenths of the way from log 2.57 to log 2.58.

EXAMPLE 1 Approximate log 2573 by linear interpolation.

SOLUTION Since $2573 = 2.573 \times 10^3$, the characteristic is 3. The mantissa lies between the mantissa for 2.570 and 2.580. From Table B, the entry for 2.570 is found to be .4099, and for 2.580 is .4116.

		Number	Mantissa		
		2.570	.4099		
10	3	2.573	. . .	c	17
		2.580	.4116		

We can establish the proportion

$$\frac{c}{17}=\frac{3}{10}$$

or

$$c=\frac{3}{10}(17)=5.1\approx 5$$

The required mantissa of log 2.573 is then $.4099+.0005=.4104$. Therefore, log 2573 $=3.4104$.

Note that in the preceding example we rounded the number c to four digits after the decimal point. Otherwise, it would seem that interpolation was increasing the accuracy of the table, and this it is not possible to do.

EXAMPLE 2 Find the antilog of 2.4059.

SOLUTION Since the mantissa is .4059 and the characteristic is 2, we must find the number corresponding to the mantissa .4059 and then multiply by 10^2. Using Table B, we arrange the work as in the accompanying chart.

		Number	Mantissa		
		2.540	.4048		
10	n	. . . .	.4059	11	17
		2.550	.4065		

$$\frac{n}{10}=\frac{11}{17}$$

$$n=\frac{11}{17}(10)=6.4\approx 6$$

Therefore, the desired number is $2.546\times 10^2=254.6$.

EXAMPLE 3 Find $\log \sqrt[3]{0.7568}$.

SOLUTION From the properties of logarithms,

$$\log \sqrt[3]{0.7568} = \frac{1}{3} \log 0.7568$$

Using Table B, we arrange the work as

		Number	*Mantissa*		
10	8	7.560	.8785	c	6
		7.568	....		
		7.570	.8791		

We have the following proportion:

$$\frac{c}{6} = \frac{8}{10}$$

from which

$$c = 4.8 \approx 5$$

Therefore, $\log 0.7568 = 9.8790 - 10$. To get the desired logarithm, divide $9.8790 - 10$ by 3. Since 3 does not divide evenly into 10, and since we want to keep the answer consistent with the standard form for negative characteristics, we write

$$\log 0.7568 = 29.8790 - 30$$

Dividing by 3 gives

$$\log \sqrt[3]{0.7568} = 9.9597 - 10$$

Exercises for Sections 6.5 and 6.6

Evaluate Exercises 1–10.

1. (a) log 5.41 (b) log 5410 (c) log 0.00541

2. (a) log 1.25 (b) log 125 (c) log 0.125

3. (a) log 9.03 (b) log 0.000903 (c) log 903,000

4. (a) log 8.89 (b) log 88.9 (c) log 0.0889

5. (a) log (0.255)(85.6) (b) log (0.255/85.6)

6. (a) log (0.295)(3.11) (b) log (0.295/3.11)

7. (a) antilog 3.2601 (b) antilog 0.2601
(c) antilog $8.2601 - 10$

8. (a) antilog 0.9258 (b) antilog 5.9258
(c) antilog $9.9258 - 10$

9. (a) antilog 1.1818 (b) antilog $3.1818 - 10$
(c) antilog 6.1818

10. (a) antilog 2.8692 (b) antilog 7.8692
(c) antilog $3.8692 - 10$

Use linear interpolation to approximate the common logarithm of the numbers in Exercises 11–25.

11. 2.361 **12.** 5842 **13.** 0.009573

14. 3.142 **15.** 2.718 **16.** 49,990

17. 642,300 **18.** 5.011 **19.** 1005

20. 62.45 **21.** $\sqrt[3]{19.79}$ **12.** $(158{,}800)^4$

23. $\sqrt{0.003965}$ **24.** $\sqrt[5]{0.1113}$ **25.** $(0.1975)^3$

In Exercises 26–35, find the number x to 4 significant digits by using linear interpolation.

26. $\log x = 2.1110$ **27.** $\log x = 8.1284 - 10$

28. $\log x = 7.814$ **29.** $\log x = 3.4141$

30. $\log x = 7.7228 - 10$

31. $\log x = \frac{1}{2}$ **32.** $\log x = 0.25$

33. $\log x = \frac{1}{3}$ **34.** $\log x = \pi$

35. $\log x = -\pi$

Find the number x in Exercises 36–49.

36. $10^x = e$ **37.** $10^x = \pi$

38. $10^x = e^2$ **39.** $10^x = 0.1441$

40. $10^x = \sqrt{2.169}$ **41.** $10^x = \sqrt{20}$

42. $10^{x+2} = 17^4$ **43.** $10^{x-1} = 5^{10}$

44. $3^x = 15$ **45.** $e^{3x} = 195$

46. $5^{-x} = 0.045$ **47.** $2^{2x} = 0.009$

48. $3^x = (15.3)^5$ **49.** $e^{x+2} = 10^{-3}$

50. Using $\log 3 = 0.4771$ and $\log 4 = 0.6021$, approximate $\log 3.5$ by linear interpolation and compare with the value in Table B.

51. Knowing that $\sqrt{4} = 2$ and $\sqrt{9} = 3$ and using linear interpolation, approximate $\sqrt{7}$. Is the approximation high or low?

52. Let $f(x) = 2x - 3$. Use the values of $f(2)$ and $f(3)$ along with linear interpolation to approximate $f(2.5)$. How accurate is your result?

Use a calculator to evaluate each of the common logarithms in Exercises 53–60. In each case, indicate the mantissa and the characteristic. *Note:* For numbers less than 1, calculators carry the characteristic and the mantissa in combined form.

53. $\log 55.8$ **54.** $\log 677{,}400$ **55.** $\log 134$

56. $\log 4690$ **57.** $\log 2.11$ **58.** $\log 0.998$

59. $\log 0.00391$ **60.** $\log 0.000025$

In Exercises 61–68, use a calculator to evaluate the given antilog.

61. antilog 3.1160 **62.** antilog 1.0555

63. antilog 0.7623 **64.** antilog $8.3324 - 10$

65. antilog $9.1414 - 10$

66. antilog $7.4444 - 10$

67. antilog -5.8709 **68.** antilog -2.1602

6.7 Computations with Logarithms

Since high-speed electronic computing devices have come into being, common logarithms have only a minor role in computation. Even so, a basic understanding of logarithmic computation remains important. Further, the arithmetic of logarithmic computation increases your appreciation for some of the background theory. We shall not explore all the short cuts and conventions related to logarithmic computation but only the fundamental techniques.

Some of the following examples could as easily have been worked by conventional arithmetic. Our aim here is just to show how to use logarithms; so do not be too concerned if the preferred technique might not be logarithms.

EXAMPLE 1 Use logarithms to evaluate

$$M = \frac{2158 \times 0.512}{0.00042}$$

SOLUTION Find log M and use the rules of logarithms to write

$$\log M = \log 2158 + \log 0.512 - \log 0.00042$$

Log 2158 is found by interpolation to be 3.3341. Log 0.512 is found directly to be $-10 + 9.7093$. Log 0.00042 is found directly to be $-10 + 6.6232$. Hence,

$$\begin{aligned}\log M &= 3.3341 + (-10 + 9.7093) - (-10 + 6.6232)\\ &= 3 + 0.3341 - 1 + 0.7093 + 4 - 0.6232\end{aligned}$$

Adding the characteristics yields

$$\begin{aligned}\log M &= 6 + (0.3341 + 0.7093 - 0.6232)\\ &= 6 + 0.4202\\ &= 6.4202\end{aligned}$$

By interpolation, antilog $0.4202 = 2.631$, and hence

$$M = 2.631 \times 10^6$$

EXAMPLE 2 Use logarithms to approximate $(25.4)^{1/4}$.

SOLUTION Let $M = (25.4)^{1/4}$. Then $\log M = \log (25.4)^{1/4} = \frac{1}{4} \log 25.4$. From Table B, $\log 25.4 = 1.4048$ and hence,

$$\log M = 0.3512$$

By interpolation, $M = 2.245$.

EXAMPLE 3 Approximate $M = (85.1 + \sqrt[4]{25.4})/\log 3$.

SOLUTION There is no formula for the logarithm of a sum, so the two terms of the numerator must be added before log M can be found. From the previous example, $\sqrt[4]{25.4} = 2.245$, so that

$$M = \frac{85.1 + 2.245}{0.4771} = \frac{87.345}{0.4771}$$

$$\log M = \log 87.34 - \log 0.4771$$

$$= 1.9412 - (9.6786 - 10)$$

$$\log M = 2.2626$$

$$M = 183.1$$

Exercises for Section 6.7

Use logarithms to approximate the computations in Exercises 1–15 to four significant digits.

1. $(65.7)(0.00411)$

2. $\dfrac{365 \times 1.423}{120 \times 0.00133}$

3. $(2563^{1/3})(0.5792^{1/2})$

4. $\dfrac{\log 23.7}{\log 0.00505}$

5. $\sqrt[3]{0.9711} \cdot \sqrt[5]{8}$

6. $(5.6)^4 \cdot (0.16)^3$

7. $3.1^{3.1}[2.112]$

8. $\dfrac{21.2+46.33}{\sqrt{57.12}}$

9. $[e^2+2^e][176+(51.3)(0.756)]$

10. $\dfrac{76.1+\log(8 \times 10^{40})}{\log 5.3}$

11. $\dfrac{-52.3 \times 64}{\log 3.83}$

12. $\dfrac{0.6003(0.0237)}{0.0114(0.00934)}$

13. $\dfrac{1.123(0.6975)}{0.0075(1278)}$

14. $3^{1.212}$

15. $\dfrac{(\sqrt{2})(\sqrt{3})}{\sqrt{5}}$

Solve for x in Exercises 16–28.

16. $3^x=\pi$

17. $2^x=\pi$

18. $7.1^x=4.667$

19. $2^x=\dfrac{2015}{3143}$

20. $(0.5)^x=5.555$

21. $2.6^x=5.26$

22. $21^x=213$

23. $1.4^x=3.665$

24. $9.2^x=100$

25. $4.1^{x+1}=15.9$

26. $2.3^{x-1}=12$

27. $25.3^{x+2}=17$

28. $6.54^{x-2}=8.35$

The calculator has allowed using complicated combinations of the logarithm and exponential even at an elementary level. The following computations would not have been in a textbook of this level just a few years ago. With a calculator, evaluate each of the following.

29. Given $f(t)=71.68 \log(7.990t)+\log(554.23)$, find $f(23.43)$.

30. Given $G(x)=3.0709 \log(15.66-x)+111.59$, find $G(9.82)$.

31. Given $F(x)=[\ln(x+34.952)]^2$, find $F(1.553)$.

32. Given $f(t)=e^{3.32t} \ln(16.07t)$, find $f(0.09879)$.

33. Given $h(x)=\dfrac{1775 \log(x^2-34.33)}{\log(767.39)}$, find $h(22.76)$.

34. Given $s(x)=\dfrac{1450.6+3.75 \log(x-11.15)}{988.5-0.972 \log(0.753x)}$, find $s(20.38)$.

35. Given $f(x)=(53.96)^{\log(22.02x-0.45)}$, find $f(0.771)$.

36. Given $H(t)=e^{[1.87+\ln(2.58t-2.02)]}$, find $H(1.105)$.

6.8 Exponential and Logarithmic Equations

Equations in which the variable occurs as an exponent are called **exponential equations**. To solve these equations, we use the fact that the logarithm is a one-to-one function. Thus, if $\log x=\log y$, then $x=y$. Hence, taking the logarithm of both sides is an admissible operation if both sides can be assumed positive.

EXAMPLE 1 Solve the exponential equation $3^x=2^{2x+1}$.

SOLUTION Taking the common logarithm of both sides,

$$\log 3^x=\log 2^{2x+1}$$

and using Rule 2 for logarithms, we get

$$x \log 3 = (2x+1) \log 2$$

Solving for x gives

$$x(\log 3 - 2 \log 2) = \log 2$$

$$x = \frac{\log 2}{\log 3 - 2 \log 2}$$

$$= \frac{0.3010}{-0.1249} = -2.41$$

EXAMPLE 2 The expression $(e^x - e^{-x})/2$ is called the hyperbolic sine of x and is denoted by $\sinh x$. Solve the equation $\sinh x = 3$.

SOLUTION We must solve the exponential equation

$$\frac{e^x - e^{-x}}{2} = 3$$

Thus, $e^x - e^{-x} = 6$. Multiplying both sides by e^x and bringing all terms to one side gives

$$e^{2x} - 6e^x - 1 = 0$$

Let $u = e^x$. Then this equation becomes $u^2 - 6u - 1 = 0$, which is quadratic in u. From the quadratic formula,

$$u = \frac{6 \pm \sqrt{40}}{2} = 3 \pm \sqrt{10}$$

Since $u = e^x$ is always positive, discard the root $3 - \sqrt{10}$, which is negative. Thus,

$$e^x = 3 + \sqrt{10}$$

Then the desired solution is

$$x = \ln (3 + \sqrt{10}) \approx 1.82$$

Equations involving logarithms are called **logarithmic equations**. Using one of the rules of logarithms often simplifies things enough for you to solve such an equation.

EXAMPLE 3 Solve the logarithmic equation

$$\log (x^2 - 1) - \log (x - 1) = 3$$

SOLUTION Simplifying the left-hand side by combining the two logarithm terms, we get

$$\log \frac{x^2-1}{x-1}=3$$

$$\log (x+1)=3$$

Thus, $x+1=10^3$. (Can you tell why this is true?) Or

$$x=-1+10^3=999$$

EXAMPLE 4 Solve the equation $x^{\log x}=\dfrac{x^3}{100}$

SOLUTION Taking the common logarithm of both sides, we have

$$(\log x)(\log x)=\log x^3-\log 100$$

or

$$(\log x)^2-\log x^3+\log 100=0$$

Note that $(\log x)^2$ does not equal $2 \log x$ but $\log x^3$ equals $3 \log x$. Hence, the equation becomes

$$(\log x)^2-3 \log x+2=0$$

Factoring gives $(\log x-2)(\log x-1)=0$, from which

$$\log x=2 \text{ or } \log x=1$$

Thus, $x=100$ or 10.

EXAMPLE 5 A certain power supply has a power output in watts governed by the equation $P=50e^{-t/250}$, where t is the time in days. If the equipment aboard a satellite requires 10 watts of power to operate properly, what is the operational life of the satellite?

SOLUTION Solving the equation $10=50e^{-t/250}$ for t gives

$$\frac{-t}{250}=\ln \frac{10}{50}$$

$$=\ln 0.2=\frac{\log 0.2}{\log e}=\frac{-1+0.3010}{0.4343}=\frac{-0.6990}{0.4343}$$

$$=-1.609$$

or $$t=250 \times 1.609$$

$$=402 \text{ days}$$

Hence, the operational life of the satellite is 402 days.

Exercises for Section 6.8

Solve for x in Exercises 1–16.

1. $7^{x+1}=2^x$

2. $3^x 2^{2x+1}=10$

3. $10^{x^2}=2^x$

4. $8^x=10^x$

5. $2^{1+x}=3$

6. $\left(\frac{1}{2}\right)^x>3$

7. $2^{\log x}=2$

8. $\log\log\log x=1$

9. $(\log x)^{1/2}=\log\sqrt{x}$

10. $(\log x)^2=\log x^2$

11. $(\log x)^3=\log x^3$

12. $x^{\log x}=10$

13. $\log(x+15)+\log x=2$

14. $\log 3x-\log 2x=\log 3-\log x$

15. $\log(x-2)-\log(2x+1)=\log\frac{1}{x}$

16. $\log(x^2+1)-\log(x-1)-\log(x+1)=1$

17. The expression $(e^x+e^{-x})/2$ is called the hyperbolic cosine of x and is denoted by $\cosh x$. Solve the equation $\cosh x=2$.

18. Solve the two equations $y=50e^{-2x}$ and $y=2^x$ simultaneously by graphing each equation on the same coordinate system.

19. Explain why if $\log u(x)=v(x)$, then $u(x)=10^{v(x)}$.

20. Show that if $y=e^{\ln f(x)}$, then $y=f(x)$.

21. The radioactive chemical element strontium 90 has a half-life of approximately 28 years. The element obeys the radioactive decay formula $A(t)=A_0e^{-kt}$, where A_0 is the original amount and t is the time in years. Find the value of k.

22. Repeat Exercise 21 for the element iodine, whose decay formula is the same type. Express t in days. The half-life is 8 days.

23. What is the half-life of the power supply of Example 5?

24. The difference in intensity level of two sounds with intensities I and I_0 is defined by $10\log(I/I_0)$ decibels. Find the intensity level in decibels of sound, from an electric motor, that is 175.6 times greater than I_0.

25. The population growth curve is given by $P=P_0e^{kt}$, where P_0 is the initial size, t is time in hours, and k is a constant. If $k=0.0132$ for a culture of bacteria, how long does it take for the culture to double in size?

Test 1 for Chapter 6

Answer *True* or *False* to Exercises 1–10.

1. If $f(x)=2^x$, then $f(x+y)=f(x)\cdot f(y)$.

2. If $f(x)=2^x$, then if $x_1<x_2, f(x_1)>f(x_2)$.

3. If $a>b$, then $a^x>b^x$.

4. $\text{Log}\ 0=1$.

5. $\text{Log}\ (x+y)=\log x+\log y$.

6. Interpolation is a way of estimating the values of a function between tabulated values.

7. $2^{\log_2 7}=7$.

8. The domain of $\log(-x)$ is the empty set.

9. The domain of $\log(-x^2)$ is the empty set.

10. The logarithm is the reciprocal of the exponential.

11. Carefully sketch $y=2^x$, $y=2^{x+3}$, and $y=2^x+3$ on the same coordinate axes.

12. Sketch $y=\log x$, $y=-\log x$, and $y=\log(-x)$ on the same coordinate axes.

13. If $\log v=5.4371$, what is the value of v?

14. What is x if $10^{x^2-4}=7$?

15. (a) Solve for x in $2^{x+1}>0$.
(b) Solve for x in $\log(x+1)>0$.

16. Find x for $x^{4.3}=2.1$.

Test 2 for Chapter 6

1. Sketch the graph of (a) $y=2^{x-3}$; (b) $y=\log_2 (x-5)$.

2. What is x if (a) $\log_x 8=-3$; (b) $\log_3 x=4$?

3. Express $3 \log x-\log (x^2-2)+2 \log (x+1)$ as a single logarithmic term.

4. Evaluate (a) $\log_5 (125 \cdot 625)$; (b) $\log_2 (128 \cdot 64)$.

5. Use Table B to evaluate (a) log 3956; (b) log 0.00075; (c) antilog 6.7838−10.

6. Using the results in Exercise 5, evaluate $N=3956/0.00075$.

7. What is x if $3^x=2^{x+1}$?

8. What is x if $\log x+\log (x-3)=1$?

9. Given $m=ce^{-kt}$, find k if $m=50$, $c=20$, and $t=2$.

10. Use logarithms to find $\sqrt[5]{0.0596}$.

7

Systems of Equations and Inequalities

In the previous chapters you learned that a solution to an equation in two variables consists of any ordered pair of numbers that satisfies the equation, and, in fact, that there are infinitely many such ordered pairs in the solution set. For example, the ordered pairs (1, 2), (2, −1), (3, −4), (0, 5), and so on are solutions to the equation $3x+y=5$. Similarly, (8, 1), (−1, −8), (10, 3), (3, −4), and so on are solutions to $x-y=7$. Inspection of the two solution sets reveals the interesting fact that (3, −4) satisfies both equations. When the two equations are considered together like this, they are referred to as a **system of equations**, and any common solution is a **simultaneous solution** to the system of equations. In this chapter you will learn how to solve certain types of systems of equations.

By graphing each equation of a system on the same set of coordinates, you can get an idea of what it means to solve a system of equations. Figure 7.1 shows the graphs of $3x+y=5$ and $x-y=7$. The ordered pair (3, −4) appears as the point

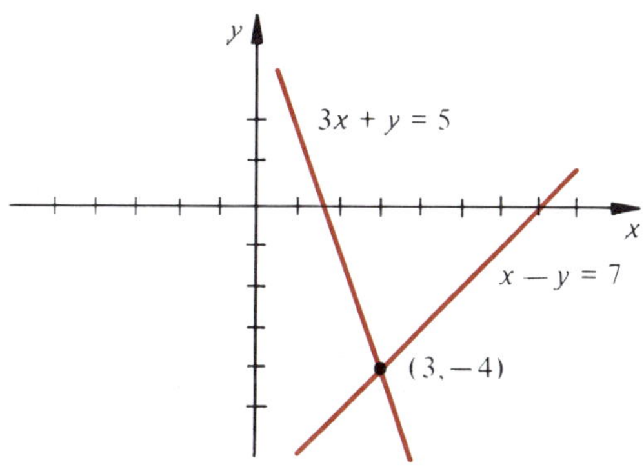

FIGURE 7.1

of intersection of the two curves. Simultaneous solutions of systems of two equations in two unknowns can always be represented like this. We sometimes use this knowledge to solve a system of equations "graphically," especially when the system contains one or more nonlinear equations.

A **linear** system of equations is one in which each equation is linear. A system that is not linear is called **nonlinear**. For example, the equations

$$\begin{aligned} 3x+y&=5 \\ x-y&=7 \end{aligned}$$

form a linear system, whereas the equations

$$\begin{aligned} x+\ 5y&=10 \\ x^2-13y&=\ 4 \end{aligned}$$

form a nonlinear system. Most of our present effort is directed toward linear systems, although some of the techniques can also be applied to nonlinear systems. Our study is further limited to systems for which the number of unknowns and the number of equations are the same.

Although any system of two equations in two unknowns can be solved graphically by sketching the curve corresponding to each equation and determining the coordinates of the points of intersection, these graphical methods are tedious; also, they generally yield only approximate results unless they are done very carefully. Hence, we shall concentrate on certain algebraic methods, using the graphs of the equations to support and check our solution.

7.1 Eliminating a Variable

Perhaps the most common method of solving systems of equations involves altering the given system algebraically until it is reduced to a single equation in one unknown. Some preliminary ideas are relevant to the procedure.

Two systems of equations are **equivalent** if both systems have precisely the same solutions. Thus, the systems

$$\text{I}\quad \begin{cases} 2x-\ y=5 \\ x-2y=4 \end{cases} \quad\text{and}\quad \text{II}\quad \begin{cases} x+y=1 \\ 3x-y=7 \end{cases}$$

are equivalent systems, since $x=2$, $y=-1$, is the solution of both. Figure 7.2(a) shows the graph of System I and Figure 7.2(b) shows the graph of System II. The point $(2, -1)$ is the point of intersection in both cases.

Operations that transform a system of equations into an equivalent system are said to be **allowable**. Three allowable operations for systems of equations are:

(1) The position of any two equations can be changed. Thus,

$$\begin{aligned} 3x+5y&=6 \\ x-3y&=4 \end{aligned} \quad\text{and}\quad \begin{aligned} x-3y&=4 \\ 3x+5y&=6 \end{aligned}$$

are equivalent systems.

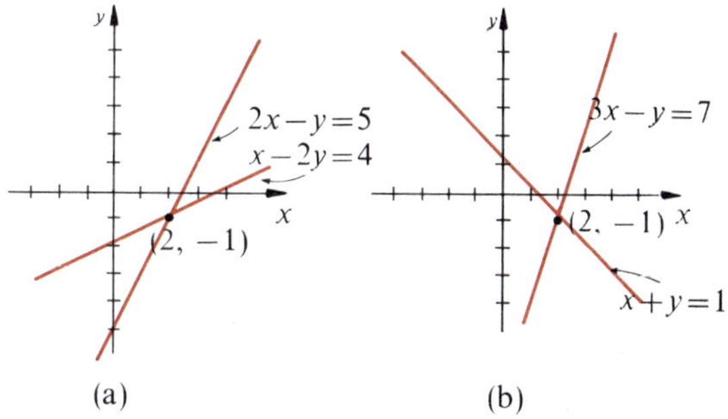

FIGURE 7.2

(2) Any equation of a system can be replaced by a nonzero multiple of itself. Thus, the systems

$$\text{I}\quad \begin{cases} x-5y=3 \\ 2x+3y=1 \end{cases} \qquad \text{II}\quad \begin{cases} x-5y=3 \\ 4x+6y=2 \end{cases} \qquad \text{III}\quad \begin{cases} x-5y=3 \\ -6x-9y=-3 \end{cases}$$

are equivalent since the lower equation in each system is a constant multiple of the others.

(3) Any equation of a system can be replaced by the sum of any two equations in the system. For example, the systems

$$\text{I}\quad \begin{cases} 3x+5y=2 \\ x-3y=1 \end{cases} \quad \text{and} \quad \text{II}\quad \begin{cases} 3x+5y=2 \\ 4x+2y=3 \end{cases}$$

are equivalent systems, since the upper equation in each system is the same and the lower equation in System II is obtained by adding together the equations in System I.

Eliminating a Variable by Addition or Subtraction

Operations corresponding to (2) and (3) are used in a technique for solving systems of equations that is called **elimination of a variable by addition or subtraction.*** The "trick" is to multiply each of the equations by just the right constants, so that after the two equations are added, the coefficients of one of the variables vanishes. Thus the number of variables is essentially reduced by one. If there are only two variables initially, this process will yield one equation in one unknown, which can be solved.

EXAMPLE 1 Use the method of elimination of a variable by addition or subtraction to solve the system of equations

$$3x+2y=5$$
$$x-2y=3$$

Sketch a figure showing the solution.

*This method is often called "Gaussian elimination."

SOLUTION First, decide which variable can most easily be eliminated by adding or subtracting the two given equations. In this case, the y variable can be eliminated simply by adding the two equations. Thus, the system

$$\begin{cases} 3x+2y=5 \\ x-2y=3 \end{cases} \quad \text{is equivalent to} \quad \begin{cases} 3x+2y=5 \\ 4x \qquad =8 \end{cases}$$

Solving $4x=8$ for x yields $x=2$. Find the value of y by substituting $x=2$ in $3x+2y=5$. Thus,

$$3(2)+2y=5$$

and $$y=-\tfrac{1}{2}$$

Therefore, the solution of the given system of equations is $x=2$, $y=-\frac{1}{2}$. Figure 7.3 shows the graph of the system.

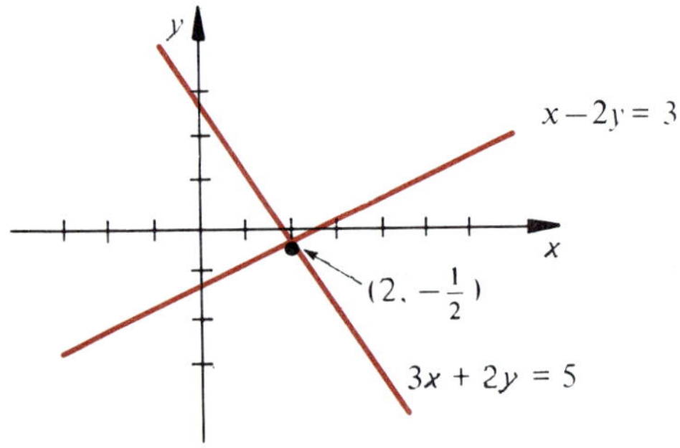

FIGURE 7.3

EXAMPLE 2 By eliminating a variable, solve the system

$$3a+2b=-4$$
$$4a+9b=1$$

SOLUTION A close look at this system of equations shows that neither of the variables will be eliminated by adding the corresponding members. However, in converting to an equivalent system in which either the a or the b coefficients are equal, it will be possible to eliminate one of the variables. If the first equation is multiplied by 4 and the second by 3, the numerical coefficient of the variable a will be 12 in each equation. Thus, writing these equivalent equations, we have

$$\begin{cases} 12a+\ 8b=-16 \\ 12a+27b=\quad 3 \end{cases}$$

Now subtracting the second equation from the first yields the equivalent system

$$\begin{cases} 12a+\ 8b=-16 \\ \qquad -19b=-19 \end{cases}$$

Thus, $b=1$. Substituting this value in one of the original equations yields

$$3a+2(1)=-4$$

Solving for a, we get $a=-2$. Therefore, the solution of the given system is $a=-2$, $b=1$.

Eliminating a Variable by Substitution

Another way of eliminating a variable from a system involves substituting a quantity for its equal. If one of the equations of a system is solved for one variable in terms of the other and the known variable substituted in the remaining equation, the number of variables is reduced by one. This method of eliminating a variable is called the **method of substitution**.

EXAMPLE 3 Solve the following system of equations by substitution.

(1) $3x+2y=5$

(2) $x-2y=3$

SOLUTION Begin with either equation, solving for either variable in terms of the other. We choose to solve the second equation for x. Thus, Equation (2) becomes

(2) $x=2y+3$

Substituting this result in Equation (1) gives

$$3(2y+3)+2y=5$$

Solving for y,

$$y=-\tfrac{1}{2}$$

and substituting this result in Equation (2), we get

$$x=2(-\tfrac{1}{2})+3=2$$

Therefore, the desired solution is $x=2$, $y=-\frac{1}{2}$.

Solutions of Linear Systems

Since all the linear systems presented so far have had precisely one solution, you may think that this is always the case. As Example 4 shows, however, a linear system can have infinitely many solutions or no solution, depending on the given equations.

EXAMPLE 4 Find the solution to each of the following *linear* systems.

$$\text{I}\ \begin{cases} x+y=2 \\ x+2y=3 \end{cases} \qquad \text{II}\ \begin{cases} x-2y=6 \\ 2x-4y=12 \end{cases} \qquad \text{III}\ \begin{cases} 2x+y=4 \\ 6x+3y=-6 \end{cases}$$

SOLUTION Figure 7.4 shows a sketch of each of the three systems.

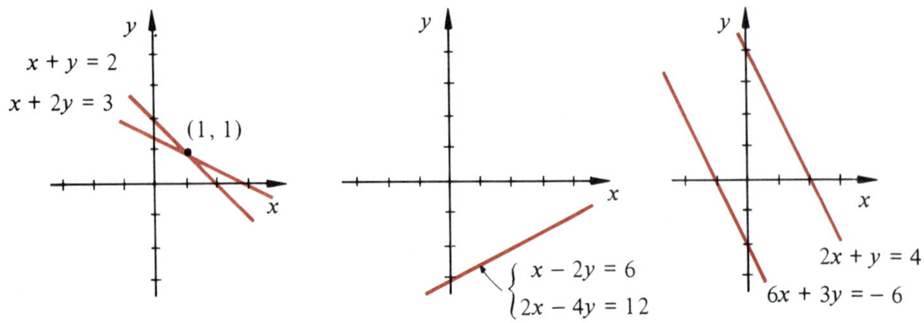

FIGURE 7.4

The two straight lines of System I cross at (1, 1). Thus, the solution is $x=1$, $y=1$, which can easily be checked by direct substitution. Since the equations of System II have the same intercepts, the two lines coincide, and the solution set is the infinite set of ordered pairs given by the coordinates of the points on the line. In System III, the lines do not intersect; that is, they are parallel. Hence the system has no solutions.

The same conclusions can be reached by algebraic methods if we proceed as follows:

System I Subtract the first equation from the second, to get

$$\begin{array}{r} x+2y= \ \ 3 \\ -x- \ y=-2 \\ \hline y= \ \ 1 \end{array}$$

This leads to the solution $x=1$, $y=1$.

System II Multiply the first equation by -2 and add it to the second equation, to get

$$\begin{array}{r} -2x+4y=-12 \\ 2x-4y= \ \ 12 \\ \hline 0 \ = \ \ \ 0 \end{array}$$

Since $0=0$ is true for all values of x and y, we conclude that this system has infinitely many solutions. Particular solutions can be obtained by letting y be a value and finding $x=6+2y$. For example, if $y=-1$, $x=6+2(-1)=4$ so $(4, -1)$ is a solution. Show that (10, 2) and (0, −3) are also solutions.

System III Multiply the first equation by -3 and add it to the second equation, to get

$$\begin{array}{r} -6x-3y=-12 \\ 6x+3y=- \ \ 6 \\ \hline 0 \ =-18 \end{array}$$

Since $0=-18$ can never be true, there is, we conclude, no solution to this system.

Example 4 shows the only three possibilities for linear systems of two equations in two unknowns. Since the graph of each of the equations is a straight line, one of the following statements must be true.

(1) The lines intersect at one point. (The system is said to be **consistent**.)

(2) The lines are parallel. (The system is said to be **inconsistent**.)

(3) The lines coincide. (The system is said to be **dependent**.)

More generally, linear systems are said to be consistent, inconsistent, or dependent as their simultaneous solution sets consist of a single element, no elements, or infinitely many elements. *Caution:* This terminology applies only to *linear* systems and cannot be extended to nonlinear systems.

Applications of Linear Systems

Examples 5 through 7 show how systems arise. No general technique or set of rules exists for converting a written problem to a set of equations.

EXAMPLE 5 The graph of $ax^2+by=1$ passes through the points (1, 4) and (2, 7). Determine the values of a and b.

SOLUTION Since both the given points must satisfy the equation $ax^2+by=1$, we can write

(1) $a+4b=1$

(2) $4a+7b=1$

Thus, the coefficients a and b can be found by solving Equations (1) and (2) simultaneously. Using substitution, write Equation (1) in the form

(1) $a=1-4b$

Substituting Equation (1) in (2) yields

$$4(1-4b)+7b=1$$

so

$$4-16b+7b=1$$

$$b=\frac{1}{3}$$

Now using Equation (1), we get

$$a=1-4\left(\frac{1}{3}\right)=-\frac{1}{3}$$

Therefore, the equation whose graph passes through points (1, 4) and (2, 7) is

$$-\frac{1}{3}x^2+\frac{1}{3}y=1$$

or

$$x^2-y=-3$$

EXAMPLE 6 A computer company markets two card readers. The 110 Card Reader reads 500 cards a minute and the 120 Card Reader reads 1000 cards a minute. The two readers processed 8000 cards in an accumulated time of 10 min. How many cards did each machine process?

SOLUTION In this problem, let A = number of cards processed by the 110 Card Reader, and B = number of cards processed by the 120 Card Reader. Therefore,

(1) $A+B=8000$

Also,

$$\frac{A}{500} = \text{time required to process } A \text{ cards on the 110 Card Reader}$$

and

$$\frac{B}{1000} = \text{time required to process } B \text{ cards on the 120 Card Reader}$$

Consequently,

$$\frac{A}{500}+\frac{B}{1000}=10$$

or equivalently,

(2) $2A+B=10000$

Solving Equation (1) for B and substituting the result in Equation (2) gives

$$2A+(8000-A)=10000$$
$$A=2000 \text{ cards}$$

Therefore, $B=6000$ cards.

EXAMPLE 7 A chemist has two acid solutions, one containing 10% acid and the other 4% acid. How many liters of each solution are needed to make 200 liters of a solution that is 6% acid?

SOLUTION Let x = number of liters of 10% acid solution and y = number of liters of 4% acid solution. Since the amount of the 10% solution plus the amount of the 4% solution must equal 200 liters, we have the equation

(1) $x+y=200$

A second equation is generated by observing that the amount of acid in x liters of the 10% solution is $0.10x$, the amount of acid in y liters of the 4% solution is $0.04y$, and the amount of acid in the final 6% solution is $200(0.06)=12$. Since the amount of acid from the two sources must equal the amount of acid in the final solution, the appropriate equation is

(2) $0.10x+0.04y=12$

The solution to this system of equations is found by solving Equation (1) for y and substituting the result in Equation (2). Thus,

$$0.10x+0.04(200-x)=12$$
$$0.10x+8-0.04x=12$$
$$x=66\tfrac{2}{3} \text{ liters}$$

From (1), we get $y=133\frac{1}{3}$ liters.

Exercises for Section 7.1

Solve the systems of equations in Exercises 1–14 by the method of elimination of a variable by addition or subtraction. Indicate whether the system is consistent, inconsistent, or dependent. In each case, graph the system.

1. $2x-4y=2$, $-2x+y=4$
2. $x+5y=4$, $4x+3y=6$
3. $3x+4y=23$, $x-3y=-1$
4. $x+y=4$, $x-y=2$
5. $x-4y=8$, $2x-8y=16$
6. $3x+2y-6=0$, $x-3y-3=0$
7. $3x+4y=10$, $6x+8y=-2$
8. $3x+5y=2$, $6x+10y=4$
9. $3x-y=2$, $2x+y=6$
10. $2x+3y=7$, $6x-y=1$
11. $3x+y=0$, $2x-2y=2$
12. $x+5y=0$, $3x+15y=3$
13. $x+7y-7=0$, $8x-7y-3=0$
14. $3x+5y=4$, $\frac{3}{2}x+\frac{5}{2}y=2$

Solve the systems of equations in Exercises 15–29 by eliminating a variable by substitution. Indicate whether the system is consistent, inconsistent, or dependent.

15. $r-s=3$, $r-2s=5$
16. $2x-y-2=0$, $x+y+5=0$
17. $x+3y=7$, $2x+6y=9$
18. $2s-t=5$, $6s+2t=-5$
19. $5y-x=2$, $2y+3x=11$
20. $x+y=8$, $-19x+8y=10$
21. $4x+5y=-6$, $3x-2y=-16$
22. $4x=10-5y$, $7x=41+3y$
23. $6x-8y=14$, $3x-4y=7$
24. $4x+3y-6=0$, $2x+5y+4=0$
25. $3x+5y=5$, $x+4y=11$
26. $4x-7y=29$, $6x+5y=-3$
27. $9z-13w+3=0$, $6z-7w=3$
28. $5x+4y=0$, $3x+5y+13=0$
29. $2x+3y=-10$, $3x-2y=-2$

In the remainder of the exercise set, use the most convenient method of eliminating variables to solve the systems of equations.

30. The velocity-time equation of car A is $v=3t+20$ ft/sec, where t is the time in seconds. If the velocity-time equation for car B is $v=6t+5$ ft/sec, at what time do both cars have the same velocity? What is this velocity?
31. A company buys a total of 25 grinders from two different companies. It buys 3 more Brand A grinders than Brand B. The total number of each brand of grinder can be found by solving the system of equations

    $A+B=25$
    $A-B=3$

    Can you explain how these equations are derived? Solve for A and B.
32. Two integers have a sum of 30 and a product of 225. What are the two integers?
33. The sum of two numbers is 12 and the difference is 4. What are the two numbers?
34. The area of a field is to be 540 yd^2 with a perimeter of 128 yd. Find the dimensions of the field.

35. The graph of $ax+by^2=2$ passes through the points (4, −2) and (−2, 2). Determine the coefficients a and b.

36. The resistance of a wire is given by $R=aT+K$, where a and K are constants and T is the temperature. Find a and K if $R=25$ ohms (Ω) when $T=10\,^\circ\text{C}$ and $R=30\ \Omega$ when $T=100\,^\circ\text{C}$.

37. A boat travels 60 miles upstream in 10 hr. If it returns in 8 hr, find the rate of the current and the boat. Assume that both rates are constant during the trip.

38. A truck enters a freeway traveling 50 mph. One hour later a car enters the freeway by the same ramp, traveling 80 mph. Determine the length of time before the car overtakes the truck.

39. A technician needs 10 lb of an alloy that is 60% lead and 40% zinc by weight. If he has an alloy that is 80% lead and 20% zinc and another that is 50% lead and 50% zinc, how many pounds of each must he use to get the desired alloy?

40. Two water pipes running at the same time can fill a swimming pool in 6 hr. If both pipes run for 2 hr and the first is then shut off, the second pipe will take 5 more hours to fill the pool. How long does it take each pipe to fill the pool by itself?

41. A collection of nickels and dimes amounts to $2.50. If there are 35 coins in all, what would the number of dimes be?

42. An airplane flying with a tail wind takes 2 hr to make a trip of 1240 miles. If the return trip against the wind takes 2.5 hr, what is the speed of the airplane relative to the air? What is the wind speed?

43. Two computers can complete a series of calculations in 10 min when they are used together. After 6 min, one of the computers breaks down and the other takes 9 min to finish the job. How long does it take each computer alone to do the series of calculations?

7.2 Nonlinear Systems

Sometimes eliminating a variable can be used for solving systems of nonlinear equations too. The approach is practically the same as in solving systems of linear equations.

EXAMPLE 1 Solve by substitution the system

$$2x-y=4$$
$$x^2-y=4$$

SOLUTION Solving the linear equation for y, we get $y=2x-4$. Substituting this value in the quadratic yields

$$x^2-(2x-4)=4$$

Expanding and collecting like terms, we have

$$\begin{aligned} x^2-2x &= 0 \\ x(x-2) &= 0 \end{aligned}$$

from which $x=0$ and $x=2$. Substituting these values in the given linear equation, we have $x=0$, $y=-4$, and $x=2$, $y=0$, as the solutions. A picture of the system is shown in Figure 7.5.

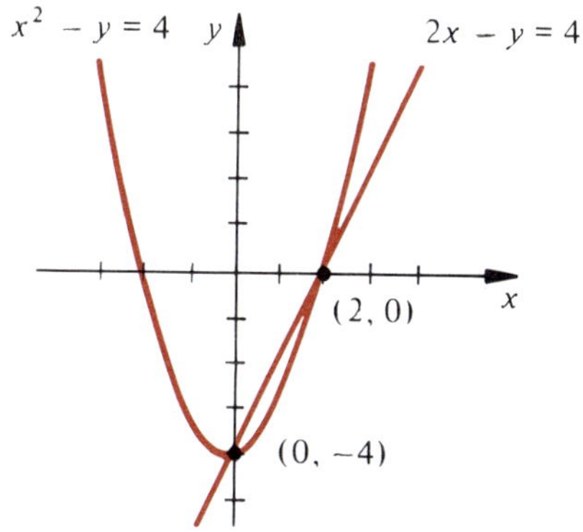

FIGURE 7.5

EXAMPLE 2 Solve the system

$$x^2 + y^2 = 4$$
$$x^2 + 4y^2 = 8$$

SOLUTION Multiplying the first of these equations by -1 and adding gives $3y^2 = 4$, from which

$$y = \pm\frac{2}{\sqrt{3}} = \pm\frac{2\sqrt{3}}{3}$$

Using this in the first equation gives $x^2 = 4 - \frac{4}{3} = \frac{8}{3}$, from which

$$x = \pm\sqrt{\frac{8}{3}} = \pm\frac{2\sqrt{6}}{3}$$

The solution set consists of the four ordered pairs

$$\left(\frac{2\sqrt{6}}{3}, \frac{2\sqrt{3}}{3}\right), \left(\frac{2\sqrt{6}}{3}, \frac{-2\sqrt{3}}{3}\right), \left(\frac{-2\sqrt{6}}{3}, \frac{2\sqrt{3}}{3}\right), \left(\frac{-2\sqrt{6}}{3}, \frac{-2\sqrt{3}}{3}\right)$$

The graph of this system is shown in Figure 7.6, in which you can see the four points of intersection, corresponding to the four ordered pairs of the solution set.

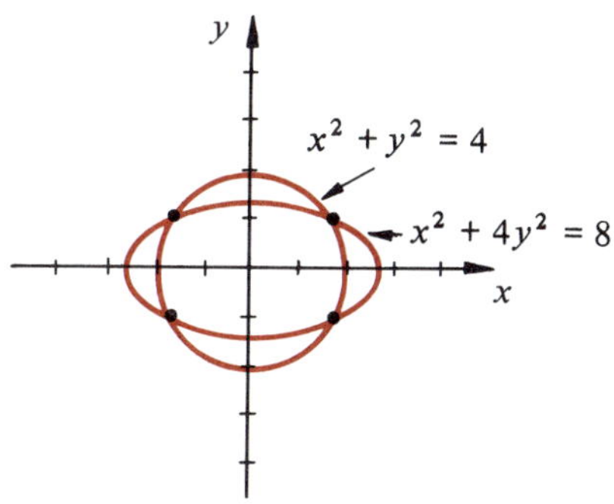

FIGURE 7.6

EXAMPLE 3 Solve the system

$$2x + y = 3$$
$$x^2 + y + x = 2$$

SOLUTION Solving the first equation for y and substituting into the second, we have

$$x^2+(3-2x)+x=2$$

or $$x^2-x+1=0$$

Using the quadratic formula we get

$$x=\frac{1\pm\sqrt{-3}}{2}$$

which is not real. Therefore, we conclude that there is no solution to this system.

Exercises for Section 7.2

Solve the nonlinear systems of Exercises 1–11. In each case, sketch both equations of the system, showing graphically the simultaneous solution.

1. $y-x=0$, $y-x^2=0$

2. $x-x^2+y=0$, $x-y+3=0$

3. $x^2-x-y=-1$, $x-y=0$

4. $x^2-3x-y=-2$, $x-y=1$

5. $2x-y=1$, $2x^2+x-y=1$

6. $x+y=0$, $-y+x^2+3x=-3$

7. $x+2y=3$, $x^2+y+x=0$

8. $x-y+2=0$, $2x^2+3y-2=0$

9. $2x+y-5=0$, $x^2-y+3=0$

10. $x^2-4x+y=2$, $2x-y=2$

11. $x^2-y-4=0$, $2y+x^2=1$

Solve the nonlinear systems in Exercises 12–32 for all *real* simultaneous solutions.

12. $x^2+y^2=1$, $x^2-y^2=1$

13. $x^2+y^2=4$, $4x^2+y^2=4$

14. $2x^2+3y^2=4$, $x^2+2y^2=10$

15. $x^2+4y^2=2$, $x^2-y^2=3$

16. $x^2+y^2=1$, $x^2+2y^2=2$

17. $x^2+y^2=1$, $2x^2-y^2=2$

18. $2y^2+x^2=9$, $y^2=x^2+3$

19. $y-x=1$, $x^2-y^2=10$

20. $x^2-4y^2=25$, $x-2y=-1$

21. $y^2+2x^2=2$, $x^2+y^2=2$

22. $y-x^2=2$, $x^2+y=10$

23. $2a^2-4b^2=7$, $a^2+2b^2=3$

24. $xy=4$, $2x+y=0$

25. $ab=16$, $a-b=0$

26. $x^2+4y^2=1$, $4x^2+y^2=1$

27. $xy=1$, $x^2+y^2=1$

28. $x^2-y^2=1$, $xy=1$

29. $x^2+y^2=16$, $y=2$

30. $x+y=7$, $x^2+y=4$

31. $x+y=1$, $x^2-y^2=1$

32. $x^2-y^2=1$, $y^2-x^2=1$

7.3 Higher-Order Linear Systems

Some systems of equations have more than two variables. We shall restrict our discussion to linear systems, and for simplicity, concentrate on systems of three equations in three unknowns.

An equation of the form $ax+by+cz=d$ is called a linear equation in three variables x, y, and z, where a, b, c, and d are constants. A *solution* to this equation is an ordered triple of numbers corresponding to values of x, y, and z that make the equation true. A *system* of three linear equations in three unknowns will be of the form

$$\begin{aligned} a_1x+b_1y+c_1z&=d_1\\ a_2x+b_2y+c_2z&=d_2\\ a_3x+b_3y+c_3z&=d_3 \end{aligned}$$

A *solution* to such a system is any ordered triple that simultaneously satisfies each of the equations of the system.

That the concepts of linear systems in two variables can be extended to higher orders should now be obvious. The techniques of solving systems of two equations can be used for higher-order systems. Consider, for example, a system of three linear equations. Choose any pair of equations and eliminate one of the variables by one of the standard techniques, usually by adding multiples of two of the equations together. This yields an equivalent equation in two variables. Now eliminate the same variable in a different pair of equations. In this way, the original three equations are essentially reduced to two linear equations in two variables—a system we know how to solve.

Like systems of two linear equations, higher-order linear systems are:

- **Consistent** if there is a unique simultaneous solution.
- **Inconsistent** if there is no simultaneous solution.
- **Dependent** if there are infinitely many simultaneous solutions.

When a third-order system in the variables x, y, and z has been reduced to an equivalent system in which the third equation involves only the z variable, the second equation involves only the y and perhaps the z variables, the system is said to be in **triangular** form. A system in triangular form is easy to solve in a systematic way. Further, a system that is not reducible to a triangular form either has no solution or has infinitely many, that is, it is either inconsistent or dependent. The next example shows a typical reduction to triangular form.

EXAMPLE 1 Solve the system

$$\begin{aligned} x-6y+3z&=-2\\ 2x-3y+z&=-2\\ 3x+3y-2z&=2 \end{aligned}$$

SOLUTION The solution to this system is obtained using the method of Gaussian elimination. Since the manipulations required to reduce three equations to triangular form are more involved than for two equations, we introduce some special notation.

First label each equation in the system. For instance, suppose we number the equations in the given system.

$$\begin{aligned} &① & x-6y+3z&=-2\\ &② & 2x-3y+z&=-2\\ &③ & 3x+3y-2z&=2 \end{aligned}$$

Then the notation

$$a① + ② \to ②$$

means "multiply Equation ① by a, add it to Equation ②, and replace Equation ② with this sum." Notice that this transformation affects only Equation ②, while Equations ① and ③ are unaltered. We show the use of this notation in the context of the given system.

$$\left.\begin{aligned} x-6y+3z&=-2\\ 2x-3y+z&=-2\\ 3x+3y-2z&=2\end{aligned}\right\}\xrightarrow{-2①+②\to②}\left\{\begin{aligned} x-6y+3z&=-2\\ 9y-5z&=2\\ 3x+3y-2z&=2\end{aligned}\right\}\xrightarrow{-3①+③\to③}$$

$$\left.\begin{aligned} x-6y+3z&=-2\\ 9y-5z&=2\\ 21y-11z&=8\end{aligned}\right\}\xrightarrow{-\frac{7}{3}②+③\to③}\left\{\begin{aligned} x-6y+3z&=-2\\ 9y-5z&=2\\ \tfrac{2}{3}z&=\tfrac{10}{3}\end{aligned}\right\}$$

From Equation ③ of the triangular form, we get $z=5$. Substituting $z=5$ in $9y-5z=2$, yields $y=3$. Finally, substituting $y=3$ and $z=5$ in $x-6y+3z=-2$, we get $x=1$. Hence, the solution is $x=1$, $y=3$, $z=5$.

The procedures used in reducing a system to an equivalent triangular form are useful even if the system is dependent or inconsistent.

EXAMPLE 2 Solve the system

$$\begin{aligned} r\quad\;\; + t&=-1\\ 3s+2t&=5\\ 3r-3s+t&=-8\end{aligned}$$

SOLUTION

$$\left.\begin{aligned} ①\quad r\quad\;\; + t&=-1\\ ②\quad 3s+2t&=5\\ ③\quad 3r-3s+t&=-8\end{aligned}\right\}\xrightarrow{③-3①\to③}\left\{\begin{aligned} r\quad\;\; + t&=-1\\ 3s+2t&=5\\ -3s-2t&=-5\end{aligned}\right\}$$

$$\xrightarrow{②+③\to③}\left\{\begin{aligned} r\quad + t&=-1\\ 3s+2t&=5\\ 0&=0\end{aligned}\right\}$$

In this form we see the *dependent* nature of the system. Any of the infinitely many solutions can be found by choosing a value for t and finding the corresponding values for r and s in the first two equations. Thus, the solutions can be written $t=c$, $r=-1-c$, and $s=\frac{1}{3}(5-2c)$, where c is any real number. For example, if $t=0$, then $r=-1$ and $s=\frac{5}{3}$; if $t=1$, then $r=-2$ and $s=1$.

EXAMPLE 3 Find the dimensions of a triangle whose perimeter is 54 cm, if the longest side is

8 cm longer than the next longest side, which in turn is 5 cm longer than the shortest side.

SOLUTION Let

$x =$ longest side

$y =$ next longest side

$z =$ shortest side

Since the perimeter is the sum of the side lengths we have the equation $x+y+z=54$. Also, by stipulation $x-y=8$ and $y-z=5$. Therefore, we wish to solve the system

$$\begin{aligned} x+y+z&=54 \\ x-y\quad\;&=\;8 \\ y-z&=\;5 \end{aligned}$$

Labelling these equations, we proceed as follows.

$$\left.\begin{array}{lr} ① & x+y+z=54 \\ ② & x-y\quad\;=\;8 \\ ③ & y-z=\;5 \end{array}\right\} \xrightarrow{②-①\to②} \left\{\begin{array}{r} x+\;y+z=\;\;54 \\ -2y-z=-46 \\ y-z=\;\;\;5 \end{array}\right.$$

$$\xrightarrow{2③+②\to③} \left\{\begin{array}{r} x+\;y+z=\;\;54 \\ -2y-z=-46 \\ -3z=-36 \end{array}\right.$$

From the equation $-3z=-36$, we get $z=12$. Using this in $-2y-z=-46$ yields $y=17$. Finally, substituting $y=17$ and $z=12$ in $x+y+z=54$, we get $x=25$. Therefore, the sides of the triangle are 12 cm, 17 cm, and 25 cm.

Systems of four equations in four unknowns are handled similarly. Three distinct pairs of equations are used to eliminate one of the variables, thus essentially reducing the original system to one of three equations in three unknowns.

EXAMPLE 4 Solve the 4th order linear system:

$$\begin{array}{lr} ① & 2x-\;y+3z+\;w=\;\;\;0 \\ ② & x+\;y-\;z-\;w=-1 \\ ③ & x-2y+3z-2w=-9 \\ ④ & 5x-5y+\;z-3w=-1 \end{array}$$

SOLUTION We first eliminate the variable w from three of the equations.

$$\left.\begin{array}{r} 2x-\;y+3z+\;w=\;\;\;0 \\ x+\;y-\;z-\;w=-1 \\ x-2y+3z-2w=-9 \\ 5x-5y+\;z-3w=-1 \end{array}\right\} \xrightarrow{①+②\to②} \begin{array}{r} 2x-\;y+3z+\;w=\;\;\;0 \\ 3x\qquad+2z\qquad\;=-1 \\ x-2y+3z-2w=-9 \\ 5x-5y+\;z-3w=-1 \end{array}$$

$$\xrightarrow{2①+③\to③}\left.\begin{aligned}2x-y+3z+w&=0\\3x+2z&=-1\\5x-4y+9z&=-9\\5x-5y+z-3w&=-1\end{aligned}\right\}\xrightarrow{3①+④\to④}\begin{aligned}2x-y+3z+w&=0\\3x+2z&=-1\\5x-4y+9z&=-9\\11x-8y+10z&=-1\end{aligned}$$

In the final system shown, ②, ③, and ④ form a system of three equations in three unknowns, which may be solved as before. The solution to the system is $x=1, y=-1, z=-2$, and $w=3$.

To close this section we note that the operations used in solving linear systems are totally arithmetic and hence easily programmable. The most general approach is by reduction to triangular form. This algorithm* is aided tremendously by the use of matrix notation, which we introduce in the next section.

Exercises for Section 7.3

For the systems of Exercises 1–22, tell which are consistent, which are inconsistent, and which are dependent. For those which are consistent, give the unique simultaneous solution and give an equivalent triangular system.

1. $\begin{aligned}x+y+z&=8\\x+y-z&=12\\x-y+z&=2\end{aligned}$

2. $\begin{aligned}x+3y+2z&=2\\3x+2y-z&=-1\\x-3y-3z&=0\end{aligned}$

3. $\begin{aligned}x+2y-z&=1\\y+z&=3\\x-y&=2\end{aligned}$

4. $\begin{aligned}2r-3s+4t&=5\\5r-7s+4t&=4\\-2r+3s-5t&=2\end{aligned}$

5. $\begin{aligned}3x+4y-7z&=12\\6x+3y+z&=0\\6x+8y-14z&=3\end{aligned}$

6. $\begin{aligned}x+y-z&=-3\\2x-y+3z&=12\\x+2y-3z&=-10\end{aligned}$

7. $\begin{aligned}x+2y+z&=3\\2x-y+3z&=-2\\4x+3y+5z&=4\end{aligned}$

8. $\begin{aligned}3a+b&=2\\-2a+c&=1\\3a+3b-c&=1\end{aligned}$

9. $\begin{aligned}x+2y-z&=0\\x-y+z&=0\\2x-3y-2z&=0\end{aligned}$

10. $\begin{aligned}x+y+2z&=4\\2x+2y+4z&=1\\2y-z&=5\end{aligned}$

11. $\begin{aligned}3a-5b-5c&=3\\a+b-2c&=7\\2a-3b-2c&=0\end{aligned}$

12. $\begin{aligned}9a-4b+2c&=8\\3a+2b+c&=1\\12a-18b&=17\end{aligned}$

13. $\begin{aligned}6x-8y+3z&=-5\\15x+12y+7z&=12\\9x+20y-4z&=25\end{aligned}$

14. $\begin{aligned}3r+5s-3t&=31\\2r-3s+2t&=13\\5r+2s-5t&=20\end{aligned}$

15. $\begin{aligned}3r-7s&=1\\3r-3s+8t&=2\\s+2t&=-1\end{aligned}$

16. $\begin{aligned}5x-y+3z&=3\\-2x+y-2z&=2\\x+4y-5z&=-1\end{aligned}$

17. $\begin{aligned}3x-2y+z&=2\\x+y-z&=-1\\2x+y+z&=0\end{aligned}$

18. $\begin{aligned}x+y&=2\\x+z&=-1\\y-z&=3\end{aligned}$

19. $\begin{aligned}x+z&=2\\x+y-3z&=8\\x-y+2z&=3\end{aligned}$

20. $\begin{aligned}x+2y-3z&=1\\y-z&=4\\-x+y+z&=5\end{aligned}$

21. $\begin{aligned}x+4y-3z&=8\\2x-3y-z&=2\\x+y+z&=10\end{aligned}$

22. $\begin{aligned}x-y-z&=4\\2x+y+z&=-2\\3x-2y-z&=1\end{aligned}$

Find the unique simultaneous solution to each of the consistent linear systems in Exercises 23–28.

23. $\begin{aligned}x+y-z+w&=10\\3x-y+2z-5w&=-1\\x-2y+z+3w&=0\\x+4y-3z+3w&=24\end{aligned}$

*An **algorithm** is a patterned sequential approach to a particular problem. Algorithms are useful, indeed necessary, for programming instructions to a computer.

24.
$$\begin{aligned} 4x+y+2z-w&=0\\ -3x+z-w&=-2\\ x-4y-w&=-1\\ 5x+3y+z&=0 \end{aligned}$$

25.
$$\begin{aligned} 2x+3y-z-w&=3\\ x-4y+3z+5w&=-1\\ x+5y+z+w&=3\\ -3x-4y+2z+w&=-1 \end{aligned}$$

26.
$$\begin{aligned} x+y+z+w&=4\\ x-y+z-w&=6\\ x+2y-z-w&=1\\ 2x-y+3z+4w&=5 \end{aligned}$$

27.
$$\begin{aligned} x+2y+z+s+2t&=2\\ x-y-z+2s-t&=7\\ 2x+3y+3z-s+4t&=-3\\ x-4y+z-s-3t&=6\\ 3x+5y-2z-3s-5t&=0 \end{aligned}$$

28.
$$\begin{aligned} 2x+y+z-3s+t&=-1\\ -3x+4z+s-t&=-6\\ 5x+2y-5s+5t&=0\\ x+y+2z+3t&=-2\\ -7x-y+z-4s&=-11 \end{aligned}$$

In the remaining problems, determine the system of equations from the statement of the problem.

29. The total number of nickels, dimes, and quarters in a box is 900. There are 20 more nickels than there are dimes, and the number of nickels and dimes together is 44 more than the number of quarters. How many coins of each denomination are in the box?

30. A triangle has a perimeter of 50 in. The longest side is 3 in. longer than the next longest and 10 in. longer than the shortest. Find the length of each side.

31. A company runs three production lines that together have an output of 45 parts an hour. Twice the production of the first line equals the sum of the other two lines, and the output of the second line is four parts an hour more than the third line. Find the production rate of each line.

32. A laboratory produces an alloy of copper, tin, and zinc having a weight of 37 grams (g). If the copper in the alloy weighs 3 g more than the zinc and the zinc weighs 8 g more than the tin, how much of each element is in the alloy?

7.4 Matrix Methods

In mathematics, groupings of numbers such as

$$\begin{array}{c} 7\\ 8\\ 3\\ -1 \end{array} \qquad \begin{array}{ccc} & 4 & \\ & 2 \quad -1 & \\ 7 & 6 & 0 \end{array} \qquad \begin{array}{cccc} 2 & 3 & 5 & 2\\ 6 & 7 & 8 & 9\\ 3 & -2 & 0 & 6 \end{array}$$

(a) (b) (c)

are called **arrays**. A rectangular array is called a **matrix**. Thus, of the arrays shown, only (a) and (c) are matrices. Do not think of a matrix as something having a numerical *value*; it is a scheme for conveniently listing or cataloging numbers. In fact applications of matrices range from use as a filing system to the solution of difficult systems of equations.

The **dimension**, or size, of a matrix is indicated by giving the number of rows followed by the number of columns. Thus an $m \times n$ (read "m by n") matrix is a rectangular array having m rows and n columns. A rectangular array $m \times n$ and an array $n \times m$ are not considered the same size unless $m = n$.

Of the arrays in the preceding paragraph (a) is a 4×1 matrix, (b) is nonrectangular and thus not a matrix, and (c) is a 3×4 matrix. We denote matrices by capital letters and display them by enclosing them in brackets. Thus,

$$A = \begin{bmatrix} 7 \\ 8 \\ 3 \\ -1 \end{bmatrix} \quad \text{and} \quad B = \begin{bmatrix} 2 & 3 & 5 & 2 \\ 6 & 7 & 8 & 9 \\ 3 & -2 & 0 & 6 \end{bmatrix}$$

describe the two matrices (a) and (c). Although it need not be the case, all matrices we shall study have real number entries.

A **square matrix** has the same number of rows and columns; then the number of rows (or columns) is called the **order** of the matrix. Thus,

$$\begin{bmatrix} 2 & 1 \\ 0 & -3 \end{bmatrix} \quad \text{and} \quad \begin{bmatrix} 0 & 5 & -2 \\ 2 & 3 & 0 \\ -1 & 7 & 9 \end{bmatrix}$$

are matrices of orders 2 and 3.

To identify specific elements within a matrix, we use a double subscript notation to indicate the row and column of the element. The first number of the subscript refers to the row and the second to the column. Thus, the notation a_{ij} means the element in the ith row and the jth column of a matrix A. The scheme in the following 3×4 matrix is typical.

$$A = \begin{bmatrix} a_{11} & a_{12} & a_{13} & a_{14} \\ a_{21} & a_{22} & a_{23} & a_{24} \\ a_{31} & a_{32} & a_{33} & a_{34} \end{bmatrix}$$

Elements such as a_{11}, a_{22}, and a_{33}, which have the same row and column number, are called **main diagonal elements**.

The theory of matrices includes algebraic operations between two or more matrices. Our study of matrices, however, will be restricted to showing how they are used to describe and solve linear systems of equations.

The work required to reduce a linear system to triangular form can be simplified if we observe that the symbols used for the variables are not required in the process itself. That is, reducing a system to triangular form really involves only manipulating the coefficients of the variables. A systematic way of avoiding the needless rewriting of variables is to use a matrix to keep track of the coefficients of a system of equations. For example, the system

$$\begin{aligned} 3x - y + 2z &= 3 \\ x + y - z &= -2 \\ 2x - 3y - 5z &= -5 \end{aligned}$$

can be represented by the matrix

$$\left[\begin{array}{ccc:c} 3 & -1 & 2 & 3 \\ 1 & 1 & -1 & -2 \\ 2 & -3 & -5 & -5 \end{array}\right]$$

This matrix is called the **matrix of the system**, or the **augmented matrix**. We have inserted the dashed line to emphasize the matrix of the coefficients of the variables, which alone is called the **coefficient matrix**. Notice that *the form of the augmented matrix assumes that each equation in the system is written in the form* $ax+by+cz=d$. If an equation of a system is not given in this form, it must be so written before it can be used in the augmented matrix.

Once the matrix of a system of equations is written, we treat the rows of the matrix just as we treat the equations they represent. Of course, we must make a mental note that the numbers in the first column represent the coefficients of x; those in the second column, the coefficients of y; and those in the third column, the coefficients of z.

A system of equations written in matrix form is solved using the same operations as in the method of Gaussian elimination. The use of *equivalent matrices* corresponds to equivalent systems. Thus, two matrices are equivalent if

(1) The position of any two rows is interchanged.

(2) A row of one matrix is a nonzero multiple of the other.

(3) A row of one matrix is the sum of two rows of the other.

The following examples show the use of the method of elimination along with matrix notation to solve a system of equations.

EXAMPLE 1 Use matrix notation to solve the system

$$\begin{aligned} 3x+2y &= 10 \\ x-3y &= -4 \end{aligned}$$

SOLUTION This system is represented by the matrix

$$\left[\begin{array}{cc:c} 3 & 2 & 10 \\ 1 & -3 & -4 \end{array}\right]$$

The following transformation yields the desired solution.

$$\left[\begin{array}{cc:c} 3 & 2 & 10 \\ 1 & -3 & -4 \end{array}\right] \xrightarrow{① - 3② \rightarrow ②} \left[\begin{array}{cc:c} 3 & 2 & 10 \\ 0 & 11 & 22 \end{array}\right]$$

The resulting matrix is the matrix for the system

$$\begin{aligned} 3x+2y &= 10 \\ 11y &= 22 \end{aligned}$$

We get $y=2$ from the equation $11y=22$. Substituting $y=2$ into $3x+2y=10$, we get $x=2$. The solution is then $x=2$, $y=2$.

Notice in Example 1 that row ① remained unchanged although we used it to operate on row ②. In this case, we call line ① the *operative line* of the operation. In matrices of 3 or more rows we can perform more than one operation at a time *provided* that the operative line is the same for all uses of the operation. This is shown in the next example.

EXAMPLE 2 Use matrix notation to solve the system

$$\begin{aligned} x-6y+3z&=-2\\ 2x-3y+z&=-2\\ 3x+3y-2z&=2 \end{aligned}$$

This is the same system that was solved in Example 1, Section 7.3.

SOLUTION The given system is represented by the augmented matrix

$$\left[\begin{array}{rrr|r} 1 & -6 & 3 & -2\\ 2 & -3 & 1 & -2\\ 3 & 3 & -2 & 2 \end{array}\right]$$

The following steps yield the desired form.

$$\left[\begin{array}{rrr|r} 1 & -6 & 3 & -2\\ 2 & -3 & 1 & -2\\ 3 & 3 & -2 & 2 \end{array}\right] \xrightarrow[(-3)①+③\to③]{(-2)①+②\to②} \left[\begin{array}{rrr|r} 1 & -6 & 3 & -2\\ 0 & 9 & -5 & 2\\ 0 & 21 & -11 & 8 \end{array}\right]$$

$$\xrightarrow{7②-3③\to③} \left[\begin{array}{rrr|r} 1 & -6 & 3 & -2\\ 0 & 9 & -5 & 2\\ 0 & 0 & -2 & -10 \end{array}\right]$$

Notice that this is the matrix of the triangular form of the given system, that is,

$$\begin{aligned} x-6y+3z&=-2\\ 9y-5z&=2\\ -2z&=-10 \end{aligned}$$

Solving the last equation, we get $z=5$. Substituting $z=5$ into the middle equation yields $y=3$. Finally, we find that $x=1$ by substituting $y=3$ and $z=5$ into the upper equation. The solution $x=1$, $y=3$, $z=5$ agrees with that in Example 1 of Section 7.3.

EXAMPLE 3 Solve the system

$$\begin{aligned} x+y-z&=-1\\ 2x-y+z&=2\\ x-5y+5z&=7 \end{aligned}$$

SOLUTION The given system is represented by the augmented matrix

$$\left[\begin{array}{rrr|r} 1 & 1 & -1 & -1\\ 2 & -1 & 1 & 2\\ 1 & -5 & 5 & 7 \end{array}\right]$$

The following steps lead to the solution.

$$\left[\begin{array}{rrr|r} 1 & 1 & -1 & -1 \\ 2 & -1 & 1 & 2 \\ 1 & -5 & 5 & 7 \end{array}\right] \xrightarrow[(-1)①+③\to③]{(-2)①+②\to②} \left[\begin{array}{rrr|r} 1 & 1 & -1 & -1 \\ 0 & -3 & 3 & 4 \\ 0 & -6 & 6 & 8 \end{array}\right]$$

$$\xrightarrow{(-2)②+③\to③} \left[\begin{array}{rrr|r} 1 & 1 & -1 & -1 \\ 0 & -3 & 3 & 4 \\ 0 & 0 & 0 & 0 \end{array}\right]$$

The third row of all zeros shows that x, y, and z are related only by the equations that the first two rows represent. The system is therefore *dependent*. Any of the infinitely many solutions may be found by assigning a value to z and solving for x and y. For example, if we let $z=1$, then $x=\frac{1}{3}$ and $y=-\frac{1}{3}$. If $z=0$, then $x=\frac{1}{3}$ and $y=-\frac{4}{3}$. In general, if $z=c$, the second row yields $y=c-\frac{4}{3}$ and the first row yields $x=\frac{1}{3}$.

EXAMPLE 4 Solve the system

$$\begin{aligned} x-\ y+\ z&=\ \ 8 \\ 2x+\ y-\ z&=-2 \\ -x-5y+5z&=\ 10 \end{aligned}$$

SOLUTION The matrix of the system is

$$\left[\begin{array}{rrr|r} 1 & -1 & 1 & 8 \\ 2 & 1 & -1 & -2 \\ -1 & -5 & 5 & 10 \end{array}\right]$$

The reduction of this matrix to triangular form follows.

$$\left[\begin{array}{rrr|r} 1 & -1 & 1 & 8 \\ 2 & 1 & -1 & -2 \\ -1 & -5 & 5 & 10 \end{array}\right] \xrightarrow[①+③\to③]{(-2)①+②\to②} \left[\begin{array}{rrr|r} 1 & -1 & 1 & 8 \\ 0 & 3 & -3 & -18 \\ 0 & -6 & 6 & 18 \end{array}\right]$$

$$\xrightarrow{(2)②+③\to③} \left[\begin{array}{rrrr} 1 & -1 & 1 & 8 \\ 0 & 3 & -3 & -18 \\ 0 & 0 & 0 & -18 \end{array}\right]$$

The last row represents

$$0x+0y+0z=-18$$

which is impossible for any x, y, and z. Hence, the system is *inconsistent*.

The matrix method can be used for linear systems of any order. The following example is typical of fourth order systems.

EXAMPLE 5 Solve the system

$$\begin{aligned} x - y + 2z + w &= -3 \\ 2x + y - 3z - w &= 4 \\ 4x + 3y - z + 2w &= 9 \\ 3x - y + 5z - 2w &= -9 \end{aligned}$$

SOLUTION This system is represented by the matrix

$$\left[\begin{array}{rrrr|r} 1 & -1 & 2 & 1 & -3 \\ 2 & 1 & -3 & -1 & 4 \\ 4 & 3 & -1 & 2 & 9 \\ 3 & -1 & 5 & -2 & -9 \end{array}\right]$$

The following sequence of operations leads to the solution.

$$\left[\begin{array}{rrrr|r} 1 & -1 & 2 & 1 & -3 \\ 2 & 1 & -3 & -1 & 4 \\ 4 & 3 & -1 & 2 & 9 \\ 3 & -1 & 5 & -2 & -9 \end{array}\right] \xrightarrow[\substack{-3\text{①}+\text{④}\to\text{④}}]{\substack{-2\text{①}+\text{②}\to\text{②} \\ -4\text{①}+\text{③}\to\text{③}}}$$

$$\left[\begin{array}{rrrr|r} 1 & -1 & 2 & 1 & -3 \\ 0 & 3 & -7 & -3 & 10 \\ 0 & 7 & -9 & -2 & 21 \\ 0 & 2 & -1 & -5 & 0 \end{array}\right] \xrightarrow{\substack{-\frac{7}{3}\text{②}+\text{③}\to\text{③} \\ -\frac{2}{3}\text{②}+\text{④}\to\text{④}}}$$

$$\left[\begin{array}{rrrr|r} 1 & -1 & 2 & 1 & -3 \\ 0 & 3 & -7 & -3 & 10 \\ 0 & 0 & \frac{22}{3} & 5 & -\frac{7}{3} \\ 0 & 0 & \frac{11}{3} & -3 & -\frac{20}{3} \end{array}\right] \xrightarrow{-\frac{1}{2}\text{③}+\text{④}\to\text{④}} \left[\begin{array}{rrrr|r} 1 & -1 & 2 & 1 & -3 \\ 0 & 3 & -7 & -3 & 10 \\ 0 & 0 & \frac{22}{3} & 5 & -\frac{7}{3} \\ 0 & 0 & 0 & -\frac{11}{2} & -\frac{11}{2} \end{array}\right]$$

The last row gives the solution $w=1$. Using this value for w, the third row yields $z=-1$. Continuing, $y=2$ and $x=0$.

In each of the preceding examples, we used a triangular form in which the zeros are below the main diagonal. We could just as easily have used a triangular form with the zeros above the main diagonal. Either way, the solution is the same.

Exercises for Section 7.4

In Exercises 1–20, use matrix methods to determine whether the given linear system is consistent, dependent, or inconsistent. If the system is consistent, give the unique solution.

1. $\begin{aligned} x+2y&=10 \\ x-2y&=-6 \end{aligned}$

2. $\begin{aligned} 5x+2y&=6 \\ 3x+4y&=12 \end{aligned}$

3. $\begin{aligned} 3x-y&=5 \\ x+4y&=6 \end{aligned}$

4. $\begin{aligned} x-2y+z&=3 \\ 2x+y-z&=7 \\ 3x-y+2z&=6 \end{aligned}$

5. $\begin{aligned} 8x+5y &= 4 \\ 5y-3z &= 1 \\ 12x+ 5z &= 6 \end{aligned}$

6. $\begin{aligned} 2x+y-z &= 2 \\ x-y+z &= 4 \\ y-z &= -2 \end{aligned}$

7. $\begin{aligned} 3x+5y-3z &= 31 \\ 2x-3y+2z &= 13 \\ 5x+2y-5z &= 20 \end{aligned}$

8. $\begin{aligned} -x+3y-z &= 0 \\ 2x+y+2z &= 2 \\ -x+7y-z &= -2 \end{aligned}$

9. $\begin{aligned} 2x- 5z &= 2 \\ 9y-4z &= 7 \\ 3x-12y &= -2 \end{aligned}$

10. $\begin{aligned} x-2y+z &= 7 \\ 2x+ 3z &= 4 \\ y+2z &= 1 \end{aligned}$

11. $\begin{aligned} x+y+2z &= 3 \\ 3x-y-z &= 1 \\ x+5y+9z &= 11 \end{aligned}$

12. $\begin{aligned} 6x-8y+3z &= -5 \\ 15x+12y+7z &= 12 \\ 9x+20y-4z &= 25 \end{aligned}$

13. $\begin{aligned} 2x+y-z &= 9 \\ x-y+z &= 0 \\ -x+3y-2z &= 5 \end{aligned}$

14. $\begin{aligned} x+2y-z &= 1 \\ y+z &= 3 \\ x-y &= 2 \end{aligned}$

15. $\begin{aligned} x-y &= 2 \\ y-z &= -3 \\ 3x+2y-5z &= 1 \end{aligned}$

16. $\begin{aligned} x-2y+3z &= 7 \\ 2x-y-z &= -4 \\ -x+3y-4z &= -5 \end{aligned}$

17. $\begin{aligned} x-3y+2z &= 0 \\ 2x+y-z &= -3 \\ 10x-2y &= -12 \end{aligned}$

18. $\begin{aligned} -x-2y+z &= -7 \\ x-3y+4z &= 2 \\ 2x-5y-7z &= 5 \end{aligned}$

19. $\begin{aligned} x+y+z &= 3 \\ x-2y-3z &= 4 \\ 2x+y+2z &= 9 \end{aligned}$

20. $\begin{aligned} 3x+y-2z &= 0 \\ x-y+z &= 1 \\ -2x+6y-7z &= -2 \end{aligned}$

The systems in Exercises 21–25 are consistent. Find their unique solution by matrix methods.

21. $\begin{aligned} 3x-2y+z+w &= 10 \\ x+y-z-w &= -5 \\ 2x+y+z+5w &= 18 \\ -x-2y+5z-2w &= 5 \end{aligned}$

22. $\begin{aligned} 2x-5y+z-w &= -5 \\ x+4y-z-2w &= -7 \\ -x-7y+3z+w &= -5 \\ -3x+8y-6z-3w &= -6 \end{aligned}$

23. $\begin{aligned} 3x+y+z-w &= -2 \\ -5x+2y-3z-3w &= -1 \\ 7x+5y+4z-w &= -1 \\ 4x-y-2z+3w &= 7 \end{aligned}$

24. $\begin{aligned} 2x+5y-2z+w &= -11 \\ x-8y-z+3w &= -18 \\ -3x-9y+3z+2w &= -1 \\ 6x+7y+5z+8w &= 8 \end{aligned}$

25. $\begin{aligned} 2x-y+4z+w+u &= -2 \\ x-8y-2z+2w-u &= 10 \\ -3x+7y-z-3w+2u &= -8 \\ -x+5y+3z+2w-3u &= -4 \\ x+9y-2z+w+6u &= 14 \end{aligned}$

7.5 Determinants

Addition and multiplication of matrices can be defined, but we shall not do so. Of primary interest in this section is an operation on square matrices that yields a real number called the **determinant** of the matrix. To display the determinant of a matrix, replace the brackets with vertical bars. We denote the determinant of a matrix A by **det** $\boldsymbol{A}$. Thus, if

$$A=\begin{bmatrix} 3 & 7 \\ -1 & 2 \end{bmatrix}$$

then det A is symbolized by

$$\det A = \begin{vmatrix} 3 & 7 \\ -1 & 2 \end{vmatrix}$$

Restricting ourselves for the moment to 2×2 matrices, we make the following definition for the value of det A.

DEFINITION

Let

$$A = \begin{bmatrix} a_1 & b_1 \\ a_2 & b_2 \end{bmatrix}$$

Then the determinant of A is given by

$$\det A = \begin{vmatrix} a_1 & b_1 \\ a_2 & b_2 \end{vmatrix} = a_1b_2 - a_2b_1$$

Practically, we get det A by multiplying the element in the upper left by the element in the lower right and then subtracting the product of the element in the lower left and the element in the upper right. Remember, det A is a real number and not an array, whereas A itself is an array and not a number.

EXAMPLE 1 If

$$A = \begin{bmatrix} 2 & 3 \\ -1 & 4 \end{bmatrix}$$

then

$$\det A = \begin{vmatrix} 2 & 3 \\ -1 & 4 \end{vmatrix} = 2(4) - (-1)(3) = 11$$

EXAMPLE 2 $\begin{vmatrix} 3 & 2 \\ 5 & 3 \end{vmatrix} = 3(3) - (5)(2) = -1$

The definition of determinants of higher-order matrices depends on the definition of 2×2 determinants. More precisely, a third-order determinant is defined in terms of a second-order determinant, a fourth-order determinant in terms of a third-order determinant, and generally, an nth-order determinant in terms of a determinant of order $(n-1)$.

Recall that an element of a matrix is conveniently located with double subscript notation, in which the first number of the subscript refers to the row of the matrix and the second to the column. Thus, a_{ij} means the element in the ith row and the jth column.

DEFINITION

The determinant obtained from A by deleting the ith row and jth column, called the **minor** of the element a_{ij}, is denoted by M_{ij}.

EXAMPLE 3 Find the minors of a_{12} and a_{33}, given the matrix

$$A=\begin{bmatrix} 3 & 2 & 1 \\ 7 & 6 & -2 \\ 0 & 5 & 9 \end{bmatrix}$$

SOLUTION Find the minor of a_{12} by eliminating the first row and the second column of the given matrix. Thus,

$$M_{12}=\begin{vmatrix} 3 & 2 & 1 \\ 7 & 6 & -2 \\ 0 & 5 & 9 \end{vmatrix}=\begin{vmatrix} 7 & -2 \\ 0 & 9 \end{vmatrix}=63$$

Similarly, the minor of a_{33} is

$$M_{33}=\begin{vmatrix} 3 & 2 & 1 \\ 7 & 6 & -2 \\ 0 & 5 & 9 \end{vmatrix}=\begin{vmatrix} 3 & 2 \\ 7 & 6 \end{vmatrix}=4$$

Using the definition of a minor, we can give the following definition of the value of an ***n*th-order determinant in terms of minors**

DEFINITION

An nth-order determinant is the sum of the n products

$$(-1)^{i+j}a_{ij}M_{ij}$$

in any row or column. The value obtained is independent of the column or row used.

In particular, the definition says that a third-order determinant is the sum of the *three products* formed by multiplying each element of any row (or column) by its corresponding minor and assigning to each product a plus sign if the sum of the number of the column and the number of the row in which the element lies is even, and a minus sign if it is odd.

If we use the jth column to expand the determinant, we say that we have "expanded the determinant about the jth column." An expansion about the ith row can also be described in similar words.

A third-order determinant can be expanded by minors in six different ways. By elements of the first row,

$$\begin{aligned}\det A &= (-1)^{1+1}a_{11}M_{11}+(-1)^{1+2}a_{12}M_{12}+(-1)^{1+3}a_{13}M_{13}\\ &= a_{11}M_{11}-a_{12}M_{12}+a_{13}M_{13}\end{aligned}$$

By elements of the second row,

$$-a_{21}M_{21}+a_{22}M_{22}-a_{23}M_{23}$$

By elements of the third row,

$$a_{31}M_{31}-a_{32}M_{32}+a_{33}M_{33}$$

By elements of the first column,

$$a_{11}M_{11}-a_{21}M_{21}+a_{31}M_{31}$$

By elements of the second column,

$$-a_{12}M_{12}+a_{22}M_{22}-a_{32}M_{32}$$

By elements of the third column,

$$a_{13}M_{13}-a_{23}M_{23}+a_{33}M_{33}$$

Each of these expansions yields exactly the same value. Completing the evaluation in each case is essentially computing three second-order determinants. The following expansion shows this in only the first case, that is, by elements of the first row:

$$\det A=a_{11}\begin{vmatrix} a_{22} & a_{23} \\ a_{32} & a_{33} \end{vmatrix}-a_{12}\begin{vmatrix} a_{21} & a_{23} \\ a_{31} & a_{33} \end{vmatrix}+a_{13}\begin{vmatrix} a_{21} & a_{22} \\ a_{31} & a_{32} \end{vmatrix}$$

That the value of the determinant is independent of the column or row used for the expansion is a theorem proved in more advanced mathematics courses. Example 4 demonstrates that independence.

EXAMPLE 4 Using expansion of minors (a) about the first column (b) about the third row, find the value of the third-order determinant

$$\begin{vmatrix} 2 & -1 & 5 \\ 1 & 3 & -3 \\ 4 & 0 & 1 \end{vmatrix}$$

SOLUTION (a)

$$\begin{vmatrix} 2 & -1 & 5 \\ 1 & 3 & -3 \\ 4 & 0 & 1 \end{vmatrix}=(-1)^{1+1}(2)\begin{vmatrix} 3 & -3 \\ 0 & 1 \end{vmatrix}+(-1)^{2+1}(1)\begin{vmatrix} -1 & 5 \\ 0 & 1 \end{vmatrix}$$

$$+(-1)^{3+1}(4)\begin{vmatrix} -1 & 5 \\ 3 & -3 \end{vmatrix}$$

$$=2\begin{vmatrix} 3 & -3 \\ 0 & 1 \end{vmatrix}-\begin{vmatrix} -1 & 5 \\ 0 & 1 \end{vmatrix}+4\begin{vmatrix} -1 & 5 \\ 3 & -3 \end{vmatrix}$$

$$=2(3-0)-(-1-0)+4(3-15)$$

$$=-41$$

(b) $$\begin{vmatrix} 2 & -1 & 5 \\ 1 & 3 & -3 \\ 4 & 0 & 1 \end{vmatrix} = (-1)^{3+1}(4)\begin{vmatrix} -1 & 5 \\ 3 & -3 \end{vmatrix} + (-1)^{3+2}(0)\begin{vmatrix} 2 & 5 \\ 1 & -3 \end{vmatrix}$$

$$+(-1)^{3+3}(1)\begin{vmatrix} 2 & -1 \\ 1 & 3 \end{vmatrix}$$

$$=4\begin{vmatrix} -1 & 5 \\ 3 & -3 \end{vmatrix} - 0 + \begin{vmatrix} 2 & -1 \\ 1 & 3 \end{vmatrix}$$

$$=4(3-15)+(6-(-1))$$

$$=-41$$

Note the advantage of choosing to expand a determinant about rows or columns containing some zeros, since then fewer calculations are needed.

EXAMPLE 5 Evaluate the determinant of the matrix A.

$$A = \begin{vmatrix} 3 & -7 & 6 \\ 4 & 9 & 0 \\ 0 & -1 & 0 \end{vmatrix}$$

SOLUTION Both the third column and the third row contains two zeros. Evaluating by the elements of the third row gives

$$\det A = -(-1)\begin{vmatrix} 3 & 6 \\ 4 & 0 \end{vmatrix} = -24$$

and expanding using the third column,

$$\det A = 6\begin{vmatrix} 4 & 9 \\ 0 & -1 \end{vmatrix} = 6(-4) = -24$$

In either case, the result is the same.

In working with determinants, you will find that knowing the following properties is helpful. We assume in stating these properties that A is a square matrix of any order.

(1) If a row (or column) of A has all zeros, then $\det A = 0$.

(2) If A is obtained from B by interchanging two rows (or columns), then $\det A = -\det B$.

(3) If any two rows (or columns) of A are equal, then $\det A = 0$.

(4) If A and B are identical except that a row (or column) of B is k times the same row (or column) of A, then $\det A = k \det B$.

(5) The value of a determinant is unchanged if the rows and the columns of the matrix are interchanged.

(6) The value of a determinant is unchanged if a multiple of one row (or column) is added to another row (or column).

Property (6) is particularly useful when expanding by minors. The idea is to apply this property repeatedly until all the elements in a given row (or column) are zero except one. Then, when expanding by minors, only one (lower-order) determinant need be evaluated. In applying Property (6) we can use the familiar notation employed in linear systems and matrices (Sections 7.3 and 7.4).

EXAMPLE 6 Evaluate

$$\det A = \begin{vmatrix} 3 & -4 & 3 \\ 5 & 1 & 2 \\ 2 & 7 & -5 \end{vmatrix}$$

SOLUTION The following **row** operations are used to replace a_{12} and a_{32} by zeros:

$$\begin{vmatrix} 3 & -4 & 3 \\ 5 & 1 & 2 \\ 2 & 7 & -5 \end{vmatrix} \begin{array}{c} \textcircled{1}+4\textcircled{2}\to\textcircled{1} \\ \hline \textcircled{3}-7\textcircled{2}\to\textcircled{3} \end{array} \begin{vmatrix} 23 & 0 & 11 \\ 5 & 1 & 2 \\ -33 & 0 & -19 \end{vmatrix}$$

Expanding by elements of the second column, we have

$$\det A = \begin{vmatrix} 23 & 11 \\ -33 & -19 \end{vmatrix} = 23(-19) - 11(-33) = -74$$

The complete expansion by minors of the third-order determinant

$$A = \begin{vmatrix} a_{11} & a_{12} & a_{13} \\ a_{21} & a_{22} & a_{23} \\ a_{31} & a_{32} & a_{33} \end{vmatrix}$$

yields

$$\det A = a_{11}a_{22}a_{33} + a_{12}a_{23}a_{31} + a_{13}a_{21}a_{32} \\ - a_{13}a_{22}a_{31} - a_{11}a_{23}a_{32} - a_{12}a_{21}a_{33}$$

This same result can be obtained by using the following device: *Rewrite the first two columns to the right of the determinant and form the indicated products, adding or subtracting as shown.*

(7.1)

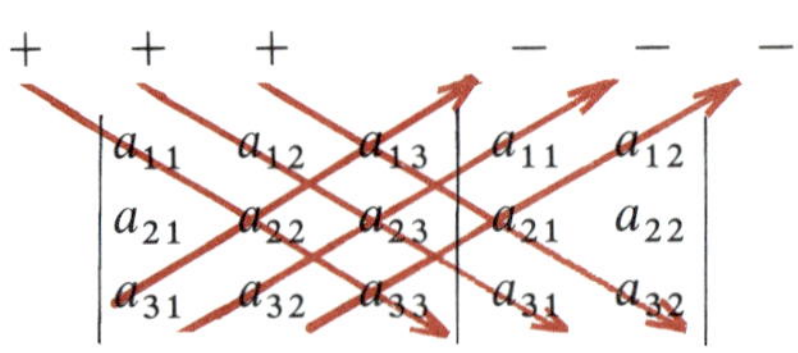

This device is not used to define a determinant because it cannot be generalized to higher-order determinants; that is, **the device works only for 3 × 3 determinants.**

EXAMPLE 7 Evaluate the determinant in Example 6 using (7.1).

SOLUTION

$$\begin{vmatrix} 3 & -4 & 3 \\ 5 & 1 & 2 \\ 2 & 7 & -5 \end{vmatrix} \begin{matrix} 3 & -4 \\ 5 & 1 \\ 2 & 7 \end{matrix} = (-15)+(-16)+(105)-(6)-(42)-(100) = -74$$

EXAMPLE 8 Evaluate

$$\det A = \begin{vmatrix} 6 & 2 & 9 & 1 \\ 8 & 4 & 1 & 4 \\ 2 & 0 & 3 & 5 \\ 2 & 5 & 6 & 2 \end{vmatrix}$$

SOLUTION Notice that we cannot use the method shown in Example 7, since det A is a 4×4 determinant. However, we can use property (6) of determinants for the following row operations.

$$\begin{vmatrix} 6 & 2 & 9 & 1 \\ 8 & 4 & 1 & 4 \\ 2 & 0 & 3 & 5 \\ 2 & 5 & 6 & 2 \end{vmatrix} \xrightarrow[\textcircled{4}-2\textcircled{1}\to\textcircled{4}]{\textcircled{2}-4\textcircled{1}\to\textcircled{2},\ \textcircled{3}-5\textcircled{1}\to\textcircled{3}} \begin{vmatrix} 6 & 2 & 9 & 1 \\ -16 & -4 & -35 & 0 \\ -28 & -10 & -42 & 0 \\ -10 & 1 & -12 & 0 \end{vmatrix}$$

Therefore

$$\det A = -\begin{vmatrix} -16 & -4 & -35 \\ -28 & -10 & -42 \\ -10 & 1 & -12 \end{vmatrix}$$

At this point we can evaluate this 3×3 determinant, either by the procedure in Example 7 or by transforming all but one element in a row or column to zeros. Either way the result is

$$\det A = -1552$$

Exercises for Section 7.5

In Exercises 1–8, evaluate each of the determinants.

1. $\begin{vmatrix} 2 & 4 \\ 3 & 5 \end{vmatrix}$ **2.** $\begin{vmatrix} -1 & 7 \\ 2 & 1 \end{vmatrix}$ **3.** $\begin{vmatrix} 0 & 1 \\ 2 & 6 \end{vmatrix}$

4. $\begin{vmatrix} -3 & -4 \\ -2 & -7 \end{vmatrix}$ **5.** $\begin{vmatrix} 2 & -6 \\ 5 & -3 \end{vmatrix}$ **6.** $\begin{vmatrix} 4 & 4 \\ -3 & 2 \end{vmatrix}$

7. $\begin{vmatrix} 5 & 10 \\ 2 & 4 \end{vmatrix}$ **8.** $\begin{vmatrix} -3 & \frac{1}{2} \\ 6 & 4 \end{vmatrix}$

In Exercises 9–17, evaluate the given third-order determinant in two different ways. Compare the results.

9. $\begin{vmatrix} 2 & 0 & 0 \\ 3 & 2 & 4 \\ 1 & -3 & 5 \end{vmatrix}$ **10.** $\begin{vmatrix} -1 & 6 & 4 \\ -2 & 0 & 5 \\ 3 & 0 & -7 \end{vmatrix}$

11. $\begin{vmatrix} 2 & 3 & -4 \\ 2 & 0 & 3 \\ -1 & 6 & 5 \end{vmatrix}$

12. $\begin{vmatrix} 7 & 2 & 5 \\ 4 & -1 & 8 \\ 3 & 6 & 3 \end{vmatrix}$ **13.** $\begin{vmatrix} 3 & 1 & 9 \\ 6 & 1 & -2 \\ -4 & 1 & 5 \end{vmatrix}$

14. $\begin{vmatrix} \frac{1}{2} & 6 & 5 \\ 1 & 10 & 2 \\ -2 & 3 & -1 \end{vmatrix}$

15. $\begin{vmatrix} 6 & 7 & 6 \\ 2 & 5 & 2 \\ 3 & 1 & -3 \end{vmatrix}$

16. $\begin{vmatrix} 10 & -6 & 4 \\ 2 & -1 & 2 \\ 5 & 0 & 7 \end{vmatrix}$

17. $\begin{vmatrix} 9 & -6 & 0 \\ -1 & -3 & -1 \\ -5 & -4 & -7 \end{vmatrix}$

18. Solve the equation $\begin{vmatrix} 3 & 1 \\ -1 & x \end{vmatrix} = 0.$

19. Solve the equation $\begin{vmatrix} x & 1 \\ 1 & x \end{vmatrix} = 0.$

20. Solve the inequality $\begin{vmatrix} x & 2 \\ 3 & 5 \end{vmatrix} < \begin{vmatrix} 5 & 2x \\ 4 & 1 \end{vmatrix}.$

21. Compute the following determinant:

$$\begin{vmatrix} \begin{vmatrix} 3 & 1 \\ -1 & 2 \end{vmatrix} & \begin{vmatrix} 7 & 0 \\ 1 & 2 \end{vmatrix} \\ \begin{vmatrix} 5 & 0 \\ 0 & 1 \end{vmatrix} & \begin{vmatrix} 0 & 0 \\ 0 & 1 \end{vmatrix} \end{vmatrix}$$

22. Show that a third-order determinant with two identical rows is zero.

23. Solve the equation

$$\begin{vmatrix} x & 2 & 1 \\ -1 & 3 & 0 \\ 1 & x & 0 \end{vmatrix} = 0$$

24. For what values of k is the following determinant equal to zero?

$$\begin{vmatrix} k & 1 & 0 \\ 0 & 1 & 1 \\ 1 & 0 & 1 \end{vmatrix}$$

25. Evaluate

$$\begin{vmatrix} 1 & 4 & -2 & 0 \\ 0 & 1 & 1 & 2 \\ -3 & -2 & 1 & 0 \\ -1 & 1 & 2 & 3 \end{vmatrix}$$

26. Evaluate

$$\begin{vmatrix} 3 & -2 & 0 & 5 \\ -1 & 3 & 6 & 2 \\ 0 & 1 & -1 & 4 \\ 2 & -5 & 1 & 0 \end{vmatrix}$$

27. Prove the six properties of determinants for second-order determinants.

7.6 Solution by Determinants (Cramer's Rule)

Determinants have an immediate and important application to solving systems of n equations in n unknowns. For two equations in two unknowns, consider the following system of equations in x and y.

(7.2) $a_1x + b_1y = c_1$

(7.3) $a_2x + b_2y = c_2$

To solve this system algebraically for the x variable, multiply Equation (7.2) by b_2 and Equation (7.3) by $-b_1$, to get

$$a_1b_2x + b_1b_2y = c_1b_2$$
$$-a_2b_1x - b_1b_2y = -c_2b_1$$

Adding the two equations, we get

$$(a_1b_2 - a_2b_1)x = c_1b_2 - c_2b_1$$

Solving for x on the assumption that $a_1b_2-a_2b_1\neq 0$ gives

$$x=\frac{c_1b_2-c_2b_1}{a_1b_2-a_2b_1}$$

Likewise,

$$y=\frac{a_1c_2-a_2c_1}{a_1b_2-a_2b_1}$$

By the definition of a second-order determinant, the numerator and denominator of both expressions can be written in the form

(7.4) $$x=\frac{\begin{vmatrix} c_1 & b_1 \\ c_2 & b_2 \end{vmatrix}}{\begin{vmatrix} a_1 & b_1 \\ a_2 & b_2 \end{vmatrix}} \qquad y=\frac{\begin{vmatrix} a_1 & c_1 \\ a_2 & c_2 \end{vmatrix}}{\begin{vmatrix} a_1 & b_1 \\ a_2 & b_2 \end{vmatrix}}$$

Formula (7.4) is called **Cramer's rule** for the solution of two linear equations in two unknowns. Good practice dictates that in solving a system by this rule, one should express all equations in the **standard form** $ax+by=c$.

The following observations make it easy to remember Cramer's rule when the equations are written in standard form.

- The denominator in both cases is the determinant of the coefficient matrix of the system.
- The determinant in the numerator is found by replacing the column of coefficients of the variable being solved by the corresponding constant terms.

One of the advantages of Cramer's rule is that a system of equations can be solved with a minimum of algebraic manipulation. If the determinant of the coefficient matrix is nonzero, the system has a unique solution given by Formula (7.4).

EXAMPLE 1 Solve the following system of equations by Cramer's rule:

$$2x+3y=11$$
$$-x+7y=-2$$

SOLUTION

$$x=\frac{\begin{vmatrix} 11 & 3 \\ -2 & 7 \end{vmatrix}}{\begin{vmatrix} 2 & 3 \\ -1 & 7 \end{vmatrix}}=\frac{11(7)-(-2)(3)}{2(7)-(-1)(3)}=\frac{77+6}{14+3}=\frac{83}{17}$$

$$y=\frac{\begin{vmatrix} 2 & 11 \\ -1 & -2 \end{vmatrix}}{\begin{vmatrix} 2 & 3 \\ -1 & 7 \end{vmatrix}}=\frac{2(-2)-(-1)(11)}{17}=\frac{-4+11}{17}=\frac{7}{17}$$

Therefore, the solution is $x=\frac{83}{17}$, $y=\frac{7}{17}$.

EXAMPLE 2 Solve the following system of equations by Cramer's rule:

$$-3x+5y=2$$
$$7x-4y=0$$

SOLUTION

$$x=\frac{\begin{vmatrix}2 & 5\\ 0 & -4\end{vmatrix}}{\begin{vmatrix}-3 & 5\\ 7 & -4\end{vmatrix}}=\frac{2(-4)-(0)(5)}{-3(-4)-(5)(7)}=\frac{8}{23}$$

$$y=\frac{\begin{vmatrix}-3 & 2\\ 7 & 0\end{vmatrix}}{\begin{vmatrix}-3 & 5\\ 7 & -4\end{vmatrix}}=\frac{-3(0)-(2)(7)}{-23}=\frac{14}{23}$$

Therefore, the solution is $x=\frac{8}{23}$, $y=\frac{14}{23}$.

The concepts of consistent, inconsistent, and dependent linear systems can be interpreted in terms of the determinants used in Cramer's rule. If the determinant of the coefficient matrix is zero and the determinants in the numerators of Formula (7.4) are nonzero, the system is *inconsistent*; that is, it has no solution. If both the determinant of the coefficient matrix and the determinants in the numerator are zero, the equations have the same solution set, and the system is *dependent*. Otherwise, Cramer's rule gives a unique solution, and the system is *consistent*.

EXAMPLE 3 The system

$$3x+\ 5y=2$$
$$6x+10y=1$$

is inconsistent, since the determinant of the coefficient matrix is

$$\begin{vmatrix}3 & 5\\ 6 & 10\end{vmatrix}=30-30=0$$

whereas the determinants in the numerator expressions are easily shown to be nonzero. The system

$$-2x+2y=0$$
$$x-\ y=0$$

is dependent, since not only is the determinant of the coefficient matrix equal to zero but also the determinants in both numerators are zero. Indeed, the solution set in the second case is the infinite set of ordered pairs of the form (t, t) where t is any real number.

Cramer's rule can easily be extended to systems of n equations in n unknowns, but it is enough for you to know the result of this extension for $n=3$. The general system of three linear equations in three unknowns,

$$\begin{aligned}a_1x+b_1y+c_1z&=d_1\\ a_2x+b_2y+c_2z&=d_2\\ a_3x+b_3y+c_3z&=d_3\end{aligned}$$

has a solution given by Cramer's rule.

$$x=\frac{\begin{vmatrix}d_1 & b_1 & c_1\\ d_2 & b_2 & c_2\\ d_3 & b_3 & c_3\end{vmatrix}}{\begin{vmatrix}a_1 & b_1 & c_1\\ a_2 & b_2 & c_2\\ a_3 & b_3 & c_3\end{vmatrix}};\quad y=\frac{\begin{vmatrix}a_1 & d_1 & c_1\\ a_2 & d_2 & c_2\\ a_3 & d_3 & c_3\end{vmatrix}}{\begin{vmatrix}a_1 & b_1 & c_1\\ a_2 & b_2 & c_2\\ a_3 & b_3 & c_3\end{vmatrix}};\quad z=\frac{\begin{vmatrix}a_1 & b_1 & d_1\\ a_2 & b_2 & d_2\\ a_3 & b_3 & d_3\end{vmatrix}}{\begin{vmatrix}a_1 & b_1 & c_1\\ a_2 & b_2 & c_2\\ a_3 & b_3 & c_3\end{vmatrix}}. \tag{7.5}$$

Notice that Formula (7.5) has the same form as the rule developed for two equations in two unknowns. As in that rule, if the equations are written in standard form, $ax+by+cz=d$, the determinants in the denominator are formed from the coefficients of the variables. Similarly, the determinants in the numerator are formed by replacing the column of coefficients corresponding to the variable of interest by the constant terms on the right side of the equations.

EXAMPLE 4 Solve the following system using Cramer's rule:

$$\begin{aligned}2x-y+2z&=3\\ x+y-z&=-3\\ x+\quad z&=2\end{aligned}$$

SOLUTION The equations are given in standard form, so Cramer's rule can be used. The determinant in the numerator of y is expanded about the first column; all others are expanded about the second column to take advantage of the zero element.

$$x=\frac{\begin{vmatrix}3 & -1 & 2\\ -3 & 1 & -1\\ 2 & 0 & 1\end{vmatrix}}{\begin{vmatrix}2 & -1 & 2\\ 1 & 1 & -1\\ 1 & 0 & 1\end{vmatrix}}$$

$$=\frac{(-1)(-1)^3\begin{vmatrix}-3 & -1\\ 2 & 1\end{vmatrix}+(1)(-1)^4\begin{vmatrix}3 & 2\\ 2 & 1\end{vmatrix}+(0)(-1)^5\begin{vmatrix}3 & 2\\ -3 & -1\end{vmatrix}}{(-1)(-1)^3\begin{vmatrix}1 & -1\\ 1 & 1\end{vmatrix}+(1)(-1)^4\begin{vmatrix}2 & 2\\ 1 & 1\end{vmatrix}+(0)(-1)^5\begin{vmatrix}2 & 2\\ 1 & -1\end{vmatrix}}$$

$$=\frac{(-3-(-2))+(3-4)-(0)}{(1-(-1))+(2-2)-(0)}=\frac{-2}{2}=-1$$

$$y=\frac{\begin{vmatrix}2 & 3 & 2\\ 1 & -3 & -1\\ 1 & 2 & 1\end{vmatrix}}{\begin{vmatrix}2 & -1 & 2\\ 1 & 1 & -1\\ 1 & 0 & 1\end{vmatrix}}=\frac{2(-3+2)-(3-4)+(-3+6)}{2}=\frac{2}{2}=1$$

$$z=\frac{\begin{vmatrix}2 & -1 & 3\\ 1 & 1 & -3\\ 1 & 0 & 2\end{vmatrix}}{\begin{vmatrix}2 & -1 & 2\\ 1 & 1 & -1\\ 1 & 0 & 1\end{vmatrix}}=\frac{(2+3)+(4-3)}{2}=\frac{6}{2}=3$$

Thus, the solution is the ordered triple $(-1, 1, 3)$.

EXAMPLE 5 A company buys three kinds of fasteners. Brand A fasteners are \$0.30 each, Brand B fasteners \$0.15 each, and Brand C fasteners \$0.05 each. The total order of 150 fasteners amounts to \$20.50. How many of each kind of fastener were purchased if there are twice as many of Brand C as there are of Brand A?

SOLUTION Let N_A=number of Brand A fasteners; N_B=number of Brand B fasteners; and N_C=number of Brand C fasteners. Now form three linear equations in N_A, N_B, and N_C; that is,

$$N_A+N_B+N_C=150 \qquad \text{(total number of fasteners)}$$
$$30N_A+15N_B+5N_C=2050 \qquad \text{(total value of the fasteners in cents)}$$
$$N_C=2N_A \qquad \text{(relative number of Brands } A \text{ and } C\text{)}$$

Before using Cramer's rule, write the third equation in the standard form

$$-2N_A+N_C=0$$

Then

$$N_A=\frac{\begin{vmatrix}150 & 1 & 1\\ 2050 & 15 & 5\\ 0 & 0 & 1\end{vmatrix}}{\begin{vmatrix}1 & 1 & 1\\ 30 & 15 & 5\\ -2 & 0 & 1\end{vmatrix}}=\frac{0-0+(2250-2050)}{-2(5-15)-(0)+(15-30)}=\frac{200}{5}=40$$

$$N_B=\frac{\begin{vmatrix}1 & 150 & 1\\ 30 & 2050 & 5\\ -2 & 0 & 1\end{vmatrix}}{5}=\frac{-2(750-2050)-(0)+(2050-4500)}{5}=\frac{150}{5}=30$$

$$N_C=\frac{\begin{vmatrix}1 & 1 & 150\\ 30 & 15 & 2050\\ -2 & 0 & 0\end{vmatrix}}{5}=\frac{-2(2050-2250)-(0)+(0)}{5}=\frac{400}{5}=80$$

Consequently, there are 40 Brand A, 30 Brand B, and 80 Brand C fasteners.

Exercises for Section 7.6

Solve each of the systems of equations in Exercises 1–21 by determinants.

1. $x-4y=2$, $-2x+y=-4$
2. $2x+5y=-4$, $4x+3y=6$
3. $3x+4y=23$, $x-3y=-1$
4. $x+y=4$, $x-y=2$
5. $x-4y=8$, $2x-8y=16$
6. $3x+2y-6=0$, $x-3y-3=0$
7. $3x+4y=10$, $6x+8y=-2$
8. $3x+5y=2$, $6x+10y=4$
9. $3x-y-2=0$, $2x+y-6=0$
10. $r-s=3$, $r-2s=5$
11. $2x-y-2=0$, $x+y+5=0$
12. $2s=t+s$, $6s+2t=-5$
13. $x+y+z=8$, $x+y-z=12$, $x-y+z=2$
14. $x+2y-z=1$, $y+z=3$, $x-y=2$
15. $r+3s+2t=2$, $3r+2s-t=-1$, $r-3s-3t=0$
16. $3x+4y-7z=12$, $6x+3y+z=0$, $6x+8y-14z=3$
17. $2x-3y+4z=5$, $5x-7y+4z=4$, $-2x+3y-5z=2$
18. $x+y-z+3=0$, $2x-y+3z-12=0$, $x+2y-3z+10=0$
19. $x+2y-z=0$, $x-y+z=0$, $2x-3y-2z=0$
20. $3m-5n-5p=3$, $m+n-2p=7$, $2m-3n-2p=0$
21. $9u-4v+2w=8$, $3u+2v+w=1$, $12u-18w=17$

22. A company receives a shipment of 25 motors, both ac and dc. There are three more ac than dc motors. How many ac motors are in the shipment?

23. The width of a rectangle is 6 cm less than its length. If the perimeter of the rectangle is 40 cm, what are its dimensions?

24. A chemist mixes 12 lb of one chemical with 10 lb of another and finds the cost to be \$54. If a mixture of 8 lb of the first chemical and 15 lb of the second costs \$61, what is the price per pound of each chemical used in the mixtures?

25. A rocket traveling at a constant rate of 50 ft/sec passes a checkpoint. One second later a rocket traveling 60 ft/sec passes the same checkpoint. How long does it take for the second rocket to overtake the first?

26. The total number of teeth on two gears is 94. If twice the number of teeth on the small gear is five more than the larger, how many teeth are on each gear?

27. A box contains \$19 in nickels, dimes, and quarters. The total number of coins is 100 and there are three times as many nickels as dimes. How many dimes are in the box?

28. The power output of three amplifiers totals 45 W. Twice the power of the first amplifier equals the total power of the other two amplifiers, and the output of the second amplifier is 4 W more than the third. Find the power output of each amplifier.

29. Three antenna wires total 50 m in length. The longest wire is 5 m longer than the next longest and 10 m longer than the shortest. Find the length of each antenna wire.

30. An airplane made a 2000-mile trip with the wind in 3 hr; the return trip against the wind took 4 hr. If the speed of the wind was the same throughout the trip, what was the speed of the wind, and what was the average speed of the airplane in still air?

31. The following two linear equations arise in the solution of certain differential equations. Solve for u' and v', using Cramer's rule.

$$R_1(x)u' + Q_1(x)v' = F(x)$$
$$R_2(x)u' + Q_2(x)v' = 0$$

Functions are sometimes specified by their general form and certain points through which their graphs must pass. The following exercises are typical.

32. The graph of $y = ax^2 + bx$ passes through $(4, -2)$ and $(-2, 2)$. Find the values of a and b.

33. Determine the cubic $y = x^3 + ax^2 + bx + c$ whose graph passes through $(1, 0)$, $(3, 0)$, and $(0, 1)$.

34. Determine the cubic $y = x^3 + ax^2 + bx + c$ whose graph passes through $(-1, 0)$, $(0, 0)$, and $(1, 0)$.

35. Determine the specific form of the rational function $y = ax/(x + c)$ that passes through $(1, 2)$ and $(2, 1)$.

36. Determine the function with general form $y = a(x - b)/(x - c)$ that passes through $(2, 0)$, $(0, -1)$, and $(-1, 1)$.

7.7 Linear Inequalities in Two Unknowns

Having studied inequalities in one variable (Section 3.6), we now examine inequalities containing two variables. The operations established for inequalities in one variable also apply to inequalities in two variables, and therefore we need not duplicate them here. The solution set of an inequality in x and y is the set of all ordered pairs (x, y) that satisfy the inequality.

The concept of an inequality in two variables is closely related to the corresponding idea of an equality in two variables. This relation is best explained graphically. Consider the linear equation

$$2x + 3y = -5$$

The graph of this equation is the straight line shown in Figure 7.7. The line L separates the plane into three disjoint subsets, namely,

(1) $L = \{(x, y) | 2x + 3y = -5\}$; the set of points making up the line L

(2) $A = \{(x, y) | 2x + 3y > -5\}$; the set of points *above* the line L

(3) $B = \{(x, y) | 2x + 3y < -5\}$; the set of points *below* the line L

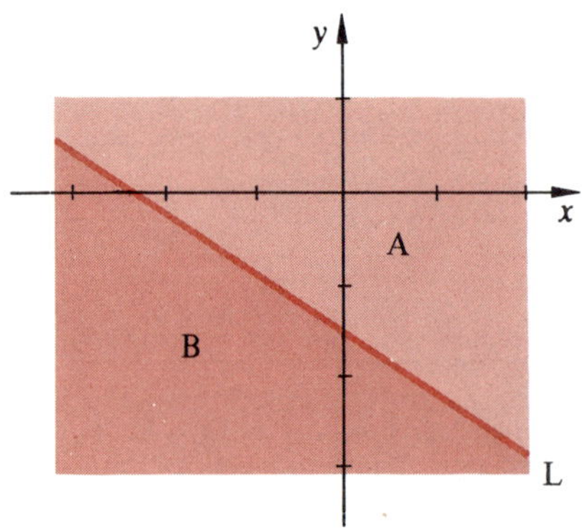

FIGURE 7.7

Thus, we see that the graph of a linear inequality in two variables corresponds to a *half-plane* composing the region on one side of a straight line. Usually then, to graph a linear inequality, first draw the line of the corresponding linear equation and decide which half-plane you need.

EXAMPLE 1 Graph the solution set for the inequality $2x+y<1$.

SOLUTION First draw the line $2x+y=1$. Figure 7.8 pictures it as a dashed line, to indicate its exclusion from the solution set. We must decide which side of this line represents the solution set. One way to do this is to select some point (x_1, y_1) not on the line and substitute it in the inequality. If (x_1, y_1) satisfies the inequality, this point lies in the desired region, and the half-plane containing (x_1, y_1) is the graph. In this problem let us use the check point (0, 0). Since $2(0)+0=0<1$ satisfies the inequality, we can conclude that the graph of $2x+y<1$ is the half-plane below the line. The shading in the figure indicates the set that is intended.

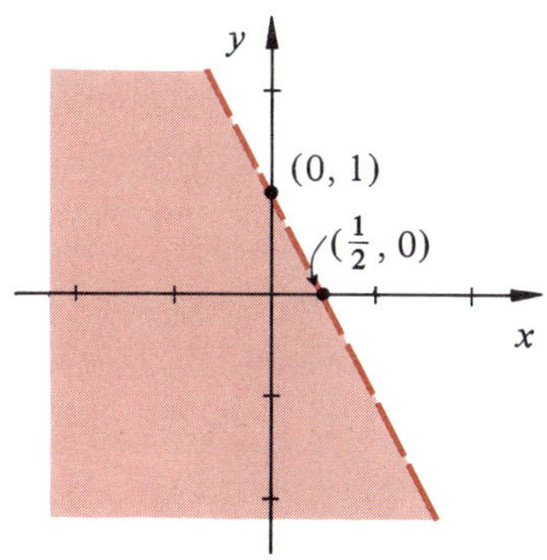

FIGURE 7.8

EXAMPLE 2 Find and graph the solution set for the inequality

$$2x+y-1\leqslant x+3y+4$$

SOLUTION By adding $-x-3y+1$ to both sides of the inequality, we get the equivalent inequality $x-2y\leqslant 5$. We then graph $x-2y=5$, shown as a solid line in Figure 7.9 to indicate its

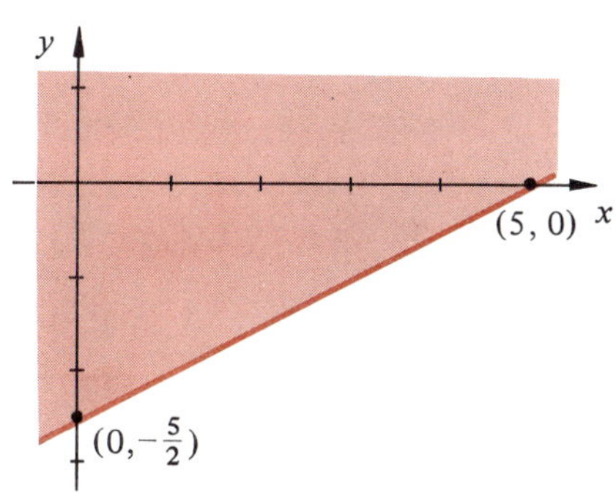

FIGURE 7.9

inclusion in the solution set. At the origin, we find that $x-2y=0$, which is <5. Therefore, all points on the same side of the line as the origin satisfy $x-2y\leqslant 5$. Figure 7.8 graphs the solution set.

EXAMPLE 3 A warehouse has 10,000 ft^2 of usable storage space in which to store refrigerators and air conditioners. Each refrigerator requires 6 ft^2 and each air conditioner requires 4 ft^2. Write an inequality describing the number of each that can be stored in the warehouse. Draw the graph.

SOLUTION The total area needed to store x refrigerators is $6x$ ft^2 and the total area for y air conditioners is $4y$ ft^2. Therefore, the sum $6x+4y$ must be less than or equal to 10,000 ft^2, that is,

$$6x+4y\leqslant 10{,}000$$

Figure 7.10 shows the graph of the line $6x+4y=10{,}000$. The desired half-plane lies below this line, as indicated by the shading. Some restrictions are implied for this problem due to its physical nature. Since the number of both refrigerators and air conditioners must be greater than zero, the solution is restricted to the triangular region indicated by the dark shading.

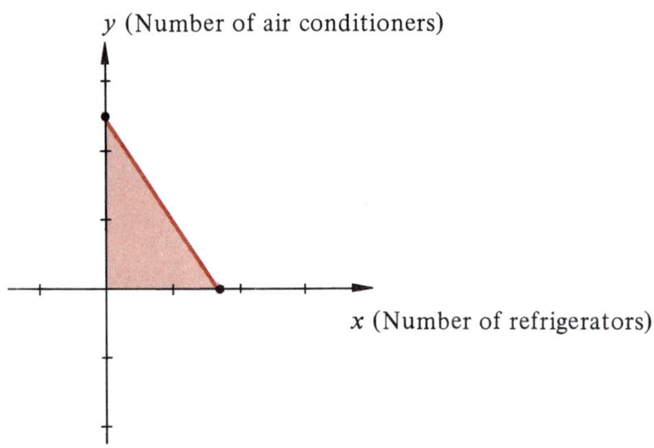

FIGURE 7.10

Exercises for Section 7.7

Graph the solution sets for the inequalities in Exercises 1–16.

1. $x>y$

2. $x\leqslant y$

3. $3x-y+7\leqslant 0$

4. $y<x+4$

5. $x+y>x-y$

6. $y-x\geqslant y+x$

7. $2x-y-8>x-2y+8$

8. $6x-2<y+1$

9. $3(x-1)+4x\geqslant 7x+y+2(x-y)$

10. $3x+y+2x-3y\leqslant 2$

11. $|y|<3$

12. $|x|<1$

13. $|x|<y$

14. $x\leqslant |y|$

15. $|x|+|y|<1$

16. $|x|+|y|>1$

17. The formula relating Fahrenheit degrees to degrees Celsius is $F=(\frac{9}{5})C+32$. For what values of C is $F<0$?

18. The current in a certain electrical network is given by $i = 25 - 0.5t$ A, where t is the elapsed time. For what time t is $i \geqslant 0$?

19. A retailer wishes to purchase two types of football shoes: a premium brand for \$35 a pair and a lesser quality pair for \$15. Make a sketch showing the possibilities if the capital outlay is limited to \$1000.

20. In designing lighting for a given outdoor area, you are given two different kinds of light bulbs, one 100 W and the other 200 W. The light will normally be on an average of 6 hr per night. Make a sketch showing the possible combinations of the two different kinds of light bulbs if you are limited to a cost of \$1.00 a night and the cost per kilowatt hour is \$0.05.

7.8 Systems of Linear Inequalities

Recall that the graph of a linear inequality in two variables is a half-plane that consists of the region on one side of a straight line. To solve a system of two or more inequalities, graph each inequality on the same plane. The solution set is then the region common to all the given inequalities. Another way of stating this is: The solution set of the system is the intersection of the solution sets of the given inequalities.

EXAMPLE 1 Solve the system

$$6x + 3y < 12$$
$$x + 2y \leqslant 2$$

SOLUTION The shaded area in Figure 7.11 represents the solution set.

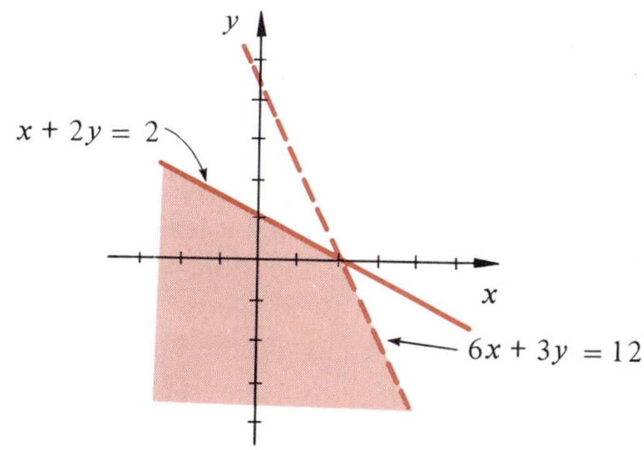

FIGURE 7.11

EXAMPLE 2 Solve the system

$$6x + y \leqslant 24$$
$$2x + 3y \leqslant 24$$
$$x \geqslant 0$$
$$y \geqslant 0$$

SOLUTION If each of the four indicated half-planes is drawn, their intersection is the shaded area shown in Figure 7.12.

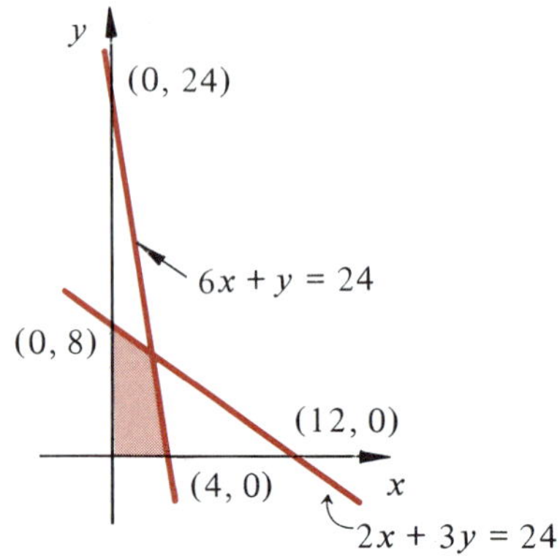

FIGURE 7.12

7.9 Linear Programming

In the 1940s mathematicians developed several techniques for evaluating decisions in fields such as business and economics under conditions of certain limitations, or constraints. One of these methods is called **linear programming**.

In this context, the term *programming* refers to the allocation of some usually limited resources to give the best results—say, the least cost, the most profit, or the least distance. (Thus, the word *programming* in this sense is *not* synonymous with the *programming* of computer terminology.) These allocation problems are linear in the sense that the equations describing the constraints are linear. Thus, linear programming is a method for optimizing some linear function, called the **objective function**, subject to constraints given in linear form.

For example, a typical objective function might be $P=3x-2y$ and some typical constraints are $x \geqslant 0$, $y \geqslant 0$, and $x+y \leqslant 2$. Any pair of numbers that satisfies the linear constraints is called a **feasible solution**. Thus, $x=1$, $y=1$ is a feasible pair for the example given. If a feasible solution gives the maximum value for P, then it is said to be **optimal**. For the objective function $P=3x-2y$, $P(1, 1)=1$. For the feasible point $(2, 0)$, $P(2, 0)=6$. How do we know when a feasible point will give the maximum value for P? Certain linear programming problems in two variables can be solved graphically. The technique is explained in the context of Example 1.

EXAMPLE 1 Determine the optimal solution for $P=3x-2y$ with constraints $x \geqslant 0, y \geqslant 0, x+y \leqslant 2$.

SOLUTION We begin by graphing the set of feasible solutions as shown in Figure 7.13(a). From this we want to obtain a point for which P is maximum. Observe that for any particular value P_0 of P, $3x-2y=P_0$ is a straight line with slope $\frac{3}{2}$ and y-intercept $-P_0/2$ (See Figure 7.13(b)). Thus, P will be a maximum when the y-intercept of the line $y=(3/2)x-P/2$ has the smallest value. (We need the smallest value since the intercept is the *negative* of $P/2$.) Since our optimal solution must be feasible, the line $y=(3/2)x-(1/2)P$ must intersect the feasible region. Hence, we need only select the line from the family of parallel lines with slope $\frac{3}{2}$ that has the smallest y-intercept *and* intersects the feasible region. This is shown in Figure 7.13(b) to be the line with x-intercept (2, 0). Thus the optimal feasible solution is (2, 0) and the maximum value is 6.

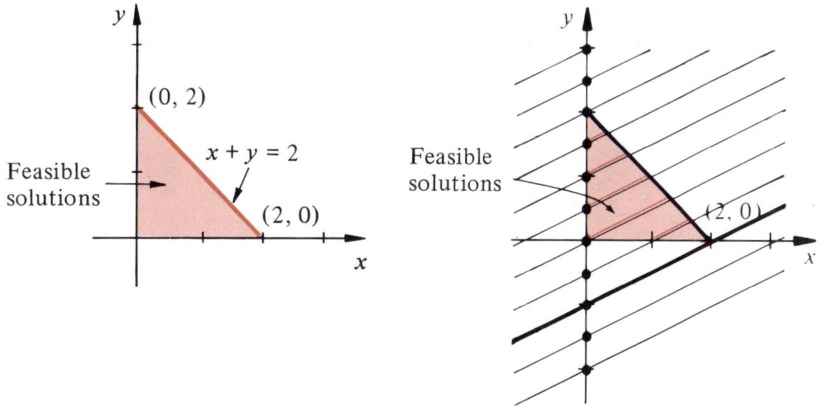

FIGURE 7.13

The solution to Example 1 was a corner point of the feasible region. This is no accident for there is a theorem proven in most elementary linear programming texts that states:

The optimal solution is found at a corner point of the graph of the set of feasible solutions.

In summary, the graphical method of solving a linear programming problem in two variables is:

(1) Find the solution set to the set of linear inequality constraints. This is the set of feasible solutions to the problem.

(2) Determine the coordinates of all corner points.

(3) Test all corner points to find the optimal solution.

EXAMPLE 2 Determine the lowest value for $y-5x$ subject to $x\geqslant 0$, $y\geqslant 0$, $2x+y\leqslant 4$, $x+2y\leqslant 4$.

SOLUTION Figure 7.14 shows the set of feasible solutions. The corner points are (0, 0), (0, 2), (2, 0), and $(\frac{4}{3}, \frac{4}{3})$. The value of $y-5x$ at each of these four points is 0, -10, 2, and $-\frac{16}{3}$. Hence, the optimum feasible solution is $x=2$, $y=0$ and the minimum value is -10.

Simple linear programming has numerous practical applications. Establishing the constraint inequalities is usually the hardest part.

EXAMPLE 3 A small company produces two styles of the same book, paperback and hardback. Producing the paperback yields a profit of \$400 per day while the hardback yields \$700 per day. The company can employ only one printing crew, which will work only

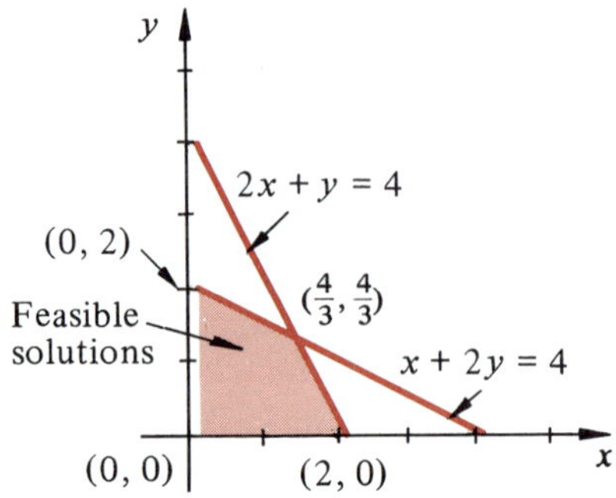

FIGURE 7.14

a five-day week. Paper limitations demand that the paperback be produced no more than three days per week, whereas hardback production is limited to four days or less per week. If you owned the company, how would you assign the printing crew?

SOLUTION First, establish the constraints in mathematical terms. Let x equal the number of days the paperback is printed and y the number of days the hardback is printed. Then we want to maximize the profit given by $P = 400x + 700y$. The constraints are

$x + y \leqslant 5$ (restriction to 5-day week)

$x \leqslant 3$ (restriction on paperback supplies)

$y \leqslant 4$ (restriction on hardback supplies)

$x \geqslant 0,\ y \geqslant 0$ (number of days must be positive or zero in either case)

Figure 7.15 displays the graph of the constraint inequalities. The corner points are (0, 4), (1, 4), (3, 2), (3, 0), and (0, 0). The value of the profit at each of these points is

at (0, 4), $P = 2800$

at (1, 4), $P = 3200$

at (3, 2), $P = 2600$

at (3, 0), $P = 1200$

at (0, 0), $P = 0$

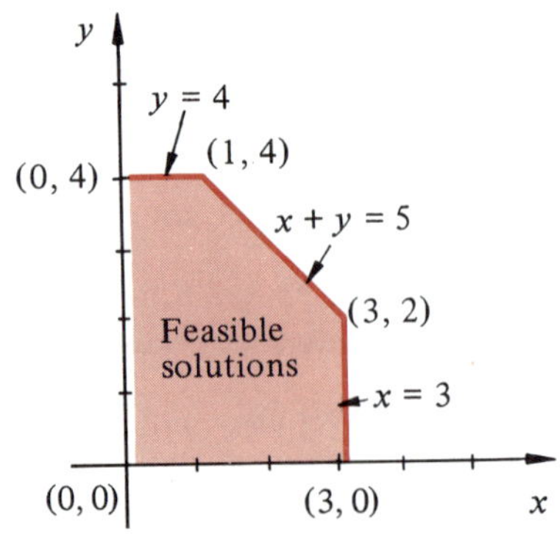

FIGURE 7.15

Hence the crew should devote one day to paperbacks and four to hardbacks to realize the greatest profit, \$3200 per week.

EXAMPLE 4 A small shop makes both drum brakes and disk brakes. Both types require a lathe and a grinder in production. Drum brakes require 2 hr on the lathe and 4 hr on the grinder, and disk brakes require 3 hr on the lathe and 2 hr on the grinder. The company makes a profit of \$14 on drum brakes and \$18 on disk brakes. Assuming that the company has one lathe and one grinder and that both machines are used 16 hr a day, how many of each should the company make to derive the largest profit?

SOLUTION List the required hours in a table.

	Hours, lathe	*Hours, grinder*
Drum brake	2	4
Disk brake	3	2

Let x equal the number of drum brakes produced and y the number of disk brakes. Then we want to maximize the profit, $P = 14x + 18y$. The constraints are

$2x + 3y \leqslant 16$ (restriction of lathe to 16-hr day)

$4x + 2y \leqslant 16$ (restriction of grinder to 16-hr day)

$x \geqslant 0,\ y \geqslant 0$ (the number of pieces made must be positive or zero)

Figure 7.16 pictures the graph of the constraint inequalities. We leave it for you to show that the corner point, (2, 4), gives the maximum profit. Hence, the company makes the greatest profit by producing 2 drum brakes and 4 disk brakes. What is the maximum profit?

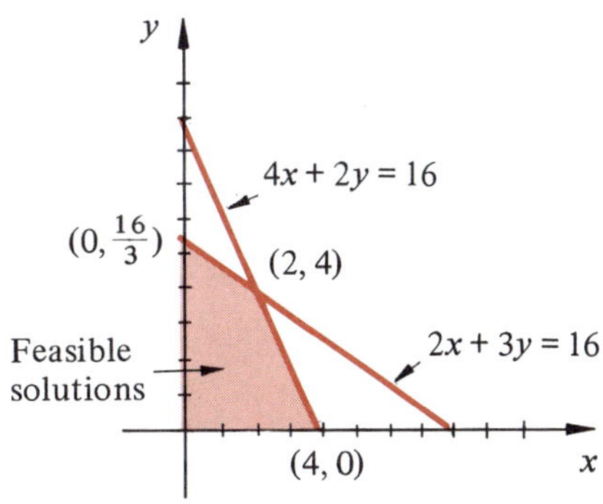

FIGURE 7.16

Exercises for Sections 7.8 and 7.9

In Exercises 1–15, use a graphical method to solve the given systems of inequalities.

1. $x \leqslant 2$
   $y > 6$

2. $x > -1$
   $x + y > 0$

3. $x \leqslant 0$
   $y \leqslant x$

4. $x - 4y \leqslant 2$
   $-2x + y \geqslant -4$

5. $2x + 5y > -4$
   $4x + 3y \leqslant 6$

6. $x + y < 4$
   $x - y > 2$

7. $x \geq 0$
$y \geq 0$
$2x+3y \leq 6$

8. $x \geq 0$
$y \geq 0$
$2x+y \geq 2$

9. $y \geq 0$
$x-3y \geq -3$
$-2x+y \geq -4$

10. $x>2$
$x<5$
$y>-1$

11. $x+y<4$
$2x+y<8$
$x-y>2$

12. $4x+3y \leq 6$
$2x+5y \geq -4$
$-2x+y \leq 2$

13. $x>0, y>0$
$x+y<2$
$2x+y>1$

14. $2x-y>1$
$x-y<1$
$x>0, y>0$

15. $x+2y>2$
$x-y<1$
$x>0, y>0$

In Exercises 16–30, solve the indicated linear programming problems.

16. Given $x \geq 0$, $y \geq 0$, $x \leq 10$, $x+y \leq 15$, maximize $P=3x+2y$.

17. Given $x \geq 0$, $y \geq 0$, $y \leq 2$, $x+2y \leq 8$, maximize $P=x+3y$.

18. Given $x \geq 0$, $y \geq 0$, $x+y \leq 2$, $x+3y \leq 4$, maximize $P=2x+5y$.

19. Given $x \geq 0$, $y \geq 0$, $x+y \leq 5$, $x+2y \leq 8$, maximize $P=5x+y$.

20. Given $x \geq 0$, $y \geq 0$, $3x+5y \leq 30$, $2x+y \leq 8$, maximize (a) $P=3x+2y$, and (b) $P=x+5y$.

21. Given $x \geq 0$, $y \geq 0$, $x+3y \leq 8$, $2x+y \leq 6$, maximize (a) $P=2x+3y$, and (b) $P=3x+y$.

22. Given $x \geq 0$, $y \geq 0$, $x+y \geq 5$, $x+2y \geq 8$, minimize $P=7x+2y$.

23. Given $x \geq 0$, $y \geq 0$, $x \leq 2$, $x+2y \geq 3$, $x+3y \geq 4$, minimize $P=5x+3y$.

24. Given $x \geq 0$, $y \geq 0$, $z \geq 0$, $x+2y+z \leq 10$, $3x+y+z \leq 9$, maximize $P=2x+y+5z$.

25. Given $x \geq 0$, $y \geq 0$, $z \geq 0$, $2x+y+z \leq 5$, $x+2y+3z \leq 4$, maximize $P=3x+y+3z$.

26. A businessman finds that his profit equation is $P=5x+2y$, where x is the number of desks and y is the number of book cabinets he sells. How many of each should he sell each day for the greatest profit if $x \geq 0$, $y \geq 0$, $x+\frac{1}{2}y \leq 8$, $\frac{1}{2}x+y \leq 7$ are the constraints on the profit equation?

27. A company makes \$10 on steel castings and \$8 on aluminum castings. If x is the number of steel castings and y the number of aluminum, the profit equation is $P=10x+8y$. Company engineers have found that they must satisfy the following constraints for each day's production: $x \geq 0$, $y \geq 0$, $2x+3y \leq 120$, $4x+y \leq 90$. How many of each kind of casting should the company make each day for the greatest profit?

28. A company produces deluxe and economy model radios. Each economy model yields a profit of \$4, whereas each deluxe model yields \$10 profit. Both models require approximately the same amount of material to build, and the supply allows for a total of 500 radios a day. The sales force reports that while the demand for the economy model does not exceed 350 a day, the deluxe radio can be sold as fast as it is made. The deluxe model requires twice as much time to build as the economy, and if only the economy model were built there would be enough time to produce 700 radios per day. How many economy models and deluxe models should the company construct to maximize their profit?

29. A woman wants to invest \$20,000, part at 7 percent and part at 8.5 percent. The amount she invests at 8.5 percent cannot be more than twice the amount she invests at 7 percent. How much should she deposit at each rate to maximize total income?

30. A company produces two kinds of vacuum cleaners with profits of \$5 and \$8, respectively, on the economy and the deluxe models. No more than 1000 of the economy model can be made in one day and the deluxe model takes three times as long to make. If only the economy model were made, time would allow for production of 1300 a day. What number of each should be built to realize the greatest profit? What is this maximum profit?

Test 1 for Chapter 7

Answer *True* or *False* to Exercises 1–10.

1. The solution set to a system of two linear equations in two unknowns is empty if the graphs of the two equations do not intersect.
2. Cramer's rule gives the solution to a system of linear equations in terms of determinants.
3. An equivalent system is obtained if both sides of any one equation are squared.
4. Another name for *matrix* is *determinant*.
5. Equivalent systems of equations have the same solution set.
6. The determinant of a matrix is always positive or zero.
7. A minor of a 2×2 matrix is any element of the matrix.
8. If every element in one row of a square matrix A is 0, then $\det A = 0$.
9. If every element in one row of a square matrix A is 1, then $\det A = 0$.
10. Solutions to a system of equations in two unknowns are displayed graphically as the points at which the graphs of the equations cross the x-axis.
11. Solve the system

$$\begin{aligned} x+y+z&=1\\ x+y-z&=2\\ x-y+z&=3 \end{aligned}$$

12. Solve the following system using matrix notation:

$$\begin{aligned} x+\ y+\ z&=\ 2\\ 3x-\ y-2z&=\ 4\\ 5x-2y+3z&=-6 \end{aligned}$$

13. Solve the following system using Cramer's rule:

$$\begin{aligned} x-5y&=7\\ 3x+2y&=4 \end{aligned}$$

14. Solve the system

$$\begin{aligned} x^2+y^2&=9\\ x^2-y^2&=1 \end{aligned}$$

15. Graphically solve the system

$$\begin{aligned} x+y&>1\\ 2x-y&\leq 5 \end{aligned}$$

16. Tickets for a special benefit dinner sold at \$4.50 for adults and \$1.75 for children. The 1573 tickets sold brought in \$5340.50. How many of each kind were sold?

Test 2 for Chapter 7

1. Solve the following system graphically:

$$\begin{aligned} 2x+3y&=2\\ 5x-4y&=-1 \end{aligned}$$

2. Solve the system in Exercise 1 by elimination.
3. Solve by elimination by substitution

$$\begin{aligned} 3x^2+y^2&=2\\ x^2+2y^2&=4 \end{aligned}$$

4. Evaluate the minor of a_{31} if

$$A=\begin{bmatrix} 1 & -2 & 1\\ 0 & -1 & 4\\ 3 & 5 & 2 \end{bmatrix}$$

5. Evaluate using expansion by minors

$$\begin{vmatrix} 3 & 2 & -1\\ 5 & 8 & 2\\ -4 & 1 & 0 \end{vmatrix}$$

6. Solve the system

$$\begin{aligned} 2r+s+2t&=5\\ 4r-s-3t&=1\\ 8r+s-\ t&=5 \end{aligned}$$

7. Use Cramer's rule to solve the system

$$\begin{aligned} 3x-4y&=\ 5\\ -2x+\ y&=-3 \end{aligned}$$

8. Use Cramer's rule to solve the system

$$x+z=5$$
$$y-x=0$$
$$z-y=-1$$

9. Graphically solve the system of inequalities

$$3x-4y\leqslant 5$$
$$x+2y>2$$

10. Maximize the function $P=3x+4y$, subject to the constraints $x\geqslant 0$, $y\geqslant 0$, $x+3y\leqslant 6$, $2x+y\leqslant 7$.

8

The Algebra of Polynomials

8.1 Polynomials

A polynomial function is a function that can be written in the form

$$P(x) = a_n x^n + a_{n-1} x^{n-1} + \cdots + a_1 x + a_0$$

where n is a nonnegative integer, called the **degree** of the polynomial, and the coefficients $a_n, a_{n-1}, \ldots, a_1, a_0$ are constants. The "leading" coefficient, a_n, is understood to be nonzero (otherwise the degree of the polynomial would not be n) while some or all of the other coefficients may be zero. Zero coefficients are ordinarily not written. For example,

$$P(x) = 4x^3 + 0x^2 + 7x + 0$$

is written as

$$P(x) = 4x^3 + 7x$$

In general, the coefficients of a polynomial can be real numbers or complex numbers. A polynomial having only real number coefficients is called a *real polynomial*; otherwise, it is called a *complex polynomial*. Examples of real polynomials are

$$3x^5 + 3x - 4 \quad \text{and} \quad x^{15} - 24x^9 + 11x$$

Examples of complex polynomials are

$$2x^5 + ix^2 - (4+i)x - 1 \quad \text{and} \quad (3-2i)x^2 + 2x - 32$$

For the simplicity, we restrict our discussion of the algebra of polynomials to real polynomials. In particular, we will explain some of the theory concerned with finding zeros of real polynomials (or, equivalently, the roots of real polynomial equations.)

The zeros of a real polynomial function can be either real numbers or complex numbers, so *the permissible values for real polynomial functions are the set of complex numbers.*

EXAMPLE 1 (a) The zeros of x^2+4 are $\pm 2i$.

(b) The zeros of $2x^3+x^2+5x-3$ are $\frac{1}{2}$ and $\frac{1}{2}(-1 \pm i\sqrt{11})$.

When we discuss polynomials, we use the degree of a polynomial for its name. For instance, $P(x)=4x^2-5x+4$ is called a second degree or quadratic polynomial. Other common names are listed in Table 8.1

TABLE 8.1

Degree	Name	Form
0	constant	$P(x)=a_0$
1	linear	$P(x)=a_1x+a_0$
2	quadratic	$P(x)=a_2x^2+a_1x+a_0$
3	cubic	$P(x)=a_3x^3+a_2x^2+a_1x+a_0$
4	quartic	$P(x)=a_4x^4+a_3x^3+a_2x^2+a_1x+a_0$

Polynomial functions may be combined with other polynomials by the operations of addition, subtraction, multiplication or division. If $p_1(x)$ and $p_2(x)$ are two polynomials of lesser degree than $P(x)$ and if $p_1(x)p_2(x)=P(x)$, then $P(x)$ is said to be **factorable** and $p_1(x)$ and $p_2(x)$ are called **factors** of $P(x)$. The factors are called real or complex according to whether the coefficients are real or complex numbers. Any polynomial factor of degree one is called a **linear** factor.

EXAMPLE 2 (a) $x^3-(2+i)x+3i$ is a cubic polynomial with complex coefficients.

(b) (x^2+2x+3) and (x^3+x-2) are factors of the fifth degree polynomial $x^5+2x^4+4x^3-x-6$.

(c) $(x-1)$ is a real linear factor of x^3-1.

(d) $(x-2i)$ is a complex linear factor of x^2+4.

The sum, difference, or product of two polynomials is a polynomial. The quotient of two polynomials is not necessarily a polynomial, however. For instance, if the polynomial

$$f(x)=2x^3+x^2-5x-17$$

is divided by $g(x)=x^2-2x+3$, the result is

$$\begin{array}{r|l}
 & 2x \ +5 \\ \hline
x^2-2x+3 & 2x^3+x^2-5x-17 \\
 & 2x^3-4x^2+6x \\ \cline{2-2}
 & \quad 5x^2-11x-17 \\
 & \quad 5x^2-10x+15 \\ \cline{2-2}
 & \qquad -x-32 \quad \text{Remainder}
\end{array}$$

The quantity $-x-32$ is called the **remainder** in the division process and $2x+5$ is the **quotient**. The division can be written in the form

$$\frac{2x^3+x^2-5x-17}{x^2-2x+3}=2x+5-\frac{x+32}{x^2-2x+3}$$

which is not a polynomial.

The theory of polynomials relative to division is lengthy and tedious. If we restrict our attention, however, to dividing a polynomial by a linear polynomial, the theory is relatively simple.

8.2 The Remainder and the Factor Theorems

Polynomials can be divided by linear factors of the form $(x-r)$, where r is a real or a complex number. Consider the division of

$$P(x)=2x^3-3x^2-4x-17$$

by the linear term $x-3$.

$$\begin{array}{r|l}
 & 2x^2+3x \ +5 \\ \hline
x-3 & 2x^3-3x^2-4x-17 \\
 & 2x^3-6x^2 \\ \cline{2-2}
 & \quad 3x^2-4x \\
 & \quad 3x^2-9x \\ \cline{2-2}
 & \qquad 5x-17 \\
 & \qquad 5x-15 \\ \cline{2-2}
 & \qquad\quad -2 \quad \text{Remainder}
\end{array}$$

Thus,

$$\frac{2x^3-3x^2-4x-17}{x-3}=2x^2+3x+5+\frac{-2}{x-3}$$

An interesting and useful observation that we can make about the remainder obtained in this division problem is that if we evaluate the function $P(x)$ at $x=3$, we get

$$\begin{aligned}
P(3)&=2(3)^3-3(3)^2-4(3)-17 \\
&=54-27-12-17 \\
&=-2
\end{aligned}$$

This is the same number as the remainder obtained by dividing $P(x)$ by $x-3$. This observation motivates the following theorem, known as the **remainder theorem**.

THEOREM 1

> **The Remainder Theorem** If $P(x)$ is divided by the linear factor $(x-r)$ until a constant remainder R is obtained, then $P(r)=R$; that is, the remainder is always a constant equal to the value of the function at $x=r$.

PROOF

By assumption

$$\frac{P(x)}{x-r}=q(x)+\frac{R}{x-r}$$

Therefore,

(8.1) $$P(x)=(x-r)q(x)+R$$

Letting $x=r$ gives

$$P(r)=(r-r)q(r)+R=R$$

EXAMPLE 1 What is the remainder when x^5+x^4+6 is divided by $x+2$?

SOLUTION In this case $P(x)=x^5+x^4+6$ and $r=-2$. Thus,

$$R=P(-2)=(-2)^5+(-2)^4+6=-32+16+6=-10$$

EXAMPLE 2 Find the remainder when $P(x)=x^3-1$ is divided by $x-1$.

SOLUTION Here $r=1$, so $R=P(1)=1^3-1=0$.

The remainder theorem leads directly to another important theorem known as the **factor theorem**. If, in particular, r is a zero of $P(x)$, then $P(r)=R=0$. Thus, from Equation (8.1),

$$P(x)=(x-r)q(x)$$

which means that $(x-r)$ is a factor of $P(x)$.

THEOREM 2

> **The Factor Theorem** The polynomial $P(x)$ has a zero $x=r$ if and only if $(x-r)$ is a factor of the polynomial.

More generally, $x=r$ is a zero of *multiplicity* k if and only if $(x-r)^k$ is a factor of the polynomial.

EXAMPLE 3 Show that $(x-2)$ is a factor of $P(x)=x^3+2x^2-5x-6$.

SOLUTION Here $r=2$ and $P(2)=2^3+2(2)^2-5(2)-6=0$. Hence, by the factor theorem, $(x-2)$ is a factor.

EXAMPLE 4 Use the factor theorem to determine whether or not $(2x+1)$ is a factor of $P(x)=4x^3-5x+2$.

SOLUTION First, note that $(2x-1)$ is not of the form $x-r$, and therefore we may not use the value of $P(1)$ to determine whether $(2x-1)$ is a factor. However, $(2x-1)=2(x-\frac{1}{2})$, which means that if $x-\frac{1}{2}$ is a factor, so is $(2x-1)$. Thus, we evaluate $P(\frac{1}{2})$, to find that

$$P\left(\frac{1}{2}\right)=4\left(\frac{1}{2}\right)^3-5\left(\frac{1}{2}\right)+2=0$$

which shows that $(2x-1)$ is a factor of $4x^3-5x+2$.

If the zeros (and their multiplicities) are known, the polynomial itself is completely determined except for a multiplicative constant.

EXAMPLE 5 Find the fourth-degree polynomial with zeros at -2, -2, 1, and 3, and y-intercept 2. (Repeating -2 is the usual way of indicating that -2 is a zero of multiplicity 2.)

SOLUTION Since each of the zeros gives a linear factor, we write

$$P(x)=a(x+2)^2(x-1)(x-3)$$

To evaluate the constant, note that the graph must pass through (0, 2). Substituting these values for x and $P(x)$ gives

$$12a=2 \quad \text{or} \quad a=\frac{1}{6}$$

Hence, $P(x)=(\frac{1}{6})(x+2)^2(x-1)(x-3)$ is the desired polynomial.

As the degree of a polynomial function $P(x)$ increases, the problem of evaluating $P(r)$ becomes tedious, since we must find and combine high powers of r. In most cases, synthetic division (explained in Section 1.6, can be used in conjunction with the remainder theorem to help in finding $P(r)$. The mechanics of synthetic division are repeated here for your convenience.

Synthetic Division

To divide a polynomial $P(x)$ by $x-r$, do the following:

(1) Arrange the coefficients of $P(x)$ in order of descending powers of x, remembering that missing terms have the coefficient zero.

(2) Replace $x-r$ by r.

(3) Bring down the coefficient of the highest power of x, multiply it by r, and add the result to the coefficient of the next-highest power of x. Multiply this sum by r and add to the next coefficient. Continue this process until there is a product added to the constant term.

(4) The last number in the bottom row is the remainder. The numbers to the left of the remainder are the respective coefficients of the quotient. The degree of the quotient is one less than the degree of $P(x)$.

EXAMPLE 6 Divide $5x^4-18x^2-6x+3$ by $x+2$, using synthetic division.

SOLUTION The divisor is $x+2$, so $r=-2$. Thus,

$$\begin{array}{r|rrrrr} -2 & 5 & 0 & -18 & -6 & 3 \\ & & -10 & 20 & -4 & 20 \\ \hline & 5 & -10 & 2 & -10 & 23 \end{array}$$

The quotient is $5x^3-10x^2+2x-10$ with remainder 23.

EXAMPLE 7 Given $P(x)=8x^5-x^3+2$, find $P(\frac{1}{2})$, using synthetic division and the remainder theorem.

SOLUTION

$$\begin{array}{r|rrrrrr} \frac{1}{2} & 8 & 0 & -1 & 0 & 0 & 2 \\ & & 4 & 2 & \frac{1}{2} & \frac{1}{4} & \frac{1}{8} \\ \hline & 8 & 4 & 1 & \frac{1}{2} & \frac{1}{4} & \frac{17}{8} \end{array}$$

The remainder is $\frac{17}{8}$. Therefore, by the remainder theorem, $P(\frac{1}{2})=\frac{17}{8}$.

EXAMPLE 8 Show that $x+2i$ is a factor of $x^3-5x^2+4x-20$.

SOLUTION We use synthetic division, with $r=-2i$.

$$\begin{array}{r|rrrr} -2i & 1 & -5 & 4 & -20 \\ & & -2i & -4+10i & 20 \\ \hline & 1 & -5-2i & 10i & 0 \end{array}$$

We conclude that since the remainder is zero, $x+2i$ is a factor.

Exercises for Sections 8.1–8.2

Perform the divisions in Exercises 1–12 using synthetic division. Check the remainder, using the remainder theorem.

1. $(x^2-5x+4)\div(x-4)$
2. $(x^2+7x+12)\div(x-3)$
3. $(x^3-2x^2+3x-1)\div(x-2)$
4. $(x^3+2x^2-3x+4)\div(x+1)$
5. $(x^3-3x+9)\div(x+\frac{1}{2})$
6. $(x^3-1)\div(x+1)$
7. $(x^4-x^3+2x^2-72)\div(x-3)$
8. $(2x^3-4x^2+x-1)\div(x+2)$

9. $(8+7x-3x^5)\div(x-2)$

10. $(4x^3+x^2-16x-4)\div(x+1)$

11. $(3x^3-20x^2+23x+10)\div(x+\frac{1}{3})$

12. $(3x^3+2x^2-4x+1)\div(x-\frac{1}{3})$

Use the remainder theorem and synthetic division to find the functional values in Exercises 13–18.

13. $P(3)$ for $P(x)=x^3+2x-5$

14. $P(-4)$ for $P(x)=x^4+3x^3-x^2-2x+6$

15. $P(\frac{1}{2})$ for $P(x)=3x^3+2x^2-4x+1$

16. $P(-\frac{1}{2})$ for $P(x)=3x^3+2x^2-4x+1$

17. $P(0.1)$ for $P(x)=-6x^4+5x^2-2$

18. $P(0.2)$ for $P(x)=3+x-4x^2-7x^3$

19. $P(3i)$ for $P(x)=x^2-2x-1$

20. $P(-i)$ for $P(x)=x^3+1$

21. $P(-2+i)$ for $P(x)=x^3+5x^2+9x+5$

22. $P(-1+i)$ for $P(x)=x^2+2x+2$

Use synthetic division and the factor theorem to determine whether or not the second expression is a factor of the first in Exercises 23–36.

23. $x^3-1, x+1$

24. $x^4-x^3+2x^2-72, x-3$

25. $4x^3+x^2-16x-4, x-2$

26. $2x^4-7x^3-x^2+8, x-5$

27. $3x^5+7x-8, x-2$

28. $x^3+27, x+3$

29. $x^4+2x^3-15x^2-32x-16, x-4$

30. $x^4-8x^3+6x^2+40x-56, x-2$

31. $3x^3+2x^2-4x+1, 3x-1$

32. $3x^3-20x^2+23x+10, 3x+1$

33. $x^4-16, x+2i$

34. $x^6+1, x-i$

35. $x^3-2x^2+4x-8, x-2i$

36. $x^3+3x-5, x+3i$

Write expressions for the polynomials in Exercises 37–42 with the indicated properties.

37. Third-degree polynomial with zeros at 1, -1, and 2. Has y-intercept of -2.

38. Fifth-degree polynomial with zeros of -3, 0, 0, 2, and 1. Passes through $(3, -1)$.

39. Fourth-degree polynomial with zeros of 1, 1, 2, and 2. Has y-intercept of 3.

40. Sixth-degree polynomial with zeros of -2, -1, 0, 1, i, and $-i$. Passes through $(2, 1)$.

41. Third-degree polynomial with zeros of $(2-i)$, $(2+i)$, and $\sqrt{2}$. Has y-intercept of 2.

42. Fourth-degree polynomial with zeros of $(1\pm\sqrt{2})$, $\sqrt{3}$, and π. Has y-intercept of $1/\pi$.

In Exercises 43–46, determine k such that:

43. $x-1$ is a factor of $x^3-2x^2+3kx-3$.

44. $x+i$ is a factor of x^4-3x^2+kx-2.

45. $x-\sqrt{2}$ is a factor of $x^3-2kx+5$.

46. $x+\pi$ is a factor of kx^2-2x+5.

In Exercises 47–50, assume that n is a positive integer and use the factor theorem to show that:

47. $(x-1)$ is a factor of x^n-1.

48. $(x-1)$ is never a factor of x^n+1.

49. $(x+1)$ is a factor of x^n-1 if n is even.

50. $(x+1)$ is a factor of x^n+1 if n is odd.

8.3 The Zeros of a Polynomial Function

The factor theorem is useful only if we know something about the zeros of the polynomial. The **fundamental theorem of algebra** tells us that even the most complicated looking polynomial has a zero, although not necessarily a real zero, and hence can be factored.

THEOREM 3

The Fundamental Theorem of Algebra Every polynomial function has at least one zero.

While Theorem 3 does not give us any clue about how to find the zero, it assures us that the polynomial

$$P(x) = 49x^9 - \sqrt{2}x^6 + \pi^2 x^4 + (2+\pi)x^3 - 6.235x^2 + 10^5$$

has at least one zero.

Generally, if $P(x)$ is a polynomial of degree n, where n is greater than zero, there must be at least one zero, say $x = r_1$. By the factor theorem, we write

$$P(x) = P_1(x)(x - r_1)$$

where $P_1(x)$ is the polynomial found by dividing $P(x)$ by $x - r_1$. The degree of $P_1(x)$ is one less than the degree of $P(x)$. If the degree of $P_1(x)$ is greater than zero, then this polynomial must have at least one zero, say $r = r_2$. Therefore, $(x - r_2)$ is a factor of $P_1(x)$, and we can write

$$P(x) = (x - r_1)(x - r_2)P_2(x)$$

Similarly, if the degree of $P_2(x)$ is greater than zero, the polynomial must have at least one zero. Continuing this reasoning for n steps, we shall get a zero r_n and hence a factor $(x - r_n)$. If $P_{n-1}(x)$ is divided by $(x - r_n)$, a constant c is obtained. Thus,

$$P(x) = c(x - r_1)(x - r_2)(x - r_3)\cdots(x - r_n)$$

In words, we have the following theorem:

THEOREM 4

If $P(x)$ is a polynomial of degree n, then there are such n numbers $r_1, r_2, \ldots, r_n$, not necessarily distinct, that

$$P(x) = a_n(x - r_1)(x - r_2)\cdots(x - r_n)$$

where a_n is a constant.

There is an immediate corollary:

THEOREM 5

A polynomial of degree n has exactly n zeros.

Clearly, Theorem 5 must be properly interpreted. The theorem does not say that there are *n distinct* zeros; it says that the total number of zeros, counting multiple zeros, equals the degree of the polynomial. Since the zeros of a polynomial $P(x)$ correspond to the roots of the equation $P(x) = 0$, the theorems on zeros are directly applicable to roots of polynomial equations.

EXAMPLE 1 Given that $x = -4.7$ is a root of $P(x) = x^3 + 0.4x^2 - 16.49x + 17.484 = 0$, determine the other roots.

SOLUTION Since $x = -4.7$ is a root, $(x+4.7)$ is a factor of $P(x)$. Dividing $P(x)$ by $(x+4.7)$, we get $P_1(x) = x^2 - 4.3x + 3.72$. The roots of the quadratic equation $P_1(x) = 0$, are obtained using the quadratic formula.

$$x = \frac{4.3 \pm \sqrt{(4.3)^2 - 4(3.72)}}{2} = \frac{4.3 \pm 1.9}{2}$$

$$x = 3.1,\ 1.2$$

Thus, the three roots of the given cubic equation are -4.7, 3.1, and 1.2.

Complex zeros of real polynomials must occur in complex conjugate pairs of the form $a \pm bi$, that is, a real polynomial cannot have an odd number of complex zeros. Complex zeros arise from linear factors of the from $x-(a+bi)$ and $x-(a-bi)$ or, equivalently, from the real quadratic factor $x^2 - 2ax + (a^2 + b^2)$. To verify this statement, we note that

$$\begin{aligned}[(x-(a-bi)][(x-(a+bi)] &= x^2 - (a+bi)x - (a-bi)x + (a-bi)(a+bi) \\ &= x^2 - 2ax + (a^2 + b^2)\end{aligned}$$

Real quadratic polynomials that cannot be factored into the product of real linear factors are said to be **irreducible**.

THEOREM 6

> A real polynomial can be written as a product of a constant and real linear and irreducible quadratic factors.

The real linear factors of a polynomial yield the real zeros and the real irreducible factors yield the complex zeros. Example 2 shows how a polynomial can be obtained from its zeros if we know at least one additional point on its graph.

EXAMPLE 2 Write the real polynomial with zeros -2, 2, 3, $-i$, $-i$, i, i, if the graph of the polynomial passes through (1, 48).

SOLUTION The desired polynomial is of degree 7 since there are seven zeros. The complex zeros $-i$ and i lead to the quadratic factor $(x-i)(x+i) = x^2 + 1$. Because the complex zeros are repeated the polynomial will have the factor $(x^2+1)^2$ in addition to the linear factors $x-2$, $x+2$, and $x-3$. The polynomial consists of the product of these factors and a constant a_n, that is,

$$P(x) = a_n(x^2+1)^2(x-2)(x+2)(x-3)$$

Using the fact that $P(1) = 48$, we have

$$48 = a_n(2)^2(-1)(3)(-2) = a_n(24)$$

Therefore, $a_n = 2$, and the polynomial is

$$P(x) = 2(x^2+1)^2(x-2)(x+2)(x-3)$$

8.4 The Rational Zeros of a Polynomial Function

An nth-degree polynomial, then, has n zeros. How are these zeros to be found? For degree 2, the quadratic formula can be used. Similar procedures using radicals exist for finding the zeros of third- and fourth-degree polynomials, but the formulas are seldom used. For polynomials of degree greater than 4, no general method exists, although there are several good numerical approximation methods. Numerical methods usually need a computer to be of any real practical value. Besides these advanced numerical methods, we frequently resort to trial and error. The trial and error, however, should not be pure guesswork. For example, do not expect a positive zero for a polynomial of the type $y=x^3+2x^2+x+2$. Why not?

Whereas it is generally hard to find the zeros of a polynomial, the following theorem shows how to find *rational* zeros if they exist. Note the hypothesis carefully.

THEOREM 7

Theorem on Rational Zeros If the coefficients of $f(x)=a_nx^n+a_{n-1}x^{n-1}+\cdots+a_1x+a_0$ are integers, then any rational zero must be of the form of a factor of a_0 divided by a factor of a_n.

This theorem can also be applied to polynomials with rational coefficients. Merely multiply the entire polynomial by the greatest common multiple of the denominators of the coefficients to get an equivalent polynomial with integer coefficients. Thus, if

$$P(x)=\tfrac{3}{4}x^2+\tfrac{2}{3}x-\tfrac{1}{6}$$

we can consider instead the polynomial

$$P_1(x)=12P(x)=9x^2+8x-2$$

This has integer coefficients and the same zeros as $P(x)$.

The theorem on rational zeros limits the number of trials to be made in determining the possible rational solutions to an equation. You would not even bother, for example, to try $x=2$ as a possible solution of the equation $x^4-7x^3+3x^2+2x+15=0$, since 2 is not a factor of 15.

EXAMPLE 1 The possible rational roots of $2x^3-5x+6=0$ are

$$\pm\tfrac{1}{2},\ \pm\tfrac{3}{2},\ \pm 1,\ \pm 2,\ \pm 3,\ \pm 6$$

In any testing for rational roots, if a root is found, the factor corresponding to the root should be divided out of the equation. The quotient, set equal to zero, is called the **depressed equation.**

EXAMPLE 2 Find the roots of the equation

$$P(x)=3x^4+11x^3+3x^2-20x-12=0$$

SOLUTION Any rational root must be a factor of 12 divided by a factor of 3, that is, $\pm n/d$, where n is 12, 6, 4, 3, 2, or 1 and d is 3 or 1. By synthetic division, we find that $x=-2$ is a root. Hence,

$$P(x)=(x+2)(3x^3+5x^2-7x-6)$$

Any rational root of the cubic factor is a factor of 6 divided by a factor of 3, that is, ± 6, ± 3, ± 2, ± 1, $\pm\frac{2}{3}$, or $\pm\frac{1}{3}$. We find that $x=-\frac{2}{3}$ is a root. Therefore,

$$P(x)=3(x+2)\left(x+\frac{2}{3}\right)(x^2+x-3)=0$$

Since this expression has a quadratic factor, we immediately use the quadratic formula, to get

$$x=\frac{-1\pm\sqrt{1+12}}{2}$$

Therefore, the desired roots are -2, $-\frac{2}{3}$, $(-1+\sqrt{13})/2$, and $(-1-\sqrt{13})/2$.

Exercises for Sections 8.3 and 8.4

Find all the roots of the equations in Exercises 1–17.

1. $x^3+2x^2-x-2=0$ **2.** $x^3-4x^2+x+6=0$

3. $y^3+2y^2+y+2=0$ **4.** $m^3-3m^2-4m+12=0$

5. $2x^3+x^2-18x-20=0$ **6.** $x^3-9x^2+23x-15=0$

7. $z^4-2z^3-7z^2+8z+12=0$

8. $2r^4-9r^3+15r^2-11r+3=0$

9. $3p^3-10p^2+12p-3=0$

10. $y^4+2y^3-4y^2-5y+6=0$

11. $x^4-6x^2-8x-3=0$ **12.** $x^3+8=0$

13. $9w^3-w+2=0$

14. $16x^4-150x^3+381x^2-262x-39=0$

15. $2x^3-3x^2+5x=2$ **16.** $12x^3+4x^2-3x=1$

17. $x^5-x^3-8x^2+8=0$

In Exercises 18–23, factor the given polynomials into real and complex linear factors.

18. x^3-x^2+2 **19.** x^3+6x^2+2x-3

20. $2x^3+5x^2+x-2$ **21.** $4x^3+8x^2+9x+3$

22. $4x^4+2x^3-24x^2+3x-45$

23. $4x^4-4x^3-57x^2+81x$

Solve the following systems of equations. Find both real and complex solutions.

24. $y=x^3$, $y=4x-3$ **25.** $y=x^4$, $y=2x-x^3$

26. $x^2y=9$, $y=(x-1)^2$ **27.** $x^2-xy=y+1$, $xy+x=4y^2$

28. A psychologist finds that the response to a certain stimulus varies with age group according to $R=x^3-4x^2-19x-10$, where R is response in milliseconds and x is age group in years. For what age group is the response equal to 4 msec? (Ignore negative solutions.)

29. In analyzing trends in a public opinion poll, a researcher found that the percentage P of people favoring a particular issue varied with age according to $P=x^3-21x^2-45x-8$, where x is age in years. In what age group did 15 percent of the people indicate a favorable opinon? (Ignore negative solutions.)

30. A box without a top is to be constructed from a 10 in. by 16 in. metal sheet. The box is formed by cutting equal squares from each corner of the sheet and then bending up the sides. What size square should be cut to give a box with a volume of 144 in.^{3}

8.5 The Real Zeros of a Polynomial by Graphing

Although you have learned how to locate the rational zeros of a polynomial, the technique was restricted to polynomials that have integral (or rational) coefficients. Thus, the zeros of polynomials such as $P(x)=\sqrt{2}x^3-x^2+x-\pi$ or even the non-rational zeros of a polynomial with integer coefficients cannot be found from the theorem on rational zeros.

In theory, if not in practice, a simple method of approximating the real zeros of a polynomial is a graphical one, since the real zeros of the polynomial correspond to the x-intercepts of the graph. The accuracy of the estimates depends on the scale chosen and the care taken in drawing the graph.

EXAMPLE 1 Find the real roots of the equation $x^3-4x+1=0$.

SOLUTION To determine any real root for this equation, we plot the graph of $P(x)=x^3-4x+1$ (Figure 8.1). The table of values is also given. From the graph we estimate the roots to be $x_1=-2.1$, $x_2=0.2$, $x_3=1.8$.

x	$P(x)$
3	16
2	1
1	−2
0	1
−1	4
−2	1
−3	−14

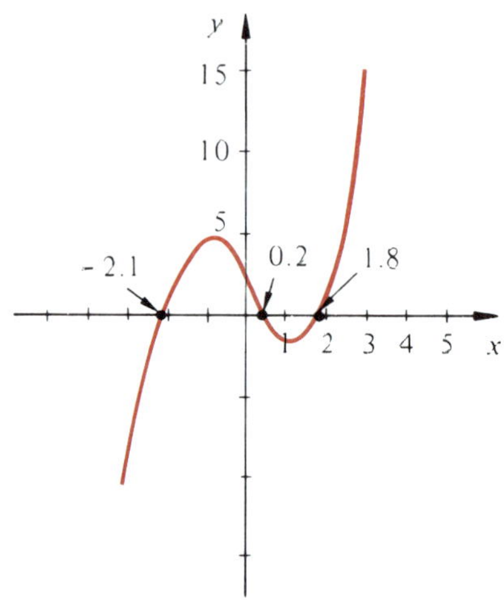

FIGURE 8.1

EXAMPLE 2 Find the real zeros of $P(x)=3x^4-4x^3+1$.

SOLUTION The graph is shown in Figure 8.2. The table of values is included. The graph and the table verify that $x=1$ is a zero. To show that it is a double zero (and that there is no other real zero in the vicinity of $x=1$), divide $3x^4-4x^3+1$ by $x-1$, to get $3x^3-x^2-x-1$. This cubic also has the zero $x=1$ and when it is divided by $x-1$, the result is $3x^2+2x+1$. The quadratic is irreducible, and hence the only real zero of the given polynomial is at $x=1$.

x	$P(x)$
2	17
$\frac{3}{2}$	$\frac{43}{16}$
1	0
$\frac{1}{2}$	$\frac{11}{16}$
0	1
$-\frac{1}{2}$	$\frac{27}{16}$
-1	8

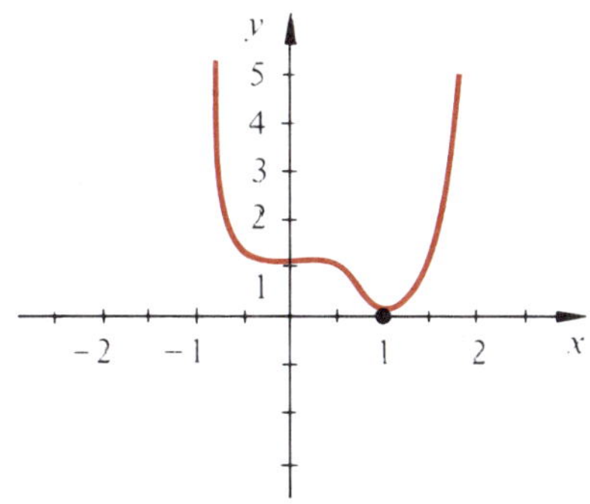

FIGURE 8.2

In general, if $P(a)$ is negative and $P(b)$ is positive, the graph of $P(x)$ must cross the x-axis somewhere between $x=a$ and $x=b$. This fact is represented graphically in Figure 8.3 and motivates the following theorem.

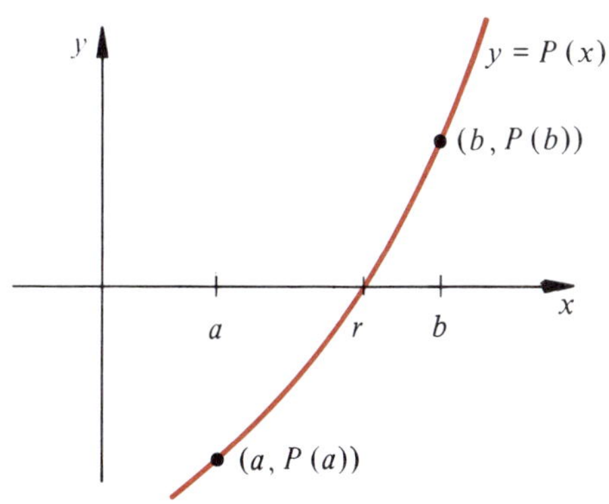

FIGURE 8.3

THEOREM 8

> Suppose that $P(x)$ is a polynomial with real coefficients and that $a<b$; then if $P(a)$ and $P(b)$ are of opposite sign, there is a number r between $x=a$ and $x=b$ for which $P(r)=0$.

A word of caution about this theorem—the theorem assures us that there will be at least one zero between $x=a$ and $x=b$ if $P(a)$ and $P(b)$ have opposite signs; it does not say that there are no zeros between $x=a$ and $x=b$ if $P(a)$ and $P(b)$ have the same sign. Figure 8.4 shows why. Further, you should be aware of the fact that there may be more than one zero between $x=a$ and $x=b$ if $P(a)$ and $P(b)$ have opposite signs.

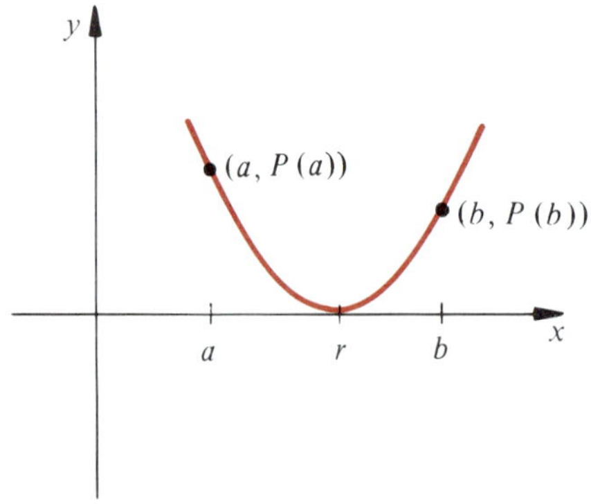

FIGURE 8.4

That is, as Figure 8.5 shows, there may be multiple crossings between $x=a$ and $x=b$.

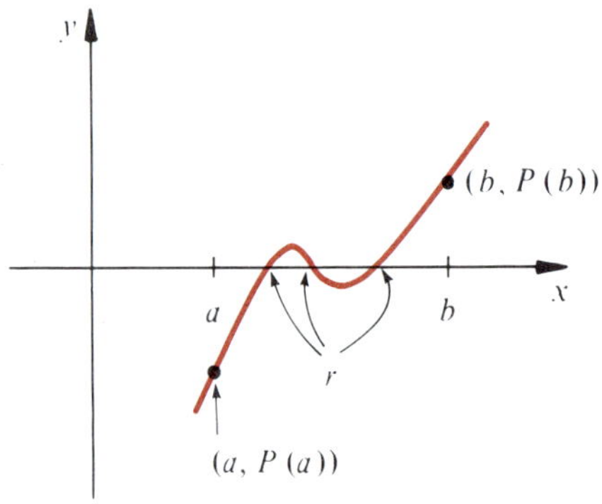

FIGURE 8.5

EXAMPLE 3 Show that the polynomial $P(x)=x^3+x^2+x-4$ has a real zero between $x=1$ and $x=2$.

SOLUTION Since $P(1)=-1$ and $P(2)=10$, it follows from Theorem 8 that there is at least one zero between $x=1$ and $x=2$.

Theorem 8 is typically used to locate the interval within which there is a zero. Because of computational difficulties, we tend to restrict a and b to integer numbers; this is not an inherent restriction of the theorem, however. Once we locate an interval that contains a zero, several techniques are available for refining our estimate of the zero. The simplest, and perhaps the most common technique, is **linear interpolation**. The assumption in this technique is that the points $(a, P(a))$ and $(b, P(b))$ are connected by a straight line. Figure 8.6 shows the difference between the actual zero and the one obtained by linear interpolation. Analytically, the equation of the straight line connecting these two points is given by

$$\frac{y-P(a)}{x-a}=\frac{P(b)-P(a)}{b-a}$$

The x-intercept of this line is $(c, 0)$. If $x=c$ and $y=0$ are used in the equation of the

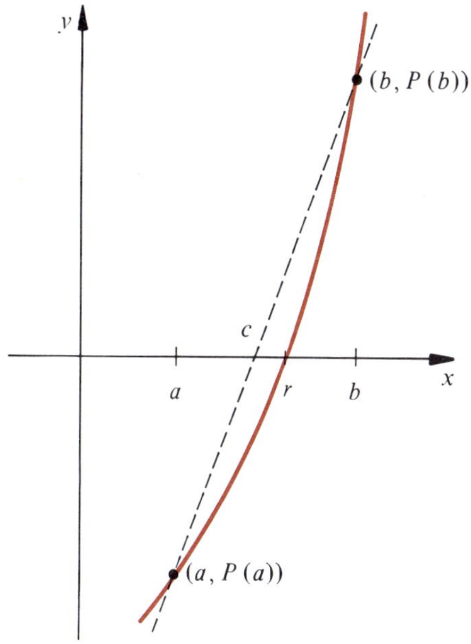

FIGURE 8.6

line, then

$$\frac{-P(a)}{c-a}=\frac{P(b)-P(a)}{b-a}$$

Solving for c gives

$$c=\frac{aP(b)-bP(a)}{P(b)-P(a)}$$

This value of c is used to approximate r.

EXAMPLE 4 Approximate the zero of the polynomial of Example 3 between 1 and 2.

SOLUTION Using the above formula with $b=2$, $a=1$, $P(2)=10$, and $P(1)=-1$, we have

$$r\approx c=\frac{10-2(-1)}{10-(-1)}=\frac{12}{11}=1.1$$

A method exists for estimating the total number of real roots of a real polynomial. The method is called **Descartes' rule of signs**, and it is based on this concept of a *variation in sign* between the terms of a polynomial:

DEFINITION

> Let $P(x)$ be a polynomial arranged in decreasing powers of x. There is then a **variation in sign** whenever two successive terms differ in sign.

Terms with zero coefficients are disregarded in considering variation in sign. For example, there are three variations in sign of

$$P(x)=2x^5-x^4+x^2+3x-2$$

from $2x^5$ to $-x^4$

from $-x^4$ to x^2

from $3x$ to -2

Descartes' rule of signs is given in Theorem 9, which is stated without proof.

THEOREM 9

> The number of positive zeros of a real polynomial is less than or equal to the variations in sign of $P(x)$. The number of negative zeros is less than or equal to the variations in sign of $P(-x)$. In both cases, the number of zeros is less than the number of variations in sign by an even number.

For example, if $P(x)$ has five variations in sign, then the number of positive zeros of $P(x)$ is 5, 3, or 1 (never 4 or 2). The theorem allows us to conclude that there is *at least one* positive real zero of such a polynomial. Notice also that a polynomial with precisely one change in sign will have precisely one positive real zero.

EXAMPLE 5 Consider the polynomial

$$P(x)=x^5-x^3-2$$

(a) Determine the number of zeros.

(b) Estimate the number of real positive and negative zeros.

(c) Estimate the number of complex zeros.

SOLUTION

(a) There are (counting multiplicities) five zeros, since the degree of $P(x)$ is 5.

(b) Since $P(x)$ has one change in sign, there is precisely one positive real zero. Since $P(-x)=-x^5+x^3-2$, $P(-x)$ has two variations in sign. Thus, $P(x)$ has zero or 2 negative zeros.

(c) Because of (b), there is either one real zero or three. Hence, there are two or four complex zeros.

Exercises for Section 8.5

In Exercises 1–10, use Theorem 9 (Descartes' rule of signs) to determine the possible number of positive and negative real roots and the possible number of complex roots.

1. $x^3+3x+2=0$

2. $4x^3-x^2-5=0$

3. $x^4+13x^2-7=0$

4. $2x^3+5x-6=0$

5. $2x^3-7x^2-3x-1=0$

6. $x^4+x^3-x^2+x-1=0$

7. $-2x^5+8x^3-7x+15=0$

8. $4x^5-2x^4+x-10=0$

9. $3x^6+x^4-8x^2+16=0$

10. $x^7-x^3+x^2-5x+7=0$

In Exercises 11–22, use Theorem 8 to show that the given polynomial equation has a real root between the indicated points. (Use linear interpolation to estimate the root.)

11. $x^3+2x^2+8x-3=0$ ($x=0$ and $x=1$)

12. $x^3-5x^2+3x+4=0$ ($x=1$ and $x=2$)

13. $x^3-3x+1=0$ ($x=1$ and $x=2$)

14. $x^3+2x+20=0$ ($x=-3$ and $x=-2$)

15. $2x^3-3x^2-12x+6=0$ ($x=3$ and $x=4$)

16. $x^3-6x^2+16=0$ ($x=5$ and $x=6$)

17. $x^4-x^3-5x^2+3x+2=0$ ($x=-1$ and $x=0$)

18. $2x^4-x^3-5x^2-3x-3=0$ ($x=-2$ and $x=-1$)

19. $x^4-x^3-2x^2-x-3=0$ ($x=2$ and $x=3$)

20. $4.3x^3-1.4x+5.6=0$ ($x=-2$ and $x=-1$)

21. $2.1x^3+3.4x^2-1.8x+5.3=0$ ($x=-3$ and $x=-2$)

22. $6.3x^4-2.5x^2-6.3=0$ ($x=1$ and $x=2$)

In Exercises 23–30, estimate the real roots of the equations by graphing the polynomial functions.

23. $x^3-4x^2+3x+1=0$

24. $x^3-3x^2+4x-5=0$

25. $2x^4-3x^3-6x^2+2x-15=0$

26. $5x^3+6x^2+6x+1=0$

27. $x^3-3x+1=0$

28. $x^3-6x^2+16=0$

29. $x^4-x^3-5x^2+3x+2=0$

30. $2x^4-3x^3-6x^2+2x-15=0$

8.6 Decomposition of Rational Functions into Partial Fractions

In calculus we sometimes find it necessary to express a rational function of the form $N(x)/D(x)$ by simpler functions. We note that if the degree of the polynomial in the numerator is less than the degree of the polynomial in the denominator, then the rational function is said to be, **proper**; otherwise it is said to be **improper**. Any improper function can be reduced to a form consisting of the sum of a polynomial and a proper rational function. The process simply involves dividing the numerator by the denominator. For example, to express the rational function

$$\frac{6x^4+7x^3+6x^2+32x-7}{3x^2+5x-2}$$

as a sum of a polynomial and a proper rational function, we divide the numerator by the denominator and get

$$2x^2-x+5+\frac{5x+3}{3x^2+5x-2}$$

A proper rational function can frequently be written as a sum of functions in which each function has but one factor in the denominator. This sum is often called the **partial fraction decomposition** of the rational function. It is literally the reverse of what we learned in Chapter 1. We learned there that

$$\frac{2}{x-3}-\frac{1}{x+1}=\frac{x+5}{x^2-2x-3}$$

Now we show how to "recover" the elements of the sum making up the proper rational function

$$\frac{x+5}{x^2-2x-3}$$

Previously you learned that every real polynomial can be factored into real linear and irreducible quadratic factors of the form $ax+b$ and ax^2+bx+c. These factors are the basis of our decomposition. The following rules govern the form of the partial fraction decomposition of a given rational function. In each case, the factors of the denominator of the rational function are being considered.

RULE 1

If the denominator has an unrepeated linear factor of the form $(ax+b)$ a fraction of the form

$$\frac{A}{ax+b}$$

must be included in the partial fraction decomposition.

RULE 2

If the denominator has a linear factor $(ax+b)$ that is repeated n times, a sum of fractions of the form

$$\frac{A}{ax+b}+\frac{A_2}{(ax+b)^2}+\cdots+\frac{A_n}{(ax+b)^n}$$

must be included in the partial fraction decomposition.

RULE 3

If the denominator has an unrepeated quadratic factor (ax^2+bx+c), a fraction of the form

$$\frac{Ax+B}{ax^2+bx+c}$$

must be included in the partial fraction decomposition.

EXAMPLE 1 Find the partial fraction decomposition of

$$\frac{x-11}{2x^2+5x-3}$$

SOLUTION The factors of the denominator are $(x+3)$ and $(2x-1)$. Hence, the given fraction can be expressed as the sum of the partial fractions

$$\frac{A}{(x+3)} \quad \text{and} \quad \frac{B}{(2x-1)}$$

where A and B are constants to be determined. Thus, we write

$$\frac{x-11}{(x+3)(2x-1)}=\frac{A}{x+3}+\frac{B}{2x-1}$$

To solve for A and B, we first clear the fractions from the equation, to get

$$x-11=A(2x-1)+B(x+3)$$

Expanding gives

$$x-11=2Ax-A+Bx+3B$$

and collecting like terms in x,

$$x-11=(2A+B)x+(3B-A)$$

For this equality to hold, the coefficients of corresponding powers of x must be equal,* so that

$$2A+B=1$$

and

$$3B-A=-11$$

Solving these equations simultaneously, we get $A=2$ and $B=-3$. Hence

$$\frac{x-11}{(x+3)(2x-1)}=\frac{2}{x+3}-\frac{3}{2x-1}$$

EXAMPLE 2 Find the partial fraction decomposition of $x^2/(x+1)^3$.

SOLUTION We assume that the partial fractions can be written

$$\frac{x^2}{(x+1)^3}=\frac{A}{x+1}+\frac{B}{(x+1)^2}+\frac{C}{(x+1)^3}$$

If we then clear the fractions,

$$x^2=A(x+1)^2+B(x+1)+C$$

expand the expression,

$$x^2=Ax^2+2Ax+A+Bx+B+C$$

collect like terms in x,

$$x^2=Ax^2+(2A+B)x+(A+B+C)$$

and equate the coefficients of like powers of x, we get finally

$$A=1$$
$$2A+B=0$$
$$A+B+C=0$$

*This is a particular case of the "identity of polynomials" theorem, which says that for two polynomials to be equal, coefficients of like powers of x must be equal.

These equations yield the values $A=1$, $B=-2$, and $C=1$. Substituting, we have

$$\frac{x^2}{(x+1)^3}=\frac{1}{x+1}-\frac{2}{(x+1)^2}+\frac{1}{(x+1)^3}$$

EXAMPLE 3 Find the partial fraction decomposition of the rational function $(2x+8)/[x(x+2)^2]$.

SOLUTION The factors of the denominator are x and $(x+2)^2$, and therefore the partial fractions are

$$\frac{2x+8}{x(x+2)^2}=\frac{A}{x}+\frac{B}{x+2}+\frac{C}{(x+2)^2}$$

Clearing the fractions yields

$$2x+8=A(x+2)^2+Bx(x+2)+Cx$$

By the method of the preceding examples, we get the simultaneous equations

$$\begin{aligned} A+B&=0\\ 4A+2B+C&=2\\ 4A&=8 \end{aligned}$$

Solving these equations yields $A=2$, $B=-2$, and $C=-2$. The desired decomposition is

$$\frac{2x+8}{x(x+2)^2}=\frac{2}{x}-\frac{2}{x+2}-\frac{2}{(x+2)^2}$$

EXAMPLE 4 Decompose $\dfrac{9x+14}{(x-2)(x^2+4)}$ into its partial fractions.

SOLUTION The denominator is composed of a linear factor $(x-2)$ and an irreducible quadratic factor (x^2+4). The required partial fractions are therefore

$$\frac{9x+14}{(x-2)(x^2+4)}=\frac{A}{x-2}+\frac{Bx+C}{x^2+4}$$

Clearing the fractions,

$$9x+14=A(x^2+4)+(Bx+C)(x-2)$$

expanding the right side and collecting like terms,

$$9x+14=(A+B)x^2+(C-2B)x+(4A-2C)$$

and equating the coefficients of corresponding powers of x, we find

$$\begin{aligned} A+B&=0\\ C-2B&=9\\ 4A-2C&=14 \end{aligned}$$

from which $A=4$, $B=-4$, and $C=1$. Substituting these values in the partial fractions, we get

$$\frac{9x+14}{(x-2)(x^2+4)}=\frac{4}{x-2}-\frac{4x-1}{x^2+4}$$

Exercises for Section 8.6

In Exercises 1–21, find the partial fraction expansion for the given rational function.

1. $\dfrac{x+1}{x^2+4x-5}$
2. $\dfrac{x-2}{x^2-x-6}$
3. $\dfrac{1}{x^2+x}$
4. $\dfrac{2}{x^3-4x}$
5. $\dfrac{2x^2+5x+5}{x(x+5)(x+2)}$
6. $\dfrac{2x^2-11x-9}{x^3-2x^2-3x}$
7. $\dfrac{x+3}{x^2+6x+5}$
8. $\dfrac{1}{(2x+3)^3}$
9. $\dfrac{x^2+3x-6}{x(x-1)^2}$
10. $\dfrac{x+1}{x^3-x^2}$
11. $\dfrac{1}{x^3+2x^2+x}$
12. $\dfrac{-x^2+11x-6}{(x+2)(x-2)^2}$
13. $\dfrac{x^2-3x+6}{(x^2-1)(3-2x)}$
14. $\dfrac{x^2}{(x+1)^3}$
15. $\dfrac{2x-1}{(x-1)^2}$
16. $\dfrac{2(x^2+1)}{x^3+2x}$
17. $\dfrac{x^2+x+2}{x^3+2x}$
18. $\dfrac{9x+14}{(x-2)(x^2+4)}$
19. $\dfrac{x^2+1}{(2x+3)(x^2+2x+4)}$
20. $\dfrac{x^3+x^2-x+1}{(x-1)^2(x^2+1)}$
21. $\dfrac{x^3-3x^2+7x-5}{(x^2+2x+3)(x^2+5)}$

Test 1 for Chapter 8

Answer *True* or *False* to Exercises 1–10.

1. Quadratic functions are polynomials.
2. A polynomial is any algebraic expression with two or more terms.
3. The only possible rational roots of $x^9+x-10=0$ are ± 10, ± 5, ± 2, and ± 1.
4. The real zeros of a polynomial are graphically represented as the x-intercepts of the graph of the polynomial.
5. A polynomial with real coefficients has real zeros.
6. Rational functions are zero only when the numerator is zero.
7. When a polynomial $f(x)$ is divided by $x+a$, the remainder is $f(a)$.
8. An nth-degree polynomial has n distinct zeros.
9. If $x=a$ is a zero of a polynomial, then $(x-a)$ is a factor of that polynomial.
10. A real cubic polynomial has at least one real zero.
11. Determine the value of k so that $x-3$ is a factor of x^3-5x^2+kx+6.
12. Use the remainder theorem to find the remainder when dividing $x^5-x^3+x^2+5$ by $x-1$.
13. Write an expression for a third-degree polynomial with roots 3, -2, and 1.
14. The graph of a fifth-degree polynomial is shown in Figure 8.7. Write the expression for the polynomial.

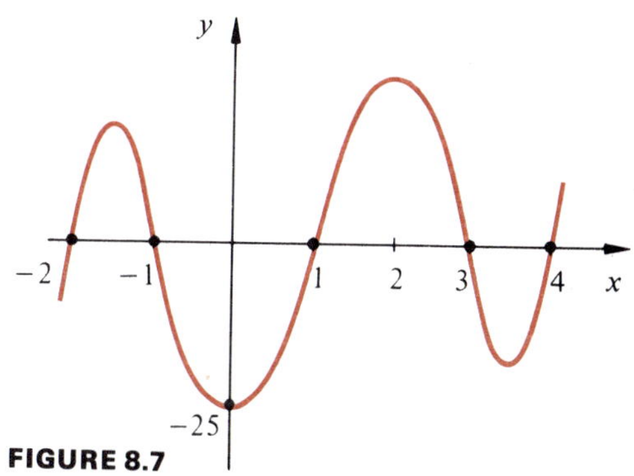

FIGURE 8.7

15. Estimate the number of positive and negative zeros of the polynomial $P(x)=x^4-x-3$. Then use linear interpolation and graphing to estimate the values of the positive zeros.

16. Determine the partial fraction decomposition of

$$\frac{2x-3}{(x^2-4)(x+3)}$$

Test 2 for Chapter 8

1. Answer *True* or *False* to the following statements.

(a) $f(x)=x^2+x^{-2}$ is a polynomial.

(b) $(x+2)$ is a factor of $(x^{15}+2^{15})$

(c) A fifth-degree polynomial has five distinct zeros.

(d) The nonreal zeros of any polynomial with real-number coefficients occur in complex conjugate pairs.

(e) $x=-\frac{3}{2}$ is a potential rational root of

$$2x^3+4x^2-x+6=0$$

2. Show that $x-2$ is a factor of $3x^4-4x^2-x-30$.

3. Use synthetic division to find the quotient and remainder of

$$\frac{2x^4+2x^3-5x^2-x+3}{2x-4}$$

4. Find all the roots of the equation

$$x^4+x^3-x^2+x-2=0$$

5. Find the partial fraction expansion of $\dfrac{8x+7}{(x-1)(x^2+4)}$

6. For a fifth-degree real polynomial that has the zeros $\sqrt{2}$, i, and $2+i$, find the remaining zeros and write the polynomial in real form.

7. Write the third-degree polynomial with zeros at 0, 2, 3, and passing through (4, 5).

8. Use a graphical method to estimate the real zeros of $y=x^3-3x^2+2$.

9 Sequences, the Binomial Theorem, and Mathematical Induction

9.1 Sequences in General

The half-life of a radioactive substance is the time required for one-half of the substance to undergo radioactive decay. Radium has a half-life of 1620 years; this means that if we start with 100 grams of radium, there will be 50 grams at the end of 1620 years, 25 grams after 3240 years, 12.5 grams after 4860 years, and so on. The mass of radium remaining after successive half-lives can then be written

$$\{100, 50, 25, 12.5, \ldots\}$$

where the first number represents the original mass of radium, the second number the mass after one half-life, the third number the mass after two half-lives, and so on. A set of numbers that is arranged in a fixed order is called a **sequence**. Thus, the half-life values of 100 grams of radium form a sequence.

The study of sequences occupies a prominent place in mathematics, and you are about to learn some of the basic concepts. We shall use the notation

$$\{a_1, a_2, a_3, \ldots\}$$

to designate a general sequence, where the subscripts indicate the position of the numbers, or **terms**, in the sequence; that is, a_1 denotes the first term, a_2 the second term, a_3 the third term, and so on.

EXAMPLE 1 In the sequence 5, 9, 13, 17, we have

$$a_1=5 \qquad a_2=9 \qquad a_3=13 \qquad a_4=17$$

A sequence that has a finite number of terms is called a finite sequence; otherwise, it is called an **infinite sequence**. The sequence in Example 1 is a finite sequence. A familiar example of an infinite sequence is the set of positive even integers $\{2, 4, 6, 8, \ldots\}$.

Sequences are usually defined by listing the terms, as in Example 1, or by specifying a formula whose domain is a subset of the positive integers. In the second case, the formula is denoted by a_n and is called the **general term** of the sequence. By this notation, the general sequence is denoted $\{a_n\}$, where it is understood that n is a positive integer. Example 2 shows a sequence in which a_n is specified by a formula.

EXAMPLE 2 The general term $a_n=2n-1$ defines a sequence whose first few terms are

$$a_1=2(1)-1=1$$
$$a_2=2(2)-1=3$$
$$a_3=2(3)-1=5$$

This sequence is denoted by $\{2n-1\}$ or by $\{1, 3, 5, \ldots\}$.

Another way of indicating the terms of a sequence is by a **recursive** formula, that is, a formula that uses the value of a known term in the sequence to calculate the succeeding terms. Normally, as illustrated in Example 3, we give the first term of the sequence and the recursive formula.

EXAMPLE 3 The formula $a_n=a_{n-1}+2$, along with the $a_1=3$, defines a sequence whose first few terms are

$$a_1=3$$
$$a_2=a_1+2=3+2=5$$
$$a_3=a_2+2=5+2=7$$
$$a_4=a_3+2=7+2=9$$

Notice that for $n=4$, $a_n=a_4$ and $a_{n-1}=a_3$.

Whether defined by a list of terms, by an explicit formula for the general term, or by a recursive formula, **a sequence is a function whose domain is some subset of the positive integers.** A sequence is sometimes called a **sequence function**. The range values of a sequence are nothing more than the terms of the sequence.

EXAMPLE 4 What is the range of the sequence function defined by $a_n = (-1)^n$?

SOLUTION Since $a_1 = -1$, $a_2 = 1$, $a_3 = -1$, and so on, the range values are merely 1 and -1.

Sometimes you are asked to try to find a reasonable or a suggested formula for the nth term of a sequence for which several successive terms are known. Example 5 shows the technique.

EXAMPLE 5 Find a formula for the general term of the sequence 5, 9, 13, 17, ...

SOLUTION Notice that the rule is "add 4 to each term." Therefore, the general term can be written

$$a_n = 5 + 4(n-1) = 4n + 1$$

Alternatively, a recursive formula for this sequence is $a_1 = 5$, $a_n = a_{n-1} + 4$.

We examine a few important sequences. The following sequences are included for reference with no other details given.

The sequence of natural numbers: $1, 2, 3, 4, 5, \ldots, n, \ldots$

The sequence of positive even numbers: $2, 4, 6, 8, \ldots, 2n, \ldots$

The sequence of positive odd numbers: $1, 3, 5, 7, 9, \ldots, 2n-1, \ldots$

The sequence of squares: $1, 4, 9, 16, 25, \ldots, n^2, \ldots$

The sequence of cubes: $1, 8, 27, 64, \ldots, n^3, \ldots$

The sequence of k powers: $1^k, 2^k, 3^k, 4^k, \ldots, n^k, \ldots$

The sequence of reciprocals: $1, \frac{1}{2}, \frac{1}{3}, \frac{1}{4}, \frac{1}{5}, \ldots, \frac{1}{n}, \ldots$

Exercises for Section 9.1

Write the first five terms of the sequences in Exercises 1–15.

1. $a_n = n + 1$

2. $a_n = 2^n$

3. $a_n = 1 - (-1)^n$

4. $a_n = n + 4$

5. $a_n = e^n$

6. $a_n = \log n$

7. $a_n = \dfrac{n+2}{n+1}$

8. $a_n = n - \dfrac{1}{n}$

9. $a_n = n^2$

10. $a_n = n(n+1)$

11. $a_n = 2n + 1$

12. $a_n = \dfrac{1}{n}$

13. $a_n = \dfrac{1}{n^2}$

14. $a_n = (-1)^{2n+1}$

15. $a_n = 5(\frac{1}{2})^n$

Find the first five terms of the sequences in Exercises 16–20 given the following recursive information.

16. $a_1 = 2$, $a_n = a_{n-1} + 2$

17. $a_1 = 3$, $a_n = a_{n-1}^2$

18. $a_1 = -1,\ a_n = 5a_{n-1}$ **19.** $a_1 = 10,\ a_n = 5 - a_{n-1}$

20. $a_1 = 1,\ a_2 = 1,\ a_n = a_{n-1} + a_{n-2}$

Find a formula for the general term of the sequences in Exercises 21–26.

21. $2, 4, 6, 8, \ldots$ **22.** $1, 3, 7, 15, 31, \ldots$

23. $1, 4, 7, 10, 13, \ldots$ **24.** $\frac{1}{2}, \frac{2}{3}, \frac{3}{4}, \frac{4}{5}, \ldots$

25. $1, 4, 9, 16, 25, \ldots$ **26.** $1, 8, 27, 64, \ldots$

27. Suppose that in a year's time an automobile loses one-fourth of the value it had at the beginning of the year. If a car costs \$4000 new, what is its value at the end of 4 yr?

28. A certain yeast plant matures in 1 hr; each hour thereafter it buds off one new plant, which also matures in 1 hr and starts to produce new plants, and so on. Write the terms of the sequence that describe the total number of plants at any time. This is an example of a *Fibonacci sequence*. Can you write a recursive formula for this sequence? (*Hint*: See Exercise 20.)

9.2 Sequences of Partial Sums

Corresponding to each sequence $\{a_n\}$, we can form another sequence $\{S_n\}$ by forming sums of the first n terms. Thus, if $a_1, a_2, a_3, \ldots, a_n, \ldots$ are the terms of a sequence, then

$$S_1 = a_1,\quad S_2 = a_1 + a_2,\quad S_3 = a_1 + a_2 + a_3,\quad \ldots, S_n = a_1 + a_2 + a_3 + \cdots + a_n$$

The sequence $\{S_n\}$ is called the **corresponding sequence of partial sums**.

A compact notation often used to express terms of the sequence of sums is the sigma, or summation, notation. The Greek letter Σ (sigma) means to form the sum of the indicated terms.

DEFINITION

By $\sum_{k=1}^{n} a_k$ is meant the sum

$$a_1 + a_2 + a_3 + \cdots + a_n$$

The subscript k, called the **index of summation**, is often called a "dummy" variable, since its only purpose is to indicate the steps in the summation. The choice of the letter k for the index is arbitrary—any other symbol would do. The 1 and the n indicate the range of the index. In the definition, k begins at 1 and increases in steps of 1 until it equals n. Writing a sum such as $\sum_{k=2}^{5} k^2$ in the form $4 + 9 + 16 + 25$ is called *expanding* the summation, or writing it in *expanded form*.

EXAMPLE 1 The sum of the first 100 integers can be written $\sum_{k=1}^{100} k$.

EXAMPLE 2 Write $\sum_{k=2}^{4} x^k$ in expanded form.

SOLUTION

$$\sum_{k=2}^{4} x^k = x^2 + x^3 + x^4$$

EXAMPLE 3 Write $\sum_{i=1}^{10} (-1)^{i-1}$ in expanded form and simplify.

SOLUTION

$$\sum_{i=1}^{10} (-1)^{i-1} = (-1)^0 + (-1)^1 + (-1)^2 + \cdots + (-1)^9 = 0$$

Two basic rules govern the use of the summation symbol. The first allows us to factor out a constant coefficient and the second allows us to compute the sum of sums term by term. Together these two properties are called the **linearity properties**.

PROPERTY 1

$$\sum_{k=1}^{n} ca_k = c \sum_{k=1}^{n} a_k$$

PROPERTY 2

$$\sum_{k=1}^{n} (a_k + b_k) = \sum_{k=1}^{n} a_k + \sum_{k=1}^{n} b_k$$

As an example of the use of these properties, the sum

$$\sum_{k=1}^{3} k(5k+1) = 1(6) + 2(11) + 3(16) = 76$$

can also be written

$$\sum_{k=1}^{3} (5k^2 + k) = 5 \sum_{k=1}^{3} k^2 + \sum_{k=1}^{3} k = 5(1+4+9) + (1+2+3) = 76$$

Either way, the value is the same.

A sum can be represented by more than one summation expression. One such instance is given in Example 4.

EXAMPLE 4 Show that $\sum_{k=1}^{n} \frac{1}{k} = \sum_{k=0}^{n-1} \frac{1}{k+1}$.

SOLUTION Let $j = k-1$ in $\sum_{k=1}^{n} \frac{1}{n}$, which yields $\sum_{j=0}^{n-1} \frac{1}{j+1}$. Since j is a dummy variable, this is equivalent to $\sum_{k=0}^{n-1} \frac{1}{k+1}$. Another way of showing that these two sums are equal is to write both in expanded form. The index on the left summation ranges from 1 to n, and the index on the right ranges from 0 to $n-1$. The second sum is said to be obtained by a *shift of index*.

EXAMPLE 5 Write the summation $\Sigma_{k=0}^{n} a_k(k-1)x^k$ as a summation whose index begins at 2.

SOLUTION Working this out formally, we let $k=t-2$. When $k=0$, then $t=2$, and when $k=n$, then $t=n+2$, so the summation becomes

$$\sum_{t=2}^{n+2} a_{t-2}(t-3)x^{t-2}$$

Let $k=t$ in this sum to get the desired index. Thus,

$$\sum_{k=2}^{n+2} a_{k-2}(k-3)x^{k-2}$$

EXAMPLE 6 The sum of the first six terms of the sequence of reciprocals, also called a *harmonic* sequence, is $1+\frac{1}{2}+\frac{1}{3}+\frac{1}{4}+\frac{1}{5}+\frac{1}{6}$. The first term can be written as $\frac{1}{1}$, and since the denominator varies from 1 to 6 in steps of 1, the sum can be written in summation notation as

$$\sum_{n=1}^{6} \frac{1}{n}$$

Exercises for Section 9.2

Write Exercises 1–10 in expanded form and evaluate.

1. $\sum_{m=1}^{5} \frac{1}{m+2}$ **2.** $\sum_{k=2}^{3} \frac{1-k}{1+k}$ **3.** $\sum_{n=0}^{5} n^2$

4. $\sum_{k=0}^{3} \left(-\frac{1}{2}\right)^k$ **5.** $\sum_{j=1}^{5} 3j$ **6.** $\sum_{k=0}^{5} (-1)^k$

7. $\sum_{k=1}^{5} (1+k)$ **8.** $\sum_{k=0}^{5} [2+3(k-1)]$

9. $\sum_{n=-2}^{2} n^2$ **10.** $\sum_{k=-1}^{2} (3k-1)$

Write the terms of the summation in Exercises 11 and 12.

11. $\sum_{k=1}^{5} x_k$ **12.** $\sum_{k=1}^{5} x^k$

Represent the summations in Exercises 13–17 using the sigma notation.

13. $1+2+3+4$ **14.** $\frac{1}{2}+\frac{1}{4}+\frac{1}{8}$

15. $1+4+9+16$ **16.** $x_0+x_1+x_2+x_3+x_4$

17. $2x+4x+8x+16x+32x$

Which of Exercises 18–23 are true?

18. $\sum_{k=1}^{n} ax_k = a\sum_{k=1}^{n} x_k$ **19.** $\sum_{i=1}^{n} (x_i+d) = nd + \sum_{i=1}^{n} x_i$

20. $\sum_{k=1}^{n} d = nd$ **21.** $\sum_{k=2}^{5} a_{k-2} = \sum_{0}^{3} a_k$

22. $\sum_{n=1}^{4} n^2 = \sum_{n=2}^{5} n^2$ **23.** $\sum_{n=1}^{3} x^n = \sum_{t=2}^{4} x^{t-1}$

Find the sequence whose sequence of sums is given by the formulas in Exercises 24–27.

24. $S_n = \frac{n}{n+1}$ **25.** $S_n = 2$

26. $S_n = n$ **27.** $S_n = (-1)^n$

28. Find the sum of the first six terms of the Fibonacci sequence defined by $a_1=2$, $a_2=3$, and $a_n=a_{n-2}+a_{n-1}$.

9.3 Arithmetic Sequences

A sequence of n equally spaced numbers is called an **arithmetic sequence**. Another way of saying this is, an arithmetic sequence is one in which each term differs from the preceding term by the same constant amount. This constant number is called the **common difference** and is usually denoted by the letter d. Thus, an arithmetic sequence is defined by the following recursive relation:

$$a_1 = a \qquad a_n = a_{n-1} + d$$

Like all sequences, arithmetic sequences can be finite or infinite; unless we state otherwise, assume that a sequence is infinite. Perhaps the simplest arithmetic sequence is the sequence of counting numbers 1, 2, 3, 4, . . . , whose general term is $a_n = n$.

EXAMPLE 1 Find the first few terms of the arithmetic sequence with $a_1 = 7$ and $d = -3$.

SOLUTION $a_1 = 7, \quad a_2 = 7 - 3 = 4, \quad a_3 = 4 - 3 = 1, \quad a_4 = 1 - 3 = -2, \quad a_5 = -2 - 3 = -5, \ldots$

EXAMPLE 2 A parachutist falls approximately 16 ft during the first second of free fall, 48 ft during the second second, 80 ft during the third second, and 112 ft during the fourth second. Notice that the numbers 16, 48, 80, 112 form an arithmetic sequence whose common difference is 32.

To find the general term of any arithmetic sequence with first term a_1 and common difference d, note that the succeeding terms are $a_2 = a_1 + d$, $a_3 = a_1 + 2d$, $a_4 = a_1 + 3d, \ldots$. Therefore, the general term of an arithmetic sequence is

(9.1) $$a_n = a_1 + (n-1)d$$

EXAMPLE 3 Find the hundredth term of the arithmetic sequence whose first term is 3 and whose common difference is 2.

SOLUTION By the formula for the general term with $a = 3$, $d = 2$, and $n = 100$,

$$a_{100} = 3 + (99)(2) = 201$$

The sequence of partial sums corresponding to an arithmetic sequence is

$$a_1, \quad a_1 + a_2 = 2a_1 + d, \quad a_1 + a_2 + a_3 = 3a_1 + 3d, \quad \ldots$$

The formula for the nth term of the sequence of sums is derived by a method attributed to Carl Gauss for computing the sum of the first 100 integers.* Gauss recognized the simple but significant fact that the sum is the same regardless of the order in which the terms are written. Thus, for the sum of the first 100 positive integers,

$$S_{100} = 1 + 2 + 3 + \cdots + 98 + 99 + 100$$

*Carl Friedrich Gauss (1777–1855) is usually ranked, with Archimedes and Isaac Newton, as one of the three greatest mathematicians of all time.

and in reverse order,

$$S_{100}=100+99+98+\cdots+3+2+1$$

Adding these two equalities,

$$2S_{100}=101+101+101+\cdots+101+101+101$$
$$2S_{100}=(100)(101)$$
$$S_{100}=5050$$

This technique can be used conveniently to derive a formula for the nth term of the sequence of sums corresponding to *any* arithmetic sequence.

$$\begin{aligned} S_n&=a_1 \quad +(a_1+d)+(a_1+2d)+\cdots+(a_n-2d)+(a_n-d)+a_n \\ S_n&=a_n \quad +(a_n-d)+(a_n-2d)+\cdots+(a_1+2d)+(a_1+d)+a_1 \\ \hline 2S_n&=(a_1+a_n)+(a_1+a_n)+(a_1+a_n)+\cdots+(a_1+a_n)+(a_1+a_n)+(a_1+a_n) \end{aligned}$$

(9.2) $$S_n=\frac{n(a_1+a_n)}{2}.$$

Thus, the sum of n terms of an arithmetic sequence is the average of the first and last terms times the number of terms. Since $a_n=a_1+(n-1)d$, we can express this formula in terms of a_1 and d:

(9.2′) $$S_n=\frac{n(a_1+a_1+(n-1)d)}{2}=a_1\cdot n+\frac{n(n-1)}{2}d$$

EXAMPLE 4 Find the sum of the first 50 terms of the arithmetic sequence whose first term is 2 and whose common difference is -3.

SOLUTION Here $a_{50}=2+(49)(-3)=2-147=-145$. Hence,

$$S_{50}=50\,\frac{(2-145)}{2}=25(-143)=-3575$$

EXAMPLE 5 Find $\sum_{k=1}^{25}(5k-3)$.

SOLUTION This sum can be identified with the sum of the first 25 terms of an arithmetic sequence whose first term is 2 and whose common difference is 5. Thus, from (9.2) the sum is equal to

$$\frac{25}{2}[2+(24)5]=1525$$

The arithmetic sequence and the linear function are closely related. To see this we sketch the points $(1, a_1)$, $(2, a_2)$, $(3, a_3)$, and so on, in Figure 9.1, and join these points by a straight line, as if the domain were all real numbers. The equation of the

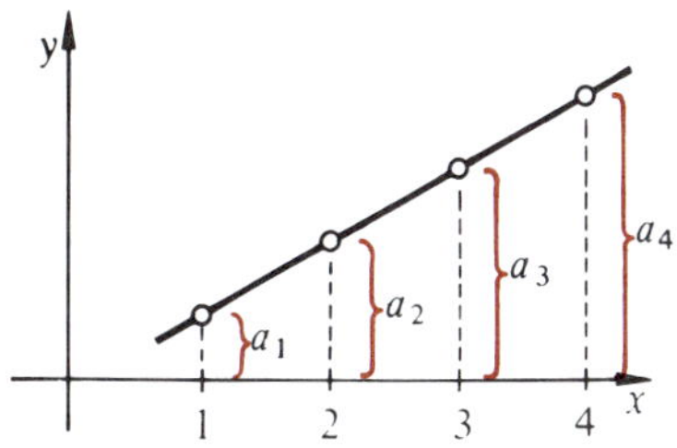

FIGURE 9.1

line is

$$\frac{y-a_1}{x-1}=\frac{a_2-a_1}{2-1}=d$$

Thus,

$$y=a_1+(x-1)d$$

which is the arithmetic sequence rule with n replaced by x. Every arithmetic sequence determines a linear function in this way. Conversely, from a linear function $f(x)$, the terms $f(1), f(2), f(3)$, and so on determine an arithmetic sequence with common difference $f(2)-f(1)$. Note that the common difference d is the slope of the line. Thus, an arithmetic sequence is really a linear function whose domain is restricted to the positive integers.

Exercises for Section 9.3

Determine which of the sequences in Exercises 1–10 are arithmetic and which are not. For those that are, give the common difference.

1. $1, 2, 3, 4, 5, \ldots$

2. $2, 5, 8, 11, \ldots$

3. $2, 4, 6, 8, 10, \ldots$

4. $3, 9, 27, 81, \ldots$

5. $3, 6, 9, 12, \ldots$

6. $1, \frac{1}{2}, 0, -\frac{1}{2}, \ldots$

7. $1, \frac{1}{2}, \frac{1}{4}, \frac{1}{8}, \ldots$

8. $1, -1, 1, 3, 1, 3, 5, 3, \ldots$

9. $\frac{1}{2}, \frac{1}{3}, \frac{1}{4}, \frac{1}{5}, \frac{1}{6}, \ldots$

10. $4, 14, 24, 34, 44, \ldots$

For Exercises 11–16, suppose that a_n is an arithmetic sequence.

11. For $a_1=2$ and $d=3$, find the general term.

12. For $a_1=3$ and $a_7=23$, find the common difference.

13. For $a_1=2$ and $S_7=35$, find the common difference.

14. For $a_1=-1$ and $d=3$, find S_8.

15. For $a_2=5$ and $d=-4$, find S_{10}.

16. For $d=3$ and $S_6=57$, find a_1.

Find the values of the sums in Exercises 17–22.

17. $\sum_{k=1}^{8}(2k-3)$

18. $\sum_{k=2}^{7}(k+4)$

19. $\sum_{k=0}^{4}(7k-4)$

20. $\sum_{n=1}^{5}\left(\frac{1}{2}n+3\right)$

21. $\sum_{k=0}^{4}7(k-4)$

22. $\sum_{k=1}^{5}(2k+1)$

23. Find a formula for the sum of the first n positive odd integers.

24. Find a formula for the sum of the first n positive even integers.

25. Show that the technique for finding the sum of an arithmetic sequence does not work for the sequence of squares.

26. A gasoline company runs a contest in which it sends the winner some money every day for a month. The amounts are \$50 the first day, \$100 the second day, \$150 the third day, and so on. In a 30-day month, how much will the winner collect?

Find the first 5 terms of the arithmetic sequence determined by the linear functions in Exercises 27–32.

27. $f(x)=x+2$

28. $f(x)=2x+1$

29. $f(x)=-x+2$

30. $f(x)=x$

31. $f(x)=2x+5$

32. $f(x)=-2x+4$

What linear function is defined by the arithmetic sequences?

33. 3, 6, 9, 12, . . .

34. $a_n=2n+3$

35. $a_1=4$, $d=3$

36. $a_1=2$, $d=-1$

37. $a_1=5$, $a_8=26$

38. $a_1=1$, $d=2$

9.4 Geometric Sequences

A number sequence made by successively multiplying by the same number is called a **geometric sequence**. The constant multiplier is called the **common ratio** and is usually denoted by r. The recursive definition of a geometric sequence is

$$a_1=a_1$$
$$a_n=ra_{n-1}$$

The number sequence giving the number of your ancestors in the preceding generations is 2, 4, 8, 16, . . . and is obviously a finite geometric sequence with common ratio 2.

A geometric sequence is completely determined by its first term a_1 and the common ratio r. If the sequence under consideration is to terminate, that is, be a finite sequence, then this should ordinarily be mentioned. To determine the general term, note that

$$a_1=a_1$$
$$a_2=a_1r$$
$$a_3=a_2r=a_1r^2$$
$$a_4=a_3r=a_1r^3$$
$$\vdots$$

$$a_n=a_{n-1}r=a_1r^{n-1} \tag{9.3}$$

EXAMPLE 1 What is the tenth term of a geometric sequence whose first term is 1 and whose common ratio is 0.1? Write an expression for the general term.

SOLUTION Using the formula for the nth term, we have

$$a_{10}=(1)(0.1)^9=10^{-9}$$

The general term for this sequence can be written

$$a_n=(1)(0.1)^{n-1}=10^{1-n}$$

EXAMPLE 2 Find the recursive form for the geometric sequence whose general formula is

$$a_n = 5\left(\frac{1}{3}\right)^{n-3}$$

SOLUTION From the given formula for a_n,

$$a_1 = 5\left(\frac{1}{3}\right)^{-2} = 45$$

Also,

$$r = \frac{a_n}{a_{n-1}} = \frac{5(\frac{1}{3})^{n-3}}{5(\frac{1}{3})^{n-4}} = \frac{1}{3}$$

Therefore, we have the recursive definition

$$a_1 = 45, \quad a_n = \frac{1}{3}a_{n-1}$$

To find the sequence of partial sums corresponding to the geometric sequence, we do the following. Let

$$S_n = a_1 + a_1 r + \cdots + a_1 r^{n-1}$$

Multiplying both sides of this equation by the common ratio r yields

$$rS_n = a_1 r + a_1 r^2 + \cdots + a_1 r^n$$

Subtracting the left side of the second equation from the first equation and equating it to the difference of the right-hand sides of the two equations, we get

$$S_n(1-r) = a_1(1-r^n)$$

from which, if $r \neq 1$,

$$S_n = \frac{1-r^n}{1-r}a_1 \tag{9.4}$$

EXAMPLE 3 Find $\sum_{k=1}^{10} (0.1)^k$.

SOLUTION This sum can be written as the sum of the first ten terms of a geometric sequence whose first term is 0.1 and whose common ratio is 0.1. Thus,

$$S_{10} = \frac{1-(0.1)^{10}}{1-0.1}(0.1) = (1.111111111)(0.1)$$
$$= 0.1111111111$$

Exponential functions of the type $f(x) = ca^x$ can be closely identified with geometric sequences of the type $a_n = ca^n$, just like the identification of arithmetic sequences and linear functions. The sequence is obtained from the function by restricting the domain of the function to the positive integers. Conversely, by sketching terms of the geometric sequence ca^n and connecting the points by a smooth curve, we get a distinctively exponential curve whose equation is $f(x)=ca^x$. (See Figure 9.2.)

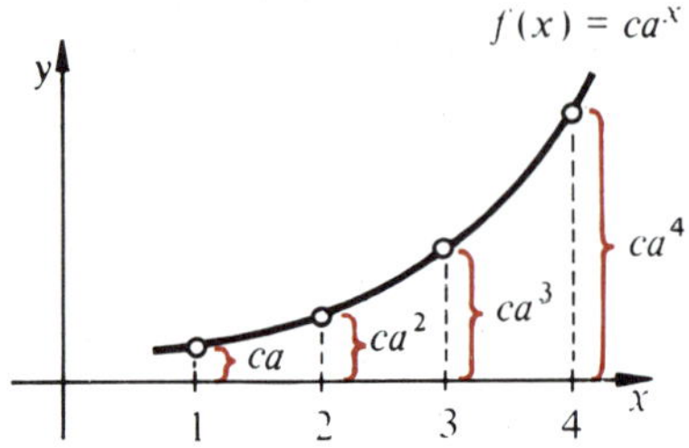

FIGURE 9.2

Geometric sequences can be used to predict population growth. If a population P_0 increases r percent each year, then the population at the end of the first year is $P_1=P_0(1+r)$, at the end of the second year is $P_2=P_1(1+r)=P_0(1+r)^2$, and so on. Hence, the population at the end of n years is

$$P_n=P_0(1+r)^n$$

EXAMPLE 4 The world population in 1978 was approximately 4 billion. If the population is increasing at 2% each year, (a) what will the world population be in 2000? (b) what is the doubling period of the population?

SOLUTION (a) Letting $P_0=4\times 10^9$, $r=0.02$, and $n=22$, we have

$$P=4\times 10^9(1.02)^{22}=6.18\times 10^9$$

(b) The doubling period is the time that it takes the population to double in size. Although we could use a specific population for P_0, we note that the doubling period depends only on the rate of growth r. Therefore, we let $P_n=2P_0$ and $r=0.02$ and solve the population equation for n. Thus,

$$2P_0=P_0(1.02)^n$$
$$2=(1.02)^n$$

Taking the logarithm of both sides, we have

$$\log 2=\log (1.02)^n$$
$$=n\log (1.02)$$

Solving for n, we get

$$n=\frac{\log 2}{\log (1.02)}=\frac{0.3010}{0.0086}=35$$

Therefore, the doubling period for $r=2\%$ is 35 years. Based on this period the world population will be approximately 8 billion in 2013 (1978+35) and 16 billion in 2048.

Exercises for Section 9.4

Determine which of Exercises 1–8 are geometric sequences; for those that are, give the value of the common ratio. For each of the geometric sequences, find the general term.

1. 4, 12, 36, 108, . . .

2. 1, −3, 7, −11, . . .

3. 3, 6, 9, 12, . . .

4. 2, 4, 16, 256, . . .

5. $\frac{1}{3}, \frac{-2}{3}, \frac{4}{3}, -\frac{8}{3}, \ldots$

6. 1, −1, 1, −1, . . .

7. $\frac{1}{2}, \frac{1}{4}, \frac{1}{8}, \frac{1}{16}, \frac{1}{32}, \ldots$

8. 6, −6, 6, −6, 6, . . .

In Exercises 9–14, assume that a_n is a geometric progression.

9. If $a_1=3$ and $r=5$, what are a_4 and S_4?

10. If $a_2=2$ and $r=2$, what is S_{10}?

11. If $a_1=5$ and $a_5=80$, what are all the possible real values of r and S_5?

12. For $a_3=\frac{1}{2}$ and $a_7=\frac{1}{32}$, find a_{20}.

13. For $a_1=-2$, $r=\frac{2}{3}$, find a_{10}.

14. For $a_4=4$ and $a_7=32$, find a_1 and r.

Find the value of the sums in Exercises 15–21.

15. $\sum_{n=1}^{5} 2(3)^n$

16. $\sum_{k=2}^{6} (3^k+2k)$

17. $\sum_{k=1}^{10} \left[\left(\frac{1}{2}\right)^k+\left(\frac{3}{2}\right)^k\right]$

18. $\sum_{n=1}^{10} 2^n$

19. $\sum_{k=1}^{6} 5\left(\frac{1}{2}\right)^k$

20. $\sum_{n=1}^{3} \left(\frac{2}{3}\right)^n$

21. $\sum_{n=1}^{3} \left(\frac{1}{4}\right)^n$

22. A certain ball, when dropped from a height, rebounds to three-fourths of the original height. At what bounce will the ball fail to reach one-fourth of the original height?

23. With each stroke, a vacuum pump removes one-fifth of the air in a container. After how many strokes will it remove three-fourths of the original air?

Find the exponential function defined by the geometric sequences in Exercises 24–26.

24. $a_1=2, r=\frac{2}{3}$

25. $a_3=4, a_4=8$

26. $a_1=1, r=3$

Give the first five terms of the geometric sequence defined by the given exponential function in Exercises 27–29.

27. $f(x)=3^x$

28. $f(x)=2(10)^x$

29. $f(x)=\left(\frac{1}{2}\right)^x$

30. What is the formula for the sum of n terms of a geometric sequence if $r=1$?

31. The speed of a chemical reaction approximately doubles each time the temperature increases by 10°C. Write a geometric sequence to show how many times faster paper will burn if the temperature rises in °C by 10, 20, 30, and 40°C.

32. A particular substance decays so that it loses half its weight each day. What is its half-life? In how many days will it have lost 90 percent of its initial weight?

33. A binary sequence is defined by the formula $a_n=2^{n-1}$. Find the formula for the sum of the first n terms of a binary sequence.

9.5 Interest

Money paid for the use of another person's money is called **interest**. An important application of geometric sequences is in computing accumulated interest. For any unit of time, the ratio of the interest to the original amount borrowed (called the **principal**) is called the **rate of interest**. Thus, the rate is given by

$$\text{Rate}=\frac{\text{interest}}{\text{principal}}$$

The rate of interest is given in percentage value and is usually quoted on an annual basis. The monthly rate is calculated by dividing the annual rate by 12, the quarterly rate by dividing the annual rate by 4, and the daily rate by dividing the annual rate by 360. Two basic forms of interest are simple interest and compound interest, which are different in the way they are computed.

Simple Interest

If the interest is computed on the original principal for the entire period, it is **simple interest** and is denoted by I_s. Simple interest is calculated by the formula

(9.5) $I_s = PRT$

where P is the original principal, R is the rate of interest, and T is the number of periods for which interest is paid (or earned).

EXAMPLE 1 Compute the interest on a \$10,000 note that is held for 5 yr at 8 percent simple interest paid quarterly.

SOLUTION The rate for each period is $R = 0.08/4 = 0.02$, since interest is paid quarterly. The number of periods is $T = 5 \cdot 4 = 20$, so

$$I_s = \$10{,}000 \times 0.02 \times 20 = \$4000$$

To compute simple interest, we could multiply the principal by the annual rate and by the number of years, to get the same total interest. In other words, aside from the convenience of receiving or paying interest quarterly, the amount of money earned or paid out is the same. Thus, the simple interest in Example 1 can also be calculated by

$$I_s = 10{,}000 \times 0.08 \times 5 = \$4000$$

Compound Interest

If the interest is allowed to accumulate so that further interest is computed not only on the original principal but also on the accumulated interest, then the difference between the original principal and the total amount is called the **compound interest**; it is denoted by I_c.

Suppose a principal P, is invested at an interest rate R for each period, compounded over n periods. The total amount A_1 at the end of the first interest period is

$$A_1 = P + RP = P(1+R)$$

At the end of the second period, the total amount is

$$A_2 = A_1 + RA_1 = P(1+R) + RP(1+R) = P(1+R)^2$$

At the end of the third period, the total amount is

$$A_3 = A_2 + RA_2 = P(1+R)^2 + RP(1+R)^2$$
$$= P(1+R)^3$$

Continuing this process for n interest periods, we have

(9.6) $A_n = P(1+R)^n$

This formula, which is the general term of a geometric sequence with first term equal to P and common ratio equal to $1+R$, is known as the **compound interest formula**. The compound interest earned or paid out for n periods is

(9.7) $I_c = A_n - P$

EXAMPLE 2 Compute the earned interest on an initial principal of \$10,000 if interest at the nominal annual rate of 8% is used to compound quarterly for 5 yr.

SOLUTION The quarterly interest rate is $0.08/4 = 0.02$, and $n = 5 \cdot 4 = 20$. By the compound interest formula,

$$A_{20} = 10{,}000(1+0.02)^{20}$$

(You'll need a calculator for this calculation.)

$$A_{20} = \$14{,}859.47$$

Hence, the interest earned is $A_{20} - P$, or

$$I_c = \$4859.47$$

By comparing this calculation with Example 1 of this section you can see the advantage (or disadvantage) of compounding.

Closely related to the concept of compound interest is present value. The **present value** of an amount A is the principal that must be invested currently to equal A after n interest periods. To determine present value, we solve for P in the compound interest formula:

(9.8) $P = (1+R)^{-n}A$

EXAMPLE 3 Find the present value of a note of \$10,000, due in 5 yr, held at 10% interest compounded monthly.

SOLUTION Here $R = 0.1/12$, $n = 60$, and $A = \$10{,}000$. Hence,

$$P = \left(1 + \frac{0.1}{12}\right)^{-60} \times 10{,}000$$
$$= \$6077.89$$

The Truth-in-Lending Act, which the U.S. Congress passed in 1974, requires financial institutions to quote an effective "annual percentage rate," or APR. A nominal annual rate of 6.5 percent, for example, is said to yield 6.81 percent annually. The APR is the ratio of interest earned (or paid) for one year to the principal at the beginning of the year, under the assumption that the principal and interest rate remain unchanged for that year. The APR is usually computed by assuming that the interest is compounded for 365 days at a rate of $R/360$, where R is the annual rate. Thus,

$$\text{APR} = \frac{P(1+R/360)^{365} - P}{P}$$

$$\textbf{(9.9)} \qquad \text{APR} = \left(1 + \frac{R}{360}\right)^{365} - 1$$

Using this formula, see if you can compute the APR's for each of the popular nominal annual rates listed in Table 9.1

TABLE 9.1

Nominal Rate	APR
5.5	5.73
6.5	6.81
7.0	7.35
7.5	7.90
8.0	8.45
9.5	10.10
10.0	10.67
10.5	11.23

Exercises for Section 9.5

In Exercises1–10, compute (a) the simple interest and (b) the compound interest for 1 yr if the compounding is done monthly.

1. \$3000 at 6%

2. \$500 at 8%

3. \$4000 at 6.5%

4. \$4200 at 8.5%

5. \$10,000 at 9.462%

6. \$10,000 at 11.012%

7. \$2000 at 8.432%

8. \$5000 at 9.423%

9. \$242.62 at 10.013%

10. \$12,500 at 11.326%

Determine present value for the amounts given in Exercises 11–15. Assume that each note is due in 1 yr.

11. \$10,000 at 9% compounded monthly.

12. \$10,000 at 9% compounded daily. (Assume that 365 days = 1 yr and the rate is $R/360$.)

13. \$500 at 7% compounded quarterly.

14. \$2000 at 8% compounded monthly.

15. \$2000 at 8% compounded daily. (See Exercise 12).

16. Using Formula (9.9), compute the APR's for the indicated nominal rates given in Table 9.1.

17. If \$500 is invested at 4.5% compounded annually, how much interest will be earned in 5 yr?

18. If a bank pays 5% interest compounded annually, how much interest is earned on \$2000 invested for 3 yr?

19. Suppose \$1000 is invested at 7% compounded semiannually. How much will be the amount of the investment at the end of 5 yr?

20. Mr. and Mrs. Profitt, at the birth of their son, invested \$100 for him at 8% compounded quarterly. When the son is 21 years old, how much will their investment have amounted to?

21. A man invests \$5000 at 6% compounded quarterly. How much interest will the man earn in 2 yr?

22. A principal of \$2000 is invested at 10% interest for 5 yr. Compare the annual to the semiannual method of compounding interest.

23. What amount should be invested at 8% compounded quarterly to assure a \$40,000 return in 20 yr?

24. What amount should be invested at 7% compounded semiannually to be worth \$25,000 in 10 yr?

25. What minimum rate of interest should a person receive if he wants his present principal of \$10,000 to double in 8 yr? Assume semiannual compounding.

26. What annual interest rate will double your investment in 10 yr if the interest is compounded quarterly?

9.6 Annuities and Mortgage Payments

One of the most popular methods of saving money is by making a sequence of equal periodic payments, called an **annuity**. The usual way of repaying a large loan (like the kind made for buying a house) is by making a similar sequence of equal periodic payments, called **mortgage payments**. Annuities and mortgage payments are closely related concepts, and in fact, one is the inverse of the other. The "value" of an annuity or the "monthly payment" of a mortgage is computed by a systematic application of the compound interest and present value formulas, along with the sum of the terms of a geometric sequence.

Annuities

The **value** of an annuity V is the sum of the principal and the compound interest that would come due if all stipulated payments should be made for the complete term of the annuity.* For example, an annuity for which monthly payments of \$100 are made for two years at 7 percent would have a quoted face value equal to the sum of each of the \$100 principals plus the accumulated compound interest. In this case, there are 24 "principals" to consider. The amount at the end of the two years for each payment is computed from the compound interest formula. Thus, the \$100 payment made at the end of the first month will earn interest for 23 months, the \$100 payment at the end of the second month will earn interest for 22 months, and so on. The scheme is shown clearly in Table 9.2.

The total value of the annuity V equals the sum of these 24 amounts; that is

$$V = \sum_{k=0}^{23} 100\left(1 + \frac{0.07}{12}\right)^k$$

*We assume that the period of compounding equals the frequency of the payments. Such an annuity is called a **simple** annuity.

TABLE 9.2

Time of Payment, Number of Months	*Value of payment plus interest after 2 yr*
1	$100\left(1+\frac{0.07}{12}\right)^{23}$
2	$100\left(1+\frac{0.07}{12}\right)^{22}$
3	$100\left(1+\frac{0.07}{12}\right)^{21}$
$\vdots$	$\vdots$
23	$100\left(1+\frac{0.07}{12}\right)^{1}$
24	$100\left(1+\frac{0.07}{12}\right)^{0}$

This sum is the sum of 24 terms of a geometric sequence with $a_1 = 100$ and $r = (1 + 0.07/12) = 1.0058333$. Therefore, using Formula (9.4), we have

$$V = 100\left(\frac{1 - 1.0058333^{24}}{1 - 1.0058333}\right) = \$2568.09$$

The generalization to annuities with n payment periods is easy. If R is the nominal annual interest rate and there are q payments a year of P dollars each, then the amount of an annuity at the end of n payment periods is given by

$$V = \sum_{k=0}^{n-1} P\left(1 + \frac{R}{q}\right)^k$$

This is evaluated using (9.4), so

$$\text{(9.10)} \quad V = P\,\frac{1-(1+R/q)^n}{1-(1+R/q)} = P\,\frac{(1+R/q)^n - 1}{R/q}$$

EXAMPLE 1 Find the amount of an annuity for which quarterly payments of \$150 are made for 5 yr at a nominal annual interest rate of 8%.

SOLUTION Using (9.10) with $P = \$150$, $R = 0.08$, $q = 4$, and $n = 5 \cdot 4 = 20$, we have

$$V = 150\,\frac{(1 + 0.08/4)^{20} - 1}{0.08/4}$$

$$= 150\,\frac{1.02^{20} - 1}{0.02} = \$3644.61$$

Mortgages

When a debt is paid off by a sequence of equal payments, the debt is said to be **amortized** by an annuity; the payments are called mortgage payments. The payments needed for amortizing a debt over a given period are determined from the concept of present value. The total debt equals the sum of the present values of each of the mortgage payments at the time the loan is made. Thus, if there are to be q payments a year of M dollars each and if there are to be n total payments, then, by (9.8), the debt D is

$$D = \sum_{k=1}^{n} M\left(1+\frac{R}{q}\right)^{-k}$$

where R is the nominal annual interest rate. To compute this sum, we use Equation (9.4), with $a_1 = M(1+R/q)^{-1}$ and $r=(1+R/q)^{-1}$, that is,

(9.11) $$D = M\frac{1-(1+R/q)^{-n}}{R/q}$$

To get the mortgage payments corresponding to this debt, we solve Equation (9.11) for M. Thus, the payment on a loan of D dollars is

(9.12) $$M = \frac{D(R/q)}{1-(1+R/q)^{-n}}$$

EXAMPLE 2 Determine the monthly payments for a \$35,000 home loan to be amortized over 30 yr at 9.5%.

SOLUTION Using (9.12) with $D = 35{,}000$, $R = 0.095$, $q = 12$, and $n = 12(30) = 360$, we have

$$M = \frac{35{,}000(0.095/12)}{1-(1+0.095/12)^{-360}} = \$294.30 \text{ per month}$$

EXAMPLE 3 A debt of \$10,000 with interest at 12% is amortized by quarterly payments over 2 yr. Determine the quarterly payments and construct an amortization schedule.

SOLUTION The quarterly payments are calculated using (9.12) with $D = 10{,}000$, $R = 0.12$, $q = 4$,

TABLE 9.3

Payment Number	*Principal at beginning of each payment period* P	*Interest due at end of period* $I = 0.03 \times P$	*Payment Amount* M	*Amount Paid on Principal* $M - I$
1	10,000.00	300.00	1424.56	1124.56
2	8875.44	266.26	1424.56	1158.30
3	7717.14	231.51	1424.56	1193.05
4	6524.09	195.72	1424.56	1228.84
5	5295.25	158.86	1424.56	1265.70
6	4029.55	120.89	1424.56	1303.67
7	2725.88	81.78	1424.56	1342.78
8	1383.10	41.49	1424.59[a]	1383.10

Note: All figures in last four columns are dollar amounts.

[a]The last payment was 3 cents higher than the others so that the debt could be paid off in 8 payments.

and $n=4(2)=8$. Thus,

$$M=\frac{10{,}000(0.12/4)}{1-(1+0.12/4)^{-8}}=\$1424.56 \text{ per quarter}$$

Finally, the amortization schedule is constructed. The principal is reduced each payment period by the amount of the payment that exceeds the interest; that is, if P_n is the principal at the beginning of a payment period, then the principal for the next payment period is

$$P_{n+1}=P_n-(M-I_n)$$

where M is the periodic payment and I_n is the interest amount for that period. The interest amount for a given period is

$$I_n=P_n\left(\frac{R}{q}\right)$$

Exercises for Section 9.6

In Exercises 1–10, find the value of the given annuity.

1. \$100 monthly for 5 years at 10%.

2. \$500 quarterly for 10 years at 9%.

3. \$5 a day for 5 yr at 10%.

4. \$360 semiannually for 20 yr at 8%.

5. \$60 monthly for 20 yr at 8%.

6. \$2 daily for 20 yr at 8%.

7. \$100 monthly for 10 yr at 10%.

8. \$600 semiannually for 10 yr at 10%.

9. \$1200 annually for 10 yr at 10%.

10. \$600 annually for 20 yr at 10%.

In Exercises 11–20 determine the monthly payment for the given home loan.

11. \$20,000 at 8% for 20 yr.

12. \$20,000 at 8% for 30 yr.

13. \$20,000 at 9% for 20 yr.

14. \$20,000 at 10% for 20 yr.

15. \$25,000 at 9% for 20 yr.

16. \$25,000 at 10% for 20 yr.

17. \$30,000 at 10% for 20 yr.

18. \$35,000 at 10% for 20 yr.

19. \$35,000 at 10% for 25 yr.

20. \$35,000 at 10% for 30 yr.

21. A debt of \$2000 with interest at 10% nominal annual interest rate is amortized by monthly payments over 1 yr. Determine the equal monthly payments and make an amortization schedule.

22. Make an amortization schedule for retiring a debt of \$3000 with 12% nominal annual interest if payments are to be quarterly, for 2 yr.

23. A woman invests \$300 a year in an ordinary annuity that pays 6% compounded annually. What is the value of the annuity at the end of 20 yr?

24. If \$500 is paid each year into an ordinary annuity that pays 5% compounded annually, what is the value of the annuity at the end of 30 yr?

25. A company proposes to build a plant that will cost \$400,000. How much money must the company invest annually at 6% so that it will have \$400,000 at the end of 10 yr?

26. How much should you invest quarterly at 6.5% annual interest to be sure that you can buy a new car in 5 yr (at approximately \$6000) without buying on credit.

27. If you want enough money for a \$10,000 down payment on a home in 5 yr, how much should you invest each month until then at 6.75% annual interest compounded monthly?

9.7 The General Power of a Binomial

The expanded product of two binomials follows a special pattern. You have already been asked to memorize the expansions of the special products of $(x+y)^2$ and $(x+y)^3$ since they are so frequently used. Now we want to show that the general expansion of $(x+y)^n$ follows a recognizable and important formula that will allow you to expand any binomial easily and also determine any particular term within that expansion.

By performing the multiplications for $n=1$ through 5, we get

$$(x+y)^1=x+y$$
$$(x+y)^2=x^2+2xy+y^2$$
$$(x+y)^3=x^3+3x^2y+3xy^2+y^3$$
$$(x+y)^4=x^4+4x^3y+6x^2y^2+4xy^3+y^4$$
$$(x+y)^5=x^5+5x^4y+10x^3y^2+10x^2y^3+5xy^4+y^5$$

Here are some pertinent observations:

(1) There are always $(n+1)$ terms; the first is x^ny^0 and the last is x^0y^n.

(2) Each intermediate term contains a product of x^jy^k, where the sum of j and k in every term is n. Thus, the product is of the form $x^{n-k}y^k$.

(3) The exponents of x decrease by 1 from one term to the next, whereas the exponent associated with y increases by 1.

(4) The coefficients follow a symmetric pattern; that is, if we follow the coefficients from either end of the expansion, we find that they match the coefficients followed from the other end in the reverse direction.

(5) It is difficult to guess the actual formula for the coefficients with such limited information. However, the coefficient of the second term is $n/1$, the coefficient of the third term is $n(n-1)/2$, the coefficient of the fourth term is $n(n-1)(n-2)/1\cdot 2\cdot 3$.

In a more formal treatment of taking the general power of a binomial, we should show that the coefficient of x^jy^k in the expansion of $(x+y)^n$ is given by

(9.13)
$$C_{n,k}=\frac{n(n-1)(n-2)\cdots(n-k+1)}{1\cdot 2\cdot 3\cdot 4\cdots k} \qquad \text{if } k\neq 0$$
$$=1 \qquad \text{if } k=0$$

Formula (9.13) is known as the **binomial coefficient**. It depends not only on the order of the expansion n, but also on the number of the term within that expansion, $(k+1)$. Thus, the general expansion of a binomial consists of terms of the form $C_{n,k}x^{n-k}y^k$, where $k=0, 1, 2, \ldots, n$.

The binomial coefficient $C_{n,k}$ can be considerably shortened, for greater ease in learning, by **factorial notation**, a mathematical abbreviation for the product of integers 1 through n. The symbol $n!$ is read "n factorial" and is defined to be

(9.14) $$n!=n(n-1)(n-2)(n-3)\cdots 3\cdot 2\cdot 1$$

The number 0! does not quite fit in this definition, so to make most mathematical formulas read more simply, **we define 0! to be 1.**

EXAMPLE 1

$$5! = 5 \cdot 4 \cdot 3 \cdot 2 \cdot 1 = 120$$

$$10! = 10 \cdot 9 \cdot 8 \cdot 7 \cdot 6 \cdot 5! = 30{,}240 \cdot 120 = 3{,}628{,}800$$

$$\frac{10!}{5!} = \frac{10 \cdot 9 \cdot 8 \cdot 7 \cdot 6 \cdot 5!}{5!} = 10 \cdot 9 \cdot 8 \cdot 7 \cdot 6 = 30{,}240$$

Most modern calculators can rapidly compute the product given by a factorial. With a calculator, you can, for example, get $60! = 8.321 \times 10^{81}$.

Here is how the factorial symbol simplifies the form of $C_{n,k}$. Since the numerator is the product of integers that begin at n and end at $(n-k+1)$, it can be written as $n!/(n-k)!$ The denominator is the product of integers from k to 1 and hence, is simply $k!$ Thus, the binomial coefficient is given by

$$\textbf{(9.15)} \quad C_{n,k} = \frac{n!}{(n-k)!k!}$$

The general term of the expansion of $(x+y)^n$ is then

$$\frac{n!}{(n-k)!k!} x^{n-k} y^k$$

The numerator is the factorial of the power of the binomial, and the denominator is the product of the factorials of each of the exponents of the $(k+1)$th term of the expansion. Thus, the sixth term (that is, $k=5$) of the binomial $(x+y)^9$ is

$$\frac{9!}{4!5!} x^4 y^5 = 126 x^4 y^5$$

EXAMPLE 2 Find the coefficient of $x^9 y^7$ in the expansion of $(2x^3 - y)^{10}$.

SOLUTION Let A be the coefficient of $x^9 y^7$; then the desired term of the expansion is

$$A x^9 y^7 = C_{10,7}(2x^3)^3(-y)^7 = C_{10,7}(2)^3(-1)^7 x^9 y^7$$

Therefore, the desired coefficient is

$$A = -8 \cdot C_{10,7} = -8 \frac{10!}{7!3!} = -8 \frac{10 \cdot 9 \cdot 8 \cdot 7!}{7!3!} = -960$$

By writing the values of $C_{n,k}$ for $k = 0, 1, 2, 3, \ldots, n$ on successive rows, as shown below, we get a triangular array of numbers, called **Pascal's triangle**.* Pascal's

*Blaise Pascal (1623–1662) was a prominent French mathematician and physicist.

triangle is perhaps the best device available for remembering the binomial coefficients, since a recognizable pattern exists between numbers in successive rows of the triangle. Each number interior to the 1s of a row is found by adding the two numbers directly above it, as indicated by the arrows. For example, the interior numbers of the fifth row are $4=1+3$, $6=3+3$, and $4=3+1$. This observation suggests Formula (9.16):

(9.16) $C_{n+1,k+1}=C_{n,k+1}+C_{n,k}$

$$
\begin{array}{llccccccccccccc}
C_{0,k} & \dashrightarrow & & & & & & & 1 & & & & & & \\
C_{1,k} & \dashrightarrow & & & & & & 1 & & 1 & & & & & \\
C_{2,k} & \dashrightarrow & & & & & 1 & & 2 & & 1 & & & & \\
C_{3,k} & \dashrightarrow & & & & 1 & & 3 & & 3 & & 1 & & & \\
C_{4,k} & \dashrightarrow & & & 1 & & 4 & & 6 & & 4 & & 1 & & \\
C_{5,k} & \dashrightarrow & & 1 & & 5 & & 10 & & 10 & & 5 & & 1 & \\
C_{6,k} & \dashrightarrow & 1 & & 6 & & 15 & & 20 & & 15 & & 6 & & 1
\end{array}
$$

EXAMPLE 3 By using the first two terms of the binomial expansion, approximate $(1.01)^{50}$.

SOLUTION Letting $x=1$ and $y=0.01$ in the binomial, we have

$$
\begin{aligned}
(1.01)^{50}=(1+0.01)^{50} &= 1^{50}+50(1^{49})(0.01) \\
&= 1+50(0.01)=1.5
\end{aligned}
$$

In fact, we have shown that $(1.01)^{50}>1.5$.

We have arrived at the results of this section intuitively. They can all be formally proved. Many textbooks choose to prove these results under the heading **binomial theorem**. In calculus, we show that this theorem can be generalized to include expansions of $(1+x)^n$, where n is not an integer and $|x|<1$. This generalized binomial expansion continues without end, but very often we are interested in only the first few terms. When the generalized form of the binomial theorem is used, the binomial coefficient cannot be computed from Equation (9.15), since the factorial of a fraction is undefined. Equation (9.13) must be used instead.

EXAMPLE 4 Using the first three terms of the generalized binomial expansion, approximate the value of $\sqrt[3]{5}$.

SOLUTION

$$\sqrt[3]{5}=(5)^{1/3}=(8-3)^{1/3}=8^{1/3}\left(1-\frac{3}{8}\right)^{1/3}=2\left(1-\frac{3}{8}\right)^{1/3}=2(1-0.375)^{1/3}$$

Using the first three terms of the generalized binomial expansion for $(1-.375)^{1/3}$ gives

$$
\begin{aligned}
\sqrt[3]{5}=2(1-0.375)^{1/3} &= 2\left[1+\frac{1}{3}(-0.375)+\frac{(\frac{1}{3})(-\frac{2}{3})}{1\cdot 2}(-0.375)^2\right] \\
&= 2[1-0.125-0.015625]=1.7188
\end{aligned}
$$

Exercises for Section 9.7

Expand the simplify Exercises 1–16, using the binomial formula.

1. $(x-y)^6$
2. $(x+7y)^4$
3. $(3x-y)^5$
4. $(x-0.1)^5$
5. $\left(x+\frac{1}{x}\right)^5$
6. $\left(x+\frac{1}{3x^2}\right)^4$
7. $(y^2+3x)^3$
8. $(\sqrt{x}+x)^5$
9. $(\sqrt{x}+\sqrt{y})^6$
10. $\left(yx^2+\frac{2}{y}\right)^4$
11. $(x^{1/3}+x^{-3/4})^4$
12. $(2y^{1/5}-3z^{1/3})^4$
13. $(2+0.1)^4$
14. $(x^{-1}+y^{-1})^5$
15. $(x^{-3/4}-x^{3/4})^3$
16. $(2x^5+5y^{-3})^4$
17. Find the 4th term in the expansion of $(\sqrt{x}+\sqrt{y})^{10}$.
18. Find the 8th term in the expansion of $(1+0.1)^{20}$.
19. Find the 12th term in the expansion of $(3x^3-7y^{1/3})^{20}$.
20. Find the 15th term in the expansion of $(2y^3-3x^4)^{20}$.
21. Find the coefficient of $x^{12}y^2$ in the expansion of $(x^3-2y^2)^5$
22. Find the coefficient of $(x/y)^5$ in the expansion of $(2x+3y^{-1})^{10}$.
23. Find the coefficient of the constant term in the expansion of $[x+(1/3x)]^8$.
24. Find the coefficient of the x term in the expansion of $(2x^{1/3}-7)^9$.

Using the first two terms of an appropriate binomial expansion, approximate Exercises 25–30.

25. $(2.3)^6$
26. $(1.99)^5$
27. $(3.99)^3$
28. $(0.99)^{100}$
29. $(1.01)^{500}$
30. 99^4

Using the first three terms of the generalized binomial expansion, approximate Exercises 31–40.

31. $\sqrt{24}$
32. $\sqrt{13}$
33. $\sqrt[3]{7}$
34. $\sqrt{66}$
35. $\sqrt[3]{30}$
36. $\sqrt{5}$
37. $\sqrt[4]{80}$
38. $\sqrt[5]{31}$
39. $\sqrt{98}$
40. $\sqrt[3]{120}$

Use a calculator to compute Exercises 41–44.

41. $40!/30!$
42. $50!$
43. $(35-5)!$
44. $54!$
45. Using the formula for $C_{n,k}$, show that $\sum_{k=0}^{n} C_{n,k}=2^n$. Notice that this formula represents the sum of the numbers in the nth row of Pascal's triangle.
46. Using the formula for $C_{n,k}$, show that $C_{n+1,k+1}=C_{n,k+1}+C_{n,k}$.

9.8 Mathematical Induction

Proving a mathematical generalization from a given set of hypotheses and axioms is called **deduction**. You are probably most familiar with this process from your plane geometry course. Another way of proving a generalization is to observe a specific event and then assume that these observations are generally true. Most scientific laws are established by this process of **incomplete induction**. Although incomplete induction is adequate for the physical sciences, it fails for mathematics.

To see why, consider this generalization: n^2-n+41 is prime if n is a positive integer. For the first 40 positive integers, this formula yields a prime number. The formula seems to work well until this point. In continuing the process, however, you will find that it fails when $n=41$. Therefore, we *cannot* generalize in mathematics using incomplete induction, since the "next" case may be the one that fails.

EXAMPLE 1 Reasoning by example (experimentally), we might conclude that all integers are less than 1,000,000,000—unless we just happen to use an example greater than 1,000,000,000.

In mathematics, a process called complete induction, or **mathematical induction**, goes from the particular to the general without the obvious problems of incomplete induction. Mathematical induction is a precise procedure for proving the validity of a generalization about some sort of sequential statement.

EXAMPLE 2 We should use mathematical induction to *prove*, not derive, the formula

$$\sum_{k=1}^{n} k^2 = \frac{1}{6}n(n+1)(2n+1)$$

Mathematical induction consists of two parts.

(1) **Verification**. The given proposition must be shown to be true for the smallest integral value of n for which the proposition holds. (Ordinarily, this step is carried out for many more than just the one case, since showing the statement true for several cases will at least make the given proposition plausible.)

(2) **Implication**. Demonstrate that if the proposition is true for $n=k$, then it must hold for $n=k+1$. The assumption that the proposition is true for $n=k$ is sometimes called the *inductive hypothesis*.

If the conditions of both the verification and implication steps have been met, then the statement is true for all positive integers greater than or equal to that number used in the verification step.

The logic behind a proof by mathematical induction is similar to knocking over a row of dominoes by pushing the first one and letting each domino knock over the one immediately following it. (See Figure 9.3.) If the first domino cannot be knocked over or if the dominoes are not close enough (that is, the kth one will not push over the $(k+1)$st), then all the dominoes will not fall.

FIGURE 9.3 All dominos will fall in this scheme. Only one domino will fall in this scheme.

EXAMPLE 3 Prove by mathematical induction that the sum of the first n integers is given by the formula $S_n = n(n+1)/2$.

SOLUTION

(1) Verification. Verify that when $k=1$, then $S_1 = 1$. Substituting 1 in the proposed formula, we get $S_1 = 1(1+1)/2 = 1$.

(2) Implication. We assume the formula true for $n=k$; that is, $S_k = k(k+1)/2$. Then we must show that $S_{k+1} = (k+1)(k+2)/2$. By definition,

$$S_{k+1} = S_k + (k+1)$$

By the inductive hypothesis,

$$S_{k+1} = \frac{k(k+1)}{2} + (k+1) = \frac{k^2+k}{2} + \frac{2k+2}{2} = \frac{k^2+3k+2}{2}$$

$$= \frac{(k+1)(k+2)}{2}$$

Since both steps have been validated, the given statement is true for all positive integers.

Both steps in mathematical induction are absolutely necessary, as emphasized in Examples 4 and 5.

EXAMPLE 4 Consider the "formula" that the sum of the first n integers is given by $A_n = n^2 - n + 1$. That $A_1 = 1$ verifies the formula for $k=1$. Assuming that $A_k = k^2 - k + 1$, we must show that $A_{k+1} = (k+1)^2 - (k+1) + 1 = k^2 + k + 1$. However, by definition,

$$A_{k+1} = A_k + (k+1) = (k^2 - k + 1) + (k+1) = k^2 + 2$$

and since $k^2 + k + 1 \neq k^2 + 2$ (except for $k=1$), the implication step is unproved.

EXAMPLE 5 Suppose that we try to show that the sum of the first n integers is given by the formula $n(n+1)/2 + 1$. Then the implication step is to assume that $B_k = k(k+1)/2 + 1$ and to try to show that $B_{k+1} = (k+1)(k+2)/2 + 1$. By definition,

$$B_{k+1} = B_k + (k+1) = \frac{k(k+1)}{2} + 1 + (k+1)$$

$$= \frac{(k+1)(k+2)}{2} + 1$$

which proves this step. However, $B_1 = 2$, which is *not* the first integer. In fact, B_n does not represent the sum for any value of n. This example shows how important it is to check both parts of the induction.

EXAMPLE 6 Show that if n is a positive integer and $x \neq y$, then $x^n - y^n$ is always divisible by $x - y$.

SOLUTION Since $x^n - y^n = x - y$ when $n=1$, the verification step is trivial. Now assume that $x^n - y^n$ is divisible by $x-y$ if $n=k$. Thus, assume that

$$\frac{x^k - y^k}{x-y} = q(x)$$

where the division yields no remainder. Then, adding and subtracting xy^k to the numerator, we have

$$\begin{aligned}\frac{x^{k+1} - y^{k+1}}{x-y} &= \frac{x^{k+1} - xy^k + xy^k - y^{k+1}}{x-y} \\ &= \frac{x(x^k - y^k) + y^k(x-y)}{x-y} \\ &= \frac{x(x^k - y^k)}{x-y} + y^k \\ &= xq(x) + y^k\end{aligned}$$

which means that for $n=k+1$, the division does not yield a remainder. Since verification and implication have been validated, we conclude that $x^n - y^n$ is divisible by $x-y$ for all n.

Exercises for Section 9.8

Using mathematical induction, prove the statements in Exercises 1–13 valid for all $n \geqslant 1$.

1. $\sum_{k=1}^{n} 2k = n(n+1)$
2. $\sum_{k=1}^{n} 2^k = 2^{n+1} - 2$
3. $\sum_{k=1}^{n} (3k-2) = \frac{n(3n-1)}{2}$
4. $\sum_{k=1}^{n} (3k-1) = \frac{n(3n+1)}{2}$
5. $\sum_{k=1}^{n} k(k+1) = \frac{n(n+1)(n+2)}{3}$
6. $\sum_{k=1}^{n} (2k-1) = n^2$
7. $\sum_{k=1}^{n} k^2 = \frac{n(n+1)(2n+1)}{6}$
8. $\sum_{k=1}^{n} k^3 = \frac{n^2(n+1)^2}{4}$
9. $\sum_{k=1}^{n} r^{k-1} = \frac{1-r^n}{1-r}, \quad r \neq 1$
10. $\sum_{k=1}^{n} \frac{1}{k(k+1)} = \frac{n}{n+1}$
11. $n(n+1)(n+2)$ is a multiple of 6 for all $n \geqslant 1$.
12. $x^{2n-1} + y^{2n-1}$ is divisible by $x+y$ for all $n \geqslant 1$.
13. $\sum_{k=1}^{n+1} \frac{1}{n+k} \leqslant \frac{5}{6}$

Test 1 for Chapter 9

Answer *True* or *False* to Exercises 1–10.

1. No term of the sequence whose nth term is $n/(n+1)$ ever exceeds 1.
2. $\sum_{k=2}^{2} 2^k = 2$
3. If $\sum_{k=1}^{n} a_k = \frac{n}{n+1}$, then $a_n = \frac{1}{n(n+1)}$.
4. For any sequence, $\sum_{k=1}^{n} a_k < \sum_{k=1}^{n+1} a_k$.
5. $\sum_{k=1}^{20} i^k = 0, \ (i = \sqrt{-1})$.

6. $(2n)!/n! = 2!$

7. The constant term in the expansion of $[x-(1/x)]^{10}$ is $C_{10,5}$

8. The integers 2, 5, 8, 11 are the first four terms of an arithmetic sequence.

9. Mathematical induction can be used to prove $\sum_{k=1}^{n} k = n(n+1)/2$.

10. $\sum_{k=2}^{10} (k+1)(k+2) = \sum_{k=0}^{8} (k+3)(k+4)$.

11. Consider the sequence 3, 9, 15, 21, ... (a) Find a formula for the general term. (b) Find the sum of the first 60 terms.

12. Prove, by mathematical induction, that

$$\sum_{k=1}^{n} (2k+1) = n(n+2).$$

13. By using the first two terms of the generalized binomial expansion, approximate $(1.01)^{100}$.

14. Find and simplify the coefficient of $x^{12}y^6$ in the expansion of $(2x^2-3y)^{12}$.

15. Find the amount of an investment of \$4500 at the end of 5 yr at 8 percent compounded quarterly.

Test 2 for Chapter 9

1. Find a formula for the general term of the sequence 1, 3, 5, 7, 9, ...

2. Write in expanded form and evaluate $\Sigma_{k=1}^{4} k(k+2)$.

3. Write the general term and find the sum of the first 99 terms of 2, 4, 6, 8, ...

4. Write the general term and write the sum of the first 10 terms of $\frac{2}{3}, \frac{2}{9}, \frac{2}{27}, \ldots$.

5. Use the binomial theorem to expand $(x+5)^6$.

6. Use the first two terms of the binomial expansion to approximate $\sqrt{7}$.

7. Find the fourth term in the expansion of $(x^2-2y)^7$.

8. Compute the compound interest on \$10,000 if it is held for 3 yr at 9% interest paid semiannually. Compute the simple interest for this period.

9. Find the value of an annuity for which quarterly payments of \$200 are made for 10 yr at an annual rate of 7%.

10. Use mathematical induction to prove $4+7+10+\cdots+(3n+1) = \frac{1}{2}n(3n+5)$.

10 Counting and Probability

10.1 The Fundamental Principal of Counting

If a coin is tossed into the air and allowed to hit the ground, there is a 50/50 chance that it will turn up heads. The theory of chance events (like flipping a coin) is studied in a branch of mathematics called probability. The basic concepts of probability depend on our ability to count accurately the number of possible outcomes for a given event.

To illustrate the counting process, suppose that you want to choose a car that is available in orange, yellow, or blue exterior paint, with or without air-conditioning and with a radio option that includes AM, AM/FM, or AM/FM/tape. How many different choices of car could you make? The choices can be displayed graphically in a "tree" diagram, as shown in Figure 10.1. Starting at point C, we represent the three color options by lines, or branches of our tree. The next level of branches represents the air-conditioning option, and finally, the last level of branches represents the radio option. Each distinct path represents a different car, and you can see that there are clearly 18 cars from which to choose.

Tree diagrams can be used to count outcomes of events whenever the events can be considered a succession of two or more independent subevents. Thus, choosing a new car can be thought of as successive events of choosing color, air-conditioning, and radio. Similarly, tossing two coins can be considered tossing one coin twice, and drawing a poker hand of five cards can be regarded as successively drawing five cards in a row from the deck.

EXAMPLE 1 Draw a tree diagram showing the number of possible outcomes of tossing three coins.

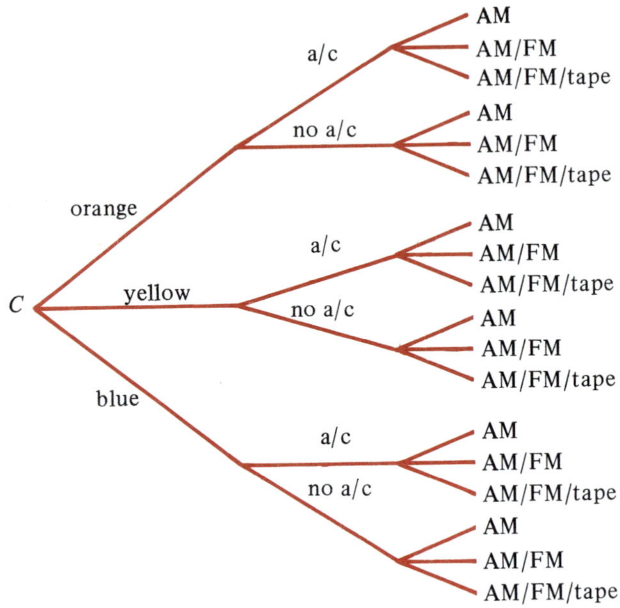

FIGURE 10.1

SOLUTION Since tossing three coins at the same time can be considered tossing the same coin three times, it is easy to draw the tree diagram shown in Figure 10.2.

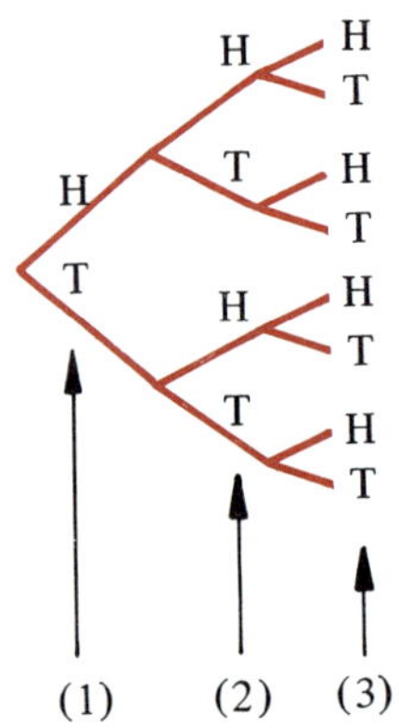

FIGURE 10.2

The tree diagrams can be used to count possible outcomes of events like those described in Figures 10.1 and 10.2, but the process is cumbersome at best. The following principle allows us to determine the number of outcomes of an event more easily than with tree diagrams.

The Fundamental Principle of Counting If one event can occur in h ways, and if after this, a second event can occur in k ways, then the two events can occur together in $h \cdot k$ ways.

The principle is stated for two events but can be generalized to include any finite number of successive events. Although the fundamental principle of counting can be used to find the number of possible outcomes of successive events, it does not tell you anything about the form of the outcomes.

EXAMPLE 2 Use the fundamental principle of counting to determine the number of different cars that can be produced from three exterior colors, air-conditioning or no air-conditioning, and three types of radios.

SOLUTION Each different car consists of a color, air-conditioning or no air-conditioning, and a radio. Since there are three color options, two air-conditioning options, and three radio options, the number of different cars is given by $3 \cdot 2 \cdot 3 = 18$. This agrees with the number indicated in Figure 10.1.

EXAMPLE 3 In how many different patterns can you answer a true-false test consisting of ten questions?

SOLUTION Assuming each answer to be independent of the others, there are two possible choices for each answer. Hence, the total is

$$2 \cdot 2 \cdot 2 \cdot 2 \cdot 2 \cdot 2 \cdot 2 \cdot 2 \cdot 2 \cdot 2 = 2^{10} = 1024$$

EXAMPLE 4 If 9 horses enter a race, in how many ways can the first 3 places be won?

SOLUTION There are 9 possible winning horses, after which 8 horses could finish in second place, and finally 7 who could finish third. By the fundamental counting principle, the number of possible ways for the top 3 to be chosen is

$$9 \cdot 8 \cdot 7 = 504$$

EXAMPLE 5 How many different five-letter "words" can be written with the letters a, b, c, d, and e if none of the letters is to be repeated? A "word" is any arrangement of letters.

SOLUTION There are 5 successive independent decisions. The first letter can be chosen in 5 ways, after which 4 letters remain. Hence, the second choice can be made in one of 4 ways. Continuing in this manner, we get as the total number

$$5 \cdot 4 \cdot 3 \cdot 2 \cdot 1 = 5! = 120$$

Example 5 is an important application of the fundamental principle, which can be generalized. The question to answer is, In how many ways can we arrange n distinct objects in a definite order? Since there are n choices for the first object, $n-1$ for the second, $n-2$ for the third, and so on, the total number of arrangements, or permutations, is given by the product $n(n-1)(n-2) \cdots 3 \cdot 2 \cdot 1 = n!$

EXAMPLE 6 How many ways can nine players on a baseball team be arranged on a ball field if each player can play every position?

SOLUTION There are 9 successive independent decisions. The first player can be positioned in 9 ways, the second player in 8 ways, the third player in 7 ways, and so on. Thus, the total number of arrangements is

$$9 \cdot 8 \cdot 7 \cdot 6 \cdot 5 \cdot 4 \cdot 3 \cdot 2 \cdot 1 = 9! = 362{,}880$$

Another typical counting problem is determining the number of groups of k elements that can be formed from a set of n objects. Suppose, for example, you want to form a committee of 2 from a group of 5 people. In how many ways can this be done? By the fundamental counting principle, there are 5 ways to pick the first member and 4 ways to pick the second to give 20 possible committees. But by this procedure, 2 committees are held different even when they have the same members.

To clarify this, let A, B, C, D, and E represent the 5 persons. Then the 20 committees are AB, AC, AD, AE, BA, BC, BD, BE, CA, CB, CD, CE, DA, DB, DC, DE, EA, EB, EC, and ED. Note that AB and BA are two distinct committees even though they have the same members. Such a distinction is often necessary, as when the first designated member is the chairman of the committee. If such a distinction is unnecessary, then there are only 10 different committees.

These two simple methods of forming committees are typical of how collections are arranged, that is, arrangements of objects in which the order of the objects is important and arrangements in which it is not. Definitions 1 and 2 formalize these two processes.

DEFINITION 1

> Any collection of objects in which the selection or arrangement has been made in a definite order is a **permutation**.

DEFINITION 2

> Any collection of objects in which the selection or arrangement has been made without regard to order is a **combination**.

EXAMPLE 7 Given the letters w, x, y, z:

(a) List the permutations of these letters taken three at a time.

(b) List the combinations of these letters taken three at a time.

SOLUTION

(a) The permutations are wxy, wxz, wyx, wyz, wzy, wzx, xwy, xyw, xyz, xzy, xzw, xwz, yxz, yzx, yzw, ywz, yxw, ywx, zxy, zyx, zxw, zwx, zyw, zwy. There are 24 permutations in all.

(b) The combinations are wxy, wxz, wyz, and xyz. There are thus four combinations in all.

10.2 Permutations

By the number of permutations of n distinct objects taken k at a time, we mean all possible distinct arrangements consisting of k objects. This number is denoted by $P_{n,k}$. To find a formula for $P_{n,k}$, note that the first object can be chosen from any one of n objects, the second from any one of the $n-1$ remaining objects, the third from any one of $n-2$ remaining objects after the second object has been chosen, and so on. Finally, the kth object is chosen from the $n-(k-1)$ objects that remain after the

$(k-1)$th object has been chosen. Using the fundamental counting principle, we get

$$P_{n,k}=\underbrace{n(n-1)(n-2)\cdots(n-k+1)}_{k\text{ numbers}}$$

This formula is usually modified by multiplying the right-hand side by the expression $(n-k)!/(n-k)!$. Thus,

$$\frac{n(n-1)(n-2)\cdots(n-k+1)(n-k)!}{(n-k)!}=\frac{n!}{(n-k)!}$$

Therefore, the number of permutations of n objects taken k at a time is given by

(10.1) $$P_{n,k}=\frac{n!}{(n-k)!}$$

In the special case in which $k=n$ (that is, the number of ways n objects can be arranged in a definite order), we have, since $0!=1$,

(10.2) $$P_{n,n}=n!$$

This result agrees with our conclusion in Section 10.1.

EXAMPLE 1 If there are 15 entries in an art contest, in how many ways can the judge award the first, second, and third prizes?

SOLUTION Since each entry and prize is distinct, we want the number of permutations of 15 things taken 3 at a time. Thus,

$$P_{15,3}=\frac{15!}{(15-3)!}=15\cdot 14\cdot 13=2730$$

EXAMPLE 2 The number of ways to deal 5 cards from a pack of 52 cards is

$$\frac{52!}{(52-5)!}=52\cdot 51\cdot 50\cdot 49\cdot 48=311{,}875{,}200$$

This number is *not* the number of distinct poker hands, since a poker player holding a hand of 5 cards ordinarily disregards the order. Our problem did not ask for the number of distinct 5-card hands; it did ask for the number of ways of dealing 5 cards, and for that number, order is a consideration.

Exercises for Sections 10.1 and 10.2

1. If there are seven boys in a race, how many ways can the second-place finish be filled once first place has been decided? How many ways can third place be filled once the first and second places have been decided?

2. List all permutations of A, B, C, D, taken three at a time. List all combinations.

3. List all permutations of the letters w, x, y, z, taken two at a time.

4. How many permutations are there of the letters in the word ANGLE?

5. Find the number of permutations of seven objects taken four at a time.

6. Evaluate $P_{9,7}$.

7. How many two-digit numbers can be formed from the numbers 1, 2, 3, 4, 5, 6 if no digits can be repeated?

8. If five women enter a 10-km run, in how many ways can they finish first, second and third?

9. In how many ways can ten people line up at a box office?

10. In how many ways can four political candidates be listed on the ballot?

11. How many batting orders can a baseball manager make up from a roster of 15 players.

12. Four persons enter a room in which there are seven chairs. In how many ways can they be seated? (Assume one chair per person.)

13. Find the total number of four-digit integers that can be formed from the digits 1, 2, 3, and 4. Assume that no digit is repeated in any one integer.

14. How many Greek-letter sorority names of two or three letters can be formed with the 24-letter Greek alphabet if letters are not repeated?

15. How many three-letter or four-letter words can be made with the letters of the word SLOPE using letters no more than once in each word? Consider a word an arrangement of letters, not necessarily an arrangement that makes sense.

16. In how many ways can nine politicians be arranged at a round table? *Hint*: Assume one position fixed and then arrange the remaining politicians.

17. In how many ways can eight cheerleaders be arranged to form a circle?

18. Given that $P_{n,2}=56$, find n.

19. Given $P_{n,3}=9P_{n,2}$, find n.

20. Show that $P_{20,2}=19P_{5,2}$.

21. Show that $\sum_{k=1}^{4} P_{4,k}=4^3$.

10.3 Combinations

Having considered choosing *ordered* arrangements of k objects from a larger set of n objects (or a set of the same size), we shall look at unordered sets having k elements chosen from the n objects. Recall that sets of this type, in which order of the objects is not considered, are called **combinations**. For the present, we denote the number of combinations of n objects taken k at a time by the symbol $N_{n,k}$. Instead of deriving the formula for $N_{n,k}$ directly from the basic definitions, we get this number indirectly by considering the number of permutations of the n objects taken k at a time, that is $P_{n,k}$.

The ordered selection of a set of k objects taken from n objects can be thought of as the following two-step procedure:

(1) Choose k objects from the n. This can be done in $N_{n,k}$ ways.

(2) Permute these k objects. This can be done in $P_{k,k}=k!$ ways.

Since the ordered selection of k objects from n objects is $P_{n,k}$, we have, from the fundamental counting principle,

$$P_{n,k}=N_{n,k}k!$$

Solving for $N_{n,k}$, we have

$$N_{n,k}=\frac{P_{n,k}}{k!}$$

or $$N_{n,k}=\frac{n!}{k!(n-k)!}$$

From (9.15), we get $N_{n,k}=C_{n,k}$; that is, *the number of combinations of n things taken k at a time is precisely the binomial coefficient*. Therefore, we write the formula for the number of combinations of n objects taken k at a time as

$$\textbf{(10.3)} \quad C_{n,k}=\frac{n!}{k!(n-k)!}$$

EXAMPLE 1 (a) The number of combinations taken 5 at a time from a set of 10 is $C_{10,5}=10!/(5!)(5!)=252$.

(b) The number of possible single bridge hands consisting of 13 cards is $C_{52,13}=52!/(13!)(39?)$.

EXAMPLE 2 How many committees of 5 can be chosen from 12 persons?

SOLUTION Order ordinarily does not count when committees are formed, so the problem is selecting combinations of 5 people from 12. That is,

$$C_{12,5}=\frac{12!}{5!7!}=\frac{12\cdot 11\cdot 10\cdot 9\cdot 8}{5\cdot 4\cdot 3\cdot 2\cdot 1}$$
$$=792$$

Now we can use the fundamental counting principle along with our knowledge of combinations to count the number of possible outcomes of a given event.

EXAMPLE 3 A baseball team has 3 catchers, 9 pitchers, 6 infielders, and 7 outfielders. In how many ways can a team of 9 be chosen?

SOLUTION The number of ways of choosing a catcher $=3$.

The number of ways of choosing a pitcher $=9$.

The number of ways of choosing an infielder $=C_{6,4}=15$

The number of ways of choosing an outfielder $=C_{7,3}=35$

Hence, by the fundamental counting principle, the total number of distinct teams $=3\cdot 9\cdot 15\cdot 35=14{,}175$.

Exercises for Section 10.3

In Exercises 1–4, evaluate each of the following:

1. $C_{4,2}$
2. $C_{5,3}$
3. $C_{7,2}$
4. $C_{9,5}$
5. How many combinations of 4 men can be selected from a roster of 15.
6. How many combinations of 3 can be selected from 12 different coins.
7. How many 5-card poker hands can be obtained from a deck of 52 cards?
8. How many ways can you get a flush (5 cards of the same suit) from a deck of 52 cards.
9. From a committee of 8, how many subcommittees of 3 can be selected?
10. Six coins are tossed onto a tabletop. In how many ways can 6 heads appear? 5 heads? 4 heads?
11. In a 10-team basketball league, how many games must be scheduled if each team plays one game with every other team in the league? How many games must be scheduled if each team plays 2 games with every other team?
12. A building has five doors. In how many ways can you enter the building and leave by a different door?
13. How many four-letter words ending in a vowel can be made from the word *vowel*?
14. How many committees of 4 men can be selected from 10 candidates if 2 of the candidates refuse to serve on a committee together?
15. In how many ways can a committee of 3 women and 2 men be chosen from 6 married couples?
16. In how many ways can a bowling team of 2 men and 2 women be selected from 8 men and 7 women?
17. A coin collector has 10 dimes and 8 quarters for sale. In how many ways can the buyer select 3 dimes and 4 quarters?
18. How many diagonals can be drawn in a 24-sided polygon?
19. Find x if $C_{x,3}=2C_{x,2}$.
20. If $P_{n,r}=120$ and $C_{n,r}=20$, what are n and r?

10.4 Probability

When a coin is flipped into the air, it is a certainty that the coin will fall. The outcome of this event can be predicted with certainty because of gravitational attraction, which is known to hold for all objects in the universe. On the other hand, we cannot state with certainty that the coin will land heads up. We know that the heads or tails landing is a matter of chance and not something that can be predicted with certainty by some law. Thus, when we flip a coin, we can be certain only that it will fall and that it will land in one of two ways, heads or tails. We usually say that there is a 50/50 chance that a coin will land heads up. The systematic study of chance events, such as flipping a coin, is called **probability**.

The probability of an event A is a number that we denote by $P(A)$. To determine $P(A)$, not only must we know the number of ways that event A can occur as the outcome of some experiment, but we must also know the number of ways in which the event A does not occur. The occurrence of A is referred to as a *success* and the nonoccurrence of A is referred to as a *failure*. Note that if s is the number of successful outcomes and f is the number of failures, then $s+f=n$ is the total number of outcomes of the operation. The probability of an event A is defined by

(10.4) $P(A)=\frac{s}{n}$

Thus to determine the probability of an event A:

(1) Determine the total number of possible outcomes of the event itself.

(2) Determine the number of possible successful outcomes.

(3) Use these numbers in Equation (10.4).

EXAMPLE 1 (a) Tossing a coin has two possible outcomes. The event "heads" is one of the two outcomes. Thus $P(\text{head})=\frac{1}{2}$.

(b) Throwing a die has six possible outcomes. The event "number less than three" can occur in two ways. Thus

$$P(\text{number less than } 3)=\frac{2}{6}=\frac{1}{3}$$

(c) Drawing a card from a deck has 52 possible outcomes. The event "ace" can occur in 4 different ways. Thus

$$P(\text{ace})=\frac{4}{52}=\frac{1}{13}$$

EXAMPLE 2 What is the probability of getting 3 heads in 5 tosses of a coin?

SOLUTION On each toss, the coin can land in 2 ways; so there are $2\cdot2\cdot2\cdot2\cdot2=2^5$ ways the coin can land in 5 tosses. The number of ways we can get 3 heads in 5 tosses is given by $C_{5,3}=5!/3!2!=10$. Hence,

$$P(3\text{ heads})=\frac{C_{5,3}}{2^5}=\frac{10}{32}=\frac{5}{16}$$

EXAMPLE 3 What is the probability that 3 balls drawn at random from a box of 5 red balls and 7 black balls will all be red?

SOLUTION The number of ways in which 3 red balls can be drawn from the 5 red balls in the box is $C_{5,3}$, and the number of ways 3 balls can be drawn from the 12 in the box is $C_{12,3}$. Therefore, the probability of a red ball is

$$P(\text{red})=\frac{C_{5,3}}{C_{12,3}}=\frac{\dfrac{5!}{3!2!}}{\dfrac{12!}{3!9!}}=\frac{10}{220}=\frac{1}{22}$$

EXAMPLE 4 What is the probability that 5 balls drawn at random from the box in Example 3 will consist of 2 red balls and 3 black balls?

SOLUTION The number of ways 2 red balls can be drawn from the 5 in the box is $C_{5,2}$ and the number of ways 3 black balls can be drawn from the 7 in the box is $C_{7,3}$. Therefore, by the counting principle, the number of ways 2 red balls and 3 black balls can be drawn is $C_{5,2} \cdot C_{7,3}$. Finally, the number of ways 5 balls can be drawn from the 12 in the box is $C_{12,5}$. Hence,

$$P(2\text{ red, }3\text{ black})=\frac{C_{5,2}\cdot C_{7,3}}{C_{12,5}}=\frac{\frac{5!}{2!3!}\ \frac{7!}{3!4!}}{\frac{12!}{5!7!}}=\frac{10\cdot 35}{792}=\frac{350}{792}=\frac{175}{396}$$

The probability of an event A cannot be less than 0 or greater than 1, that is

$$0 \leqslant P(A) \leqslant 1$$

If $P(A)=1$, the event, we say, is *certain*, and if $P(A)=0$, the event, we say, is *not possible*. (If the probability of A is $P(A)$, then the probability that A will not occur is $1-P(A)$.)

EXAMPLE 5 (a) When a coin is tossed, the probability of *not* getting heads is

$$P(\text{tail})=1-P(\text{head})=1-\frac{1}{2}=\frac{1}{2}$$

(b) The probability of *not* throwing a 5 on a roll of a die can be expressed by

$$P(\text{not }5)=1-P(5)=1-\frac{1}{6}=\frac{5}{6}$$

(c) The probability of *not* drawing an ace in drawing from a deck of 52 cards can be expressed by

$$P(\text{not an ace})=1-P(\text{ace})=1-\frac{1}{13}=\frac{12}{13}$$

Determining the number of "successful" outcomes can be tricky, particularly when the individual outcomes are not equally likely; that is, when the outcomes do not have an equal chance of occurring. Consider, for instance, the outcomes of the roll of 2 dice. There are 11 possible outcomes (that is, 2 through 12), but each of the sums is not equally likely. This is because there is more than one way to get the sums 3 through 11. To see this, note that there are 6 ways for the first die to land and 6 for the second, or a total of 36 equally likely ways of landing. The sum of 2 can be obtained with a 1 on one die and a 1 on the second die. The sum of 3 can be obtained with a 2 on the first die and 1 on the second or with a 2 on the second die and a 1 on the first. Table 10.1 shows the 36 ways of landing grouped by sums.

Table 10.1 should be used to determine the probability of events involving the throwing of two dice.

Sum	*Outcome of Pair of Dice*	*Possible ways to reach sum*
2	(1, 1)	1
3	(1, 2), (2, 1)	2
4	(2, 2), (3, 1), (1, 3)	3
5	(2, 3), (3, 2), (4, 1), (1, 4)	4
6	(3, 3), (4, 2), (2, 4), (5, 1), (1, 5)	5
7	(1, 6), (6, 1), (5, 2), (2, 5), (4, 3), (3, 4)	6
8	(4, 4), (5, 3), (3, 5), (6, 2), (2, 6)	5
9	(6, 3), (3, 6), (5, 4), (4, 5)	4
10	(5, 5), (6, 4), (4, 6)	3
11	(6, 5), (5, 6)	2
12	(6, 6)	1

EXAMPLE 6 What is the probability of obtaining a number besides 11 on a throw of two dice?

SOLUTION Since $P(11)=\frac{2}{36}=\frac{1}{18}$, it follows that

$$P(\text{not } 11)=1-\frac{1}{18}=\frac{17}{18}$$

Exercises for Section 10.4

1. What is the probability of drawing a heart from a deck of 52 cards?
2. What is the probability of drawing the king of clubs from a deck of 52 cards?
3. If a box contains 3 bolts and 5 nuts, what is the probability of drawing a bolt from the box? What is the probability of drawing a nut?
4. A piggy bank contains 20 dimes and 15 nickels. Assuming each has an equally likely chance falling out when the bank is shaken, what is the probability that the first coin to fall out will be a dime?
5. What is the probability that an odd number will turn up on one roll of a die?
6. What is the probability of getting a number greater than 4 on one roll of a die?
7. If 3 pennies are tossed simultaneously, what is the probability that exactly 2 will fall heads?
8. If girls and boys have an equally likely chance of being born, what is the probability that all three children in a family are girls?
9. A set of 3 red balls, 4 white balls, and 6 yellow balls are mixed in a box. What is the probability that a ball drawn from the box will not be red?
10. What is the probability of getting 5 heads if 7 coins are tossed onto a table top?
11. Six coins are tossed and allowed to land on the floor. What is the probability that exactly three will land tails?
12. Five boys and eight girls place their names in a hat. If 7 names are drawn from the hat, what is the probability of getting 3 boys and 4 girls?
13. A box contains 6 quarters, 5 dimes, and 15 nickels. If two coins are drawn at random from the box, what is the probability that they are both nickels?
14. In Exercise 13, what is the probability that two coins drawn from the box will have the same value?
15. Two letters are chosen at random from the word *probability*. What is the probability that they are both vowels?

16. Three cards are drawn at random from a deck of playing cards. What is the probability of getting 2 hearts and 1 spade?

17. What is the probability of rolling a 3 with a pair of dice?

18. What is the probability of rolling a 7 with a pair of dice?

19. What is the probability that a 5-card hand from a deck of playing cards will contain 2 aces and only 2?

20. What is the probability that a 7-card hand from a deck of playing cards will contain 5 cards of the same suit and only 5?

10.5 Odds

The probability of an event A is often given indirectly by two numbers, which when taken together are called the "odds in favor" of the event $O(A)$.

DEFINITION

The ratio of the probability of an event occurring to the probability that the event does not occur is called the odds (in favor) of the event A. Thus:

$$O(A)=\frac{P(A)}{1-P(A)}=\frac{\text{number of successes}}{\text{number of failures}} \quad (10.5)$$

Odds are quoted by giving both parts of the ratio. Thus, if $O(A)=\frac{3}{2}$ we say the odds are "three to two."

EXAMPLE 1 What are the odds of throwing a 2 or a 1 on a toss of one die?

SOLUTION The probability of a 2 or a 1 is $\frac{1}{3}$. Hence,

$$O(2\text{ or }1)=\frac{P(2\text{ or }1)}{1-P(2\text{ or }1)}$$

$$=\frac{\frac{1}{3}}{1-\frac{1}{3}}=\frac{1}{2}$$

Thus, the odds are one-to-two.

EXAMPLE 2 Suppose the odds in favor of an event A are 5 to 2. What is the probability of the event A?

SOLUTION

$$\frac{P(A)}{1-P(A)}=\frac{5}{2}$$

Solving for $P(A)$ yields

$$2P(A)=5-5P(A)$$

$$7P(A)=5$$

$$P(A)=\frac{5}{7}$$

10.6 Expectation

The expectation of a prize (or a game) is defined as the value of the prize times the probability of winning it. If the two distinct outcomes to a game are events A and B, and if the amounts to be paid are m_1 for A and m_2 for B, then the expectation of the game is $P(A)m_1 + P(B)m_2$. If a game has outcomes $A_1, A_2, A_3, \cdots, A_n$, with payoffs $m_1, m_2, m_3, \cdots, m_n$, then the expectation E is defined by

$$(10.6) \quad E = \sum_{i=1}^{n} m_i P(A_i)$$

In some games there is a payoff for only one event (one number, one side of a coin, and so on), in which case only one m_i is nonzero and *the expectation is the product of the payoff times the probability of that event.*

The expectation might be considered the amount you should pay for the privilege of playing the game.

EXAMPLE 1 A number wheel has the numbers 1 through 10 on it and you receive \$2 if your number comes up. What is your expectation for this game?

SOLUTION The probability that any one of the ten numbers will come up is 1/10. Since the payoff is \$2 for the winning number and nothing for the others, we have

$$E = \tfrac{1}{10} \cdot \$2 = \$0.20$$

This would be a fair price for you to pay for playing the game.

EXAMPLE 2 A box contains 1 red ball and 9 white balls. If you pick the red ball, you receive \$5; otherwise, you get \$0.50. Is a dollar a game too much to pay for this game?

SOLUTION The expectation is

$$\begin{aligned} E &= P(\text{red})(\$5) + P(\text{white})(\$0.50) \\ &= \tfrac{1}{10}(\$5) + \tfrac{9}{10}(\$0.50) \\ &= (\tfrac{1}{2} + \tfrac{9}{20}) \\ &= \$\tfrac{19}{20} = \$0.95 \end{aligned}$$

Hence, if you pay \$1 for each game you could be expected to lose \$0.05 per game—*if you played it for a long time.*

EXAMPLE 3 A contractor, before submitting a bid for the construction of an airport runway, determines that he has a 0.7 probability of making a profit of \$60,000 and a 0.3 probability of losing \$100,000. What is his expectation?

SOLUTION

$$\begin{aligned} E &= (\text{amt. of profit}) \cdot P(\text{profit}) - (\text{amt. of loss}) \cdot P(\text{loss}) \\ &= (\$60{,}000)(0.7) - (\$100{,}000)(0.3) \\ &= \$12{,}000 \end{aligned}$$

From this perspective, the contractor may decide that his expectation is too small, and he may just forget about bidding.

Exercises for Sections 10.5 and 10.6

1. What are the odds for throwing a 6 when throwing one die?
2. What are the odds for getting exactly two aces in a five-card poker hand?
3. What are the odds for getting something besides a total of 7 on the throw of two dice?
4. What are the odds for an event whose probability is 0.3?
5. What is the probability of an event A if the odds in favor of A are 8 to 3?
6. What is the probability of an event A if the odds in favor of A are 1 to 3?
7. What is the expectation of a game in which you throw a die and receive an amount of money equal to the number on the die?
8. What is the expectation of a game in which you throw a pair of dice and receive an amount of money equal to the total of the numbers on the dice?
9. A box contains 5 red balls, 3 purple balls, and 1 white ball. If you draw a red ball, you receive \$1; a purple ball, \$5; and the white ball, \$10. What is your expectation for this game?
10. A contractor knows that he stands to earn a profit of \$50,000 with 0.9 probability but with a slight chance (0.1 probability) that he will lose as much as \$200,000. What is his expectation?

10.7 Mutually Exclusive Events

Two or more events are **mutually exclusive** if one event excludes the possibility of the other. For instance, on a single roll of a die, the event E_1 that the die will turn up 2, and the event E_2 that the die will turn up 5 are mutually exclusive events, since success in rolling a 2 excludes the possibility of a 5.

Suppose we want to compute the probability of getting either a 2 or a 5 on a single roll of a die. The events E_1 (rolling a 2) and E_2 (rolling a 5) are mutually exclusive. Since success in rolling a 2 or a 5 is possible in 2 ways out of 6, we have

$$P(2 \text{ or } 5)=\frac{2}{6}=\frac{1}{6}+\frac{1}{6}=P(2)+P(5)$$

From this example, you see that the probability of mutually exclusive events is the sum of the probabilities of the individual events. This observation and its extension to more than two mutually exclusive events is stated in the following theorem.

THEOREM 1

If $E_1, E_2, \cdots, E_n$ are mutually exclusive events, then the probability of E_1 or $E_2 \cdots$ or E_n equals the sum of the probabilities of the individual events; that is,

$$P(E_1 \text{ or } E_2 \cdots \text{ or } E_n)=\sum_{i=1}^{n} P(E_i)$$

Theorem 1 applies only to mutually exclusive events. For instance, it cannot be used to compute the probability of getting an ace or a heart on a single draw from a deck of playing cards, since the ace of hearts could be drawn.

EXAMPLE 1 A box contains 5 red balls, 4 white balls, and 7 black balls. What is the probability of getting a red ball or a white ball on a single draw?

SOLUTION Drawing a red ball and drawing a white ball are mutually exclusive events. The probability of drawing a red ball is $P(\text{red})=\frac{5}{16}$ and the probability of drawing a white ball is $P(\text{white})=\frac{4}{16}$. Therefore, the probability of a red or a white ball on a single draw is

$$P(\text{red or white})=\frac{5}{16}+\frac{4}{16}=\frac{9}{16}$$

EXAMPLE 2 If a card is drawn from a deck of playing cards, what is the probability of an ace or a 2 or a face card?

SOLUTION Drawing an ace, drawing a 2, and drawing a face card are mutually exclusive events. Furthermore, their probabilities are

$$P(\text{ace})=\frac{4}{52}, \qquad P(2)=\frac{4}{52}, \qquad P(\text{face card})=\frac{12}{52}$$

Therefore,

$$P(\text{ace or 2 or face card})=\frac{4}{52}+\frac{4}{52}+\frac{12}{52}=\frac{20}{52}=\frac{5}{13}$$

10.8 Independent Events

If a box contains 3 red balls, 5 black balls, and 2 white balls, what is the probability in a drawing that a red ball will be drawn first and a black ball second? To understand the theory of compound events of this type, we divide the study into independent and dependent events.

Two events E_1 and E_2 are said to be **independent** if the probability of E_1 is not influenced by E_2, and conversely, the probability of E_2 is not influenced by E_1. Otherwise, the events are said to be **dependent**.

EXAMPLE 1 If two cards are drawn from a deck and the first card is returned to the deck before the second is drawn, the two events are independent.

EXAMPLE 2 If two cards are drawn in succession from a deck and the first card is not returned before the second is drawn, the two events are dependent.

Consider the probability of two independent successive events. Suppose one ball is to be drawn from each of the boxes shown.

Box 1	Box 2
5 red 3 black	2 red 4 black

What is the probability that two red balls will be drawn? Notice that a red ball from Box 1 can be drawn in 5 ways out of 8 possible outcomes and a red ball from Box 2 can be drawn in 2 ways out of 6 possible outcomes. Then 2 red balls can be drawn in $5 \cdot 2$ ways of $8 \cdot 6$ possible outcomes. Denoting the probability that red balls will be drawn from each box by $P(R_1 \text{ and } R_2)$, we have

$$P(R_1 \text{ and } R_2) = \frac{5 \cdot 2}{8 \cdot 6} = \frac{5}{8} \cdot \frac{2}{6} = P(R_1) \cdot P(R_2)$$

That is, the desired probability is the product of the probabilities of the independent events. We generalize this result in the following theorem.

THEOREM 2

If $E_1, E_2, \cdots, E_n$ are independent events, then the probability of all the events occurring is the product of the probabilities of the independent events; that is,

$$P(E_1 \text{ and } E_2 \cdots \text{ and } E_n) = P(E_1) \cdot P(E_2) \cdot \cdots \cdot P(E_n)$$

EXAMPLE 3 If a coin is flipped 4 times, what is the probability of getting 4 heads?

SOLUTION Since each flip is independent of all others, the probability of 4 heads is

$$\begin{aligned} P(H_1 \text{ and } H_2 \text{ and } H_3 \text{ and } H_4) &= P(H_1) \cdot P(H_2) \cdot P(H_3) \cdot P(H_4) \\ &= \tfrac{1}{2} \cdot \tfrac{1}{2} \cdot \tfrac{1}{2} \cdot \tfrac{1}{2} \\ &= \tfrac{1}{16} \end{aligned}$$

EXAMPLE 4 If a coin is flipped 4 times, what is the probability of getting 3 heads and 1 tail (in any order)?

SOLUTION The number of ways in which 3 heads can occur is $C_{4,3} = 4$. Each of these ways has the probability of occurrence

$$P(H) \cdot P(H) \cdot P(H) \cdot P(T) = \tfrac{1}{2} \cdot \tfrac{1}{2} \cdot \tfrac{1}{2} \cdot \tfrac{1}{2} = \tfrac{1}{16}$$

Therefore, the desired probability is

$$P(3H \text{ and } 1T) = 4 \cdot \tfrac{1}{16} = \tfrac{1}{4}$$

EXAMPLE 5 Consider a box with 5 red balls and 3 yellow balls. A ball is drawn from the box and replaced and then a second ball is drawn. What is the probability that a red will be drawn first and a yellow ball will be drawn second?

SOLUTION Drawing a red ball and drawing a yellow ball are independent events; since the first ball is replaced before the second is drawn. The probabilities are $P(\text{red})=\frac{5}{8}$ and $P(\text{yellow})=\frac{3}{8}$. Therefore,

$$P(\text{red and yellow})=\frac{5}{8}\cdot\frac{3}{8}=\frac{15}{64}$$

EXAMPLE 6 A number wheel with the numbers 1 to 100 is turned and allowed to stop; the number on which it stops is recorded. What is the probability that in ten turns of the wheel a number will be repeated?

SOLUTION Consider the probability of *not* turning to a duplicate number. After the first number is recorded, there are 99 numbers (of the 100) that would be permissible for the condition "no duplication." After the second number, there would be 98 remaining numbers, then 97, then 96, and so on. Then,

$$P(\text{not drawing a duplicate in 10 turns})=\frac{100}{100}\cdot\frac{99}{100}\cdot\frac{98}{100}\cdot\cdots\cdot\frac{92}{100}\cdot\frac{91}{100}=0.628$$

Thus, the probability of a duplicate is

$$\begin{aligned}P(\text{drawing a duplicate in 10 turns})&=1-P(\text{not drawing a duplicate in 10 turns})\\&=1-0.628\\&=0.372\end{aligned}$$

10.9 Dependent Events

The probability of drawing a red ball twice, in succession, from a box that contains 7 red balls and 3 black balls if the first ball is not replaced exemplifies two dependent events. In this case, the probability of a red ball on the second draw, where a red ball was drawn first, is called a *conditional* probability. The probability of E_2, given that E_1 has occurred, is generally denoted by $P(E_2|E_1)$, which is read "the probability of E_2 given that E_1 has occurred." For independent events,

$$P(E_2|E_1)=P(E_2)$$

since the probability of E_2 is not influenced by E_1. However, if E_1 and E_2 are dependent events,

$$P(E_2|E_1)\neq P(E_2)$$

In the case of the box with 7 red balls and 3 black balls, the probability of a red ball on the first draw is $\frac{7}{10}$ and the probability of a red ball on the second draw, given that a red ball was drawn first, is $\frac{6}{9}$, since only 6 of the remaining 9 balls are red. These probabilities are denoted

$$P(R_1)=\frac{7}{10} \quad \text{and} \quad P(R_2|R_1)=\frac{6}{9}$$

To see how these probabilities are used to complete the probability of drawing two red balls if the first ball is not replaced, we note that a red ball can be drawn in 7 ways on the first draw and 6 ways on the second, so 2 red balls, in succession are possible in $7 \cdot 6 = 42$ ways. Since the possible outcomes are $10 \cdot 9 = 90$, the probability of two red balls is

$$P(R_1 \text{ and } R_2) = \frac{42}{90} = \frac{7 \cdot 6}{10 \cdot 9} = \frac{7}{10} \cdot \frac{6}{9} = P(R_1) \cdot P(R_2|R_1)$$

That is, the probability that both balls will be red is the product of $P(R_1)$ and $P(R_2|R_1)$.

THEOREM 3

If E_1 and E_2 are two events, the probability of E_1 and E_2 as successive events is

$$P(E_1 \text{ and } E_2) = P(E_1) \cdot P(E_2|E_1)$$

For independent events, this theorem reduces to Theorem 2 of Section 10.8.

EXAMPLE 1 What is the probability of drawing two spades from a deck of playing cards if the first card is not replaced?

SOLUTION

$$P(S_1) = \frac{13}{52} \quad \text{and} \quad P(S_2|S_1) = \frac{12}{51}$$

Hence,

$$P(S_1 \text{ and } S_2) = \frac{13}{52} \cdot \frac{12}{51} = \frac{3}{51} = \frac{1}{17}$$

EXAMPLE 2 A box contains 7 red balls and 6 white balls. What is the probability of drawing a red ball first and a white ball second if we do not replace the first ball?

SOLUTION

$$P(R_1) = \frac{7}{13} \quad \text{and} \quad P(W_2|R_1) = \frac{6}{12}$$

The desired probability is

$$P(R_1 \text{ and } W_2) = P(R_1) \cdot P(W_2|R_1) = \frac{7}{13} \cdot \frac{6}{12} = \frac{7}{26}$$

EXAMPLE 3 Relative to Example 2, what is the probability of drawing a red ball and a white ball if we do not specify order?

SOLUTION We can draw a red ball first and a white ball second or a white ball first and a red ball second. The probability of either ordering is $\frac{7}{26}$. Since the events R and W can be permuted in 2! ways, the desired probability is

$$2!P(R \text{ and } W) = 2!\left(\frac{7}{26}\right) = \frac{7}{13}$$

Exercises for Sections 10.7–10.9

1. What is the probability of rolling a number greater than 4 on a single roll of a die?

2. If a die is rolled twice, what is the probability of rolling either a 4 or a 5?

3. A box contains 3 red balls, 2 black balls, and 2 white balls. What is the probability of drawing either a red ball or a white ball on a single draw?

4. If a penny is flipped 3 times, what is the probability of getting 2 heads and 1 tail?

5. One card is drawn from a card deck. What is the probability that it is an ace or a face card?

6. One card is drawn from a card deck. What is the probability that it is less than 4 or an ace?

7. (a) What is the probability of drawing two hearts from a card deck if the first card is replaced before the second is drawn?
   (b) What is the probability of two hearts without replacement?

8. (a) What is the probability of drawing an ace on the first draw and a face card on the second if the first card is not replaced?
   (b) What is the probability of drawing an ace and a face card in any order without replacement?

9. (a) Three cards are drawn (without replacement) from a card deck. What is the probability of a heart on the first draw, a diamond on the second draw, and a spade on the third draw?
   (b) What is the probability of 1 heart, 1 diamond, and 1 spade?

10. One card is drawn from each of three separate decks. What is the probability that (a) all three are hearts? (b) all three are the same suit?

11. Box 1 contains 7 red balls and 4 black balls, and Box 2 contains 4 red balls and 9 black balls. If one ball is drawn from each box, what is (a) the probability that they are both red? (b) the probability that one is red and one is black?

12. If a die is rolled, what is the probability of getting a 1 or a number greater than 4?

13. Two persons are chosen at random from five married couples. What is the probability that husband and wife will be chosen?

14. The names Tom, Dick, Ann, Clyde, Jane, and Sam are placed in a hat. What is the probability the first three names drawn without replacement will be in the order Ann, Dick, and Clyde?

15. A box contains 3 red balls, 7 black balls, and 5 white balls. Two balls are drawn with replacement from the box.
    (a) What is the probability of drawing a red ball or a black ball first and a white ball second?
    (b) What is the probability of drawing a red ball or a black ball first and a black ball or a white ball second?

16. A company buys one box of 100 machined parts from each of three different companies. There are 3 defective parts in the box from Company A, 2 in the box from Company B, and 2 in the box from Company C. If one part is taken from each box, what is the probability that (a) all 3 parts are defective? (b) at least 2 of the 3 will be defective?

17. Five cards are drawn without replacement from a deck of cards.
    (a) What is the probability of exactly 3 clubs?
    (b) What is the probability of at least 2 clubs?

18. Consider the three boxes shown.

Box 1	Box 2	Box 3
3 red 2 black	5 red 4 black	2 red 7 black

A ball is drawn from Box 1. If it is red, the second ball is drawn from Box 2; if it is black, the second ball is drawn from Box 3. What is the probability that (a) both balls are red? (b) both are black? (c) one is red and one is black?

19. Box A contains 2 black balls and 1 white ball, and Box B contains 1 black ball and 2 white balls. One ball is transferred from Box A to Box B and then a ball is drawn from Box B. What is the probability that it is black?

20. Box A contains 2 red balls and 2 white balls, Box B contains 3 red balls and 4 white balls, and Box C contains 2 red balls and 5 white balls. If a box is chosen at random and 1 ball drawn from it, what is the probability that it is red?

21. One hundred numbers are placed in a box. Twenty times, drawing for a number takes place, *with replacement*. (a) What is the probability that the same number will be drawn twice? (b) How many times must a drawing be made so that the probability that the same number will be drawn twice is greater than 0.5?

Test 1 for Chapter 10

Answer *True* or *False* to Exercises 1–10.

1. The number of ways to answer a 20-question true-false test is 20!.

2. Combinations are sets of elements in which the order is not considered.

3. The number of permutations of n elements is $n!$.

4. If an event is possible in n ways, the probability of any one of the ways is $1/n$.

5. If two events A and B are independent, then $P(A \text{ or } B)=P(A)\cdot P(B)$.

6. If A and B are mutually exclusive events, then $P(B)=1-P(A)$.

7. The odds in favor of an event are always positive.

8. The number of combinations of n things taken r at a time is the same as the coefficient of x^{n-r} in the expansion of $(x+y)^n$.

9. If a game pays you \$10, then your expectation is \$10.

10. If two coins are tossed, then the probability of a head and a tail is 1/3.

11. In how many ways can an all-black bridge hand be dealt? What is the probability of an all-black hand?

12. What is the probability of a 7 or an 11 in the roll of two dice?

13. A card is drawn from a deck and a prize of \$10 is awarded for an ace, \$5 for a face card or a 10, and \$1 for any other card. What is your expectation for this game?

14. What is the probability of drawing the two red aces from a deck without replacement?

15. Two numbers are chosen (with replacement) from 1, 2, 3, 4, 5. What is the probability that the two are the same?

Test 2 for Chapter 10

1. In how many ways can 10 people be arranged in teams of 3?

2. Calculate (a) $P_{5,2}$; (b) $C_{8,3}$.

3. A business firm wants to hire 6 men and 3 women. If 10 men and 7 women apply for the positions, in how many ways can the firm make the selections?

4. (a) What is the probability of drawing a five-card poker hand that contains a pair?

(b) What is the probability of drawing a five-card poker hand that contains four of a kind?

5. (a) What is the probability of getting three heads in a row in tossing a coin?

(b) What is the probability of getting in succession a head, a tail, and a head?

(c) What is the probability of getting two heads and a tail in any order?

6. Consider the two boxes shown. A ball is drawn

2 red	6 red
3 black	4 black
Box 1	Box 2

from the first box. If it is red, then a ball is drawn from the second box. What is the probability that two red balls will be drawn?

11

The Conic Sections

Analytic geometry is a branch of mathematics in which we study the geometry of curves, using algebraic equations. Analytic geometry poses two fundamental problems:

(1) To associate with an equation involving x and y a curve in the rectangular plane.

(2) To associate with a curve in the rectangular plane a relation between x and y, called the **equation** of the curve.

It is at least theoretically easy to solve the first problem. All we need to do is assign values of x (or y) and find corresponding values of y (or x). Under the assumption that we can solve for y, the totality of ordered pairs (x, y) yields the curve corresponding to the given equation. This procedure is certainly not new, since you have used it at several different places in this book. In its most basic form, it is called point plotting; and it is usually supplemented with some analysis to allow for a more comprehensive approach.

The second problem of analytic geometry is usually substantially more difficult. To write the equation of a curve, you must know the rule governing the construction of the curve. So far in this book, the only curves whose equations are specifically derived from their construction are the straight line and the circle. Some general procedures apply for curves known as the conic sections.

11.1 The Parabola

Curves that can be formed by cutting a right circular cone with a plane are called **conic sections**. Greek mathematicians ca. 300 B.C. discovered that they could get four distinct curves by cutting a right circular cone with a plane; these were the circle, the ellipse, the hyperbola, and the parabola. The properties of the conics discovered by the Greeks include those that we use as definitions.

If two right circular cones are placed vertex to vertex with a common axis, the resulting figure is referred to as a cone with two *nappes*. To see how the four conics can be generated from a cone with two nappes, refer to Figure 11.1.

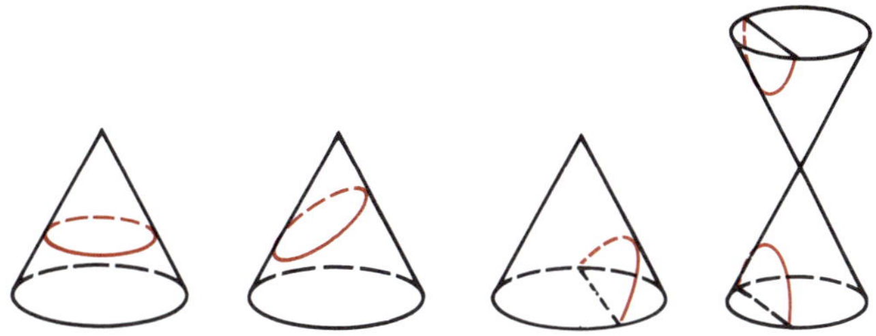

FIGURE 11.1

(a) A **circle** is obtained when the cutting plane is perpendicular to the axis, provided that it does not pass through the vertex.

(b) An **ellipse** is obtained when the cutting plane is inclined so as to cut entirely through one nappe of the cone without cutting the other nappe.

(c) A **parabola** is obtained when the cutting plane is parallel to one element in the side of the cone.

(d) A **hyperbola** is obtained when the cutting plane is inclined so that it cuts through both nappes.

Since analytic geometry is the study of the relation between algebra and geometry, we want to establish the algebraic representation of the four conic sections. The circle, which is the simplest of the conic sections, was defined and discussed in Section 4.2. Here we define and discuss the **parabola**.

DEFINITION

> A parabola is the set of all points in a plane that are equidistant from a fixed point and a fixed line.

The fixed point is the **focus** and the fixed line is the **directrix** of the parabola. The line through the focus and perpendicular to the directrix is the **axis** of the parabola. By definition, the midpoint between the focus and the directrix is a point on the parabola, known as the **vertex**. (Recall from Chapter 5 that the shape of the graph of a quadratic function is a parabola whose axis is *vertical*.)

Figure 11.2 shows a parabola with vertex located at the origin, focus at $F(a, 0)$, and directrix perpendicular to the x-axis at $D(-a, 0)$. Observe that a is the distance from the vertex to the focus, sometimes referred to as the **focal distance**.

To find the algebraic equation of this parabola, consider a point $P(x, y)$ on the parabola. Then, by definition,

(11.1) $\overline{GP} = \overline{FP}$

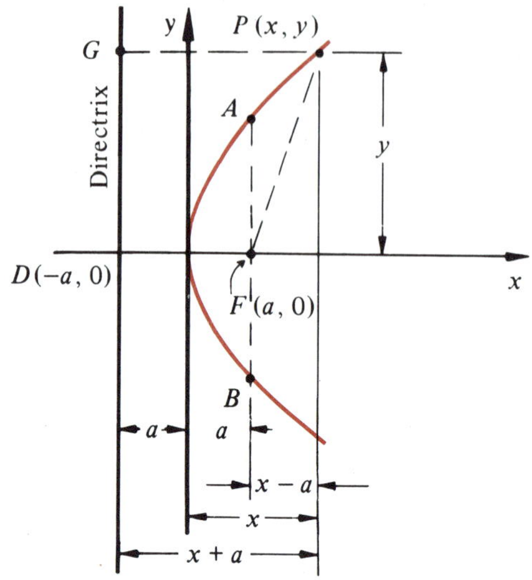

FIGURE 11.2

But from Figure 11.2, we see that

$$\overline{GP} = x + a \quad \text{and} \quad \overline{FP} = \sqrt{(x-a)^2 + y^2}$$

Substituting these expressions in Equation (11.1) yields

$$x + a = \sqrt{(x-a)^2 + y^2}$$

or when both sides are squared,

$$(x+a)^2 = (x-a)^2 + y^2$$

With like terms expanded and collected, the equation of the parabola becomes

(11.2) $y^2 = 4ax$

Equation (11.2) is referred to as the **standard form** of the equation of a parabola with vertex at the origin and focus on the x-axis. It is clearly symmetric about its axis. The focus is at $(a, 0)$ if the coefficient of x is positive. In this case, the parabola opens to the right. If the coefficient of x is negative, the focus is at $(-a, 0)$ and the parabola opens to the left.

The chord AB through the focus and perpendicular to the axis is called the **right cord**, or frequently, the latus rectum. The length of the right chord is found by letting $x = a$ in Equation (11.2). By this substitution,

$$y^2 = 4a^2$$
$$y = \pm 2a$$

The length of the right chord is therefore numerically equal to $4a$. This fact is useful because it helps to define the shape of the parabola, giving us an idea of the "opening" of the parabola.

The standard form of a parabola with vertex at the origin and focus on the y-axis is

(11.3) $x^2 = 4ay$

Deriving this formula, which parallels Equation (11.2), is left to you. The parabola represented by Equation (11.3) is symmetric about the y-axis. It opens upward if the coefficient of y is positive and downward if it is negative.

The equation of a parabola is characterized by having one linear variable and one squared variable. *The linear variable indicates the direction of the axis of the parabola.*

A unique physical property of the parabola is that it will reflect any rays emitted from the focus so that they travel parallel to the axis of the parabola. This feature makes the parabola a particularly desirable shape for reflectors in spotlights and reflecting telescopes and also radar antennas.

A rough sketch of the parabola can be drawn if the location of the vertex and the extremities of the right chord are known. As you will see in the following examples, all this information can be obtained from the standard form of the equation.

EXAMPLE 1 Discuss and sketch the graph of the equation $x^2 = 8y$.

SOLUTION This equation has the form of Equation (11.3) with $4a = 8$ or $a = 2$. The y-axis is the axis of the parabola, the focus is at (0, 2), and the directrix is the line $y = -2$. The endpoints of the right chord are then $(-4, 2)$ and (4, 2). Figure 11.3 shows the parabola.

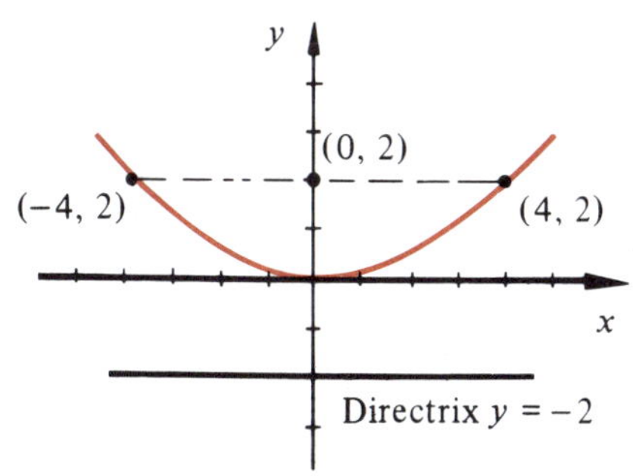

FIGURE 11.3

EXAMPLE 2 Find the equation of the parabola with focus at $(-1, 0)$ and directrix $x = 1$; sketch the curve.

SOLUTION The focus lies on the x-axis to the left of the directrix, so the parabola opens to the left. The desired equation is then the form in Equation (11.2) with $a = -1$; that is,

$$y^2 = -4x$$

To sketch the parabola, note that the length of the right chord is $\overline{AB} = 4$. Its extremities are therefore $(-1, 2)$ and $(-1, -2)$. The curve appears in Figure 11.4.

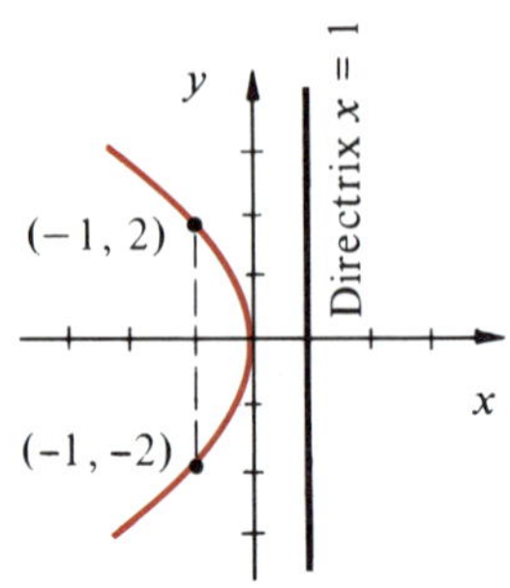

FIGURE 11.4

EXAMPLE 3 A parabolic reflector is to be built with a focal distance of 2.25 ft. If the reflector is to be 1 ft deep at its axis, what will be its diameter?

SOLUTION To solve this problem, we need the equation of the parabola used to generate the reflector. Referring to Figure 11.5 we locate the vertex at the origin and the focus on the x-axis. (We could have just as well placed the focus on the y-axis.) Then, using Equation (11.2) with $a=2.25$ yields the equation

$$y^2=4(2.25)x=9x$$

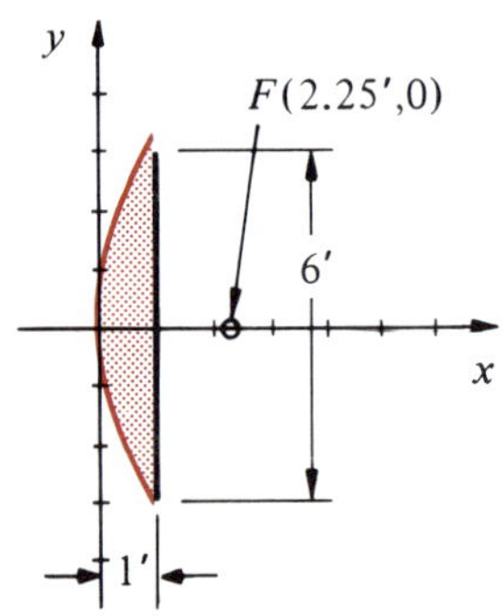

FIGURE 11.5

The diameter of the reflector can now be found by substituting $x=1$ in this equation. Thus,

$$y^2=9 \quad \text{and} \quad y=\pm 3$$

which means that the diameter of the reflector is 6 ft.

Exercises for Section 11.1

In Exercises 1–10, find the coordinates of the focus, the endpoints of the right chord, and the equation of the directrix of each of the parabolas. Sketch the graph of each parabola.

1. $y^2=-8x$
2. $x^2=12y$
3. $2x^2=12y$
4. $x^2=-24y$
5. $y^2+16x=0$
6. $y+2x^2=0$
7. $y^2=3x$
8. $3x^2=4y$
9. $y^2=-2x$
10. $y^2=10x$

In Exercises 11–18, find the equation of the parabolas having the given properties. Sketch each curve.

11. Focus at (0, 2), directrix $y=-2$.

12. Focus at $(0, -\frac{1}{2})$, directrix $y=\frac{1}{2}$.

13. Focus at $(\frac{3}{2}, 0)$, directrix $x=-\frac{3}{2}$.

14. Focus at $(-10, 0)$, directrix $x=10$.

15. Endpoints of right chord $(2, -1)$ and $(-2, -1)$, vertex at (0, 0).

16. Endpoints of right chord (3, 6) and $(3, -6)$, vertex at (0, 0).

17. Vertex at (0, 0), vertical axis, one point of the curve (2, 4).

18. Vertex at (0, 0), horizontal axis, one point of the curve (2, 4).

In Exercises 19–22, solve the given system of equations graphically.

19. $2x+4y=0$
$x^2-4y=0$

20. $y=e^x$
$y^2=-3x$

21. $y^2=12x$
$y=\log x$

22. $y=x^3$
$x^2=8y$

23. The supporting cable of a suspension bridge hangs in the shape of a parabola. Find the equation of a cable hanging from two 400-ft-high supports that are 1000 ft apart if the lowest point of the cable is 250 ft below the top of the supports. Choose the origin in the most convenient location.

24. A parabolic antenna is to be built by revolving the parabola $y^2=24x$. Sketch the cross-section of an antenna that has a circular front measuring 12 ft in diameter. Locate the focus.

11.2 The Ellipse

An **ellipse** can be constructed from a loop of string by placing two pins at F and F', as shown in Figure 11.6, and placing the loop of string over them. Pull the string taut with the point of a pencil and then move the pencil, keeping the string taut. The figure generated is an ellipse. From this construction, observe that the sum of the distances from the two fixed points to the point P is always the same since the loop of string is kept taut. This property characterizes the ellipse.

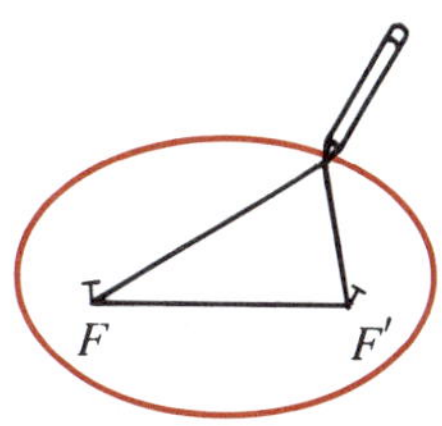

FIGURE 11.6

DEFINITION

> An ellipse is the set of all points in a plane the sum of whose distances from two fixed points in the plane is constant.

The fixed points are called the **foci** of the ellipse. The midpoint of a line through the foci is called the **center** of the ellipse. We use the center of the ellipse to locate the ellipse in the plane as we used the vertex to locate the parabola.

To get the equation of an ellipse, consider an ellipse with foci located on the x-axis so that the origin is midway between them, as in Figure 11.7. Letting the foci

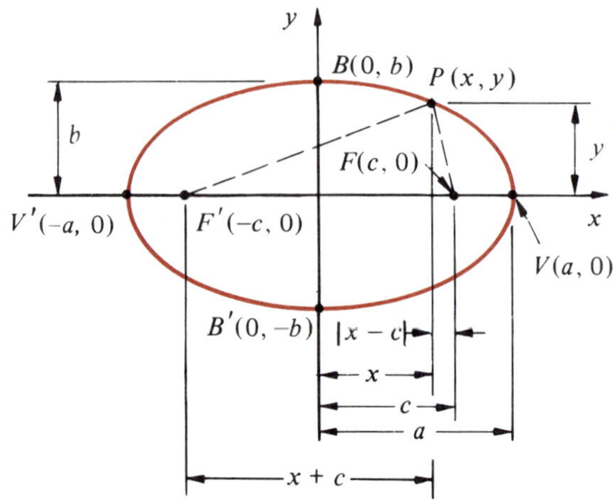

FIGURE 11.7

be the points $F(c, 0)$ and $F'(-c, 0)$ and letting the sum of the distances from a point $P(x, y)$ of the ellipse to the foci be $2a$, where $a > c$, we have

$$PF + PF' = 2a$$

From Figure 11.7, it is clear that

$$PF = \sqrt{(x-c)^2 + y^2}$$

and

$$PF' = \sqrt{(x+c)^2 + y^2}$$

so that

$$\sqrt{(x-c)^2 + y^2} + \sqrt{(x+c)^2 + y^2} = 2a$$

Transposing the first radical and squaring gives

$$(x+c)^2 + y^2 = 4a^2 - 4a\sqrt{(x-c)^2 + y^2} + (x-c)^2 + y^2$$

Then if like terms are expanded and collected—

$$a\sqrt{(x-c)^2 + y^2} = a^2 - cx$$

and the equation squared again and simplified,

$$(a^2 - c^2)x^2 + a^2y^2 = a^2(a^2 - c^2)$$

by substituting $b^2 = a^2 - c^2$, we get

$$b^2x^2 + a^2y^2 = a^2b^2$$

Finally, dividing through by the nonzero quantity a^2b^2, we can write the equation of an ellipse with its center at the origin and its foci on the x-axis as

(11.4) $$\frac{x^2}{a^2} + \frac{y^2}{b^2} = 1$$

Equation (11.4) is clearly symmetric about both axes and the origin. By letting $y = 0$ in Equation (11.4), we see that the x-intercepts of the ellipse are $(a, 0)$

and $(-a, 0)$. The segment of the line through the foci from $(a, 0)$ to $(-a, 0)$ is called the **major axis** of the ellipse. Clearly, the length of the major axis is $2a$, which is also the value chosen for the sum of the distances PF and PF'. The length a is then known as the **semimajor axis** and the endpoints of the major axis as the **vertices** of the ellipse.

The y-intercepts of the ellipse are found to be $(0, b)$ and $(0, -b)$ by letting $x=0$ in Equation (11.4). The segment of the line perpendicular to the major axis from $(0, b)$ to $(0, -b)$ is called the **minor axis**. Since the length of the minor axis is $2b$, the length of the **semiminor axis** is b. The graph of an ellipse can readily be sketched once the semimajor and semiminor axes are known. In this book, **the semimajor axis is represented by the letter *a* and the semiminor axis by the letter *b*.**

The foci of an ellipse are located by solving the equation $b^2=a^2-c^2$ for the focal distance c. Thus,

$$c=\sqrt{a^2-b^2}$$

A similar derivation will show that

$$\frac{x^2}{b^2}+\frac{y^2}{a^2}=1 \tag{11.5}$$

is the equation of an ellipse with its center at the origin and its foci on the y-axis. (The letter a is again being used to represent the semimajor axis.)

EXAMPLE 1 Find the equation of an ellipse centered at the origin with foci on the x-axis if the major axis is 10 and the minor axis is 4.

SOLUTION In this case, the major axis is on the x-axis. The semimajor axis is $a=5$ and the semiminor axis is $b=2$. Substituting these values into Equation (11.4) gives the required equation,

$$\frac{x^2}{5^2}+\frac{y^2}{2^2}=1$$

or $$\frac{x^2}{25}+\frac{y^2}{4}=1$$

EXAMPLE 2 Find the equation of the ellipse with vertices at $(0, 5)$ and $(0, -5)$ and foci at $(0, 4)$ and $(0, -4)$.

SOLUTION From the given information, we can conclude that the foci are on the y-axis, the center of the ellipse is at the origin, and the semimajor axis is $a=5$. To find the semiminor axis, use the relation $b^2=a^2-c^2$. Thus, $b=\sqrt{25-16}=\sqrt{9}=3$. Substituting $a=5$ and $b=3$ in Equation (11.5) yields

$$\frac{x^2}{9}+\frac{y^2}{25}=1$$

The ellipse is sketched in Figure 11.8.

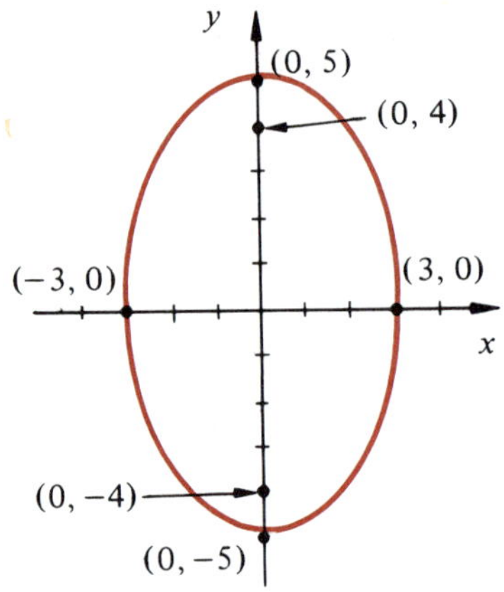

FIGURE 11.8

EXAMPLE 3 Sketch the graph of the ellipse $4x^2+16y^2=64$.

SOLUTION The given equation divided by 64 can be written in the form

$$\frac{x^2}{16}+\frac{y^2}{4}=1$$

The major axis lies along the x-axis, since the denominator of the x term in the equation is larger than the denominator of the y term. Consequently, the semimajor axis is $a=4$ and the semiminor axis is $b=2$. The vertices are then (4, 0) and (−4, 0), and the endpoints of the minor axis are (0, 2) and (0, −2). The ellipse is sketched in Figure 11.9.

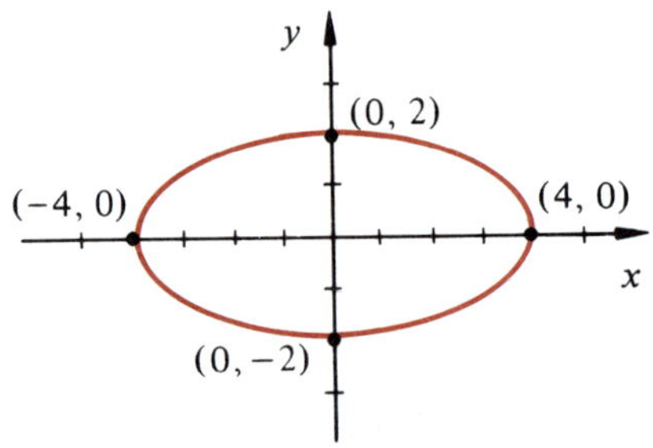

FIGURE 11.9

One of the first scientific applications of the ellipse was in astronomy. The astronomer Kepler (c. 1600) discovered that the planets moved in elliptical orbits about the sun with the sun at one focus. Artificial satellites also move in elliptical orbits about the earth. Another application is in the design of machines, in which elliptic gears are used to get a slow, powerful movement with a quick return. A third application of the ellipse is found in electricity, in which the magnetic field of a single-phase induction motor is elliptical under normal operating conditions.

Exercises for Section 11.2

In Exercises 1–10, discuss each of the given equations and sketch their graphs.

1. $5x^2+y^2=25$ **2.** $4x^2+9y^2=36$

3. $16x^2+4y^2=16$ **4.** $3x^2+9y^2=27$

5. $3x^2+y^2=9$ **6.** $25x^2+4y^2=100$

7. $2x^2+3y^2-24=0$ **8.** $5x^2+20y^2=20$

9. $9x^2+4y^2=4$ **10.** $4x^2+y^2=25$

In Exercises 11–20, find the equation of the ellipses having the given properties. Sketch each curve.

11. Vertices at $(\pm 4, 0)$, minor axis 6.

12. Vertices at $(0, \pm 1)$, minor axis 1.

13. Vertices at $(0, \pm 5)$, semiminor axis $\frac{3}{2}$.

14. Vertices at $(\pm 6, 0)$, semiminor axis 2.

15. Major axis 10, foci at $(\pm 4, 0)$.

16. Major axis 10, foci at $(0, \pm 3)$.

17. Foci at $(\pm 1, 0)$, semimajor axis 4.

18. Vertices at $(0, \pm 7)$, foci at $(0, \pm\sqrt{28})$.

19. Vertices at $(\pm\frac{5}{2}, 0)$, one point of the curve at $(1, 1)$.

20. Vertices at $(\pm 3, 0)$, one point of the curve at $(\sqrt{3}, 2)$.

In Exercises 21–24, solve the given system of equations graphically.

21. $\dfrac{x^2}{4}+y^2=4$
$2y+3x=0$

22. $\dfrac{x^2}{9}+\dfrac{y^2}{9}=1$
$y=e^{x+2}$

23. $y=x^2$
$x^2+4y^2=4$

24. $y^2-12x=0$
$y^2+9x^2=9$

25. An elliptical cam with a horizontal major axis of 10 in. and a minor axis of 3 in. is to be machined by a numerically controlled vertical mill. Find the equation of the ellipse to be used in programming the control device.

26. An elliptical cam having the equation $9x^2+y^2=81$ revolves against a push rod. What is the maximum travel of the push rod?

11.3 The Hyperbola

The final conic that we consider is the **hyperbola**.

DEFINITION

> A hyperbola is the set of all points in a plane the difference of whose distances from two fixed points in the plane is constant.

In Figure 11.10, we have drawn a hyperbola with foci at $F(c, 0)$ and $F'(-c, 0)$. The origin is then at the midpoint between the foci, which corresponds to the center of the hyperbola. The points $V(a, 0)$ and $V'(-a, 0)$ are called the **vertices**, and the segment $\overline{VV'}$ is called the **transverse axis** of the hyperbola. Clearly, the length of the transverse axis is $2a$. The segment BB', which is perpendicular to the transverse axis at the center of the hyperbola, is called the **conjugate axis** and has length $2b$. The conjugate axis has an important relation to the curve even though it does not intersect the curve.

To find the algebraic equation of a hyperbola, consider a point $P(x, y)$ on the hyperbola. Then, by definition,

$$\overline{F'P}-\overline{FP}=2a$$

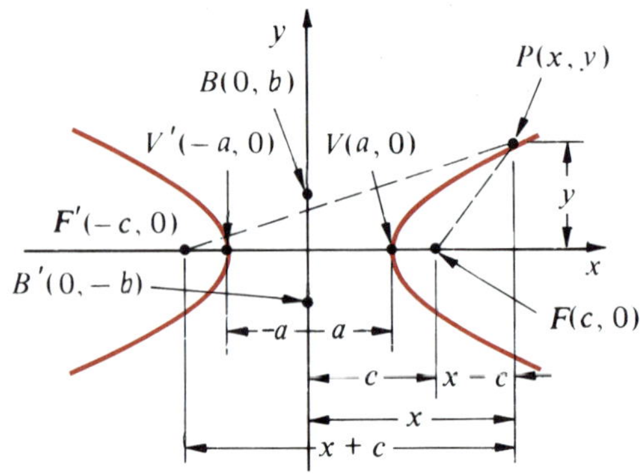

FIGURE 11.10

which in turn can be written

(11.6) $\sqrt{(x+c)^2+y^2}-\sqrt{(x-c)^2+y^2}=2a$

Using the same procedure here that we used for the ellipse, we can reduce Equation (11.6) to

$$\frac{x^2}{a^2}-\frac{y^2}{c^2-a^2}=1$$

Or if we let $b^2=c^2-a^2$, the equation of the hyperbola becomes

(11.7) $\frac{x^2}{a^2}-\frac{y^2}{b^2}=1$

which, like the ellipse, is symmetric about both axes and the origin. Letting $y=0$, we find that the x-intercepts of Equation (11.7) are $x=\pm a$. We can get additional information about the shape of the hyperbola by solving Equation (11.7) for y. Performing the necessary algebraic operations yields

$$y=\pm\frac{b}{a}\sqrt{x^2-a^2}$$

from which it is evident that the curve does not exist for $x^2<a^2$. Consequently, the hyperbola consists of two separate curves, or *branches*—one to the right of $x=a$ and a similar one to the left of $x=-a$.

The shape of the hyperbola is constrained by two straight lines, called the *asymptotes* of the hyperbola. The asymptotes of a hyperbola are the extended diagonals of the rectangle formed by drawing lines parallel to the coordinate axes through the endpoints of both the transverse axis and the conjugate axis. Referring to Figure 11.11, we see that the slopes of the diagonals of this rectangle are

$$m=\pm\frac{y}{x}=\pm\frac{b}{a}$$

and therefore the asymptotes are given by the lines

$$y=\pm\frac{b}{a}x$$

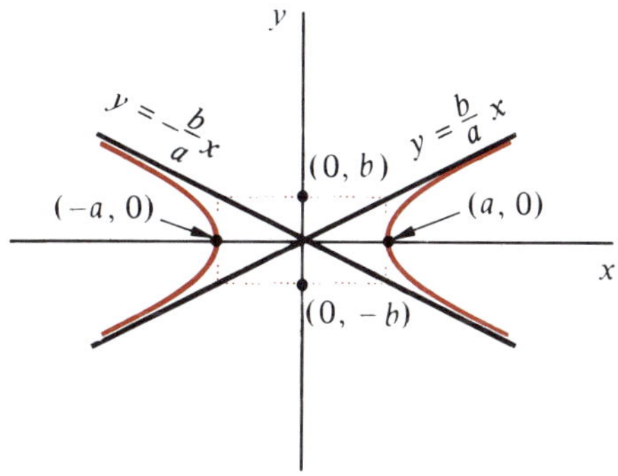

FIGURE 11.11

We now want to show that these lines are asymptotes of the hyperbola; that is, we want to show that the hyperbola approaches arbitrarily close to the lines as x increases without bound. Solving Equation (11.7) for y^2, we can write it in the form

$$y^2 = \frac{b^2x^2}{a^2}\left(1 - \frac{a^2}{x^2}\right)$$

or if we take the square root of both sides, then

$$\textbf{(11.8)} \quad y = \pm\frac{b}{a}x\sqrt{1 - \frac{a^2}{x^2}}$$

Now consider the value of the right member as x becomes large. The quantity a^2/x^2 becomes small, and therefore,

$$\pm\frac{b}{a}x\sqrt{1 - \frac{a^2}{x^2}}$$

approaches $\pm(b/a)x$, which means that the hyperbola $(x^2/a^2) - (y^2/b^2) = 1$ is asymptotic to the lines $y = \pm(b/a)x$.

If we begin with the foci of the hyperbola on the y-axis and the center at the origin, the standard form of the equation of a hyperbola becomes

$$\textbf{(11.9)} \quad \frac{y^2}{a^2} - \frac{x^2}{b^2} = 1$$

The vertices of this hyperbola are on the y-axis; **the positive term always indicates the direction of the transverse axis**. For the hyperbola, like the ellipse, standard form demands that the coefficients of x^2 and y^2 be in the denominator and that the number on the righthand side be 1. For the hyperbola, the sign of the term, *not* the magnitude of the denominator, determines the transverse axis.

To sketch the hyperbola, first draw the rectangle through the extremities of the transverse and conjugate axes and extend the diagonals of the rectangle. Then draw the hyperbola so that it passes through the vertex and comes close to the extended diagonals as x moves away from the origin.

EXAMPLE 1 Discuss and sketch the graph of $4x^2 - y^2 = 16$.

SOLUTION Dividing by 16, we have

$$\frac{x^2}{4} - \frac{y^2}{16} = 1$$

which is the equation of a hyperbola with center at the origin and foci on the x-axis. It has vertices at $(\pm 2, 0)$, and its conjugate axis extends from $(0, 4)$ to $(0, -4)$. The foci are found from the equation $b^2 = c^2 - a^2$. Thus,

$$c = \sqrt{a^2 + b^2} = \sqrt{4 + 16} = \sqrt{20} = 2\sqrt{5}$$

and the foci are located at $(\pm 2\sqrt{5}, 0)$. If these points are plotted and the rectangle and its extended diagonals drawn, the hyperbola shown in Figure 11.12 is obtained.

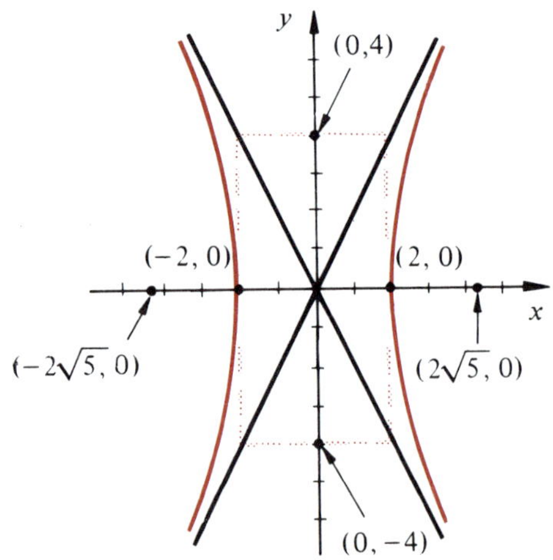

FIGURE 11.12

EXAMPLE 2 Discuss and sketch the graph of $\dfrac{y^2}{13} - \dfrac{x^2}{9} = 1$.

SOLUTION This hyperbola has a vertical transverse axis with vertices at $(0, \sqrt{13})$ and $(0, -\sqrt{13})$. The extremes of the conjugate axis are then $(3, 0)$ and $(-3, 0)$. The foci are at $(0, \sqrt{22})$ and $(0, -\sqrt{22})$. This information is used to sketch the hyperbola in Figure 11.13.

EXAMPLE 3 Determine the equation of the hyperbola centered at the origin with foci at $(\pm 6, 0)$ and a transverse axis eight units long.

SOLUTION Here $a = 4$ and $c = 6$. Since $c^2 = a^2 + b^2$, we have $b^2 = c^2 - a^2 = 36 - 16 = 20$. Substituting $a^2 = 16$ and $b^2 = 20$ in Equation (11.7) gives

$$\frac{x^2}{16} - \frac{y^2}{20} = 1$$

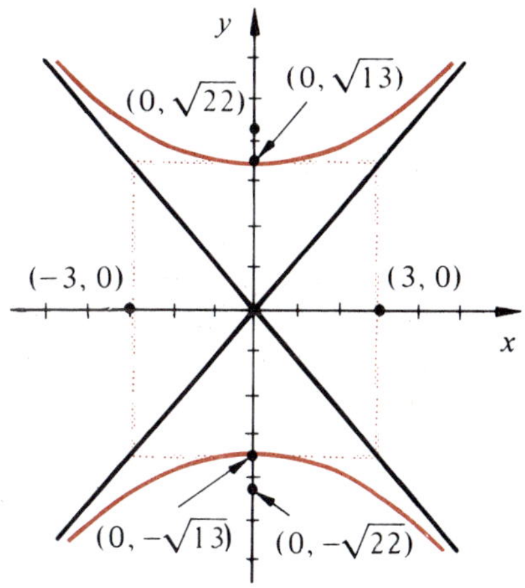

FIGURE 11.13

Exercises for Section 11.3

In Exercises 1–10, discuss the properties of the graph of each of the given equations and then sketch the graph.

1. $x^2 - y^2 = 16$
2. $y^2 - x^2 = 9$
3. $4x^2 - 9y^2 = 36$
4. $9x^2 - y^2 = 9$
5. $4y^2 - 25x^2 = 100$
6. $3x^2 - 3y^2 = 9$
7. $4x^2 - 16y^2 = 25$
8. $4y^2 - x^2 = 9$
9. $y^2 + 1 = x^2$
10. $x^2 - 25 = 5y^2$

In Exercises 11–20, find the equation of the hyperbolas having the given properties. Sketch each curve.

11. Vertices at $(\pm 4, 0)$, foci at $(\pm 5, 0)$.
12. Vertices at $(0, \pm 3)$, foci at $(0, \pm 5)$.
13. Conjugate axis 4, vertices at $(0, \pm 1)$.
14. Conjugate axis 1, vertices at $(\pm 4, 0)$.
15. Transverse axis 6, foci at $(\pm \frac{7}{2}, 0)$.
16. Transverse axis 3, foci at $(\pm 2, 0)$.
17. Vertices at $(0, \pm 4)$, asymptotes $y = \pm(\frac{1}{2})x$.
18. Vertices at $(\pm 3, 0)$, asymptotes $y = \pm 2x$.
19. Vertices at $(0, \pm 3)$, one point of the curve $(2, 7)$.
20. Vertices at $(\pm 3, 0)$, one point of the curve $(7, 2)$.

11.4 Translation of Axes

The equation of a circle centered at the origin has the form $x^2 + y^2 = r^2$ (Section 4.2). It is sometimes desirable to be able to write the equation of a circle centered at some other point in the plane besides the origin. This can be done by writing the equation of the circle with respect to a new pair of axes that are parallel to the original axes. Changing from one pair of axes to another like this is called a *translation of axes*. We shall investigate how translating axes affects the form of the standard equations of the conic sections.

Consider a point $P(x, y)$ in the xy-coordinate plane. Under a translation of axes, the coordinates of P are changed to agree with a new pair of axes. To see how

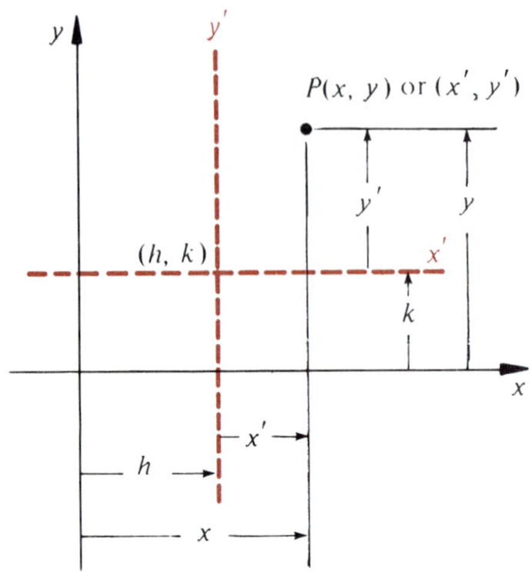

FIGURE 11.14

this happens, let us construct an x'-axis parallel to the x-axis and a y'-axis parallel to the y-axis, as shown in Figure 11.14. Let the intersection of the new axes be at the point (h, k) in the original coordinate system. Points referred to the x'-axis and y'-axis will be designated by a prime (′) notation. Thus the coordinates of the point P, with respect to the translated axes, are (x', y'). To transform (x, y) into (x', y'), one must know the relation between the new and old axes. From Figure 11.14, we see that the two coordinate systems are related by the equations

(11.10)
$$x = x' + h$$
$$y = y' + k$$

or

(11.11)
$$x' = x - h$$
$$y' = y - k$$

These equations are called *the equations for the translation of axes.*

Now let us consider a circle with its center at (h, k) in the xy-coordinate plane, as shown in Figure 11.15. If (h, k) is taken as the intersection of the $x'y'$-axes, the equation of the circle referred to these axes can be written

$$(x')^2 + (y')^2 = r^2$$

To express this equation in xy-coordinates, we apply Equation (11.11). Thus

(11.12) $(x-h)^2 + (y-k)^2 = r^2$

is the standard equation of a circle of radius r centered at a point (h, k).

The procedure for getting the standard form of the equation of a circle centered at a point (h, k) can be extended to the other conic sections when they have been displaced from the origin.

The equation of the parabola that has its vertex at the origin and a vertical axis is $x^2 = 4ay$. Hence, the equation of the same parabola with its vertex at (h, k)

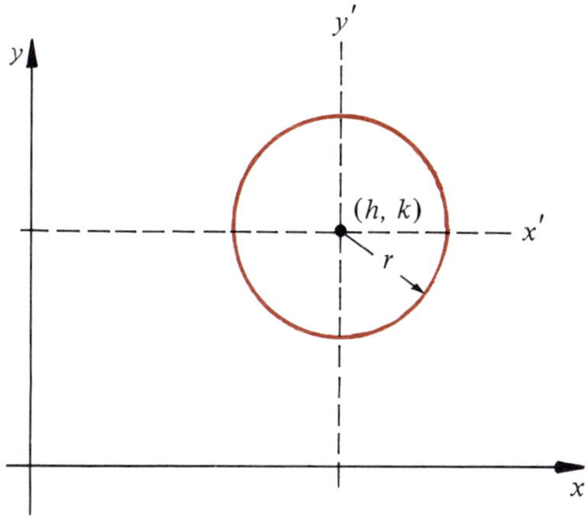

FIGURE 11.15

can be written in terms of the $x'y'$-coordinate plane as

$$(x')^2 = 4ay'$$

or by the translating Equation (11.11), as

(11.13) $(x-h)^2 = 4a(y-k)$

Similarly, the parabola with a horizontal axis and vertex at (h, k) is given by

(11.14) $(y-k)^2 = 4a(x-h)$

By a similar procedure, the standard equation of an ellipse centered at a point (h, k) with horizontal major axis is

(11.15) $\dfrac{(x-h)^2}{a^2} + \dfrac{(y-k)^2}{b^2} = 1$

and if the major axis is vertical, the equation is

(11.16) $\dfrac{(x-h)^2}{b^2} + \dfrac{(y-k)^2}{a^2} = 1$

Finally, the standard equation of a hyperbola centered at a point (h, k) with horizontal transverse axis is

(11.17) $\dfrac{(x-h)^2}{a^2} - \dfrac{(y-k)^2}{b^2} = 1$

and if the transverse axis is vertical, the equation is

(11.18) $\dfrac{(y-k)^2}{a^2} - \dfrac{(x-h)^2}{b^2} = 1$

EXAMPLE 1 Write the equation of the ellipse centered at $(2, -3)$ with horizontal axis 10 units and vertical axis 4 units. (See Figure 11.16.)

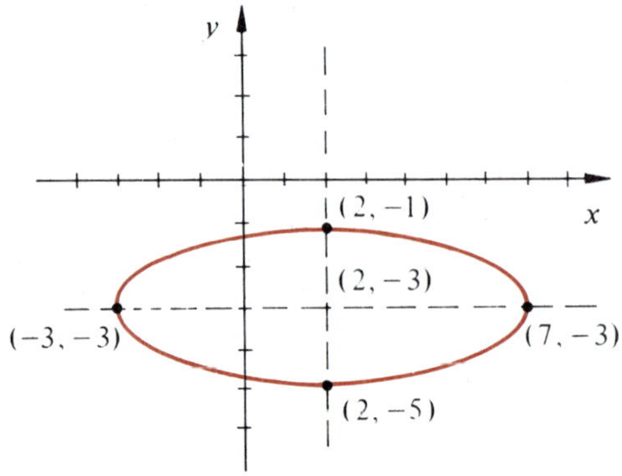

FIGURE 11.16

SOLUTION Since the longer of the two axes is the horizontal axis, use Equation (11.15), with $a=5$ and $b=2$. Also, $h=2$ and $k=-3$. Making these substitutions, we have as the equation of the ellipse.

$$\frac{(x-2)^2}{25}+\frac{(y+3)^2}{4}=1$$

EXAMPLE 2 Write the equation of the parabola whose directrix is the line $y=-2$ and whose vertex is at (3, 1). (See Figure 11.17.)

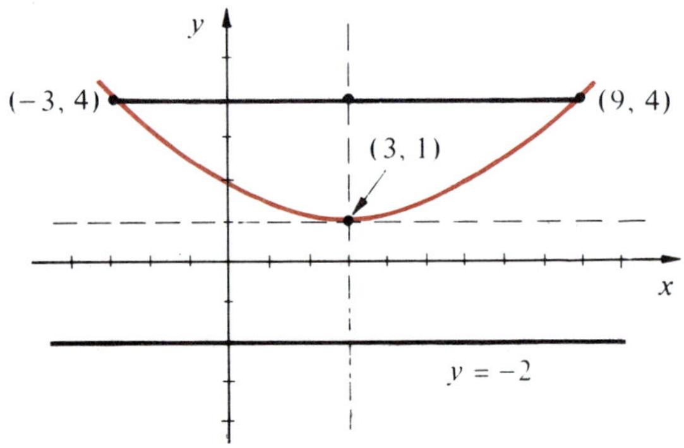

FIGURE 11.17

SOLUTION The parabola has a vertical axis and opens upward since the directrix is horizontal and lies below the vertex. The vertex lies midway between the focus and the directrix so that $a=3$. Letting $h=3$, $k=1$, and $a=3$ in Equation (11.13), the equation of the parabola is

$$(x-3)^2=12(y-1)$$

EXAMPLE 3 Discuss and sketch the graph of the hyperbola

$$\frac{(x+5)^2}{16}-\frac{(y-2)^2}{9}=1$$

SOLUTION The hyperbola is centered at $(-5, 2)$. By Equation (11.17) it has a horizontal transverse axis with vertices at $(-9, 2)$ and $(-1, 2)$. The endpoints of the conjugate axis are at $(-5, 5)$ and $(-5, -1)$. Also the foci are $(-10, 2)$ and $(0, 2)$, since $c=\sqrt{9+16}=5$. The graph appears in Figure 11.18.

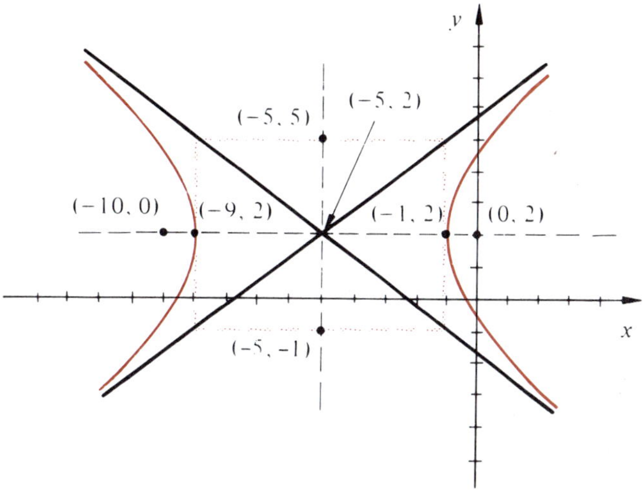

FIGURE 11.18

Exercises for Section 11.4

In Exercises 1–6, write the equations of the *parabolas* having the given properties. Sketch each graph.

1. Vertex at $(3, 1)$, focus at $(5, 1)$.

2. Vertex at $(-2, 3)$, focus at $(-2, 0)$.

3. Directrix $y=2$, vertex at $(1, -1)$.

4. Directrix at $x=-1$, vertex at $(0, 4)$.

5. Endpoints of right chord $(2, 4)$ and $(2, 0)$, opening to the right.

6. Endpoints of right chord $(-1, -1)$ and $(5, -1)$, opening upward.

In Exercises 7–12, write the equations of the *ellipses* having the given properties. Sketch each graph.

7. Major axis 8, foci at $(5, 1)$ and $(-1, 1)$.

8. Minor axis 6, vertices at $(2, -1)$ and $(10, -1)$.

9. Minor axis 2, vertices at $(\frac{1}{2}, 0)$ and $(\frac{1}{2}, -8)$.

10. Semimajor axis $\frac{3}{2}$, foci at $(1, 1)$ and $(1, -1)$.

11. Vertices at $(-6, 3)$ and $(-2, 3)$, foci at $(-5, 3)$ and $(-3, 3)$.

12. Center at $(1, -3)$, major axis 10, minor axis 6, vertical axis.

In Exercises 13–18, write the equations of the *hyperbolas* having the given properties. Sketch each graph.

13. Center at $(-1, 2)$, transverse axis 7, conjugate axis 8, vertical axis.

14. Center at $(3, 0)$, transverse axis 6, conjugate axis 2, horizontal axis.

15. Vertices at $(5, 1)$ and $(-1, 1)$, foci at $(6, 1)$ and $(-2, 1)$.

16. Vertices at $(2, \pm 4)$, conjugate axis 2.

17. Vertices at $(-4, -2)$ and $(0, -2)$, asymptotes $m=\pm\frac{1}{2}$.

18. Vertices at $(3, 3)$ and $(5, 3)$, asymptotes $m=\pm 3$.

Write the equation of the family of curves indicated in Exercises 19–24.

19. Circles with center on the x-axis.

20. Parabolas with vertical axis and vertex on the x-axis.

21. Parabolas with vertex and focus on the x-axis.

22. Ellipses with center on the y-axis and horizontal major axis.

23. Circles passing through the origin with center on the x-axis.

24. Circles tangent to the x-axis.

11.5 The General Second-Degree Equation

The general second-degree equation is of the form

(11.19) $Ax^2+Bxy+Cy^2+Dx+Ey+F=0$

where A, B, C, D, E, and F are constants. Each conic described in Sections 11.1 through 11.4 can be expressed in the form of Equation (11.19) with $B=0$. This is seen by expanding the standard form of each conic. If it is assumed that $B=0$, the following statements are easy to show.

- If $A=C=1$, Equation (11.19) represents a circle.
- If $A \neq C$ and A and C have the same numerical sign, Equation (11.19) represents an ellipse.
- If A and C have different numerical signs, Equation (11.19) represents a hyperbola.
- If A or $C=0$ (but not both), then Equation (11.19) represents a parabola.
- Special cases such as a single point or no graph may result.

If $B=0$, the general form of a conic can be reduced to one of the standard forms by completing the square on x and y. Two examples of this technique follow.

EXAMPLE 1 Discuss and sketch the graph of $x^2-4y^2+6x+24y-43=0$.

SOLUTION This is the equation of a hyperbola, since the coefficients of the x^2 and y^2 terms have unlike signs. To sketch the hyperbola, reduce the given equation to standard form by rearranging the terms and completing the square on the x-terms and the y-terms. Thus,

$$x^2-4y^2+6x+24y-43=0$$

can be written

$$(x^2+6x\quad)-4(y^2-6y\quad)=43$$

Completing the square on each variable, we get

$$(x^2+6x+9)-4(y^2-6y+9)=43+9-36$$

$$(x+3)^2-4(y-3)^2=16$$

$$\frac{(x+3)^2}{16}-\frac{(y-3)^2}{4}=1$$

The center of the hyperbola is the point $(-3, 3)$. The transverse axis is horizontal with vertices at $(1, 3)$ and $(-7, 3)$. The endpoints of the conjugate axis are at $(-3, 5)$ and $(-3, 1)$. Finally, the foci are at $(-3+2\sqrt{5}, 3)$ and $(-3-2\sqrt{5}, 3)$, since $c=\sqrt{16+4}=2\sqrt{5}$. The graph appears in Figure 11.19.

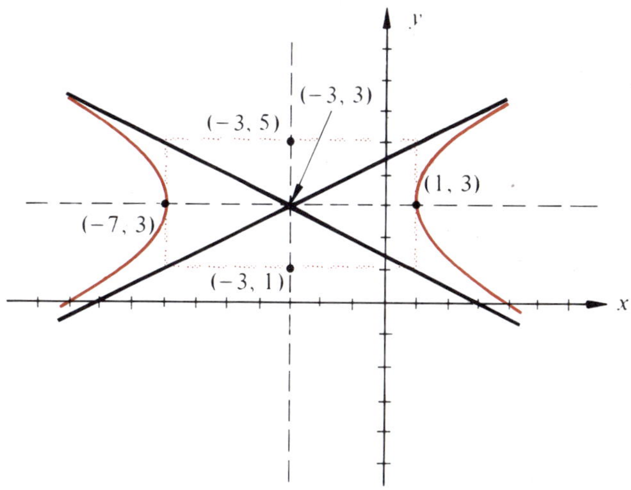

FIGURE 11.19

EXAMPLE 2 Discuss and sketch the graph of $2y^2+3x-8y+9=0$.

SOLUTION If the square on the y-variable is completed, this equation can be reduced to the form of Equation (11.14). Thus,

$$\begin{aligned} 2y^2+3x-8y+9&=0 \\ 2(y^2-4y\quad)&=-3x-9 \\ 2(y^2-4y+4)&=-3x-9+8 \\ 2(y-2)^2&=-3\left(x+\frac{1}{3}\right) \\ (y-2)^2&=-\frac{3}{2}\left(x+\frac{1}{3}\right) \end{aligned}$$

This is the standard form of the equation of a parabola, with horizontal axis and vertex at $(-\frac{1}{3}, 2)$. We see that $4a=-\frac{3}{2}$, so $a=-\frac{3}{8}$. Therefore, the focus is at $(-\frac{17}{24}, 2)$, and the endpoints of the right chord are $(-\frac{17}{24}, \frac{11}{4})$ and $(-\frac{17}{24}, \frac{5}{4})$. The parabola, which opens to the left, is shown in Figure 11.20.

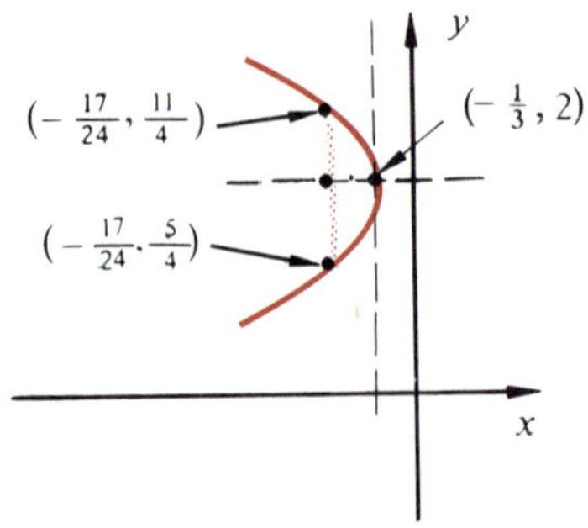

FIGURE 11.20

Exercises for Section 11.5

Transform each of the equations in Exercises 1–20 into standard form and sketch their graphs.

1. $x^2+y^2+4x+6y+4=0$

2. $x^2-2x-8y+25=0$

3. $9x^2+4y^2+18x+8y-23=0$

4. $2x^2+2y^2-4x-16=0$

5. $x^2-y^2-4x-21=0$

6. $y^2+4y+6x-8=0$

7. $x^2-6x-3y=0$

8. $x^2+4y^2+8x=0$

9. $2y^2-2y+x-1=0$

10. $3x^2+4y^2-18x+8y+19=0$

11. $4x^2-y^2+8x+2y-1=0$

12. $y^2-2x-4y+10=0$

13. $9x^2+4y^2-18x+16y-11=0$

14. $y^2-25x^2+50x-50=0$

15. $x^2-2y^2+2y=0$

16. $y^2-y-\frac{1}{2}x+\frac{1}{4}=0$

17. $x(x+4)=y^2+3$

18. $x^2+y(4+2y)=0$

19. $y=x^2+5x+7$

20. $y=2x^2+10x$

Test 1 for Chapter 11

Answer *True* or *False* to Exercises 1–10.

1. The coordinates of every point (x, y) on a curve satisfy the equation of the curve.

2. The right chord of $y^2=4ax$ is perpendicular to the x-axis.

3. $2x^2+y^2=c$ represents an ellipse for any real constant c.

4. The graph of $x^2-y^2=0$ is a hyperbola.

5. The circle $x^2+y^2=4$ has radius 2.

6. The ellipse $2x^2+y^2=1$ has a major axis of length $\sqrt{2}$.

7. The equation of an ellipse centered at the origin is of the form $ax^2+by^2=1$, where a and b are both >0.

8. The hyperbola has four branches.

9. The hyperbola $y^2=1+x^2$ has two y-intercepts, $+1$ and -1.

10. The center of the circle $(x+3)^2+(y-2)^2=1$ is located at $(3, -2)$.

11. Sketch the graph of $x^2+y^2+4y=0$.

12. Determine the equation of a parabola with vertex at $(1, -2)$ and directrix $y=6$.

13. Sketch the graph of $x^2+4y^2=4$. Locate the foci.

14. Sketch the graph of $2x^2+x+y^2+2y=1$. Determine the translation that will place the center of the curve at a new origin.

15. Sketch $x^2-4y^2=4$, being sure to give the oblique asymptotes both graphically and analytically.

Test 2 for Chapter 11

1. Sketch the parabola with focus at (3, 4) and directrix $y=-2$.

2. Find the equation of the parabola with focus at $(-2, 1)$ and directrix $x=3$.

3. Find the equation of the ellipse with center at (0, 5), a vertical major axis of 6 and a minor axis of 3.

4. Find the equation of the hyperbola with center at the origin, vertices at $(\pm 3, 0)$ and foci at $(\pm 4, 0)$.

5. Find the equation of the circle that has its center at $(-2, -3)$ and passes through (0, 2).

6. Explain how each of the conic sections is obtained from a right circular cone.

Discuss and sketch the graph of Exercises 7–10.

7. $x^2+y^2-y=0$

8. $y=(x+3)(x-2)$

9. $x(x+8)=y^2$

10. $x^2+3y^2+x-y=0$

TABLE A
Exponential Functions

x	e^x	e^{-x}	x	e^x	e^{-x}
0.00	1.0000	1.0000	2.5	12.182	0.0821
0.05	1.0513	0.9512	2.6	13.464	0.0743
0.10	1.1052	0.9048	2.7	14.880	0.0672
0.15	1.1618	0.8607	2.8	16.445	0.0608
0.20	1.2214	0.8187	2.9	18.174	0.0550
0.25	1.2840	0.7788	3.0	20.086	0.0498
0.30	1.3499	0.7408	3.1	22.198	0.0450
0.35	1.4191	0.7047	3.2	24.533	0.0408
0.40	1.4918	0.6703	3.3	27.113	0.0369
0.45	1.5683	0.6376	3.4	29.964	0.0334
0.50	1.6487	0.6065	3.5	33.115	0.0302
0.55	1.7333	0.5769	3.6	36.598	0.0273
0.60	1.8221	0.5488	3.7	40.447	0.0247
0.65	1.9155	0.5220	3.8	44.701	0.0224
0.70	2.0138	0.4966	3.9	49.402	0.0202
0.75	2.1170	0.4724	4.0	54.598	0.0183
0.80	2.2255	0.4493	4.1	60.340	0.0166
0.85	2.3396	0.4274	4.2	66.686	0.0150
0.90	2.4596	0.4066	4.3	73.700	0.0136
0.95	2.5857	0.3867	4.4	81.451	0.0123
1.0	2.7183	0.3679	4.5	90.017	0.0111
1.1	3.0042	0.3329	4.6	99.484	0.0101
1.2	3.3201	0.3012	4.7	109.95	0.0091
1.3	3.6693	0.2725	4.8	121.51	0.0082
1.4	4.0552	0.2466	4.9	134.29	0.0074
1.5	4.4817	0.2231	5	148.41	0.0067
1.6	4.9530	0.2019	6	403.43	0.0025
1.7	5.4739	0.1827	7	1096.6	0.0009
1.8	6.0496	0.1653	8	2981.0	0.0003
1.9	6.6859	0.1496	9	8103.1	0.0001
2.0	7.3891	0.1353	10	22026	0.00005
2.1	8.1662	0.1225			
2.2	9.0250	0.1108			
2.3	9.9742	0.1003			
2.4	11.023	0.0907			

TABLE B
Four-Place Logarithms of Numbers from 1 to 10

n	0	1	2	3	4	5	6	7	8	9
1.0	0.0000	0043	0086	0128	0170	0212	0253	0294	0334	0374
1.1	.0414	0453	0492	0531	0569	0607	0645	0682	0719	0755
1.2	.0792	0828	0864	0899	0934	0969	1004	1038	1072	1106
1.3	.1139	1173	1206	1239	1271	1303	1335	1367	1399	1430
1.4	.1461	1492	1523	1553	1584	1614	1644	1673	1703	1732
1.5	.1761	1790	1818	1847	1875	1903	1931	1959	1987	2014
1.6	.2041	2068	2095	2122	2148	2175	2201	2227	2253	2279
1.7	.2304	2330	2355	2380	2405	2430	2455	2480	2504	2529
1.8	.2553	2577	2601	2625	2648	2672	2695	2718	2742	2765
1.9	.2788	2810	2833	2856	2878	2900	2923	2945	2967	2989
2.0	.3010	3032	3054	3075	3096	3118	3139	3160	3181	3201
2.1	.3222	3243	3263	3284	3304	3324	3345	3365	3385	3404
2.2	.3424	3444	3464	3483	3502	3522	3541	3560	3579	3598
2.3	.3617	3636	3655	3674	3692	3711	3729	3747	3766	3784
2.4	.3802	3820	3838	3856	3874	3892	3909	3927	3945	3962
2.5	.3979	3997	4014	4031	4048	4065	4082	4099	4116	4133
2.6	.4150	4166	4183	4200	4216	4232	4249	4265	4281	4298
2.7	.4314	4330	4346	4362	4378	4393	4409	4425	4440	4456
2.8	.4472	4487	4502	4518	4533	4548	4564	4579	4594	4609
2.9	.4624	4639	4654	4669	4683	4698	4713	4728	4742	4757
3.0	.4771	4786	4800	4814	4829	4843	4857	4871	4886	4900
3.1	.4914	4928	4942	4955	4969	4983	4997	5011	5024	5038
3.2	.5051	5065	5079	5092	5105	5119	5132	5145	5159	5172
3.3	.5185	5198	5211	5224	5237	5250	5263	5276	5289	5302
3.4	.5315	5328	5340	5353	5366	5378	5391	5403	5416	5428
3.5	.5441	5453	5465	5478	5490	5502	5514	5527	5539	5551
3.6	.5563	5575	5587	5599	5611	5623	5635	5647	5658	5670
3.7	.5682	5694	5705	5717	5729	5740	5752	5763	5775	5786
3.8	.5798	5809	5821	5832	5843	5855	5866	5877	5888	5899
3.9	.5911	5922	5933	5944	5955	5966	5977	5988	5999	6010
4.0	.6021	6031	6042	6053	6064	6075	6085	6096	6107	6117
4.1	.6128	6138	6149	6160	6170	6180	6191	6201	6212	6222
4.2	.6232	6243	6253	6263	6274	6284	6294	6304	6314	6325
4.3	.6335	6345	6355	6365	6375	6385	6395	6405	6415	6425
4.4	.6435	6444	6454	6464	6474	6484	6493	6503	6513	6522
4.5	.6532	6542	6551	6561	6571	6580	6590	6599	6609	6618
4.6	.6628	6637	6646	6656	6665	6675	6684	6693	6702	6712
4.7	.6721	6730	6739	6749	6758	6767	6776	6785	6794	6803
4.8	.6812	6821	6830	6839	6848	6857	6866	6875	6884	6893
4.9	.6902	6911	6920	6928	6937	6946	6955	6964	6972	6981
5.0	.6990	6998	7007	7016	7024	7033	7042	7050	7059	7067
5.1	.7076	7084	7093	7101	7110	7118	7126	7135	7143	7152
5.2	.7160	7168	7177	7185	7193	7202	7210	7218	7226	7235
5.3	.7243	7251	7259	7267	7275	7284	7292	7300	7308	7316
5.4	.7324	7332	7340	7348	7356	7364	7372	7380	7388	7396

TABLE B
Four-Place Logarithms of Numbers from 1 to 10 (continued)

n	0	1	2	3	4	5	6	7	8	9
5.5	.7404	7412	7419	7427	7435	7443	7451	7459	7466	7474
5.6	.7482	7490	7497	7505	7513	7520	7528	7536	7543	7551
5.7	.7559	7566	7574	7582	7589	7597	7604	7612	7619	7627
5.8	.7634	7642	7649	7657	7664	7672	7679	7686	7694	7701
5.9	.7709	7716	7723	7731	7738	7745	7752	7760	7767	7774
6.0	.7782	7789	7796	7803	7810	7818	7825	7832	7839	7846
6.1	.7853	7860	7868	7875	7882	7889	7896	7903	7910	7917
6.2	.7924	7931	7938	7945	7952	7959	7966	7973	7980	7987
6.3	.7993	8000	8007	8014	8021	8028	8035	8041	8048	8055
6.4	.8062	8069	8075	8082	8089	8096	8102	8109	8116	8122
6.5	.8129	8136	8142	8149	8156	8162	8169	8176	8182	8189
6.6	.8195	8202	8209	8215	8222	8228	8235	8241	8248	8254
6.7	.8261	8267	8274	8280	8287	8293	8299	8306	8312	8319
6.8	.8325	8331	8338	8344	8351	8357	8363	8370	8376	8382
6.9	.8388	8395	8401	8407	8414	8420	8426	8432	8439	8445
7.0	.8451	8457	8463	8470	8476	8482	8488	8494	8500	8506
7.1	.8513	8519	8525	8531	8537	8543	8549	8555	8561	8567
7.2	.8573	8579	8585	8591	8597	8603	8609	8615	8621	8627
7.3	.8633	8639	8645	8651	8657	8663	8669	8675	8681	8686
7.4	.8692	8698	8704	8710	8716	8722	8727	8733	8739	8745
7.5	.8751	8756	8762	8768	8774	8779	8785	8791	8797	8802
7.6	.8808	8814	8820	8825	8831	8837	8842	8848	8854	8859
7.7	.8865	8871	8876	8882	8887	8893	8899	8904	8910	8915
7.8	.8921	8927	8932	8938	8943	8949	8954	8960	8965	8971
7.9	.8976	8982	8987	8993	8998	9004	9009	9015	9020	9025
8.0	.9031	9036	9042	9047	9053	9058	9063	9069	9074	9079
8.1	.9085	9090	9096	9101	9106	9112	9117	9122	9128	9133
8.2	.9138	9143	9149	9154	9159	9165	9170	9175	9180	9186
8.3	.9191	9196	9201	9206	9212	9217	9222	9227	9232	9238
8.4	.9243	9248	9253	9258	9263	9269	9274	9279	9284	9289
8.5	.9294	9299	9304	9309	9315	9320	9325	9330	9335	9340
8.6	.9345	9350	9355	9360	9365	9370	9375	9380	9385	9390
8.7	.9395	9400	9405	9410	9415	9420	9425	9430	9435	9440
8.8	.9445	9450	9455	9460	9465	9469	9474	9479	9484	9489
8.9	.9494	9499	9504	9509	9513	9518	9523	9528	9533	9538
9.0	.9542	9547	9552	9557	9562	9566	9571	9576	9581	9586
9.1	.9590	9595	9600	9605	9609	9614	9619	9624	9628	9633
9.2	.9638	9643	9647	9652	9657	9661	9666	9671	9675	9680
9.3	.9685	9689	9694	9699	9703	9708	9713	9717	9722	9727
9.4	.9731	9736	9741	9745	9750	9754	9759	9763	9768	9773
9.5	.9777	9782	9786	9791	9795	9800	9805	9809	9814	9818
9.6	.9823	9827	9832	9836	9841	9845	9850	9854	9859	9863
9.7	.9868	9872	9877	9881	9886	9890	9894	9899	9903	9908
9.8	.9912	9917	9921	9926	9930	9934	9939	9943	9948	9952
9.9	.9956	9961	9965	9969	9974	9978	9983	9987	9991	9996

TABLE C
Powers and Roots

n	n^2	$\sqrt{n}$	n^3	$\sqrt[3]{n}$	n	n^2	$\sqrt{n}$	n^3	$\sqrt[3]{n}$
1	1	1.000	1	1.000	**51**	2,601	7.141	132,651	3.708
2	4	1.414	8	1.260	**52**	2,704	7.211	140,608	3.733
3	9	1.732	27	1.442	**53**	2,809	7.280	148,877	3.756
4	16	2.000	64	1.587	**54**	2,916	7.348	157,464	3.780
5	25	2.236	125	1.710	**55**	3,025	7.416	166,375	3.803
6	36	2.449	216	1.817	**56**	3,136	7.483	175,616	3.826
7	49	2.646	343	1.913	**57**	3,249	7.550	185,193	3.849
8	64	2.828	512	2.000	**58**	3,364	7.616	195,112	3.871
9	81	3.000	729	2.080	**59**	3,481	7.681	205,379	3.893
10	100	3.162	1,000	2.154	**60**	3,600	7.746	216,000	3.915
11	121	3.317	1,331	2.224	**61**	3,721	7.810	226,981	3.936
12	144	3.464	1,728	2.289	**62**	3,844	7.874	238,328	3.958
13	169	3.606	2,197	2.351	**63**	3,969	7.937	250,047	3.979
14	196	3.742	2,744	2.410	**64**	4,096	8.000	262,144	4.000
15	225	3.873	3,375	2.466	**65**	4,225	8.062	274,625	4.021
16	256	4.000	4,096	2.520	**66**	4,356	8.124	287,496	4.041
17	289	4.123	4,913	2.571	**67**	4,489	8.185	300,763	4.062
18	324	4.243	5,832	2.621	**68**	4,624	8.246	314,432	4.082
19	361	4.359	6,859	2.668	**69**	4,761	8.307	328,509	4.102
20	400	4.472	8,000	2.714	**70**	4,900	8.367	343,000	4.121
21	441	4.583	9,261	2.759	**71**	5,041	8.426	357,911	4.141
22	484	4.690	10,648	2.802	**72**	5,184	8.485	373,248	4.160
23	529	4.796	12,167	2.844	**73**	5,329	8.544	389,017	4.179
24	576	4.899	13,824	2.884	**74**	5,476	8.602	405,224	4.198
25	625	5.000	15,625	2.924	**75**	5,625	8.660	421,875	4.217
26	676	5.099	17,576	2.962	**76**	5,776	8.718	438,976	4.236
27	729	5.196	19,683	3.000	**77**	5,929	8.775	456,533	4.254
28	784	5.292	21,952	3.037	**78**	6,084	8.832	474,552	4.273
29	841	5.385	24,389	3.072	**79**	6,241	8.888	493,039	4.291
30	900	5.477	27,000	3.107	**80**	6,400	8.944	512,000	4.309
31	961	5.568	29,791	3.141	**81**	6,561	9.000	531,441	4.327
32	1,024	5.657	32,768	3.175	**82**	6,724	9.055	551,368	4.344
33	1,089	5.745	35,937	3.208	**83**	6,889	9.110	571,787	4.362
34	1,156	5.831	39,304	3.240	**84**	7,056	9.165	592,704	4.380
35	1,225	5.916	42,875	3.271	**85**	7,225	9.220	614,125	4.397
36	1,296	6.000	46,656	3.302	**86**	7,396	9.274	636,056	4.414
37	1,369	6.083	50,653	3.332	**87**	7,569	9.327	658,503	4.431
38	1,444	6.164	54,872	3.362	**88**	7,744	9.381	681,472	4.448
39	1,521	6.245	59,319	3.391	**89**	7,921	9.434	704,969	4.465
40	1,600	6.325	64,000	3.420	**90**	8,100	9.487	729,000	4.481
41	1,681	6.403	68,921	3.448	**91**	8,281	9.539	753,571	4.498
42	1,764	6.481	74,088	3.476	**92**	8,464	9.592	778,688	4.514
43	1,849	6.557	79,507	3.503	**93**	8,649	9.644	804,357	4.531
44	1,936	6.633	85,184	3.530	**94**	8,836	9.695	830,584	4.547
45	2,025	6.708	91,125	3.557	**95**	9,025	9.747	857,375	4.563
46	2,116	6.782	97,336	3.583	**96**	9,216	9.798	884,736	4.579
47	2,209	6.856	103,823	3.609	**97**	9,409	9.849	912,673	4.595
48	2,304	6.928	110,592	3.634	**98**	9,604	9.899	941,192	4.610
49	2,401	7.000	117,649	3.659	**99**	9,801	9.950	970,299	4.626
50	2,500	7.071	125,000	3.684	**100**	10,000	10.000	1,000,000	4.642

Answers to Odd-Numbered Exercises

Sections 1.1 – 1.3 (page 10)

1. T **3.** F **5.** T **7.** F **9.** F **11.** $0.\overline{142\,857}$, nonterminating

13. 0.375, terminating **15.** $01.\overline{2}$, nonterminating **17.** $1.8\overline{3}$, nonterminating

19. $0.\overline{461\,538}$, nonterminating **21.** -1 **23.** 4 **25.** -6 **27.** -432 **29.** -8

31. 13 **33.** -12 **35.** 28 **37.** -19.4 **39.** -58.9166 **41.** 5489.1542 **43.** -63.732

45.

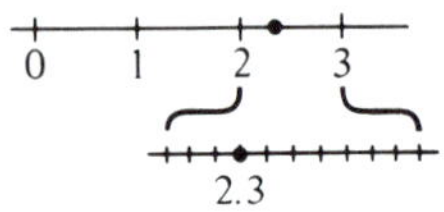

47.

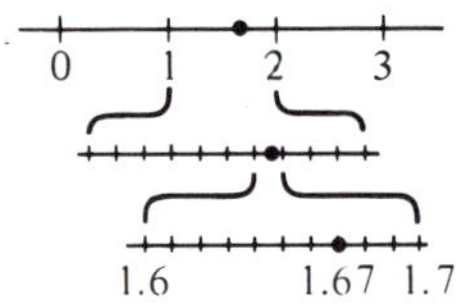

49.

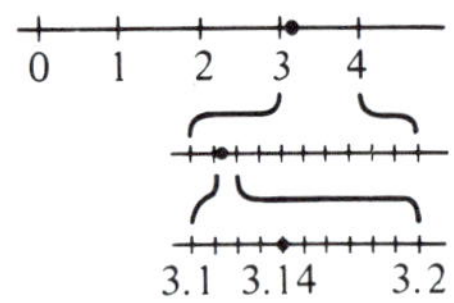

51.

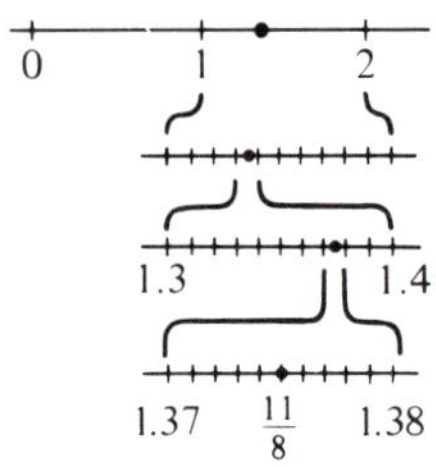

53.

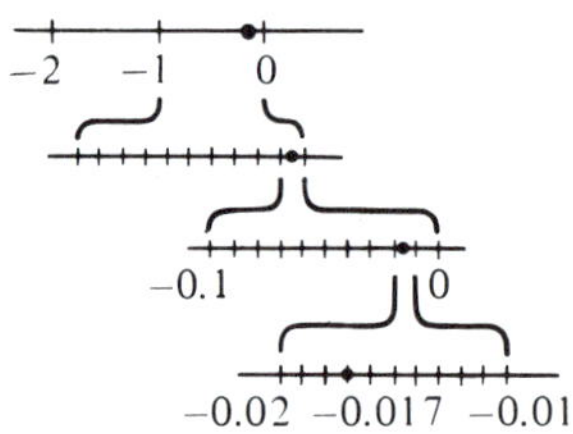

55. $17 < 25$ **57.** $-\frac{7}{6} < \frac{9}{7}$ **59.** $-0.1 < 0.002$ **61.** $-\frac{99}{7} < \frac{99}{12}$ **63.** $\sqrt{2} < \pi$ **65.** 7 **67.** 3

69. 14 **71.** 6 **73.** -2 **75.** 4 **77.** 4 **79.** 7.1 **81.** 6 **83.** 0.301

85. Try 1 and -1. True when a and b have the same sign or one or both are zero.

Section 1.4 (page 14)

1. $3x-2y+1$ **3.** x^2+4y^2-2y **5.** $-2y^2$ **7.** $2x^2$

9. $x^2y^2+3xy-x^2-y^2+12$ **11.** $y^2-4x^3+xy^3-xy^2+6x-16$

13. $-10x^2+16y^2-4$ **15.** $-5m+19n$ **17.** -2 **19.** 0

21. x^2-y^2-16 **23.** $-2x^2+2y^2$ **25.** $-8x+30$ **27.** $3+4y-4x$

29. $-10x+42y$ **31.** $-3i_1+2i_2$ **33.** $5T^2-2T-3$

Section 1.5 (page 20)

1. x^5+2x^2 **3.** x^5-5x^4 **5.** x^2-4y^2 **7.** x^4-1

9. $x^4-2x^3+2x^2-2x+1$ **11.** x^2y^2-1 **13.** $x^3+3x^2y^2+3xy^4+y^6$

15. $x^4-x^2y^2-x^2y+y^3$ 17. $x^2+y^2+z^2+2xy+2xz+2yz$

19. $6x^2-12y^2-xy-2xz+3yz$ 21. $x^2+8x+16$ 23. $x^2+30x+225$

25. b^2-144 27. $4x^2+20x+25$ 29. $81y^2-36y+4$ 31. $16x^2-25$

33. $x^3+6x^2+12x+8$ 35. $125a^3+75a^2b+15ab^2+b^3$

37. $8x^3-60x^2+150x-125$ 39. x^3+8 41. x^3-27

43. $8x^3+1$ 45. $x^2+14x+45$ 47. $6x^2-14x-40$ 49. $2y^2+9y+10$

51. $2x^2+5xy-12y^2$ 53. $36x^2-111x+40$ 55. $h^2+6h+12$

Section 1.6 (page 25)

1. $\frac{1}{2}abc^2$ 3. $4st$ 5. $\frac{xy}{3}-\frac{1}{4x^2y}$ 7. $ab^2-a^2b+\frac{1}{b^3}$

9. $\frac{xy}{4}+\frac{1}{x^2y}-\frac{y^2}{2}$ 11. $2x^2-3x+2$ 13. $3y^2+4y-2-\frac{5}{y+1}$ 15. $2x-1$

17. $x^2+7x+10$ 19. $x-1$ 21. $x^2+3+\frac{5}{x-2}$ 23. $x^2-\frac{1}{2}x-\frac{11}{4}+\frac{83}{8(x+\frac{1}{2})}$

25. $x^3+2x^2+8x+24$ 27. $-3x^4-6x^3-12x^2-24x-41-\frac{74}{x-2}$ 29. $3x^2-21x+30$

31. $\frac{1}{R_1}+\frac{1}{R_2}+\frac{1}{R_3}$

Section 1.7 (page 30)

1. $x(3x^2-5x+15)$ 3. $y^3(13y^4-27)$ 5. $(z-12)^2$ 7. $(3x+1)^2$

9. $(\frac{1}{2}x+1)^2$ 11. $(x+13)(x-13)$ 13. $4(m+4)(m-4)$ 15. $(x-\sqrt{7})(x+\sqrt{7})$

17. $(\frac{1}{2}x+\frac{1}{3})(\frac{1}{2}x-\frac{1}{3})$ 19. $(3x+2)^2$ 21. $\frac{1}{9}(x+1)^2$ 23. $(x-4)(x+3)$

25. $(y+6)(y+7)$ 27. $(x-9)(x+6)$ 29. $(2x+1)(x+3)$ 31. $(5x-3)(x-5)$

33. $[6(a+b)-5][(a+b)-1]$ 35. $(3x-5y)(x+3y)$ 37. $(3x+2y)(x-2y)$

39. $4(2x-y)(x-2y)$ 41. $(x+4)(x^2-4x+16)$ 43. $(2x+1)(4x^2-2x+1)$ 45. $(a+\frac{1}{2})(a^2-\frac{1}{2}a+\frac{1}{4})$

47. $(y-3)^3$ 49. $(2x+1)^3$ 51. $(1+ab)(a^2+b^2)$ 53. $(z-w)(z+w-3)$

55. $(a-c+d)(a+c-d)$ 57. $(x+2-\sqrt{3})(x+2+\sqrt{3})$ 59. $(a-4-\sqrt{21})(a-4+\sqrt{21})$

61. $(x+\frac{3}{2}+\sqrt{\frac{3}{2}})(x+\frac{3}{2}-\sqrt{\frac{3}{2}})$ 63. $4(y-1-\sqrt{3})(y-1+\sqrt{3})$

65. $[\sqrt{2}(b+\frac{3}{2})-\sqrt{\frac{15}{2}}][\sqrt{2}(b+\frac{3}{2})+\sqrt{\frac{15}{2}}]$

Section 1.8 (page 34)

1. $\frac{5c}{12a^2b^2}$ 3. $\frac{1}{6x^3y^2}$ 5. $\frac{6a^4}{b^2c^2}$ 7. $\frac{x-3}{2}$

9. $\frac{6y}{5x(2x-3)}$ 11. $\frac{x(x-2)}{y(x-1)}$ 13. $\frac{2y+1}{y+2}$ 15. $\frac{x(2x+3)}{x+6}$

17. $\frac{(s+1)^2}{s+3}$ 19. 1 21. $-(2y+1)(y+4)$ 23. $-\frac{(a+100)(a+400)}{2(a+40)}$

25. $\frac{2ac+ab-3bc}{abc}$ 27. $\frac{2t-3s+1}{3st}$ 29. $\frac{3+2y^2-3xy}{xy^2}$ 31. $\frac{5x-7}{(x+1)(x-3)}$

33. $\frac{x^2+5x-10}{(x-5)(x+3)}$ 35. $\frac{x+5}{x(x+2)}$ 37. $\frac{7a+4}{a(2a+3)}$ 39. $\frac{3x+32}{3x+4}$

41. $\frac{p^2+5p+2}{p(p+1)^2}$ 43. $\frac{x-4}{(x+1)(x-2)(x+3)}$ 45. $\frac{5x^2-x-3}{(x+3)(x-2)^2}$ 47. $\frac{x+5}{(x-2)(x-4)}$

49. $x+1$ 51. $\frac{x+2}{x-1}$ 53. $\frac{(x+2)(x-3)}{(x-4)(x+1)}$ 55. $\frac{2t^3+3t^2-15}{6t^3}$

57. $\frac{5s+9}{s(s+3)(s+1)}$

Section 1.9 (page 39)

1. $(x-\sqrt{3})(x+\sqrt{3})$, Domain: any set containing the real numbers.

3. $(x+3)(x+1)$, Domain: any set containing the integer numbers.

5. $\left(x-\frac{3+\sqrt{29}}{2}\right)\left(x-\frac{3-\sqrt{29}}{2}\right)$, Domain: any set containing the real numbers.

7. $(x-2+\sqrt{14})(x-2-\sqrt{14})$, Domain: any set containing the real numbers.

9. $(x+\sqrt{3}i)(x-\sqrt{3}i)$, Domain: any set containing the complex numbers.

11. $7+5i$ 13. $-2+3i$ 15. 4 17. $3+i$

19. $-7+22i$ 21. 26 23. $13+8\sqrt{3}i$ 25. $18+24i$

27. $(5-i)/2$ 29. $(6+9i)/13$ 31. $-i/5$

33. $\frac{(-4+3\sqrt{2})-(12+\sqrt{2})i}{18}$

Section 2.1 (page 44)

1. 243 3. 16 5. x^{14} 7. $6x^8$

9. c^{48} 11. $3^6=729$ 13. x^5y^5 15. $8x^9y^{21}$

17. 5 19. $\frac{4}{49}$ 21. $7xy^4/27$ 23. x^{12}/y^{30}

25. $\frac{1}{8}$ 27. $\frac{5}{x^2}$ 29. y^{20}/x^8 31. $y^3/8x^3$

33. $\frac{a^{10}}{b^6}$ 35. x^8y^4 37. $1/(x+y)$ 39. 2

41. $\frac{1}{6}$ **43.** $y/(xy+1)$ **45.** $\frac{1}{9}$ **47.** $(a^2+b^2)/ab$

49. $\dfrac{2x}{(x-y)(x+y)}$ **51.** $1/(y-x)$ **53.** $\dfrac{x-3}{2(x-1)(x+1)}$ **55.** 8.2344×10^9

57. 5.2×10^{-11} **59.** 4.6×10^{39} **61.** 3,485,200,000 **63.** 0.000 000 938 5

65. 0.02222 **67.** 3×10^{10} **69.** 1,500,000

Section 2.2 (page 47)

1. 12 **3.** -30 **5.** 3 **7.** 5

9. -3 **11.** x^2 **13.** $4^3=64$ **15.** $x^3\sqrt[4]{3}$

17. $5\sqrt{2}$ **19.** $12\sqrt{2}$ **21.** $\frac{2}{3}\sqrt[3]{4}$ **23.** $\dfrac{x^3\sqrt{x}}{y^2}$

25. $4x\sqrt{3xy}$ **27.** $y\sqrt[5]{8x^4}$ **29.** $\sqrt{15}$ **31.** $2a$

33. $\dfrac{\sqrt{21}}{7}$ **35.** $\dfrac{\sqrt[4]{54}}{3}$ **37.** $\dfrac{\sqrt[4]{xy^3}}{y}$ **39.** $\dfrac{3\sqrt{2x}}{2x}$

41. $\dfrac{X\sqrt{X-Z}}{X-Z}$ **43.** $\dfrac{\sqrt[3]{a^2b^2(b-a)}}{ab}$ **45.** $\dfrac{(a-b)\sqrt{ab}}{ab}$ **47.** $(x+y)\sqrt{x-y}$

49. $\dfrac{1}{x+y}$ **51.** a^2b **53.** Correct **55.** $\sqrt{2}+\sqrt{3}$

57. $8-\sqrt{7}$ **59.** $\sqrt{7}+3$ **61.** $\sqrt{2}+1$

63. $y-m<0$ is the problem.

Section 2.3 (page 50)

1. $6\sqrt{3}$ **3.** $2\sqrt[5]{a}$ **5.** $5\sqrt{3}$ **7.** $13\sqrt{3}$ **9.** $x\sqrt{y}-x^2y\sqrt{xy}$ **11.** $2\sqrt{2}$

13. $\dfrac{\sqrt{5}(8+5\sqrt{2})}{10}$ **15.** $\left(1+xy+\dfrac{1}{y}\right)\sqrt{xy}$ **17.** $\left(\dfrac{1}{x}+\dfrac{1}{y}\right)\sqrt{xy}$ **19.** $\sqrt{6}+\sqrt{14}$

21. $3x-5\sqrt{xy}$ **23.** -5 **25.** $a-2\sqrt{ab}+b$ **27.** $\sqrt{3}+\sqrt{2}-\sqrt{6}-2$

29. $a\sqrt{a}+2a\sqrt{b}+2b\sqrt{a}+b\sqrt{b}$ **31.** $\sqrt{2}(\sqrt{5}-\sqrt{3})$ **33.** $\dfrac{(\sqrt{x}-\sqrt{y})^2}{x-y}$

35. $\dfrac{(\sqrt{a}+\sqrt{b})^2}{a-b}$ **37.** $\dfrac{1}{x^3}+\dfrac{1}{x^2}\sqrt[5]{x^3}$ **39.** No

Section 2.4 (page 53)

1. $x^{1/3}$ **3.** $(x/y)^{1/4}$ **5.** $x^{2/3}$ **7.** $x^{5/12}$ **9.** $\sqrt[5]{x^3}$ **11.** $\sqrt[8]{a^3}$ **13.** $\sqrt[3]{a+b}$

15. 8 **17.** 27 **19.** $\frac{1}{8}$ **21.** $x^{7/6}$ **23.** $y^{5/6}$ **25.** $5ab^3$ **27.** $3x^{1/3}$

29. $\dfrac{2y^{1/2}+3x^{1/2}}{x^{1/2}y^{1/2}}$ **31.** $a+2a^{1/2}b^{1/2}+b$ **33.** $\dfrac{y^2+2y+1}{y}$

35. $\dfrac{(2x+1)(x-2)^{1/2}}{x-2}$ **37.** $\dfrac{-9}{x+6}$ **39.** $\sqrt[12]{5^4 \cdot 4^3}$ **41.** $\sqrt[6]{x^3(x+y)^2}$

43. 0.00279 **45.** 36.462 **47.** 2.35589 **49.** 0.97969 **51.** 0.0011387

53. 0.30840

Section 3.1 (page 60)

1. $-\frac{3}{2}$ **3.** $\frac{2}{3}$ **5.** 13 **7.** $\frac{18}{25}$ **9.** 7 **11.** $\frac{1}{2}$ **13.** -20

15. $\frac{1}{3}$ **17.** 6 **19.** 2 **21.** $-\frac{22}{3}$ **23.** $\frac{4}{5}$ **25.** $\frac{5}{3}$ **27.** $x=\dfrac{pl}{q}$

29. $R=\dfrac{R_1R_2}{R_1+R_2}$ **31.** $t=\dfrac{v-b}{a}$ **33.** $b=\dfrac{2A}{h}-a$ **35.** No solution

37. $\frac{1}{2}$ **39.** No solution **41.** 2 **43.** $-\frac{9}{2}$, 19

Section 3.2 (page 66)

1. $A=\frac{1}{2}bh$ **3.** $P=4s$ **5.** $a=v/t$ **7.** $x-8$ **9.** $5(x+3)$ **11.** $0.20(x-12000)$

13. $\frac{1}{3}(x+y+z)$ **15.** $I=PR$ **17.** $2(x+y)$ **19.** $A=6s^2$ **21.** 9 and 11 amps

23. 102, 104, 106 **25.** 0.7 mi **27.** 220 mi **29.** 133 mph **31.** 0.5 hr

33. 4 touchdowns, 4 extra points, 4 field goals, 2 safeties **35.** 8 ft from the 150-lb boy **37.** 1.785 hr

Section 3.3 (page 74)

1. 0, 2 **3.** $-1, 2$ **5.** $-2, 3$ **7.** $-5, 2$ **9.** $-4, \frac{3}{2}$ **11.** $-\frac{1}{3}, 2$ **13.** $0, \frac{1}{5}$

15. $-2, 5$ **17.** $-\frac{6}{7}, \frac{4}{5}$ **19.** $\frac{2}{3}, -2$ **21.** $-1, \frac{3}{5}$ **23.** $(1\pm\sqrt{17})/4$ **25.** $(1\pm\sqrt{33})/4$

27. $(5\pm\sqrt{5})/2$ **29.** $(-6\pm\sqrt{6})/5$ **31.** $(1\pm i\sqrt{5})/3$ **33.** $(1\pm 3i)/2$

35. $\dfrac{\pi}{6}\pm\dfrac{\sqrt{24-\pi^2}}{6}i=0.524\pm 0.627i$ **37.** $-1.66, 0.66$ **39.** $0.091\pm 0.848i$

41. $k=3/\pi$ **43.** $k\leq\frac{1}{32}$ **45.** $k=4, -4$, and others **47.** $k=2$ **49.** $k=-4$

Section 3.4 (page 76)

1. $-3+\sqrt{2}=-1.59$ and $-3-\sqrt{2}=-4.41$ **3.** $10+2\sqrt{10}=16.32$ and $10-2\sqrt{10}=3.68$

5. $\dfrac{-15+\sqrt{285}}{2}=0.94$ **7.** $T=87.4°$. $T=12.6°$ is unrealistic since water freezes at 32 °F

9. 0.9 km **11.** 5 m, 12 m **13.** 25 cm, 75 cm **15.** 56 ft × 42 ft

17. 2 mph **19.** 40 hr **21.** 11, 12, 13, and -11, -12, -13 **23.** 4 ft, 10 ft **25.** 8.4 hr

27. 27 and 49 both work **29.** No answer needed

Section 3.5 (page 80)

1. $\pm 2, \pm i\sqrt{3}$ **3.** $\pm i\sqrt{2}, \pm 2\sqrt{2}$ **5.** $\pm i, \pm 2\sqrt{2}i$ **7.** $-2, \frac{1}{3}$

9. $-\frac{3}{2}, \frac{1}{2}$ **11.** $-4, 2$ **13.** $\pm i\sqrt{13}, \pm i\sqrt{3}$ **15.** $(141 \pm i\sqrt{139})/70$

17. $\pm 1, \pm i\sqrt{5}/2$ **19.** ± 5 **21.** $-3, 5$ **23.** $\dfrac{32 \pm \sqrt{1636}}{34}$ **25.** $\dfrac{7 \pm \sqrt{97}}{2}$

27. ± 1.3636 **29.** 15 **31.** 11 **33.** $-5, 5$ **35.** $-7, 2$ **37.** 3 **39.** 2

41. No solution **43.** 18 **45.** 0, 1 **47.** 30 **49.** $h = \dfrac{\sqrt{s^2 - \pi^2 r^4}}{\pi r}$ **51.** 2

Section 3.6 (page 88)

1. 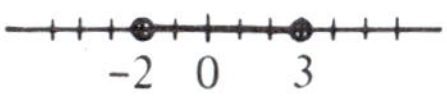

3. 

5. 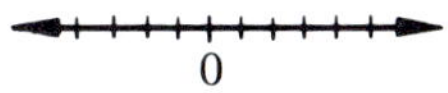

7. 

9. 

11. 

13. No graph

15. 

17. $x > -\frac{5}{2}$ **19.** $x \geqslant \dfrac{(\sqrt{2}-1)}{1+\sqrt{3}}$ **21.** $x > \frac{70}{11}$ **23.** $x > -13$ **25.** $x \leqslant -\frac{5}{3}$ **27.** $x > 0$

29. $x > \frac{4}{3}$ **31.** $x \geqslant -\frac{1}{2}$ **33.** $x > 8$ **35.** $-\frac{1}{5} \leqslant x < \frac{11}{7}$ **37.** $(-1.5, 1.5)$

39. $(-\infty, -4] \cup [4, \infty)$ **41.** $(-1, 5)$ **43.** $(-7, 3)$ **45.** $(-\infty, 4) \cup (6, \infty)$ **47.** $[-4, 16]$

49. $[\frac{1}{2}, \frac{5}{2}]$ **51.** $|x| < 5$ **53.** $|x| > 3$ **55.** $|x-4| < 2$ **57.** $|x-3| < 13$

59. $|x| > 2$ **61.** $|2x-9| \leqslant 5$ **63.** $|2x-5| \geqslant 5$ **65.** $|2x-5| \leqslant 5$

67. All reals except $x=0$. **69.** $x \geqslant -\frac{3}{2}$ **71.** $[-3, 1]$ **73.** No solution

Section 3.7 (page 93)

1. $x \leqslant -3$ or $x \geqslant 1$ **3.** $x < -\frac{1}{3}$ or $x > 2$ **5.** $-\frac{3}{2} \leqslant x \leqslant \frac{5}{2}$ **7.** All x **9.** $x < (1-\sqrt{33})/2 = -2.37$, $x > (1+\sqrt{33})/2 = 3.37$

11. $x < -5, -4 < x < -1$ **13.** $x \leqslant -2$ or $x > -\frac{3}{2}$ **15.** $-2 < x \leqslant -1, 1 \leqslant x < 2$ **17.** All x

19. $(-1-\sqrt{5})/2 < x < (-1+\sqrt{5})/2$ or $x \geqslant 2$ **21.** $\left.\begin{array}{r} x \leqslant (1-\sqrt{21})/2 \\ -\sqrt{3} < x < \sqrt{3} \\ x \geqslant (1+\sqrt{21})/2 \end{array}\right\} \begin{array}{l} x \leqslant -1.79 \\ -1.73 < x < 1.73 \\ x \geqslant 2.79 \end{array}$

23. $k \geqslant \sqrt{48}$ or $k \leqslant -\sqrt{48}$ **25.** $k = -\frac{1}{2}, \frac{1}{2}$ **27.** $0 < t < 5$ **29.** $10 < T < 20$

Section 4.1–4.2 (page 102)

1, 3, 5. 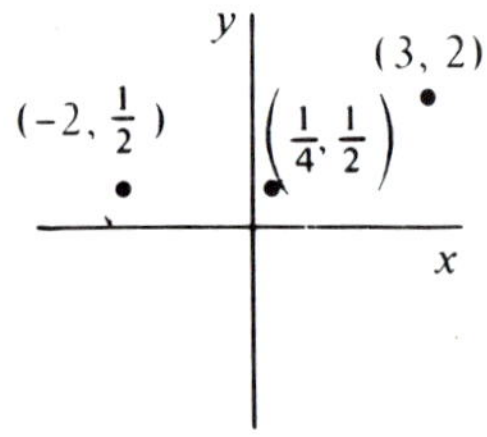

7. I, IV **9.** III **11.** 0

13. 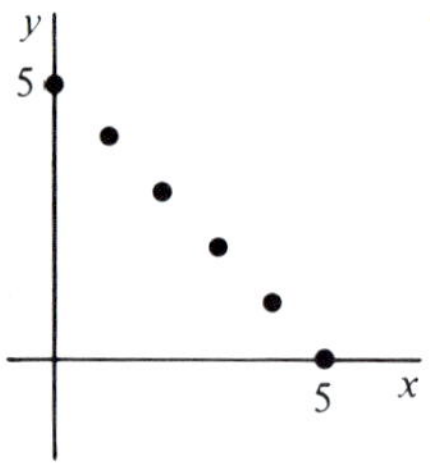

15. 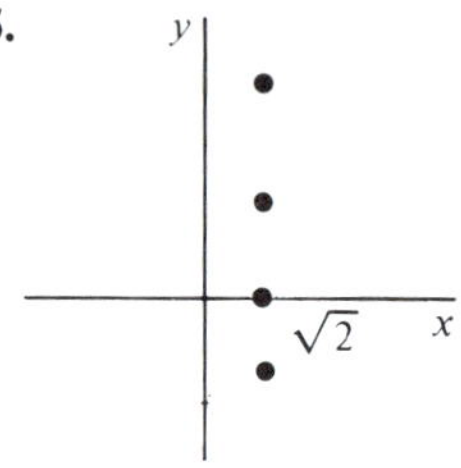

17. 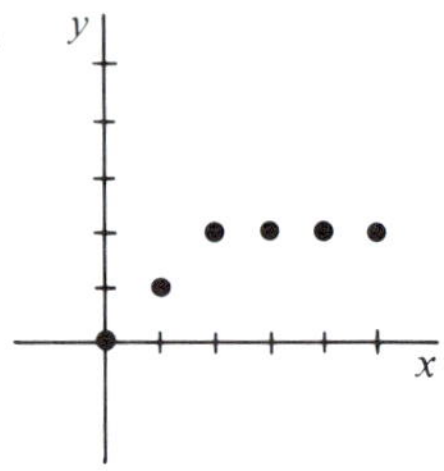

19. 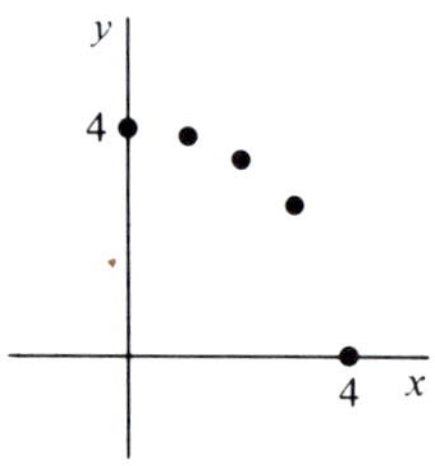

21. 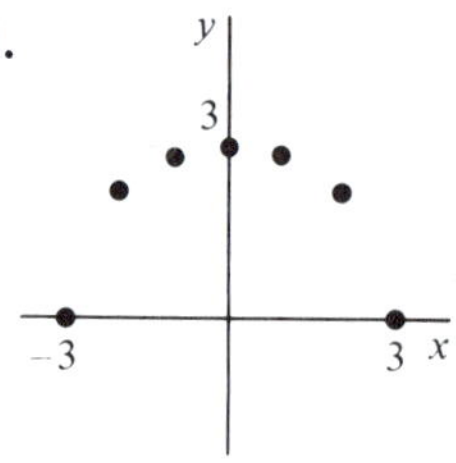

23. $\sqrt{2}$ **25.** $\frac{5}{4}$ **27.** 8.2

29. $2\sqrt{4+\sqrt{3}} \approx 4.8$ **31.** $2\sqrt{x+y}$ **33.** $2|y|$ **35.** 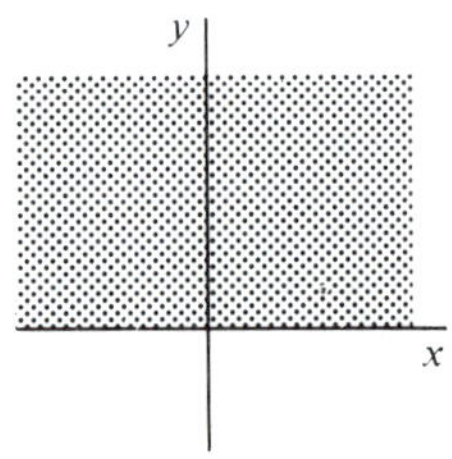

37. 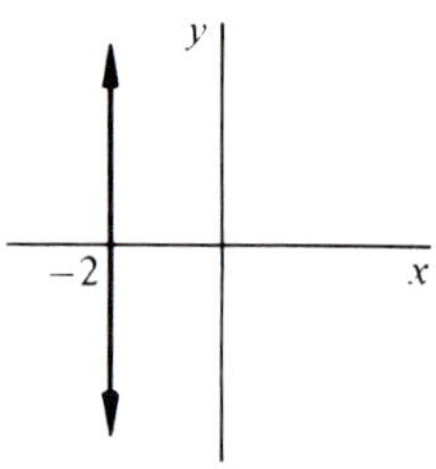

39. 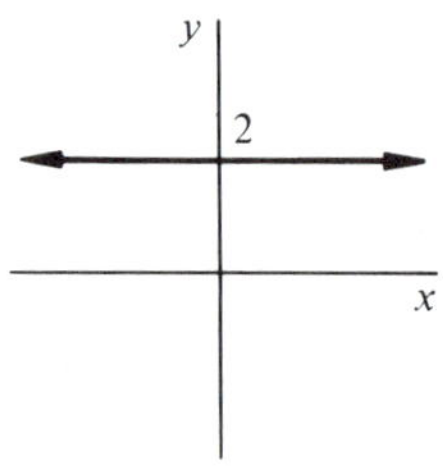

41. 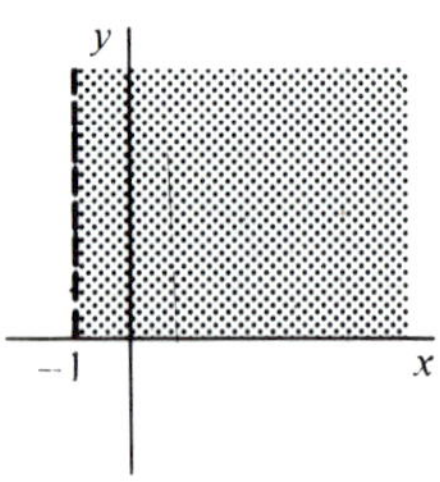

43. 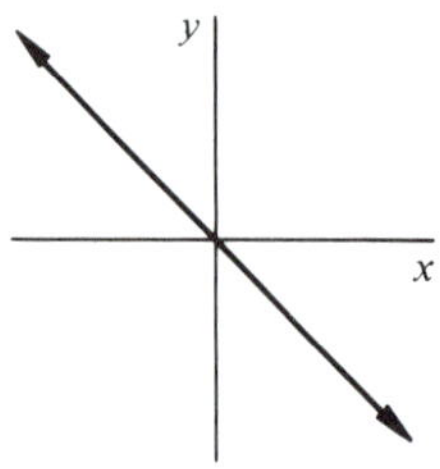

45. No graph

47. Center $(\frac{5}{4}, -\frac{1}{4})$
Radius $= \frac{5}{4}\sqrt{2}$

49. Center $(-\frac{7}{6}, \frac{1}{6})$
Radius $= \sqrt{158}/6$

51. Center $(-4, 5)$ Radius $=7$

53. Center $(0, \frac{5}{2})$ Radius $=\frac{5}{2}$

55. $x^2+y^2=26$

57. $(x+1)^2+(y-2)^2=26$

59. $(x+2)^2+(y-5)^2=3$

61. $(x-2)^2+(y-1)^2=13$

Section 4.3 (page 107)

1. $v=kt$

3. $r=k/\sqrt{p}$

5. $m=kx^2y^3$

7. $y=kh^4/g^2$

9. $n=\dfrac{k}{rt}$

11. 252 mi

13. 62.5 lb sodium, 37.5 lb chlorine

15. 4.44 and 3.56 ft

17. 400 lb/in.^{2}

19. 0.15

21. $y=6$

23. $z=1$

25. 64 ft/sec

27. $C=28.3$¢

29. $I=1$ ft-cand.

31. $P=200$ watts

33. 1 amp

35. \$2300

37. (a) Directly proportional to t^2 (b) Constant (c) Directly proportional to t

Section 4.4 (page 113)

1. Function

3. Function

5. Function

7. Not a function

9. Function

11. Not a function

13. (a) and (b)

15. D: all reals. R: all reals

17. D: $x\geqslant 0$. R: $F(x)\geqslant 0$

19. D: all reals. R: all reals

21. D: all x, except $x=-3$. R: all y except $y=0$

23. D: all t except $t=-\frac{1}{4}$. R: all $g(t)$ except $g(t)=\frac{1}{4}$.

25. D:$x\leqslant 0$. R:$f(x)\geqslant 0$.

27. D:$x\geqslant 1$. R:$f(x)\geqslant 0$.

29. (a) 10 (b) $3\pi+1$ (c) $3z+1$ (d) $3(x-h)+1$ (e) 3 (f) All reals (g) 31 (h) 13 (i) 3

31. (a) 3 (b) -3 (c) Undefined (d) $\{-3, 3, 9\}$ (e) $\{2, 5, 7\}$ (f) $H(2+5)=-3$; $H(2)+H(5)=12$

33. $2x+h$

35. $2x-2+h$

37. $3x^2+3xh+h^2$

39. $\dfrac{-1}{(x+1)(x+1+h)}$

41. (a) x_2/x_1 (b) $f(1/x)=kx$; $1/f(x)=x/k$ (c) $f(x^2)=k/x^2$; $[f(x)]^2=k^2/x^2$
(d) $-k/x(x+h)$ (e) $f(x)+1=(x+k)/x$; $f(x+1)=k/(x+1)$
(f) $f(x_1+x_2)=k/(x_1+x_2)$; $f(x_1)+f(x_2)=k(x_1+x_2)/x_1x_2$ (g) $af(x)=ak/x$; $f(ax)=k/ax$

43. In general, none are equal.

45. -11.6918

47. 2.366 amp

Section 4.5 (page 120)

1. Function

3. Not a function

5. Not a function

7.

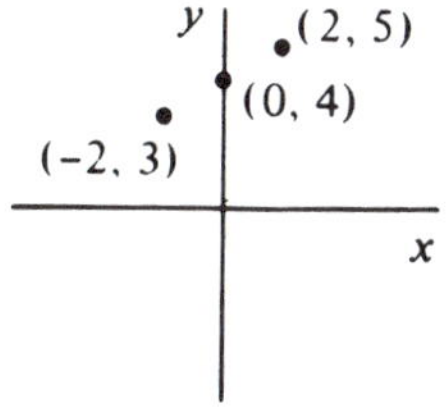

9.

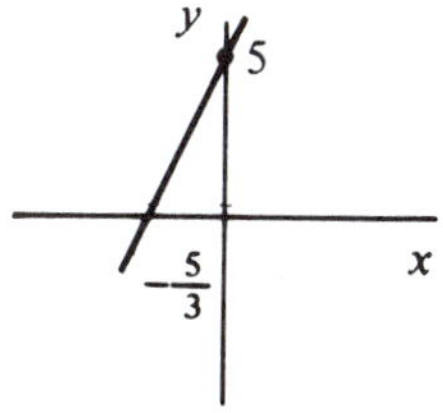

11.

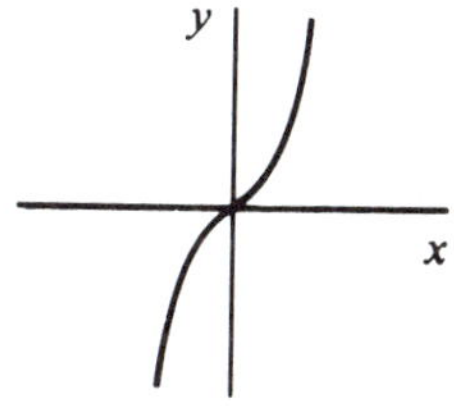

13.

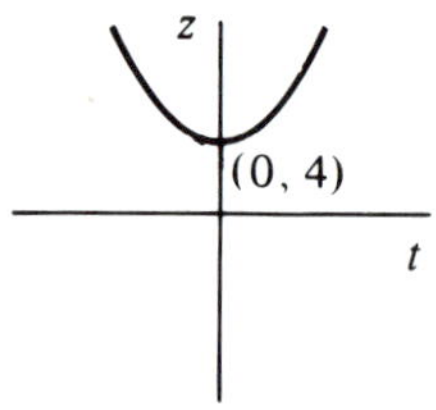

15.

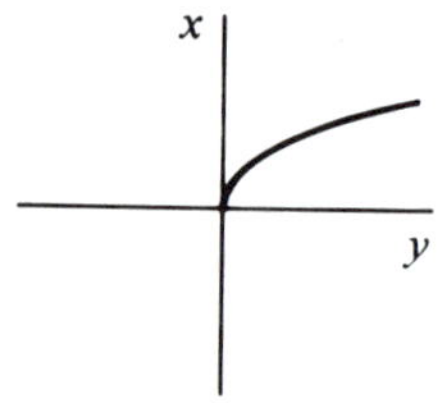

17.

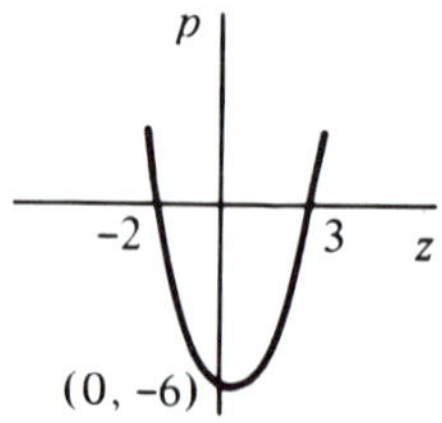

19.

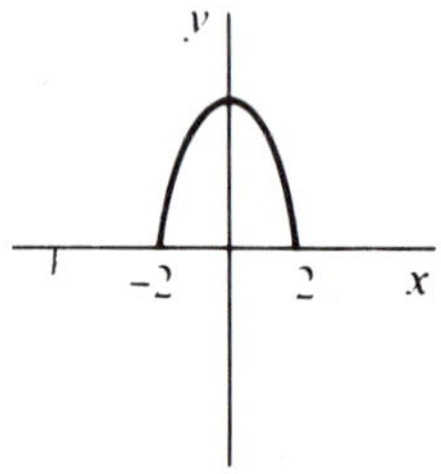

21.

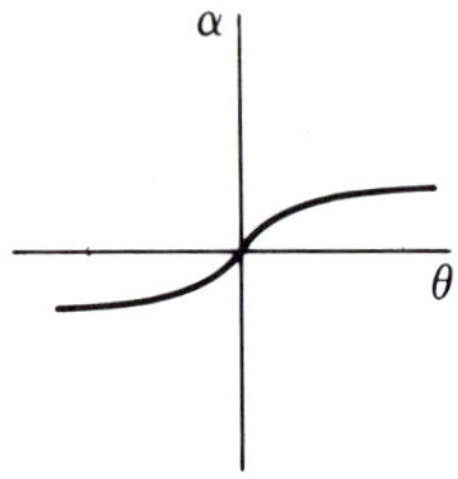

23.

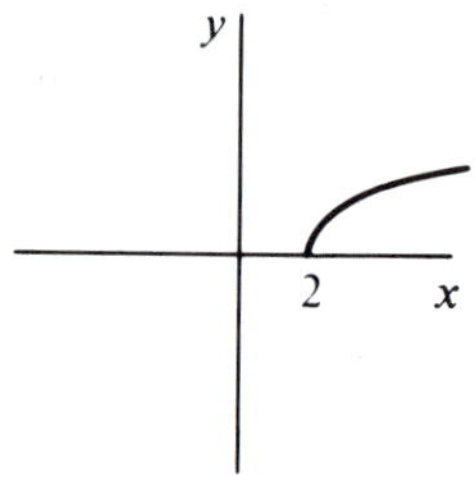

25.

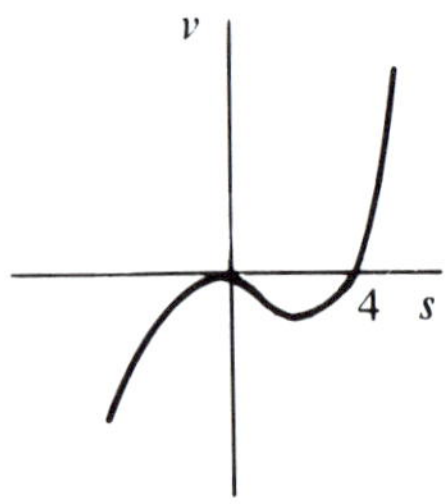

27.

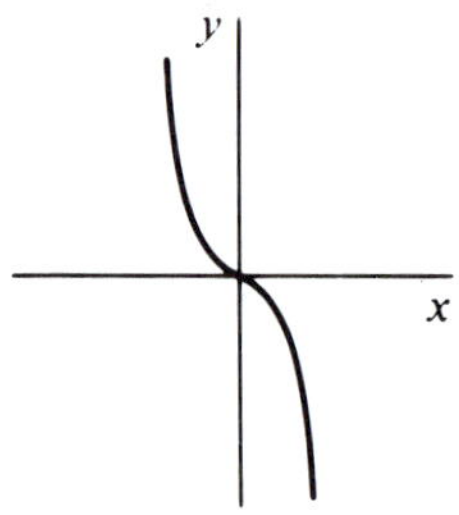

29.

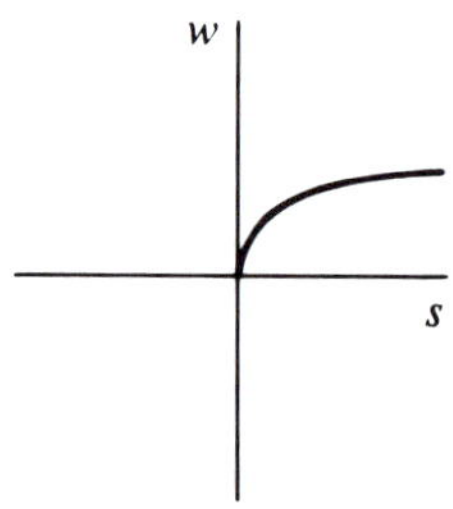

31.

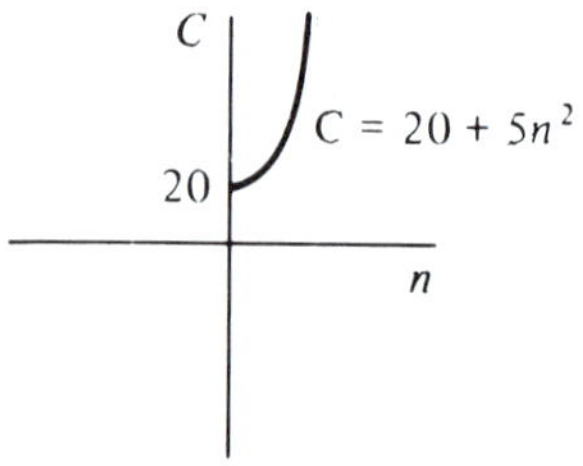

33.

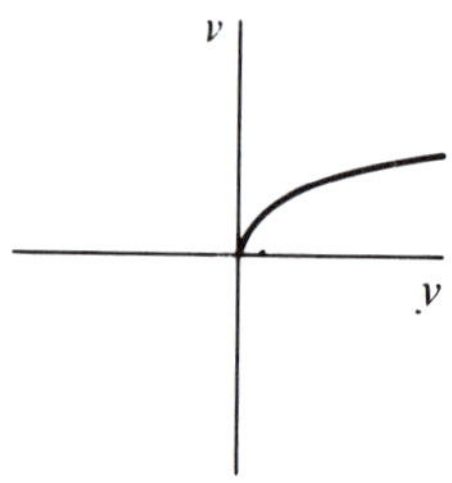

35.

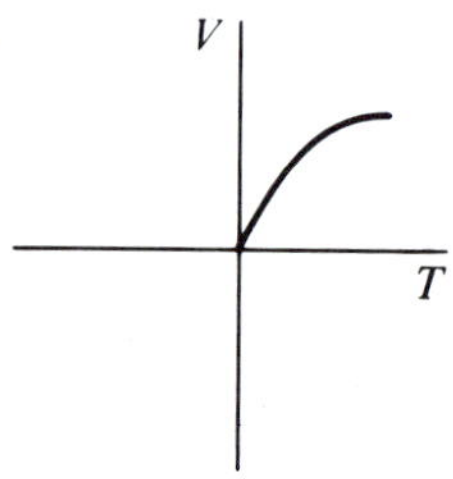

Section 4.6 (page 126)

1. $\frac{3}{2}$ **3.** $\frac{5}{2}$ **5.** 3, None **7.** $\{(7, 3), (9, 5), (3, 7), (5, 9)\}$

9. No inverse function **11.** $\{(3, -2), (4, -1), (0, 0)\}$

13. No inverse function **15.** Inverse functions

17. Not inverse functions **19.** Inverse functions

21. Inverse functions **23.** a, c, and d have inverses.

25. $6x+4; 6x+8$ **27.** $x^2; x^2$ **29.** $x-1; (x^3-1)^{1/3}$ **37.** $f^{-1}(x)=x+3$

39. $f^{-1}(x)=(1-x)/x$ **41.** $f^{-1}(x)=(1+x)/(1-x)$

Section 5.1 (page 135)

1.

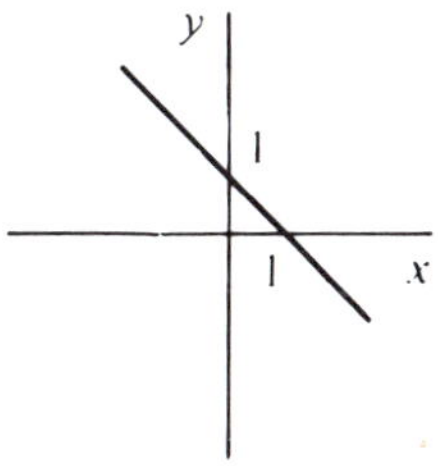

3.

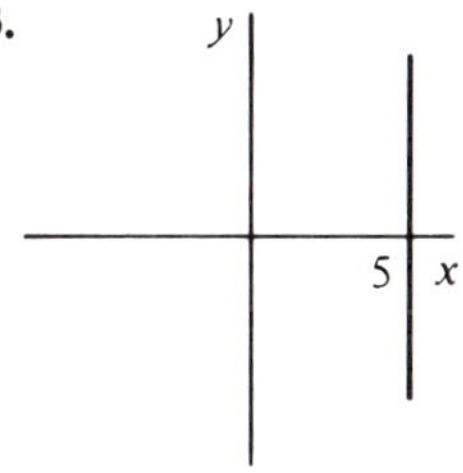

5.

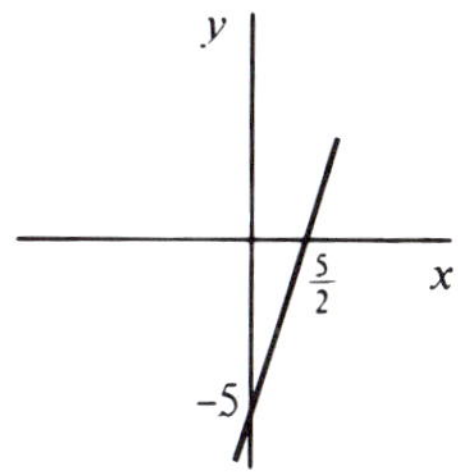

7.

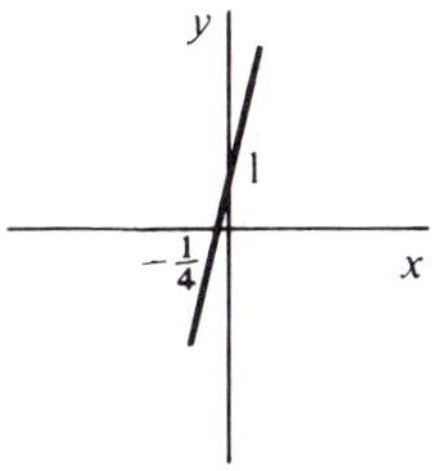

9.

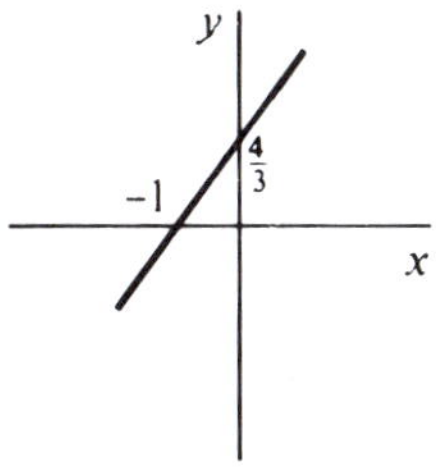

11.

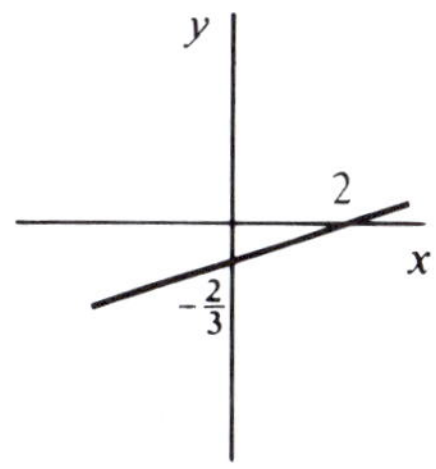

13. $m=\frac{1}{2}$ **15.** $m=-\frac{5}{4}$ **17.** $m=\frac{4}{3}$ **19.** $m=0$

21.

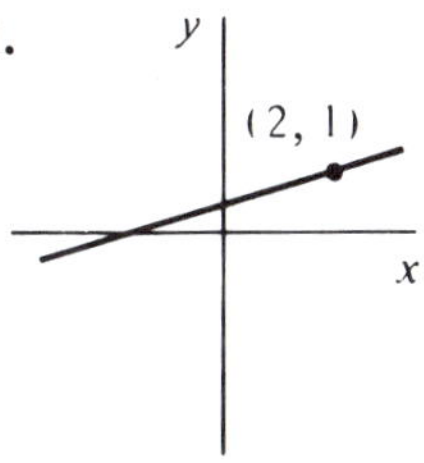

23.

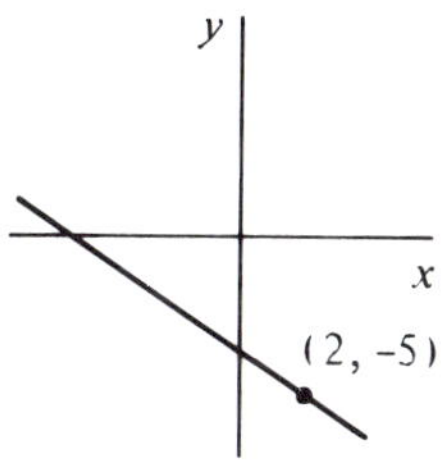

25. $m=\sqrt{481}/12 \approx 1.83$ **27.** $m=2$

29.

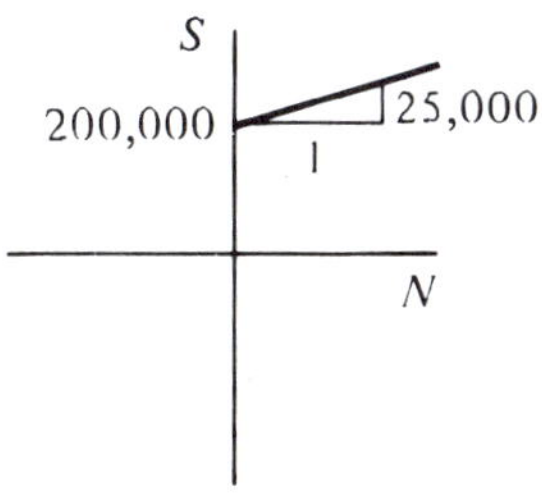

Section 5.2 (page 139)

1. $x-2y=-8$ **3.** $7x+y=33$ **5.** $x+5y=16$ **7.** $2y-3x=1$ **9.** $5y-2x=10$
11. $m=\frac{2}{3}, b=-\frac{5}{3}$ **13.** $m=-1, b=2$ **15.** $m=\frac{1}{2}, b=-2$ **17.** $m=\frac{1}{2}, b=\frac{5}{2}$ **19.** $m=0, b=5$
21. $m=\frac{1}{5}, b=\frac{7}{5}$ **23.** $2x-3y=-4$ **25.** $x+y=8$ **27.** $2x+3y=8$ **29.** $7x+5y=0$
31. $3x+y=-5$ **33.** No answer required

Section 5.3 (page 142)

1. $I = V/12$ **3.** \$225 **5.** \$151.11 **7.**

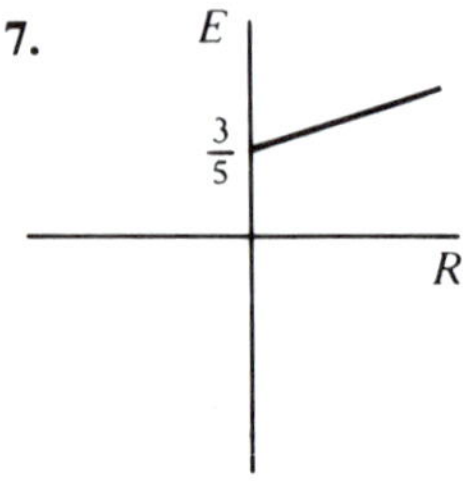

9. (a) \$1900 (b) \$6.33 (c) \$5.80 (d) \$5.00 (e) 320

11. (a) \$2.00 (b) \$200 (c) No change (d) 0, 100, 350

13; (a) \$12,100 (b) 0.637 (c) $C = 0.637x + 28{,}700$ (d) \$79,063

15. (a) \$18,500 (b) \$13,500 profit (c) \$28,571

Section 5.4 (page 151)

1. D: all x
R: $f(x) \geq -4$
Zeros $x = -2, 2$
Intercept $(0, -4)$
Vertex $(0, -4)$

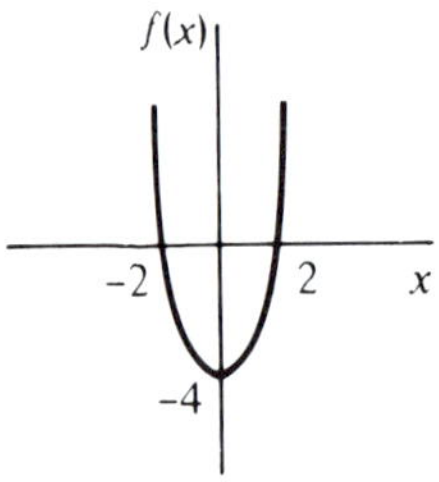

3. D: all x
R: $y \geq 1$
No zeros
Intercept $(0, 1)$
Vertex $(0, 1)$

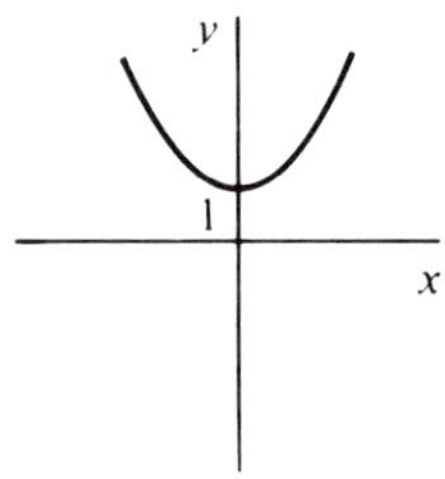

5. D: all x
R: $f(x) \leq 9$
Zeros $x = -3, 3$
Intercept $(0, 9)$
Vertex $(0, 9)$

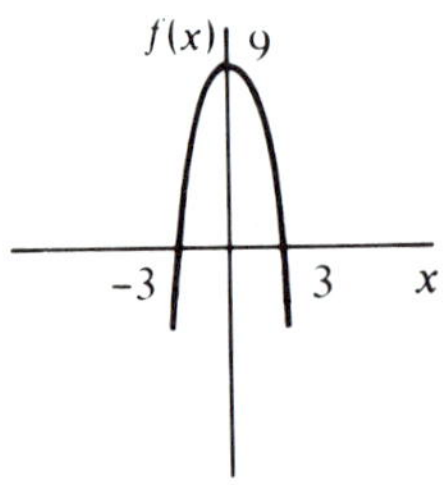

7. D: all t
R: $x(t) \geq -\frac{9}{4}$
Zeros $t = 0, 3$
Intercept $(0, 0)$
Vertex $(\frac{3}{2}, -\frac{9}{4})$

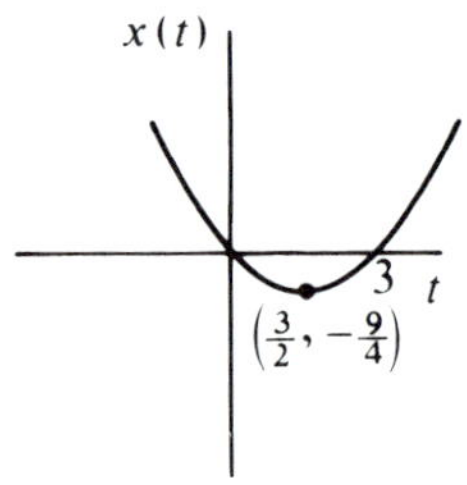

9. D: all x
R: $y \geq -\frac{1}{4}$
Zeros $x = 2, 3$
Intercept $(0, 6)$
Vertex $(\frac{5}{2}, -\frac{1}{4})$

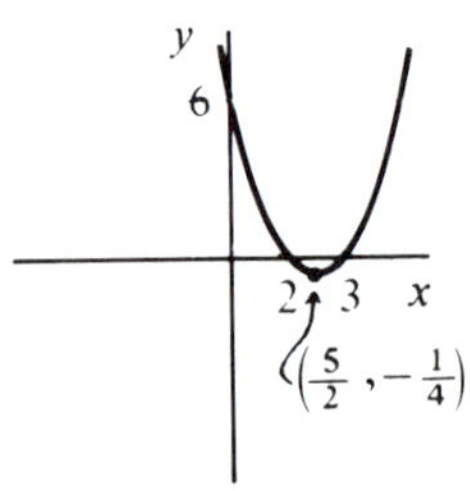

11. D: all x
R: $y \geq \frac{2}{3}$
No zeros
Intercept $(0, 1)$
Vertex $(-\frac{1}{3}, \frac{2}{3})$

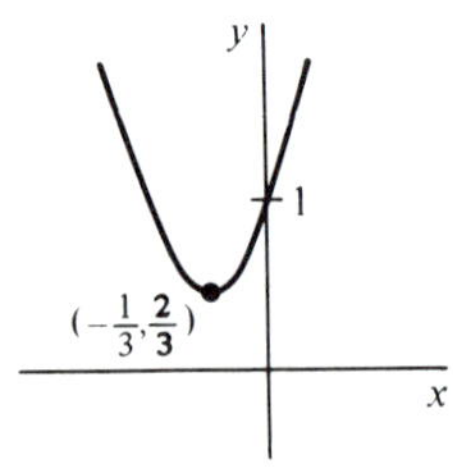

13. D: all x
R: $y \leqslant \frac{9}{4}$
Zeros $x = -1, 2$
Intercept (0, 2)
Vertex $(\frac{1}{2}, \frac{9}{4})$

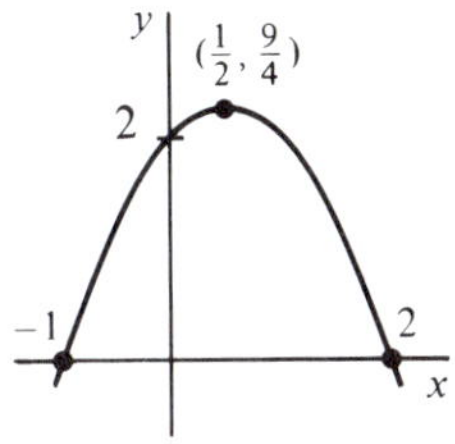

15. D: all x
R: $f(x) \geqslant -\frac{49}{8}$
Zeros $x = -1, \frac{5}{2}$
Intercept (0, −5)
Vertex $(\frac{3}{4}, -\frac{49}{8})$

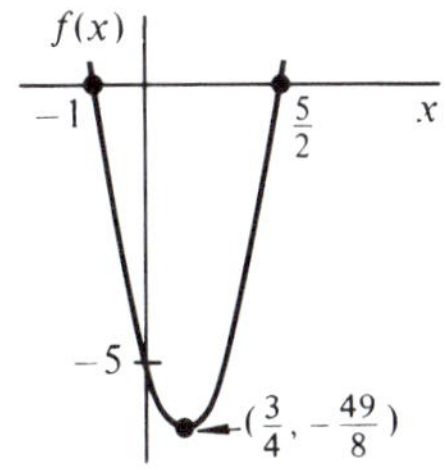

17. D: all x
R: $y \leqslant \frac{13}{4}$
Zeros $x = \dfrac{1 \pm \sqrt{13}}{2}$
Intercept (0, 3)
Vertex $(\frac{1}{2}, \frac{13}{4})$

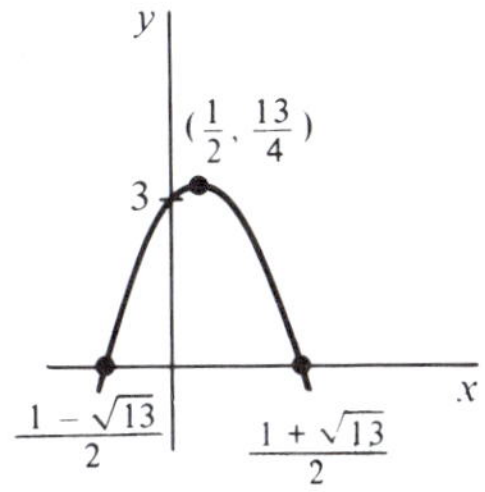

19. D: all x
R: $G(x) \geqslant \frac{3}{4}$
No zeros
Intercept (0, 1)
Vertex $(-\frac{1}{2}, \frac{3}{4})$

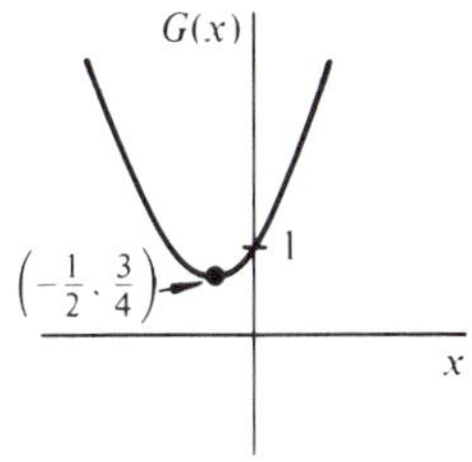

21. $y = x^2 - 2x$

23. $y = x^2 - 2x + 1$

25. $y = -x^2 - 4x - 3$

27. $y = \frac{2}{3}x^2 - \frac{4}{3}x - 2$

29. $y = -\frac{1}{10}x^2 - \frac{3}{10}x + 1$

31. Symmetric with respect to the x axis.

33.

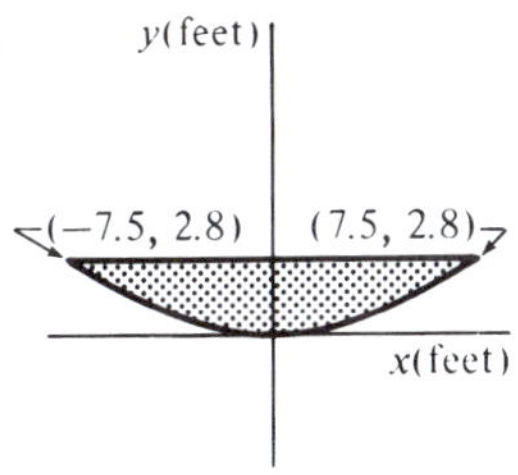

35.

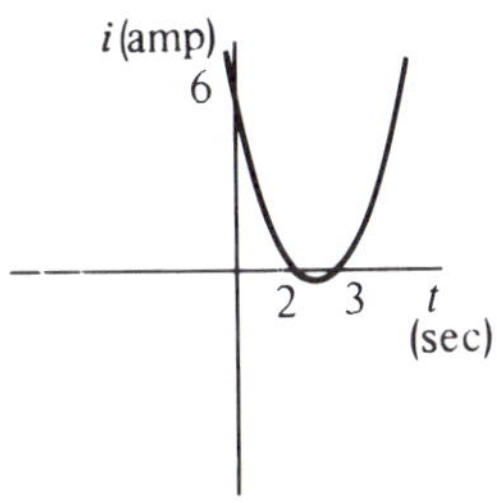

37. $v \times = 500$ fps at $t = 4$ sec
$v = 0$ at $t = 4 + 2\sqrt{5} \approx 8.47$ sec

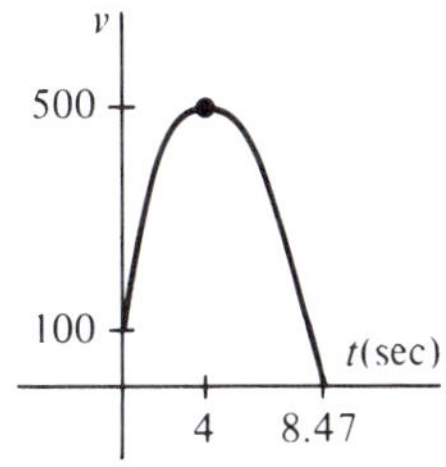

39. No

Section 5.5 (page 160)

1.

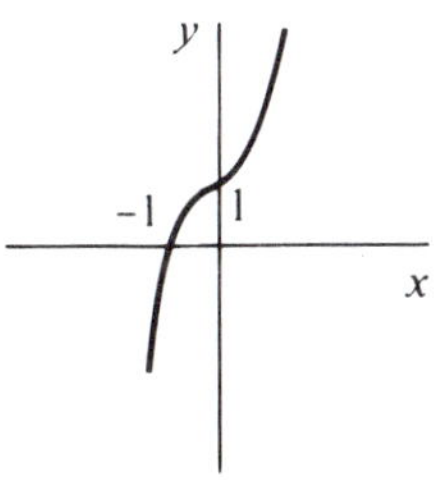

3.

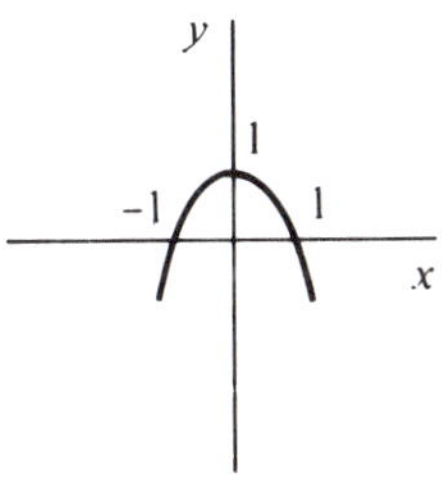

5.

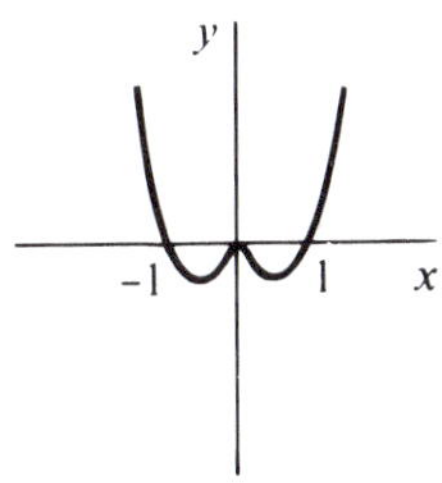

7.

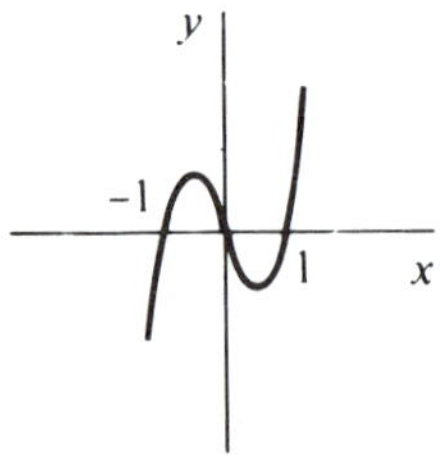

9.

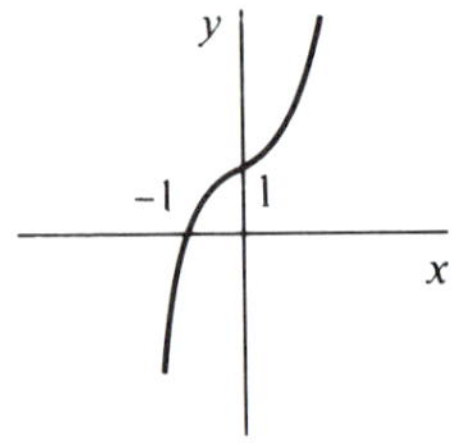

11.

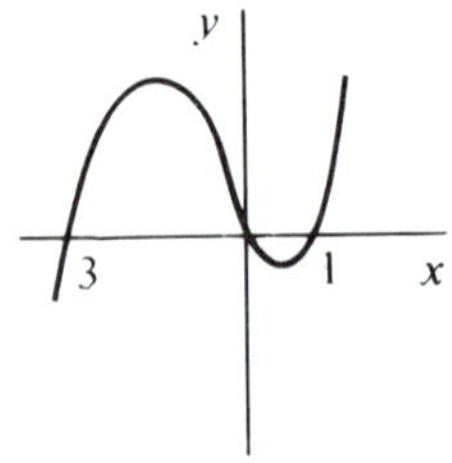

13.

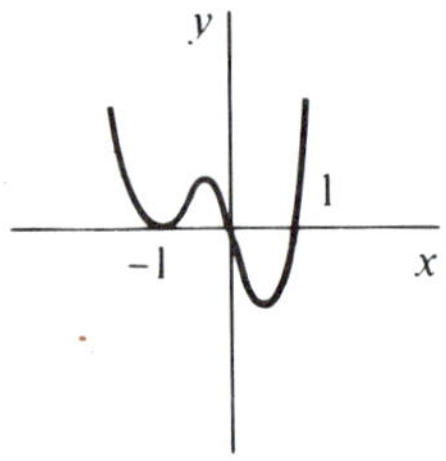

15.

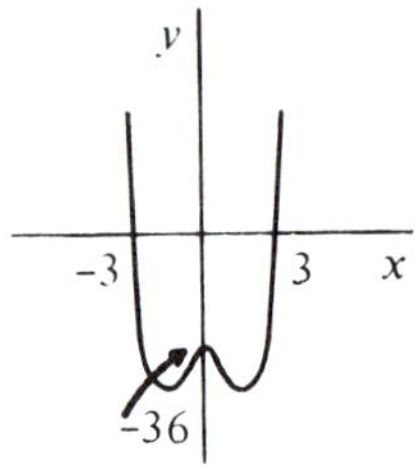

17.

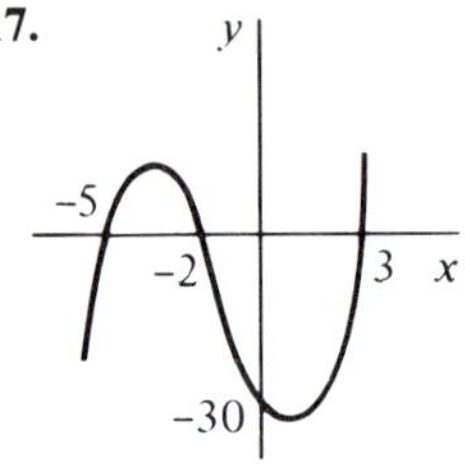

19.

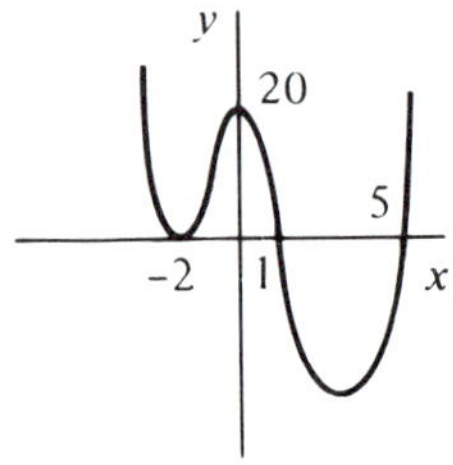

Section 5.6 (page 1 6 1)

1.

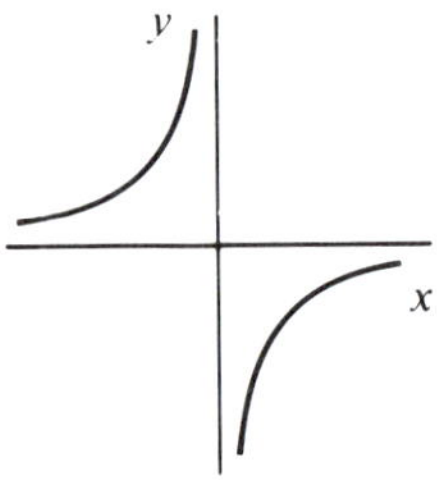

3.

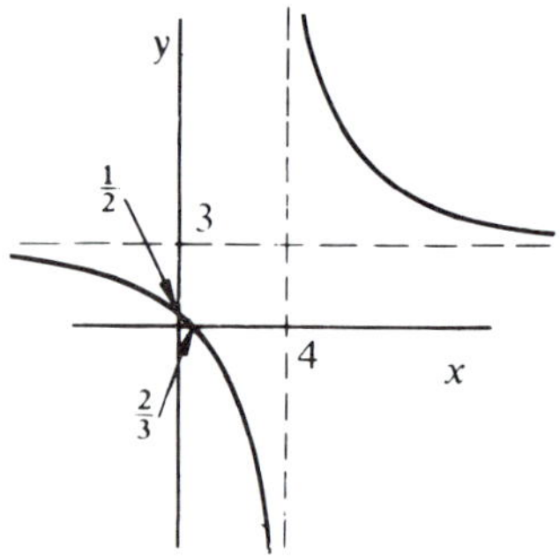

5.

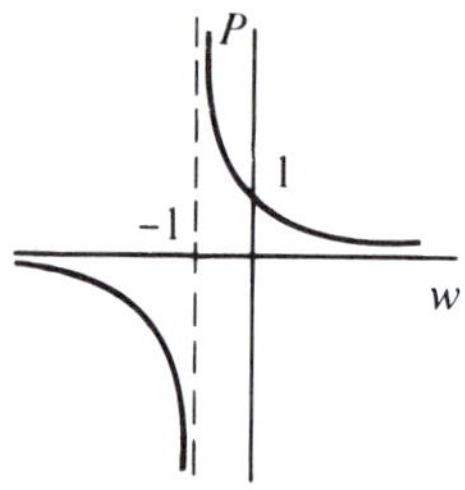

7.

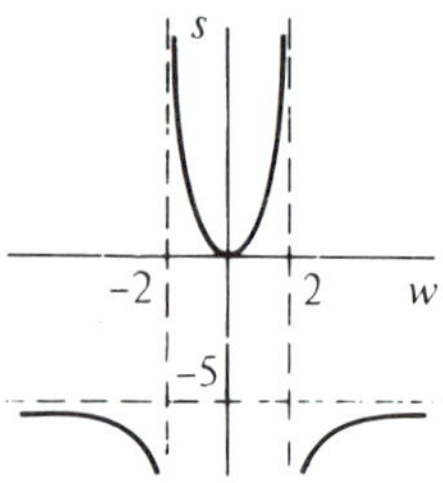

9.

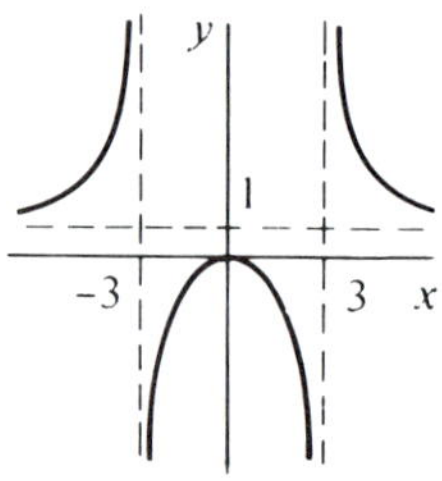

11.

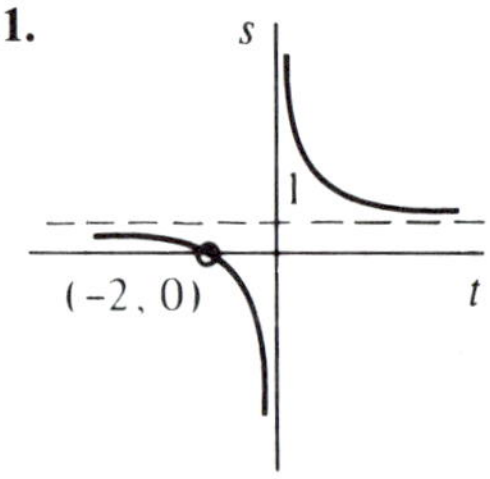

13.

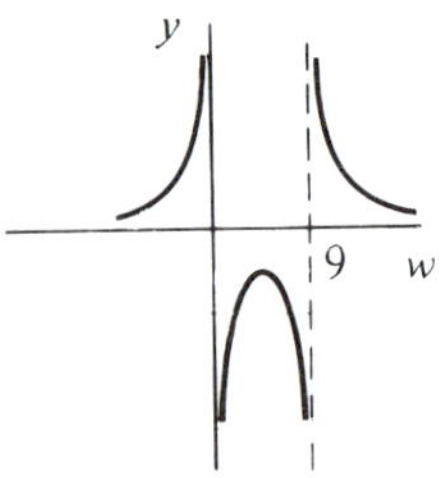

15.

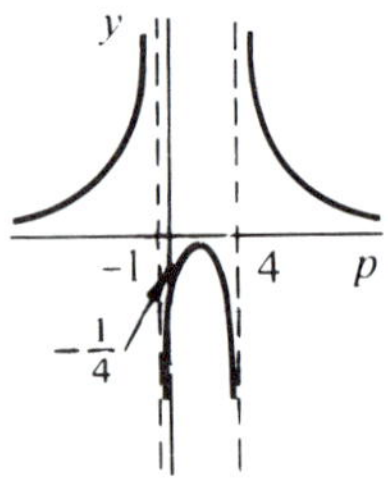

17.

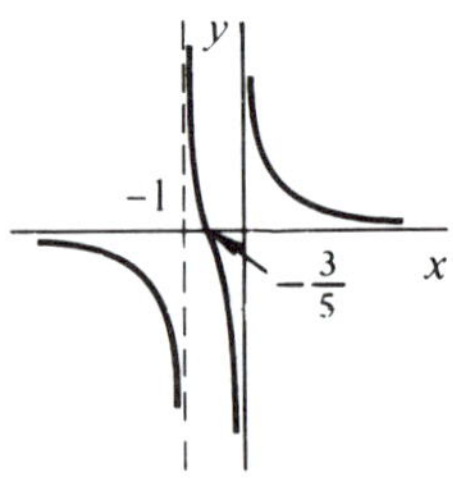

19.

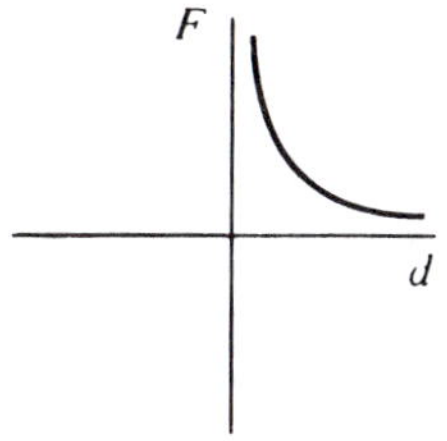

21.

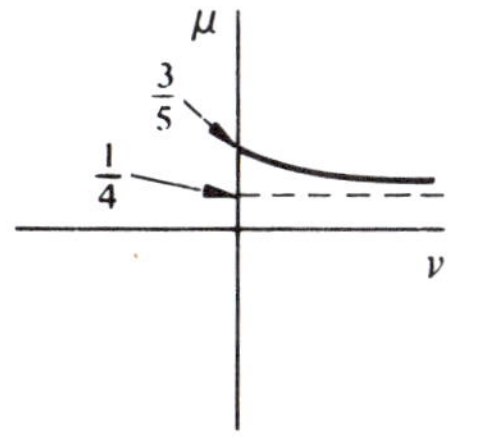

Section 5.7 (page 164)

1.

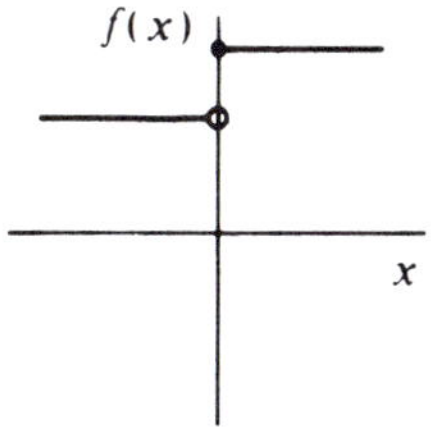

3.

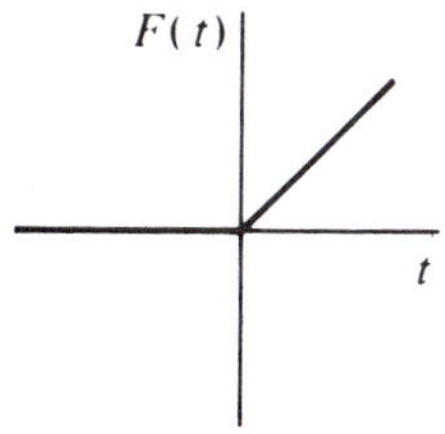

5.

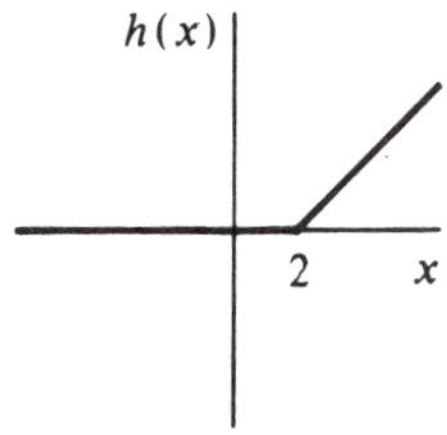

7.

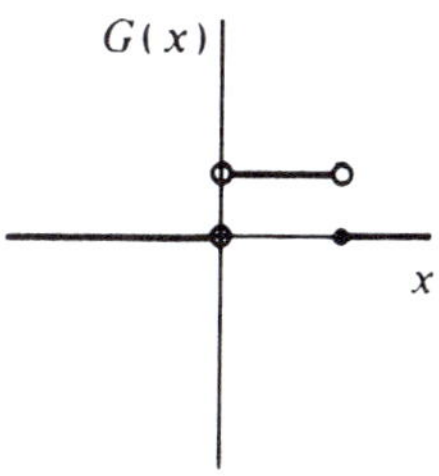

9.

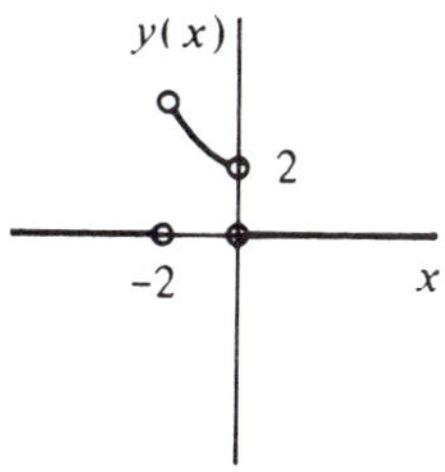

11.

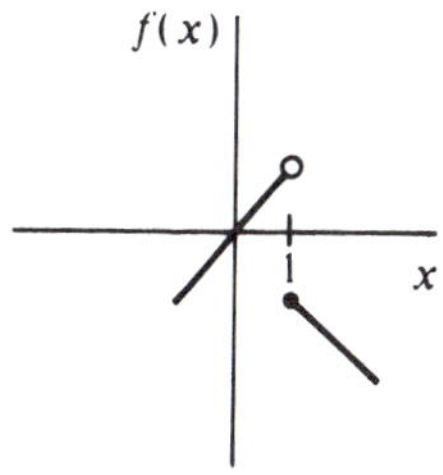

13.

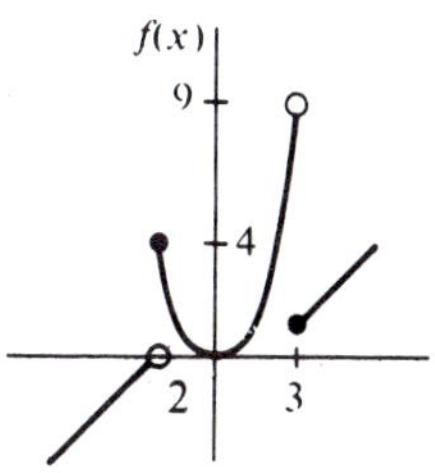

15.

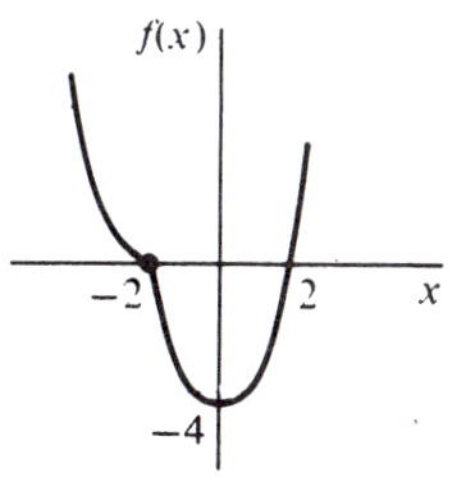

17.

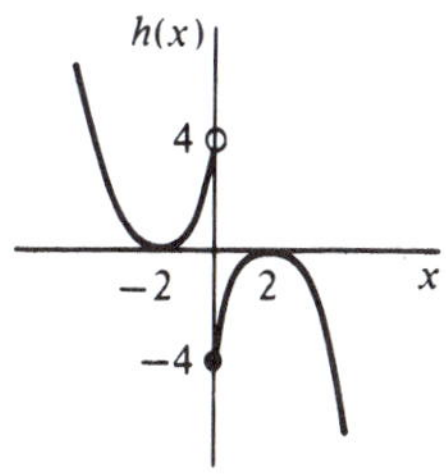

19.

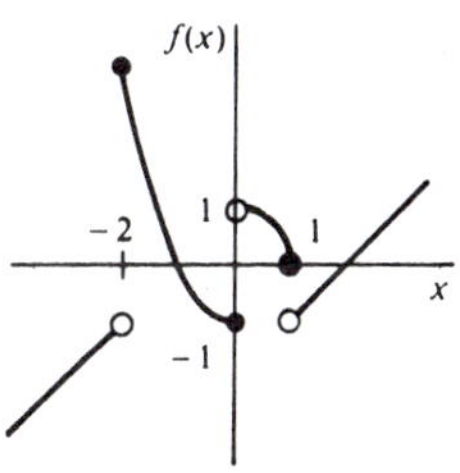

21.

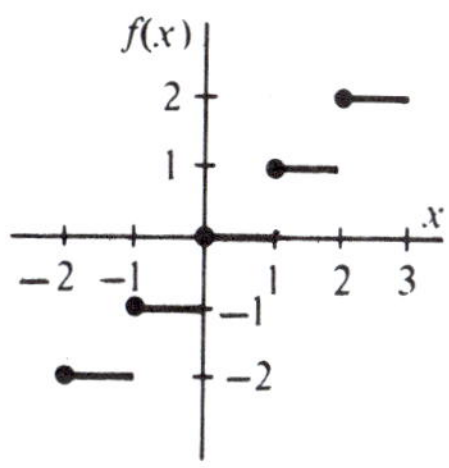

23.

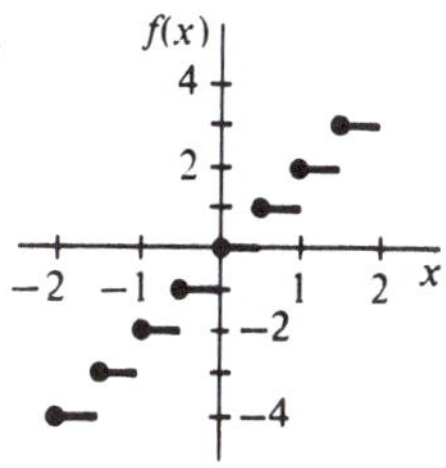

25.
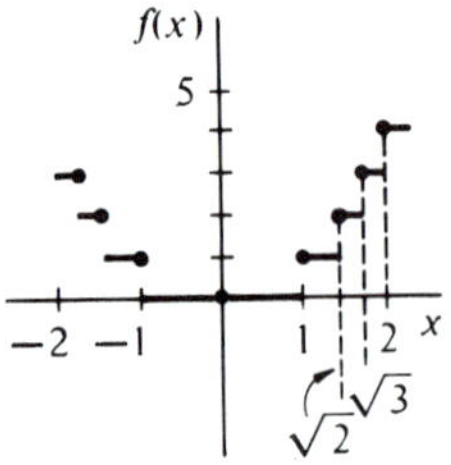

27.
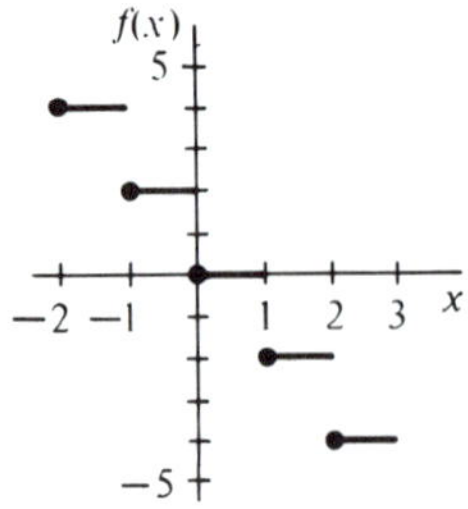

29.
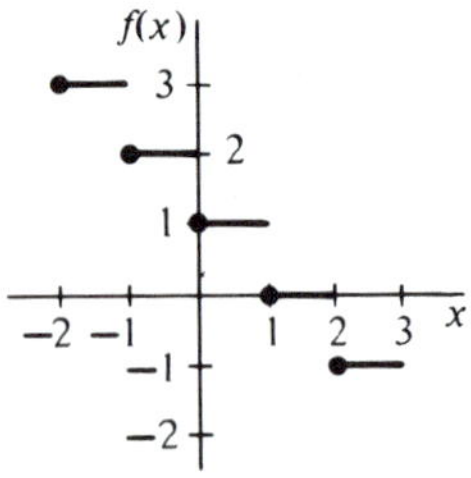

31.
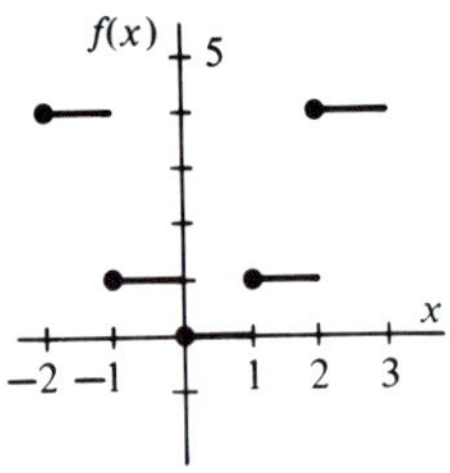

33.
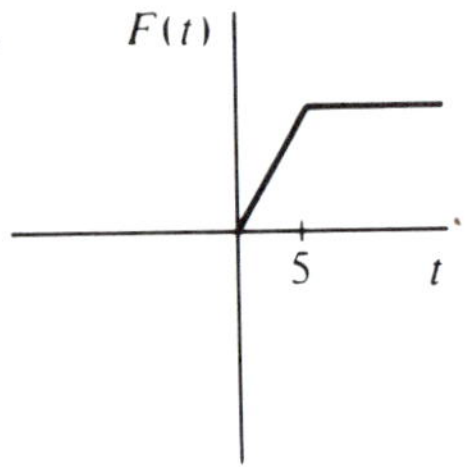

35.
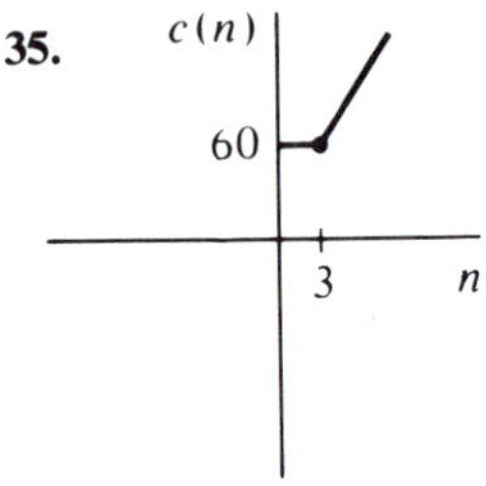

Section 5.8 (page 169)

1.
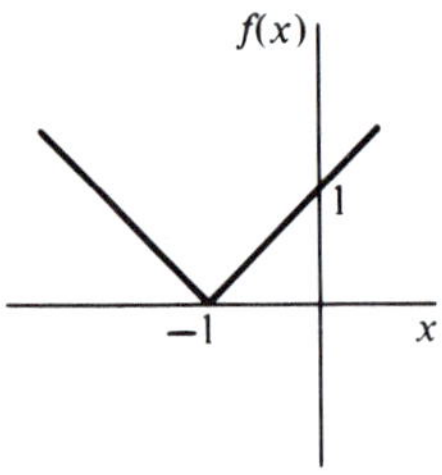

3.
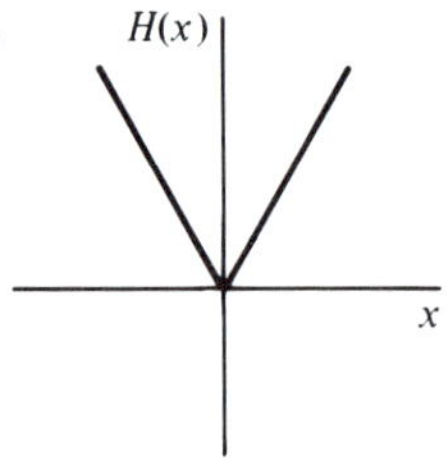

5.
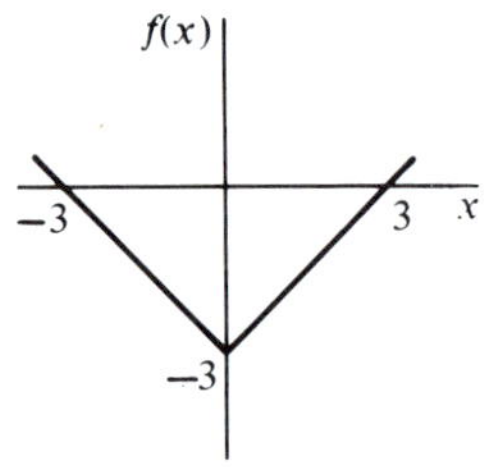

7.
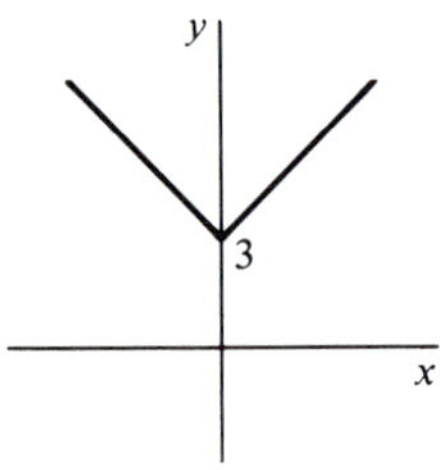

9.
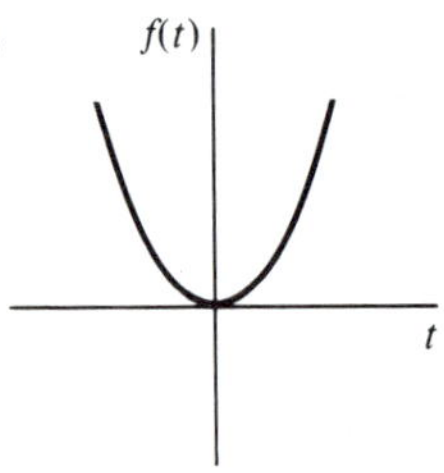

11.
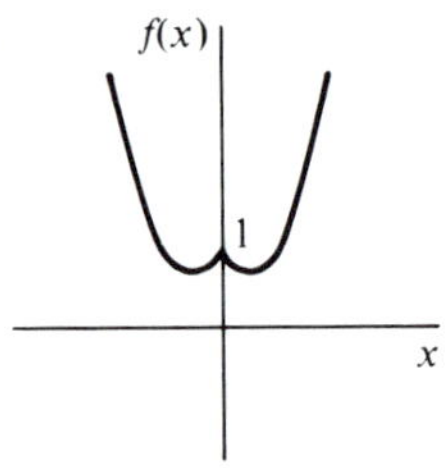

13.
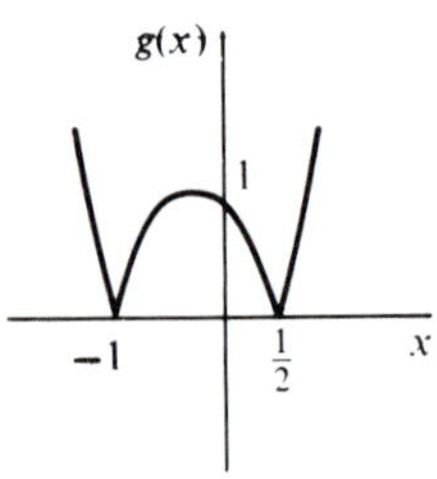

15.
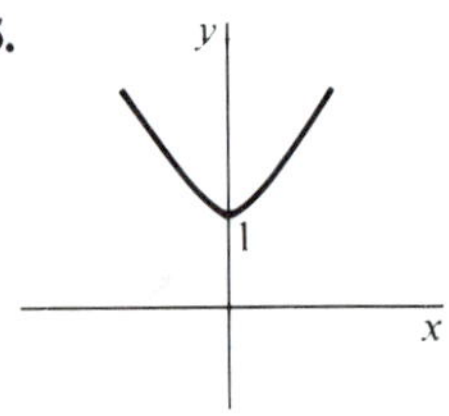

17.
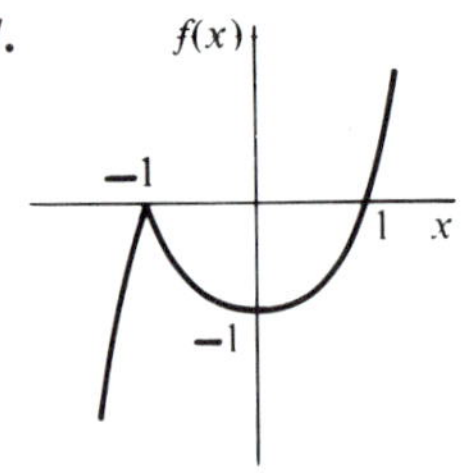

19.

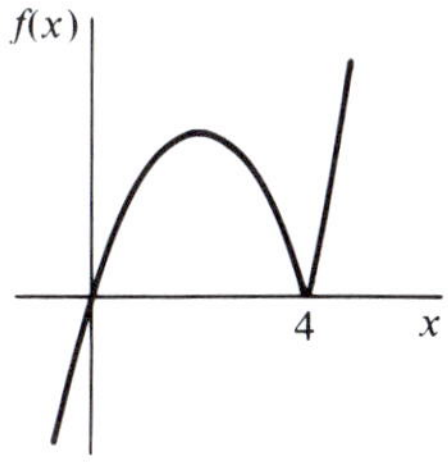

21. Let $x=2, y=-3$.

23.

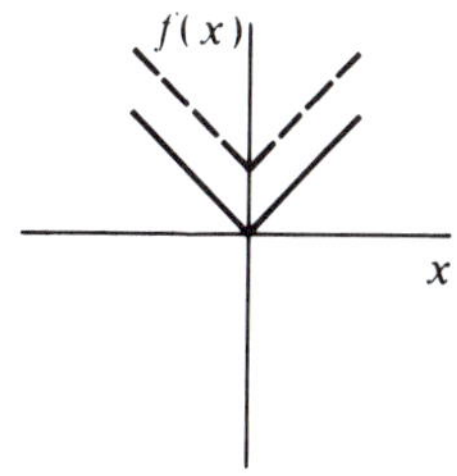

25. $x=-2, 2$ **27.** $x=-6, 4$ **29.** $x=1$ **31.** $x=\frac{3}{2}$ **33.** $x=-1, 1$ **35.** $x=-3, 1$

37. No solution **39.** $x=0$

41.

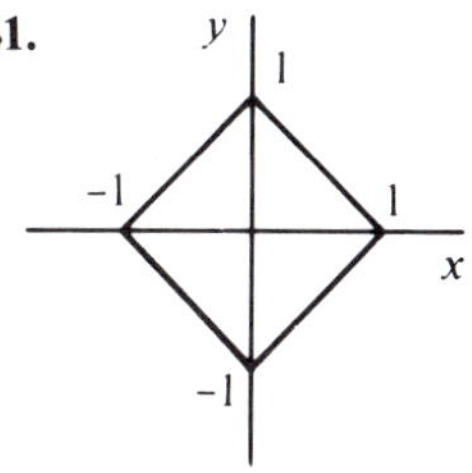

43.

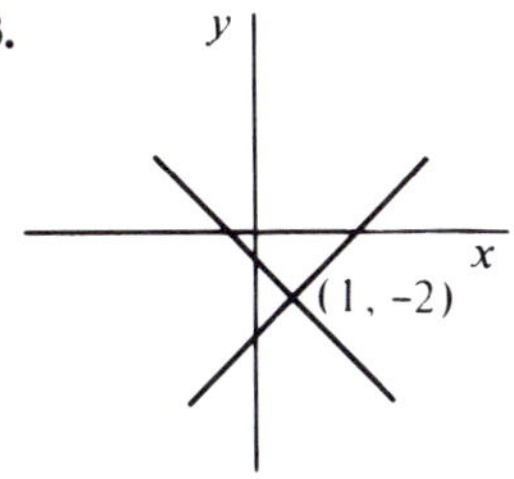

45.

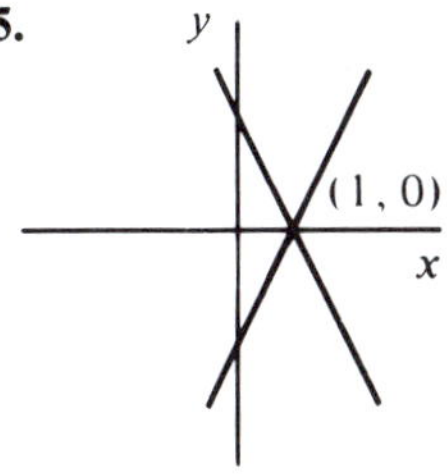

47.

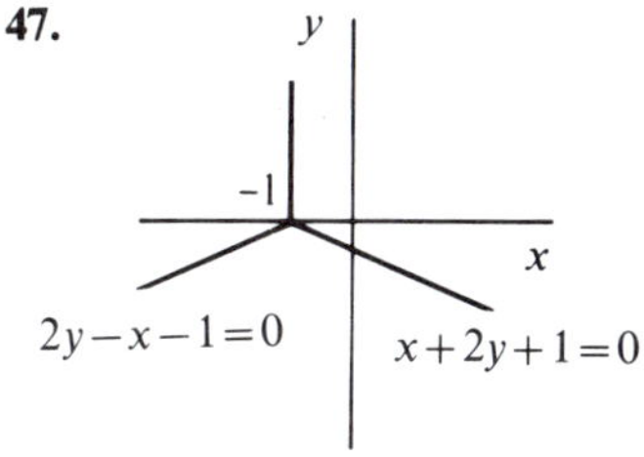

49.

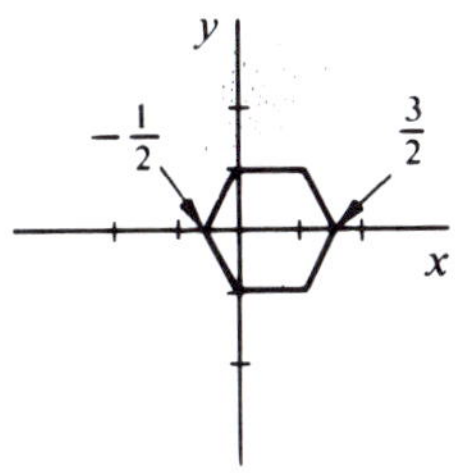

Section 5.9 (page 173)

1.

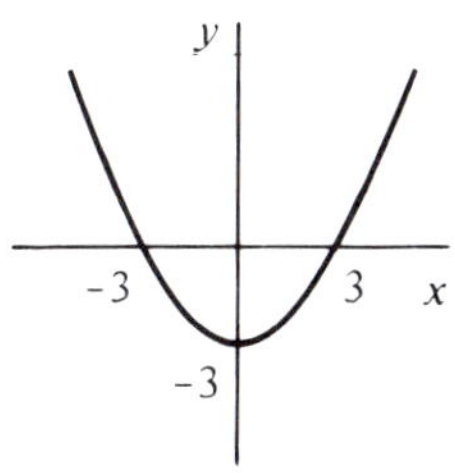

3.

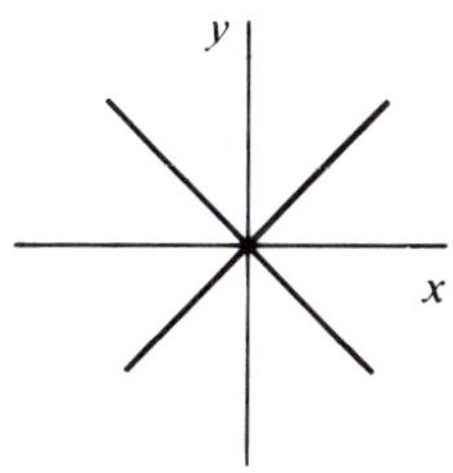

5.

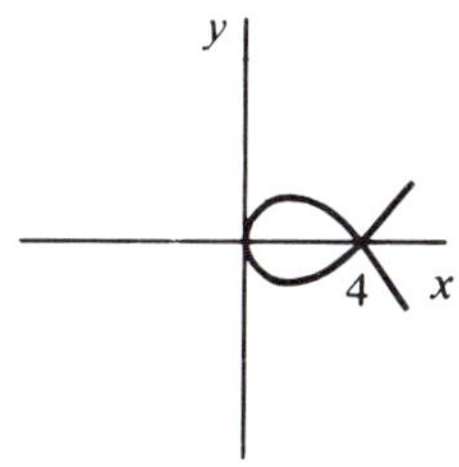

7.

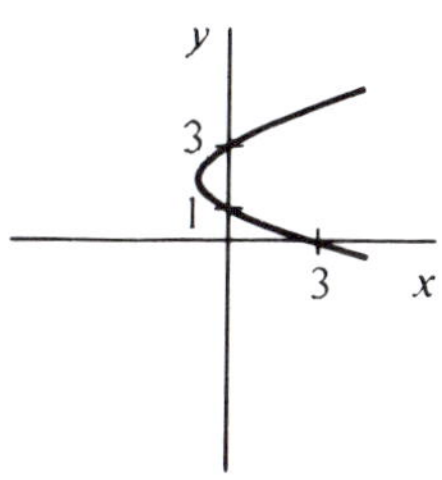

9.

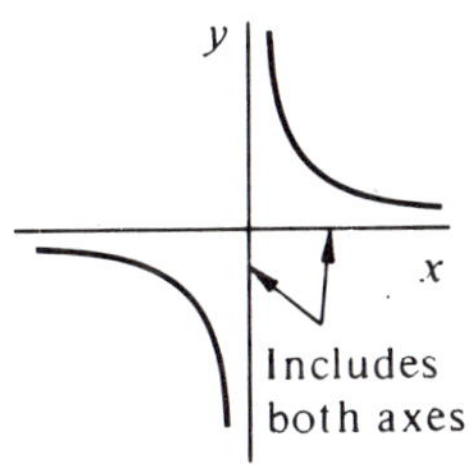

11.

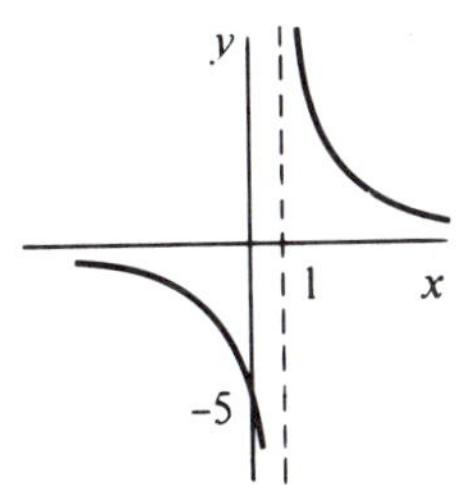

13.
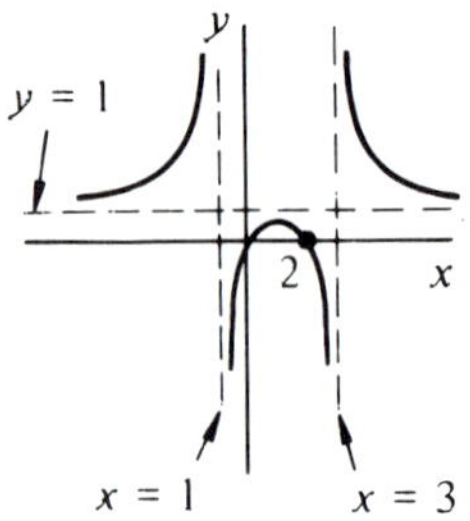

15.
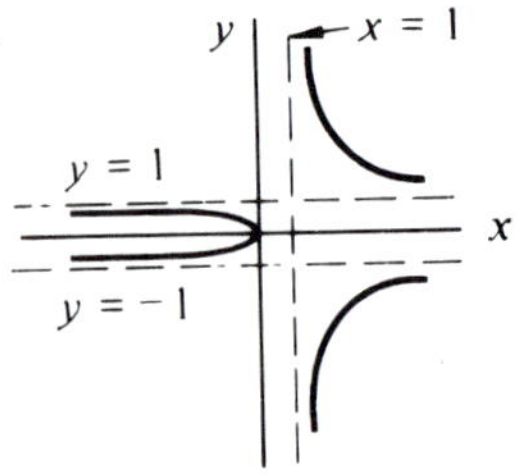

17.
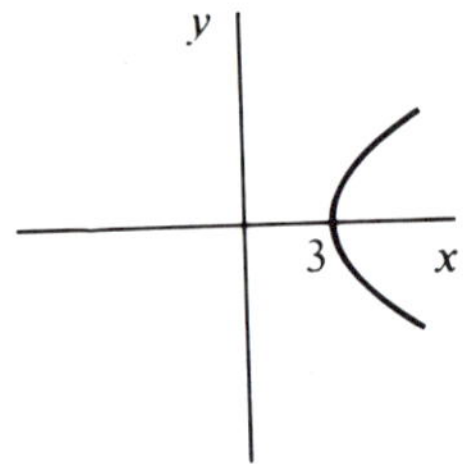

19.
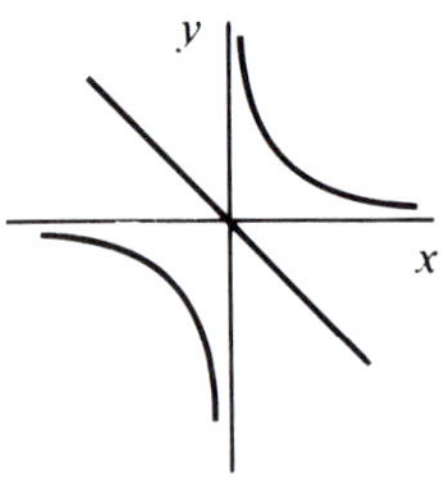

21.
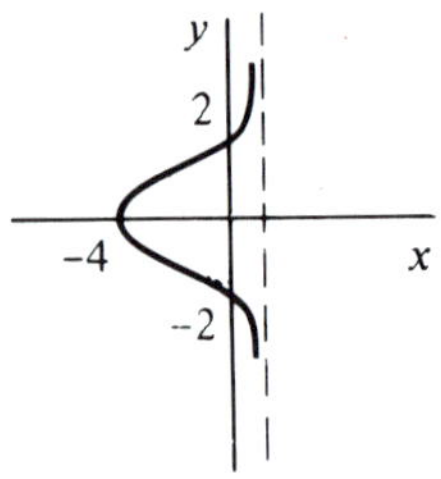

23.
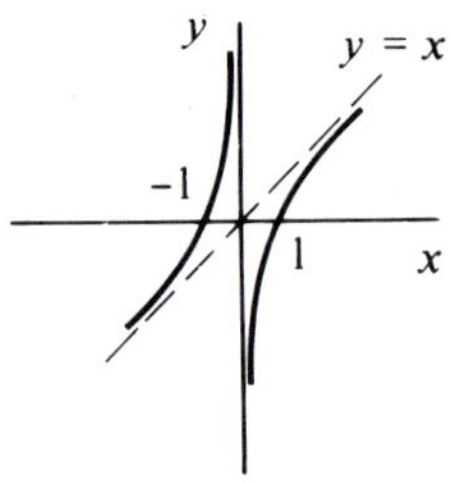

25.
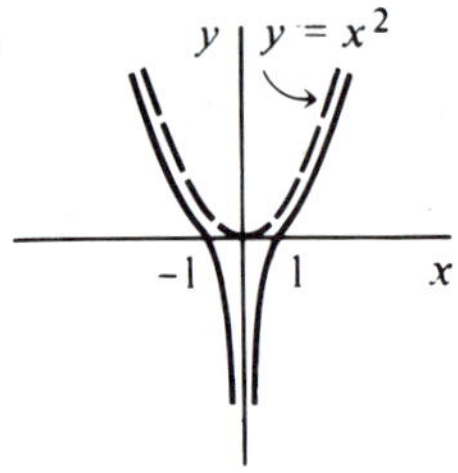

Sections 6.1–6.2 (page 182)

1. $\frac{4}{9}$ 3. $-\frac{2}{5}$ 5. 1 7. 0 9. -2

11.
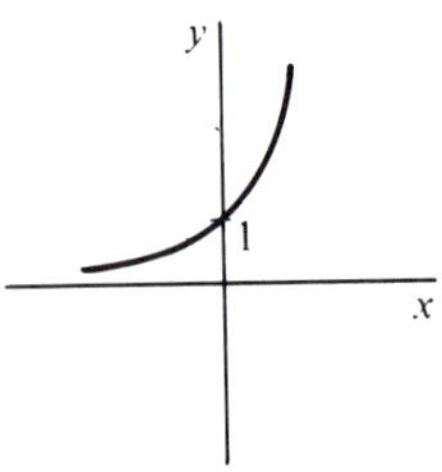

13.
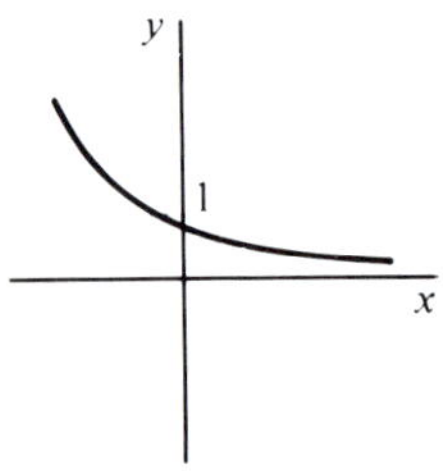

15.
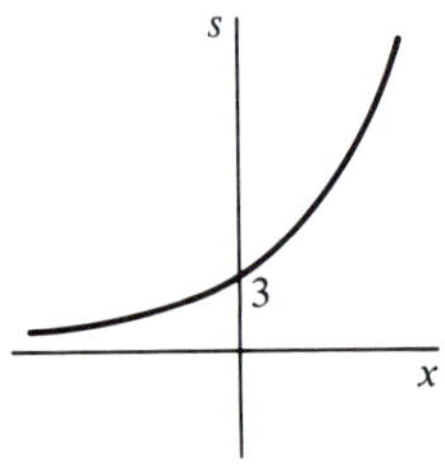

17.
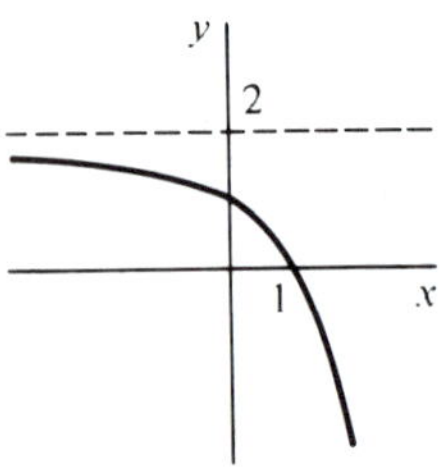

19.
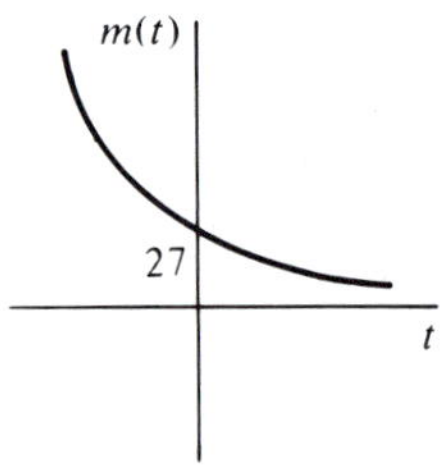

21.
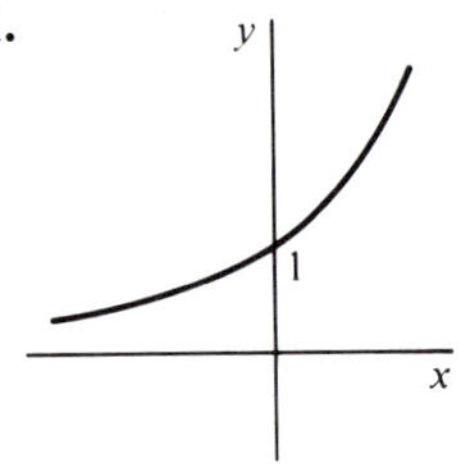

23.
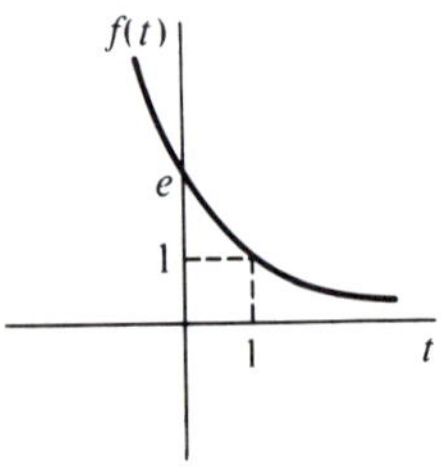

25.
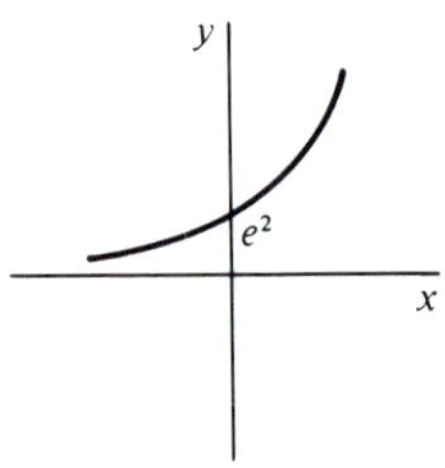

27. (a) 2.7 (b) 8.8 (c) 2.8 (d) 0.35 (e) 1.3

29. $y=1$ for all x

31. Reflection in the x-axis

33. The same curve

35. The same curve shifted 3 units in the negative x direction.

37. 7 days

39. \$1276.28

41. 2878.29 m/sec

43. 230.24 deg

45. 10.026

47. 22.68

Section 6.3 (page 188)

1. $\log_2 x=3$ **3.** $\log_5 M=-3$ **5.** $\log_7 L=2$ **7.** $b=2$ **9.** $b=10$

11. 10,000 **13.** 10 **15.** 4 **17.** 9 **19.** 2 **21.** a

23. 8 **25.** 36 **27.** 6 **29.** $-\frac{5}{2}$ **31.** $-1<x<8$ **33.** $4\leqslant x\leqslant 8$

35.

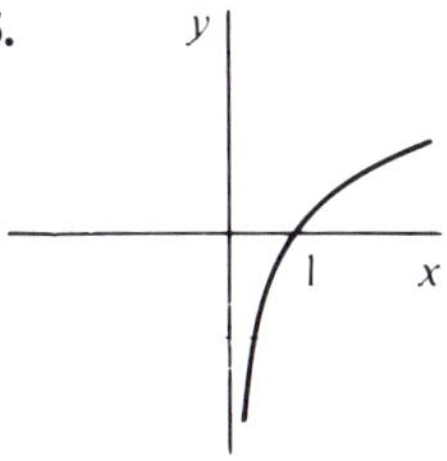

37.

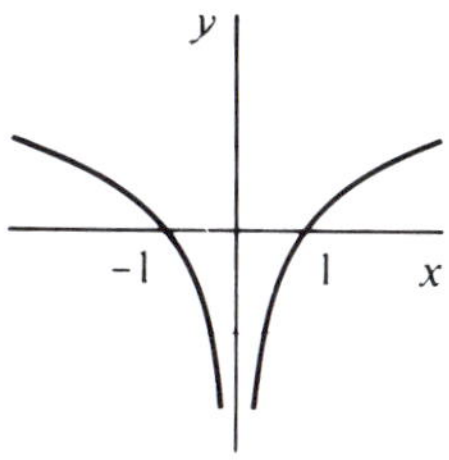

39.

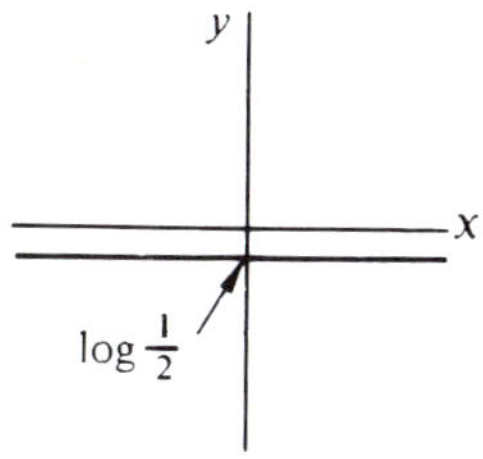

41.

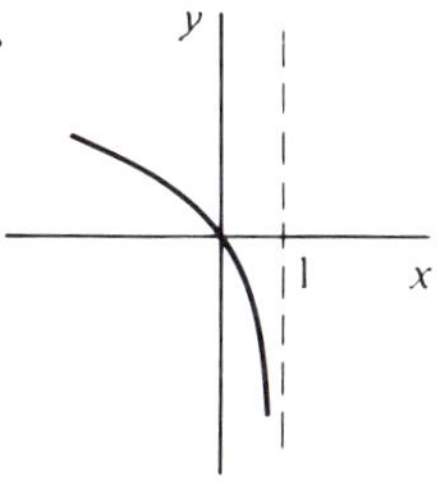

43.

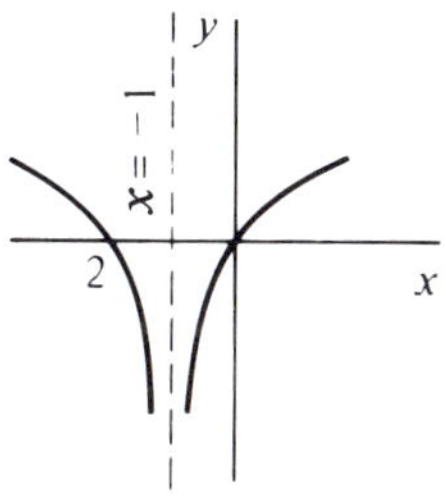

45. Reflection in the x-axis.

47. Negative portion reflected in the x-axis.

49. Curve moved 3 units in the negative x direction.

51. 2 **53.** 10 **55.** $(1, 0)$

57. No answer required

59. Reflections of each other in $y=x$

61. -0.3010

63. 1 **65.** -1 **67.** -0.69315 **69.** 3.4657 time units

Section 6.4 (page 190)

1. 9 **3.** $\frac{1}{2}$ **5.** 6 **7.** 24 **9.** 0.1761 **11.** 1.0791

13. 1.9542 **15.** 0.3495 **17.** 3.3801 **19.** -2.8539 **21.** $\log_2 x$ **23.** 0

25. $\log \dfrac{5t(t^2-4)^2}{\sqrt{t+3}}$

27. $\log \dfrac{u^3}{(u+1)^2(u-1)^5}$

31. The same for $x>0$. For $x<0$, $\log x^2$ is defined, $2 \log x$ is not.

33. Both are -2.

35. (a) Multiply each ordinate value by p. (b) Move curve up $\log p$ units.
(c) Move curve p units to the left. (d) Move curve down $\log p$ units.

Sections 6.5–6.6 (page 197)

1. (a) 0.7332 (b) 3.7332 (c) 7.7332−10
3. (a) 0.9557 (b) 6.9557−10 (c) 5.9557
5. (a) 1.3390 (b) 7.4740−10
7. (a) 1820 (b) 1.82 (c) 0.0182
9. (a) 15.2 (b) 0.000 000 152 (c) 1,520,000
11. 0.3731
13. 7.9811−10
15. 0.4343
17. 5.8077
19. 3.0022
21. 0.4321
23. 8.7991−10
25. 7.8867−10
27. 0.01344
29. 2595
31. 3.162
33. 2.155
35. 7.218×10^{-4}
37. 0.4972
39. −0.8413
41. 0.6505
43. 7.990
45. 1.758
47. −3.398
49. −8.908
51. 2.6 (low)
53. 1.7466
55. 2.1271
57. 0.3243
59. −2.4078 or −3+0.5922
61. 1306.17
63. 5.785
65. 0.13848
67. 0.00000135

Section 6.7 (page 200)

1. 0.2700
3. 10.41
5. 1.501
7. 70.46
9. 3001
11. −5739
13. 0.08172
15. 1.095
17. 1.651
19. −0.6414
21. 1.737
23. 3.860
25. 0.9606
27. −1.123
29. 165.62
31. 12.9416
33. 1651.67
35. 128.83

Section 6.8 (page 203)

1. 1.553
3. $x=0, x=\log 2=0.301$
5. 0.5850
7. $x=10$
9. $x=1, 10^4$
11. $x=1$
$x=10^{\sqrt{3}}=53.96$
$x=10^{-\sqrt{3}}=0.0185$
13. $x=5$
15. $x=2+\sqrt{5}=4.236$
17. $x=\ln(2+\sqrt{3})=1.317$
$x=\ln(2-\sqrt{3})=-1.317$
19. The logarithm is a 1 to 1 function.
21. $k=0.02476$
23. 173.3
25. 52.5 hr

Section 7.1 (page 213)

1. $(-3, -2)$
3. $(5, 2)$
5. Dependent
7. Inconsistent
9. $(\frac{8}{5}, \frac{14}{5})$
11. $(\frac{1}{4}, -\frac{3}{4})$
13. $(\frac{10}{9}, \frac{53}{63})$
15. $r=1, s=-2$
17. Inconsistent
19. $(3, 1)$
21. $(-4, 2)$
23. Dependent
25. $(-5, 4)$
27. $w=3, z=4$
29. $(-2, -2)$
31. $A=14, B=11$
33. $(8, 4)$
35. $a=0, b=\frac{1}{2}$
37. 6.75 and 0.75 mph
39. $\frac{10}{3}$ and $\frac{20}{3}$ lb
41. 20 nickels and 15 dimes
43. 18 and 22.5 min

Section 7.2 (page 216)

1. $(0, 0), (1, 1)$ **3.** $(1, 1)$ **5.** $(0, -1), (\frac{1}{2}, 0)$ **7.** No solution **9.** $(-1 \pm \sqrt{3}, 7 \mp 2\sqrt{3})$ **11.** $(\pm\sqrt{3}, -1)$ **13.** $(0, \pm 2)$ **15.** No solution **17.** $(\pm 1, 0)$ **19.** $(-\frac{11}{2}, -\frac{9}{2})$ **21.** $(0, \pm\sqrt{2})$ **23.** No solution **25.** $a=b=\pm 4$ **27.** No solution **29.** $(\pm 2\sqrt{3}, 2)$ **31.** $(1, 0)$

Section 7.3 (page 220)

1. $(7, 3, -2)$ **3.** $(\frac{5}{2}, \frac{1}{2}, \frac{5}{2})$ **5.** Inconsistent **7.** Dependent **9.** $(0, 0, 0)$ **11.** $a=1, b=2, c=-2$ **13.** $(\frac{2}{3}, \frac{3}{4}, -1)$ **15.** Inconsistent **17.** $(\frac{1}{11}, -\frac{7}{11}, \frac{5}{11})$ **19.** $(\frac{13}{3}, -\frac{10}{3}, -\frac{7}{3})$ **21.** $(\frac{160}{29}, \frac{66}{29}, \frac{64}{29})$ **23.** $(4, 2, -3, 1)$ **25.** $(2, 0, 4, -3)$ **27.** $(1, 0, 2, 3, -2)$ **29.** $n=246, d=226, q=428$ **31.** $l_1=15, l_2=17, l_3=13$

Section 7.4 (page 226)

1. $(2, 4)$ **3.** $(2, 1)$ **5.** $(-\frac{3}{4}, 2, 3)$ **7.** $(8, 5, 6)$ **9.** $(6, \frac{5}{3}, 2)$ **11.** Dependent **13.** $(3, 2, -1)$ **15.** Inconsistent **17.** Dependent **19.** $(4, -3, 2)$ **21.** $(1, -1, 2, 3)$ **23.** $(0, 1, -1, 2)$ **25.** $(1, 0, -2, 3, 1)$

Section 7.5 (page 233)

1. -2 **3.** -2 **5.** 24 **7.** 0 **19.** $x=\pm 1$ **21.** -70 **23.** $x=-3$ **25.** 5

Section 7.6 (page 239)

1. $(2, 0)$ **3.** $(5, 2)$ **5.** Dependent **7.** Inconsistent **9.** $(\frac{8}{5}, \frac{14}{5})$ **11.** $(-1, -4)$ **13.** $(7, 3, -2)$ **15.** $r=3, s=-3, t=4$ **17.** $(-135, -101, -7)$ **19.** $(0, 0, 0)$ **21.** $\left(\frac{496}{636}, -\frac{291}{636}, -\frac{270}{636}\right)$ **23.** $l=13$ cm, $w=7$ cm **25.** 5 sec **27.** 8 dimes **29.** $l_1=\frac{65}{3}$ m, $l_2=\frac{50}{3}$ m, $l_3=\frac{35}{3}$ m **31.** $u'=\dfrac{FQ_2}{R_1Q_2-R_2Q_1}$; $v'=-\dfrac{FR_2}{R_1Q_2-R_2Q_1}$ **33.** $a=-\frac{11}{3}, b=\frac{5}{3}, c=1$ **35.** $a=\frac{2}{3}, c=-\frac{2}{3}$

Section 7.7 (page 242)

1.

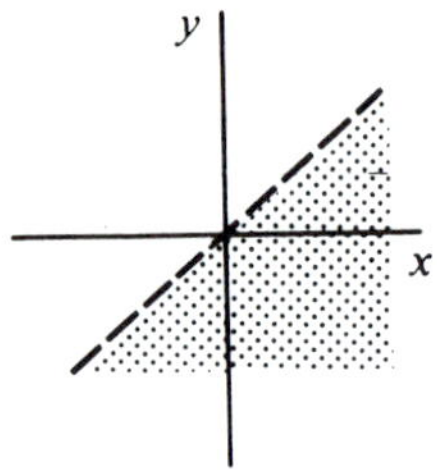

3.

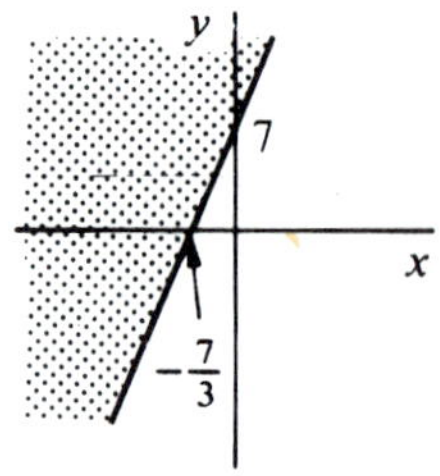

5.

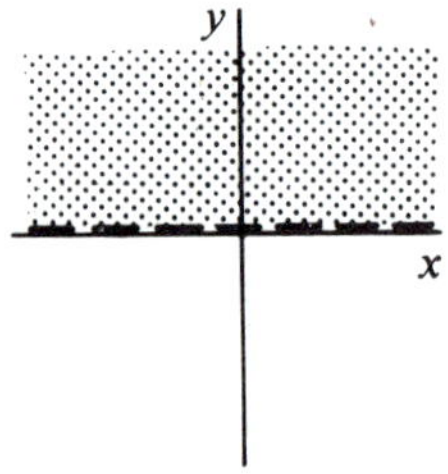

7.

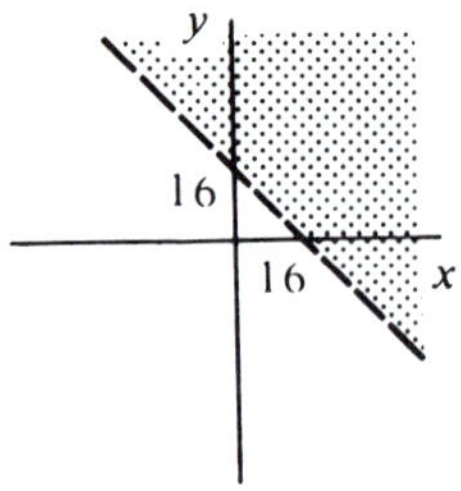

9.

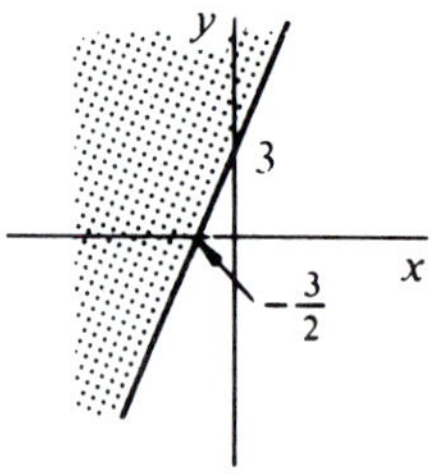

11.

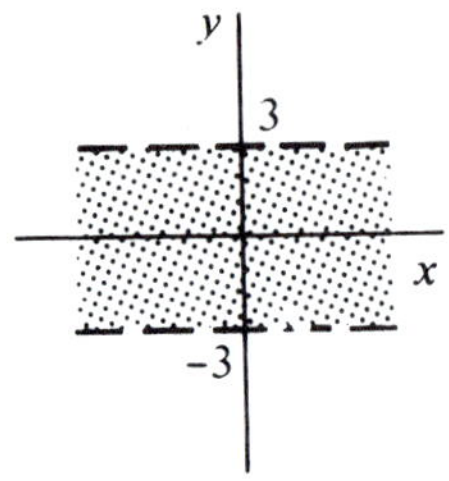

13.

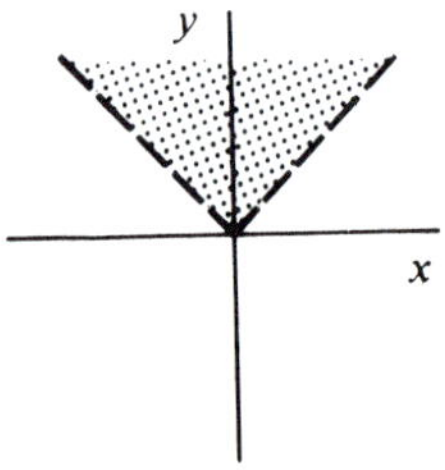

15.

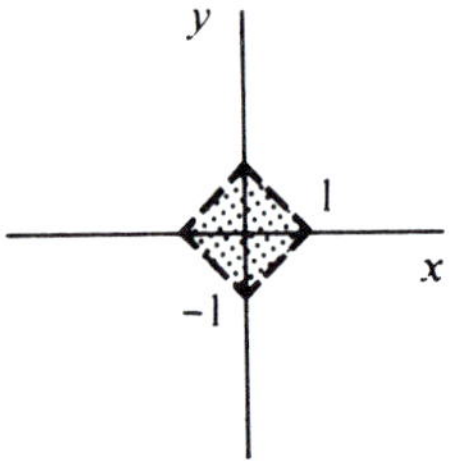

17. $C < -\frac{160}{9}$

19.

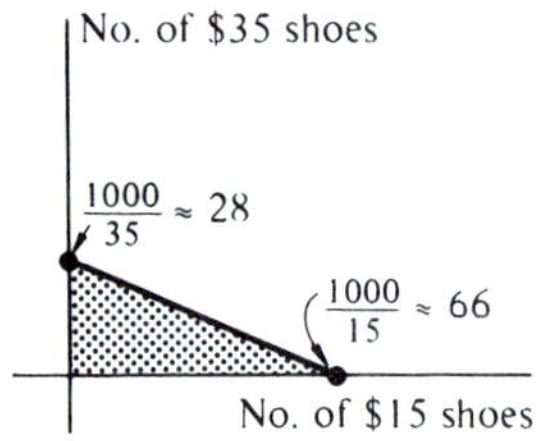

Sections 7.8–7.9 (page 247)

1.

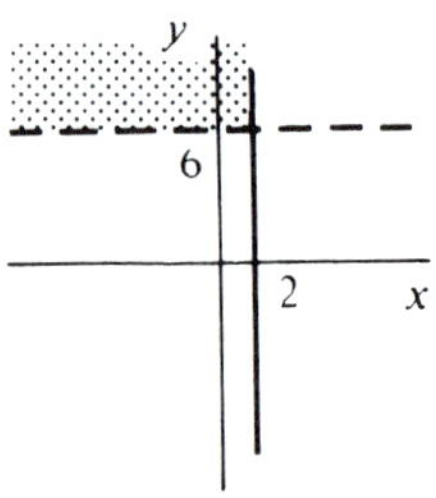

3.

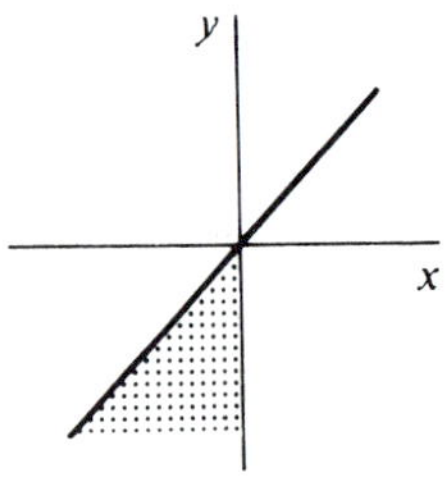

5.

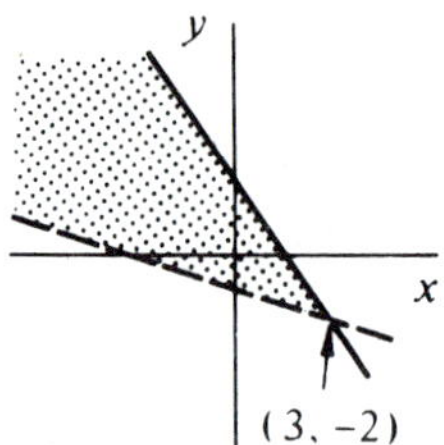

7.

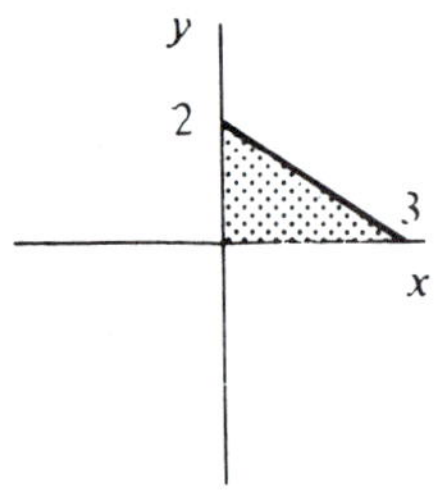

9.

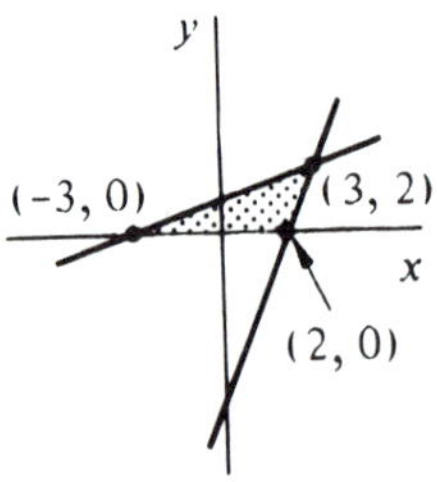

11.

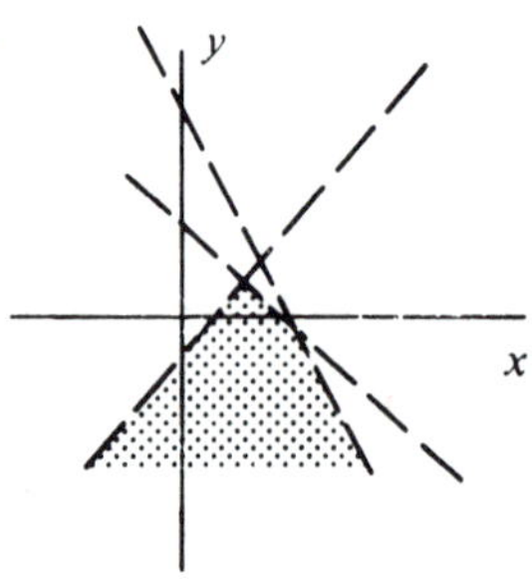

13.

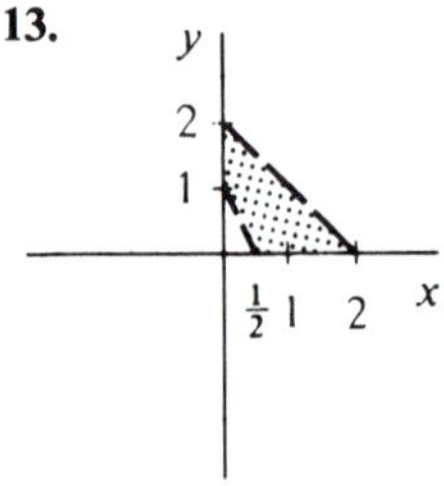

15.

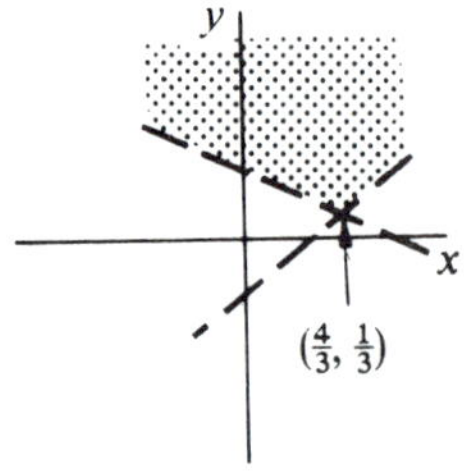

17. 10 at (4, 2) **19.** 25 at (5, 0) **21.** (a) 10 at (2, 2); (b) 9 at (3, 0) **23.** $\frac{9}{2}$ at $(0, \frac{3}{2})$ **25.** $\frac{42}{5}$ at $(\frac{11}{5}, 0, \frac{3}{5})$

27. $x=15$, $y=30$

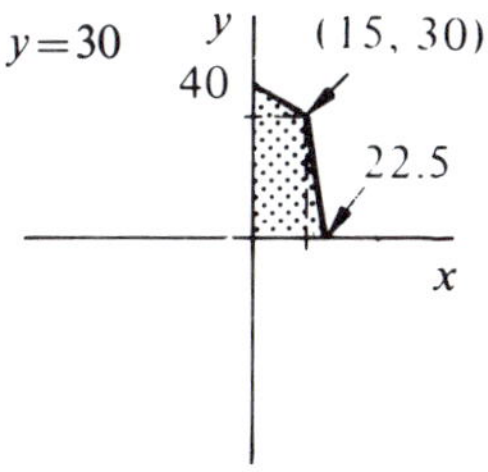

29. \$13,300 at 8.5%; \$6700 at 7.0%

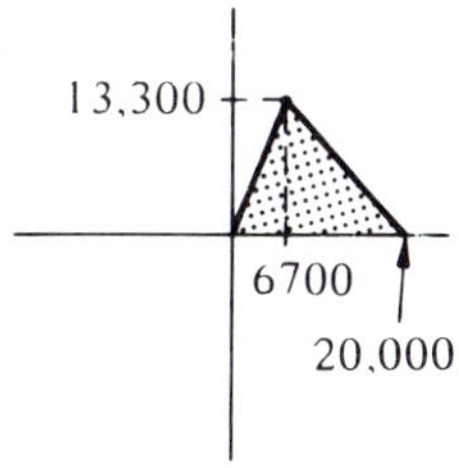

Sections 8.1–8.2 (page 256)

1. $x-1$ **3.** $x^2+3+\frac{5}{x-2}$ **5.** $x^2-\frac{1}{2}x-\frac{11}{4}+\frac{83}{8(x+\frac{1}{2})}$

7. $x^3+2x^2+8x+24$ **9.** $-3x^4-6x^3-12x^2-24x-41-\frac{74}{x-2}$

11. $3x^2-21x+30$ **13.** 28 **15.** $-\frac{1}{8}$ **17.** -1.9506 **19.** $-10-6i$

21. 0 **23.** Not a factor **25.** Factor

27. Not a factor **29.** Factor **31.** Factor **33.** Factor **35.** Factor

37. $y=-(x+1)(x-1)(x-2)$ **39.** $y=\frac{3}{4}(x-1)^2(x-2)^2$ **41.** $y=\frac{-\sqrt{2}}{5}(x-\sqrt{2})(x^2-4x+5)$

43. $\frac{4}{3}$ **45.** $1+\frac{5}{2\sqrt{2}}$

Section 8.3–8.4 (page 261)

1. 1, −1, −2 **3.** $-2, \pm i$ **5.** $-\frac{5}{2}, 1\pm\sqrt{5}$ **7.** −1, −2, 2, 3 **9.** $\frac{1}{3}, (3\pm i\sqrt{3})/2$

11. −1, −1, −1, 3 **13.** $-\frac{2}{3}, (1\pm i\sqrt{2})/3$ **15.** $\frac{1}{2}, (1\pm i\sqrt{7})/2$

17. $-1, 1, 2, -1\pm i\sqrt{3}$ **19.** $(x+1)\left(x+\frac{5}{2}-\frac{\sqrt{37}}{2}\right)\left(x+\frac{5}{2}+\frac{\sqrt{37}}{2}\right)$

21. $2(2x+1)\left(x+\frac{3}{4}+i\frac{\sqrt{15}}{4}\right)\left(x+\frac{3}{4}-i\frac{\sqrt{15}}{4}\right)$ **23.** $2x(2x-3)\left(x-\frac{1}{4}+\frac{\sqrt{217}}{4}\right)\left(x-\frac{1}{4}-\frac{\sqrt{217}}{4}\right)$

25. $(0, 0), (1, 1), (-1\pm i, -4)$ **27.** $(2, 1), \left(\frac{2}{3}, -\frac{1}{3}\right), \left(-1, \frac{-1\pm i\sqrt{15}}{8}\right)$ **29.** 23

Section 8.5 (page 266)

1. 1 negative, 2 complex **3.** 1 positive, 1 negative, 2 complex

5. 1 positive, 2 complex, or 1 positive, 2 negative

7. 3 positive, 2 complex, or 3 positive, 2 negative, or 1 positive, 4 complex, or 1 positive, 2 negative, 2 complex

9. 2 positive, 2 negative, 2 complex, or 2 positive, 4 complex, or 2 negative, 4 complex, or 6 complex

11. 0.27 **13.** 1.25 **15.** 3.07 **17.** −0.33 **19.** 2.14 **21.** −2.27

23.

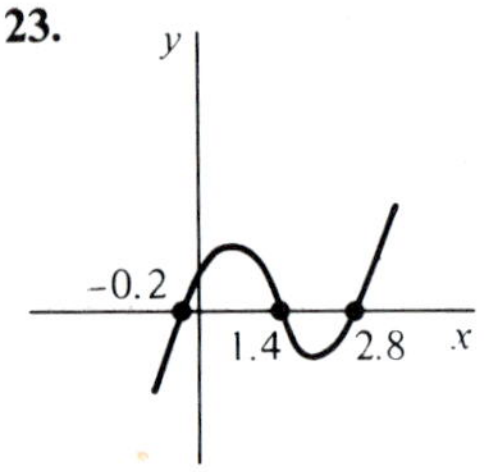

25.

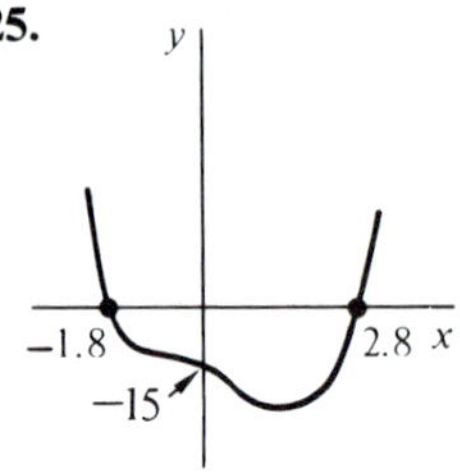

27.

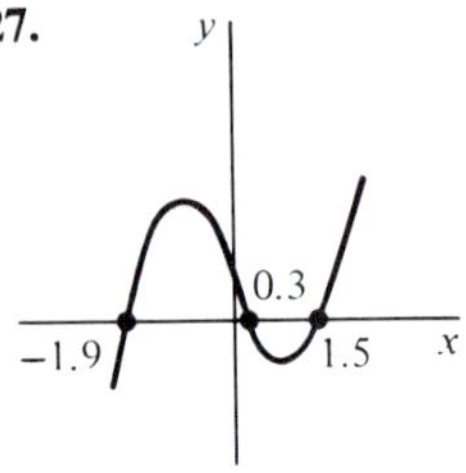

29.

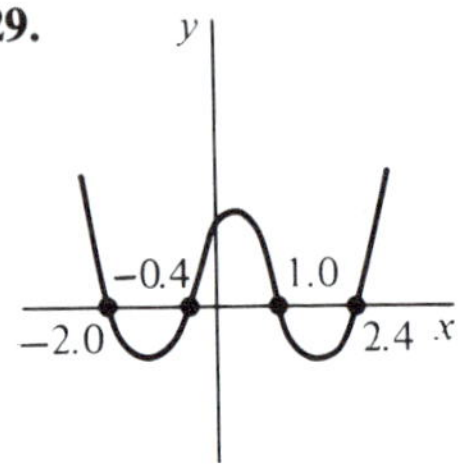

Section 8.6 (page 271)

1. $\frac{\frac{2}{3}}{x+5}+\frac{\frac{1}{3}}{x-1}$ **3.** $\frac{1}{x}-\frac{1}{x+1}$ **5.** $\frac{\frac{1}{2}}{x}+\frac{2}{x+5}-\frac{\frac{1}{2}}{x+2}$

7. $\frac{\frac{1}{2}}{x+5}+\frac{\frac{1}{2}}{x+1}$ **9.** $-\frac{6}{x}+\frac{7}{x-1}-\frac{2}{(x-1)^2}$ **11.** $\frac{1}{x}-\frac{1}{x+1}-\frac{1}{(x+1)^2}$

13. $\frac{-1}{x+1}+\frac{2}{x-1}+\frac{3}{3-2x}$ **15.** $\frac{2}{x-1}+\frac{1}{(x-1)^2}$ **17.** $\frac{1}{x}+\frac{1}{x^2+2}$

19. $\frac{1}{2x+3}-\frac{1}{x^2+2x+4}$ **21.** $\frac{2x-1}{x^2+2x+3}-\frac{x}{x^2+5}$

Section 9.1 (page 275)

1. 2, 3, 4, 5, 6 **3.** 2, 0, 2, 0, 2 **5.** e, e^2, e^3, e^4, e^5 **7.** $\frac{3}{2}, \frac{4}{3}, \frac{5}{4}, \frac{6}{5}, \frac{7}{6}$

9. 1, 4, 9, 16, 25 **11.** 3, 5, 7, 9, 11 **13.** $1, \frac{1}{4}, \frac{1}{9}, \frac{1}{16}, \frac{1}{25}$

15. $\frac{5}{2}, \frac{5}{4}, \frac{5}{8}, \frac{5}{16}, \frac{5}{32}$ **17.** $3, 3^2, 3^4, 3^8, 3^{16}$ **19.** 10, −5, 10, −5, 10

21. $2n$ **23.** $3n-2$ **25.** n^2 **27.** $(4000)(.75)^4=\$1265.63$

Section 9.2 (page 278)

1. 1.093 **3.** 55 **5.** 45 **7.** 20 **9.** 10

11. $x_1+x_2+x_3+x_4+x_5$ **13.** $\sum_{k=1}^{4} k$ **15.** $\sum_{k=1}^{4} k^2$ **17.** $\sum_{k=1}^{5} 2^k x$

19. True **21.** True **23.** True **25.** $a_1=2$, $a_n=0, n\geq 2$ **27.** $a_1=-1$, $a_n=2(-1)^n$

Section 9.3 (page 281)

1. Yes, 1 **3.** Yes, 2 **5.** Yes, 3 **7.** No **9.** No **11.** $a_n = 3n - 1$ **13.** 1 **15.** -90 **17.** 48 **19.** 50 **21.** -70 **23.** n^2 **25.** No answer required **27.** 3, 4, 5, 6, 7 **29.** 1, 0, -1, -2, -3 **31.** 7, 9, 11, 13, 15 **33.** $f(x) = 3x$ **35.** $f(x) = 3x + 1$ **37.** $f(x) = 3x + 2$

Section 9.4 (page 285)

1. Yes, 3, $a_k = (4)3^{k-1}$ **3.** No **5.** Yes, -2, $a_k = (\frac{1}{3})(-2)^{k-1}$ **7.** Yes, $\frac{1}{2}$, $a_k = (\frac{1}{2})^k$ **9.** $a_4 = 375$, $S_4 = 468$ **11.** $r = \pm 2$, $S_5 = 155$ or 55 **13.** $-1024/19{,}683$ **15.** 726 **17.** $\dfrac{3^{11} - 1}{2^{10}} - 2$ **19.** $\frac{315}{64}$ **21.** $\frac{21}{64}$ **23.** 7 strokes **25.** $f(x) = 2^{x-1}$ **27.** 3, 9, 27, 81, 243 **29.** $\frac{1}{2}, \frac{1}{4}, \frac{1}{8}, \frac{1}{16}, \frac{1}{32}$ **31.** Speed $= 2^{(T - T_0)/10}$ **33.** $2^n - 1$

Section 9.5 (page 288)

1. $I_s = \$180$, $I_c = \$185.03$ **3.** $I_s = \$260$, $I_c = \$267.89$ **5.** $I_s = \$946.20$, $I_c = \$988.33$ **7.** $I_s = \$168.64$, $I_c = \$175.31$ **9.** $I_s = \$24.29$, $I_c = \$25.44$ **11.** \$9142.38 **13.** \$466.29 **15.** \$1844.20 **17.** \$123.09 **19.** \$1410.60 **21.** \$632.46 **23.** \$8204.39 **25.** 8.855%

Section 9.6 (Round-off may cause differences in answers) page 292

1. \$7743.71 **3.** \$11,837.09 **5.** \$35,341.23 **7.** \$20,484.50 **9.** \$19,124.91 **11.** \$167.29 **13.** \$179.95 **15.** \$224.93 **17.** \$289.51 **19.** \$318.05

21. The monthly payments are \$175.83. The interest payments for each month are \$16.67, \$15.34, \$14.00, \$12.65, \$11.29, \$9.92, \$8.54, \$7.15, \$5.74, \$4.32, \$2.89, and \$1.45.

23. \$11,035.68 **25.** \$30,347.18 **27.** \$140.58

Section 9.7 (page 296)

1. $x^6 - 6x^5y + 15x^4y^2 - 20x^3y^3 + 15x^2y^4 - 6xy^5 + y^6$

3. $243x^5 - 405x^4y + 270x^3y^2 - 90x^2y^3 + 15xy^4 - y^5$

5. $x^5+5x^3+10x+\frac{10}{x}+\frac{5}{x^3}+\frac{1}{x^5}$

7. $y^6+9y^4x+27y^2x^2+27x^3$

9. $x^3+6x^2\sqrt{xy}+15x^2y+20xy\sqrt{xy}+15xy^2+6y^2\sqrt{xy}+y^3$

11. $x^{4/3}+4x^{1/4}+6x^{-5/6}+4x^{-23/12}+x^{-3}$

13. 19.4481

15. $x^{-9/4}-3x^{-3/4}+3x^{3/4}-x^{9/4}$

17. $120\sqrt{xy}x^3y$

19. $-\frac{20!3^9 7^{11}}{11!9!}x^{27}y^{11/3}$

21. -10

23. $\frac{70}{81}$

25. 121.6 **27.** 63.52 **29.** 6 **31.** 4.89975 **33.** 1.913194 **35.** 3.107

37. 2.9907 **39.** 9.8995 **41.** 3.076×10^{15} **43.** 2.653×10^{32}

Sections 10.1–10.2 (page 305)

1. 6, 5 **3.** *wx, wy, wz, xw, xy, xz, yw, yx, yz, zw, zx, zy* **5.** 840 **7.** 30

9. 10! **11.** 15!/6! **13.** 24 **15.** 180 **17.** 7! **19.** 11

Section 10.3 (page 308)

1. 6 **3.** 21 **5.** 1365 **7.** 2,598,960 **9.** 56 **11.** 45, 90

13. 48 **15.** 300 **17.** 8400 **19.** 8

Section 10.4 (page 311)

1. 0.25 **3.** 0.375, 0.625 **5.** 0.5 **7.** 0.375 **9.** $\frac{10}{13}=0.769$ **11.** $\frac{5}{16}$

13. 0.323 **15.** 0.109 **17.** $\frac{1}{18}$ **19.** 0.0399

Sections 10.5–10.6 (page 314)

1. 1 to 5 **3.** 5 to 1 **5.** $\frac{8}{11}$ **7.** $\frac{21}{6}$ **9.** $\frac{30}{9}$

Sections 10.7–10.9 (page 319)

1. $\frac{1}{3}$ **3.** $\frac{5}{7}$ **5.** $\frac{4}{13}$ **7.** (a) $\frac{1}{16}$ (b) $\frac{3}{51}$

9. (a) $\frac{13^3}{52\cdot51\cdot50}=0.017$ (b) $\frac{13^3\cdot6}{52\cdot51\cdot50}=0.1$

11. (a) $\frac{28}{143}$ (b) $\frac{79}{143}$

13. $\frac{1}{9}$

15. (a) $\frac{2}{9}$ (b) $\frac{8}{15}$

17. (a) 0.0815 (b) 0.367

19. $\frac{5}{12}$

21. (a) 0.87 (b) 13

Section 11.1 (page 326)

1. focus $(-2, 0)$
dir. $x=2$
r.c. $(-2, 4), (-2, -4)$

3. focus $(0, \frac{3}{2})$
dir $y=-\frac{3}{2}$
r.c. $(-3, \frac{3}{2}), (3, \frac{3}{2})$

5. focus $(-4, 0)$
dir. $x=4$
r.c. $(-4, 8), (4, 8)$

7. focus $(\frac{3}{4}, 0)$
dir. $x=-\frac{3}{4}$
r.c. $(\frac{3}{4}, \frac{3}{2}), (\frac{3}{4}, -\frac{3}{2})$

9. focus $(-\frac{1}{2}, 0)$
dir. $x=\frac{1}{2}$
r.c. $(-\frac{1}{2}, 1), (-\frac{1}{2}, -1)$

11. $x^2=8y$

13. $y^2=6x$ **15.** $x^2=-4y$ **17.** $x^2=y$

19. $(0, 0), (-2, 1)$ **21.** $(.09, -1.05)$ **23.** $x^2=1000y$

Section 11.2 (page 331)

1. ellipse
$a=5$
$b=\sqrt{5}$
foci $(0, \pm 2\sqrt{5})$

3. ellipse
$a=2$
$b=1$
foci $(0, \pm\sqrt{3})$

5. ellipse
$a=3$
$b=\sqrt{3}$
foci $(0, \pm\sqrt{6})$

7. ellipse
$a=\sqrt{12}$
$b=\sqrt{8}$
foci $(\pm 2, 0)$

9. ellipse
$a=1$
$b=\frac{2}{3}$
foci $(0, \pm\frac{1}{3}\sqrt{5})$

11. $9x^2+16y^2=144$ **13.** $100x^2+9y^2=225$ **15.** $9x^2+25y^2=225$ **17.** $15x^2+16y^2=240$

19. $4x^2+21y^2=25$ **21.** $(1.3, -1.9), (-1.3, 1.9)$ **23.** $(.94, .88), (-.94, .88)$

25. $9x^2+100y^2=225$

Section 11.3 (page 335)

1. $a=4$
$b=4$
foci $(\pm\sqrt{32}, 0)$

3. $a=3$
$b=2$
foci $(\pm\sqrt{13}, 0)$

5. $a=5$
$b=2$
foci $(0, \pm\sqrt{29})$

7. $a=\frac{5}{2}$
$b=\frac{5}{4}$
foci $(\pm\frac{5}{4}\sqrt{5}, 0)$

9. $a=1$
$b=1$
foci $(\pm\sqrt{2}, 0)$

11. $9x^2-16y^2=144$ **13.** $4y^2-x^2=4$ **15.** $13x^2-36y^2=117$

17. $4y^2-x^2=64$ **19.** $y^2-10x^2=9$

Section 11.4 (page 339)

1. $8x=y^2-2y+25$ **3.** $12y=-13+2x-x^2$ **5.** $4x=y^2-4y+8$

7. $7x^2+16y^2-28x-32y-68=0$ **9.** $16x^2+y^2-16x+8y+4=0$ **11.** $3x^2+4y^2+24x-24y+72=0$

13. $64y^2-49x^2-98x-256y=577$ **15.** $7x^2-9y^2-28x+18y=44$ **17.** $x^2-4y^2+4x-16y-16=0$

19. $(x-h)^2+y^2=r^2$ **21.** $y^2=4a(x-h)$ **23.** $(x-r)^2+y^2=r^2$

Section 11.5 (page 342)

1. $(x+2)^2+(y+3)^2=3^2$ **3.** $\frac{(x+1)^2}{4}+\frac{(y+1)^2}{9}=1$ **5.** $\frac{(x-2)^2}{25}-\frac{y^2}{25}=1$

7. $(x-3)^2=3(y+3)$ **9.** $(y-\frac{1}{2})^2=-\frac{1}{2}(x-\frac{3}{2})$ **11.** $\frac{(x+1)^2}{1}-\frac{(y-1)^2}{4}=1$

13. $\frac{(x-1)^2}{4}+\frac{(y+2)^2}{9}=1$ **15.** $\frac{(y-\frac{1}{2})^2}{\frac{1}{4}}-\frac{x^2}{\frac{1}{2}}=1$ **17.** $\frac{(x+2)^2}{7}-\frac{y^2}{7}=1$

19. $(x+\frac{5}{2})^2=y-\frac{3}{4}$

Index

$23.75